AFTER BABEL

AFTER BABEL

ASPECTS OF LANGUAGE
AND TRANSLATION

SECOND EDITION

GEORGE STEINER

Oxford   New York
OXFORD UNIVERSITY PRESS

Oxford University Press, Walton Street, Oxford OX2 6DP

Oxford New York
Athens Auckland Bangkok Bombay
Calcutta Cape Town Dar es Salaam Delhi
Florence Hong Kong Istanbul Karachi
Kuala Lumpur Madras Madrid Melbourne
Mexico City Nairobi Paris Singapore
Taipei Tokyo Toronto

and associated companies in
Berlin Ibadan

Oxford is a trade mark of Oxford University Press

First published 1975
Second edition 1992

British Library Cataloguing in Publication Data

Data available

Library of Congress Cataloging in Publication Data
Steiner, George, 1929-
After Babel: aspects of language and translation/George
Steiner.—2nd edn.
p. cm.
Includes bibliographical references and index.
1. Translating and interpreting. 2. Language and languages.
I. Title.
418'.02—dc20 P306.S66 1992 92-13874
ISBN 0-19-282874-6 Pbk

5 7 9 10 8 6 4

Printed in Great Britain by
Biddles Ltd
Guildford and King's Lynn

FOR ZARA

'eyn 'aḥereth

ACKNOWLEDGEMENTS

THOUGH this book has, to a great extent, had to define and map its own field, it is none the less dependent on a large body of related work. One's bibliography and footnotes are, in this respect, the most genuine act of thanks. The origin of the present study lies in the *Penguin Book of Modern Verse Translation* which I edited in 1966 (and which was later reissued under the title *Poem Into Poem*). Tony Richardson was a close collaborator in that project. His early, tragic death has left a constant void. There are deficiencies in the present book which he would have been the first to note. During the course of work, I have benefited from exchanges with translators and with the increasing number of poets and scholars concerned with translation. Let me mention only Robert Fitzgerald, Roger Shattuck, Donald Carne Ross, William Arrowsmith, Nathaniel Tarn, John Frederick Nims, Christopher Middleton, and Octavio Paz. Some of the theoretical and practical matter presented in this book first came up in the course of seminars at Harvard, Yale, and the University of Zürich. In each case, the debt which I owe to my students is considerable. It will also be obvious, at various points, how much I profited from the personal interest of Claude Lévi-Strauss and I. A. Richards. Thomas Sebeok, whose knowledge of the whole range of current language-studies may well be unrivalled, has been a good listener. Noam Chomsky has been generous in expressing his disagreements in private communication (an exchange of views is included in my earlier book, *Extraterritorial: Papers on Literature and the Language Revolution*). Mr. Robin Anderson, of Churchill College, read the first three chapters in draft and offered critical advice on technical issues. During early stages of research, I received invaluable support from the Guggenheim Memorial Foundation. Like so many other writers and scholars, I found in its Director, Prof. Gordon Ray, a vital ally. My indebtedness to my assistant, Mrs E. Southern, would be difficult to put summarily.

In a direct sense, this volume owes its existence and scope to the initiative of Jon Stallworthy and his colleagues at the Oxford University Press. Theirs have been the indispensable patience and criticism. Mr Bernard Dod and Mr Nicolas Barker have proved the most exacting and helpful of copy-editors. Jon Stallworthy is himself both a poet and a translator. The advantage has been mine.

It is, under this rubric, customary to thank one's family and immediate circle of friends for their forbearance or enthusiasm during a long spell of often obsessive work. But there is cant in doing so, for what choice had they? The dedication of this book, on the other hand, says only a fraction of what it means.

G. S.

Cambridge, October 1973

PREFACE TO THE SECOND EDITION

THIS book was written under somewhat difficult circumstances. I was at the time increasingly marginalized and indeed isolated within the academic community. This is not, necessarily, a handicap. Tenure in the academy today, the approval of one's professional peers, the assistance and laurels in their giving, are not infrequently symptoms of opportunism and mediocre conventionality. A degree of exclusion, of compelled apartness, may be one of the conditions of valid work. Scientific research and advance are in substantial measure and logic collaborative. In the humanities, in the disciplines of intuitive discourse, committees, colloquia, the conference circuit are the bane. Nothing is more ludicrous than the roll-call of academic colleagues and sponsors set out in grateful footnotes at the bottom of trivia. In poetics, in philosophy, in hermeneutics, work worth doing will more often than not be produced against the grain and in marginality.

But there are dangers. *After Babel* attempts to map a new field, a new space for argument. There has been (though it remains rare) penetrative insight into the act of translation, into the phenomenological and philosophic aspect of this act from the time of Seneca to that of Walter Benjamin and W. v. O. Quine. Practising translators (though again these are rare) have left descriptive records of their craft. The sheer volume of literary, historical, philosophic translation on which Western civilization has depended for its lineage and dissemination constitutes material for systematic analysis and reflection. But there had been, before *After Babel*, no full-scale endeavour to relate, to bring into interactive focus, the diverse areas of rhetoric, of literary history and criticism, of linguistics, and of linguistic philosophy. There had been no ordered or detailed attempt to locate translation at the heart of human communication or to explore the ways in which the constraints on translatability and the potentialities of transfer between languages engage, at the most immediate and charged level, the philosophic enquiry into consciousness and into the meaning of meaning.

Inevitably, such an attempt at innovative synthesis will be vulnerable. To an extent almost defiant of common sense, approved academic studies have fragmented into minute specialization. The parish grows smaller with every teaching appointment or research grant. The sanctioned vision is microscopic. More and more is being published in learned journals, by academic presses, about less and less. The note is one of Byzantine minutiae, of commentaries on commentaries on commentaries towering like inverted pyramids on single points often ephemeral. The specialist holds the 'generalist' or 'polymath' in vengeful disdain. And his authority and technical grasp over a given inch of ground may, indeed, exhibit a confidence, an immaculate humility, denied to the comparatist, to one who (awkwardly or with a peremptory bound) crosses stiles between fields.

To attempt a comprehensive poetics of translation was foolhardy enough. To do so in isolation from the support which might, under other circumstances, have been provided by sympathetic readers of different chapters in the university, was to incur manifest risks. The first edition of *After Babel* contained errors and imprecisions. It contained inexactitudes of phrasing, particularly in reference to what were then called transformational generative grammars. It lacked clarity in regard to the vital topic of temporality in Semitic and in Indo-European syntax. There can be no apology for these defects, only thanks to those who pointed them out (notably Professor Edward Ullendorff in a review-essay of magisterial severity). But the acerbities of the response to *After Babel* in academe did not stem from reproof over details. It betrayed a profound, worried dismay at the very concept of a larger perspective, of an alliance between philosophic concerns, poetic sensibility, and linguistics in the more formal and technical sense. To Roman Jakobson, to William Empson in his *Structure of Complex Words*, to Kenneth Burke—a neglected master in language-studies—such an alliance was the obvious imperative for hermeneutics. By the mid-1970s, the barriers ran high between specializations inebriate with a largely spurious claim to 'scientific' status. Among stamp collectors letter-writers are not always welcome.

More characteristic of the mandarin trade than direct attack was the 'passage under silence' (as French has it) of the book. Wholly representative of this strategy is the footnote in a recent (highly intelligent) monograph on philosophy and translation: *After Babel* is designated as self-evidently the most important text in the entire domain of translation studies and of the philosophic issues they entail. After which no further mention or citation occurs. Since it first appeared, *After Babel* has been drawn upon and pilfered, often without acknowledgement. A considerable secondary literature has grown up around many of the themes first stated in the book. Fascinatingly and nearly implausibly, this study of translation, with its insistence on difficulty, on the singularity of different speech-worlds and its prodigality of examples from poetry, has itself been translated into languages ranging from Romanian to Chinese. My awed thanks go to those who have undertaken this vexing task. Each translation has thrown searching light on the fundamental propositions in the original. Nevertheless, and although it has been continuously in print, *After Babel* remains to academic linguists, to those who theorize about or claim to teach translation, an irritant and the anarchic act of an outsider.

I value, therefore, the invitation from the Oxford University Press to publish this second edition. Errata have, so far as possible, been corrected. Loose or confused moments in the argument have been amended. Material published after 1974–5 has been included in new or expanded footnotes. The bibliography, which even those hostile to the enterprise found invaluable, and appropriated, has been updated. Much of this work has been made possible by the privileged context of a European university chair (the oldest in the field of comparative literature). I now can enlist the resources, the critical exchanges with colleagues, the help in research which were not available to me when I wrote the book. My particular gratitude goes to my colleague and assistant Aminadav Dyckman, a philologist, linguist, and student of Slavic poetics of passionate exactitude.

Yet even in this corrected guise, *After Babel* will, I suspect, continue to be something of a scandal or *monstrum* which the guilds of linguistic scholarship and linguistic and analytic philosophy will prefer to neglect. Central tenets in this work remain almost deliberately misunderstood or threatening. Let me set them out summarily—and without repentance.

After Babel postulates that translation is formally and pragmatically implicit in *every* act of communication, in the emission and reception of each and every mode of meaning, be it in the widest semiotic sense or in more specifically verbal exchanges. To understand is to decipher. To hear significance is to translate. Thus the essential structural and executive means and problems of the act of translation are fully present in acts of speech, of writing, of pictorial encoding inside any given language. Translation between different languages is a particular application of a configuration and model fundamental to human speech even where it is monoglot. This general postulate has been widely accepted. I try to illustrate it by considering the teeming difficulties encountered inside the same language by those who seek to communicate across spaces of historical time, of social class, of different cultural and professional sensibility. More especially, I invite consideration of the dilemmas of inadequate translation posed by the radical differences between the speech-habits, voiced and unvoiced, of men and of women. Here it is not socio-linguistics or psycholinguistics, nor even anthropology, which illuminate most. It is the intuitive probes of poets, dramatists, and novelists when they articulate the conventions of masked or failed understanding which have obtained between men and women, between women and men, in the lineaments of dialogue we call love or hatred. The subject is pivotal to our perceptions of self and society. Certain recent currents in feminism and 'women-studies' have brutalized or made trivial the complex, delicate fabric of evidence. So far as I can judge, the instigations to enquiry in this book have scarcely been followed up.

But although we 'translate' at every moment when speaking and receiving signals in our own tongue, it is evident that translation in the larger and more habitual sense arises when two languages meet. That there should be two different languages, that there should have been, at a rough estimate, more than twenty thousand spoken on this small planet, is the Babel-question. Why should *homo sapiens sapiens*, genetically and physiologically uniform in almost all respects, subject to identical biological–environmental constraints and evolutionary possibilities, speak thousands of mutually incomprehensible tongues, some of which are set only a few miles apart? The material, economic, social advantages of using a single language are blatant. The thorn-barriers posed by reciprocal incomprehension, by the need to acquire a second or third language, often of formidable phonetic and grammatical difficulty and 'strangeness', are evident. There is an elementary and elemental challenge to reflection here largely ignored as either formless or insoluble by most of academic linguistics (even as the famous question of the origins of human language had, until very recently, been ruled out of 'scientific' court).

After Babel adduces the Darwinian analogy: that of the plethora of organic species. Are there structural parallels between the ten thousand species of insects to be found in a corner of Amazonia, say, and the numbing proliferation of languages spoken on the Indian subcontinent or in those very same regions of the Amazon rain-forests? At the first level, the analogy breaks down. The Darwinian paradigm is one of evolutionary benefit. As they emerge competitively, different life-forms, however specialized, however minutely distinct, occupy different niches in the environment. Their proliferation augments the chances of precise adjustment and biological progress. No such profit accrues from the seemingly anarchic multiplicity of mutually non-communicating tongues. On the contrary: there is no mythology known to us in which the fragmentation of some initial single language (the Adamic motif) into jagged bits, into cacophony and incommunicado, has not been felt to be a catastrophe, a divine chastisement on some opaque motion of rebellion or arrogance in fallen man. Even at a glance, the disasters, be they economic, political, or social, which have attended on the thousandfold 'babbling after Babel' are palpable.

But there is, at a second level, a seminal suggestion to be found in Darwinian models. *After Babel* argues that it is the constructive powers of language to conceptualize the world which have been

crucial to man's survival in the face of ineluctable biological constraints, this is to say in the face of death. It is the miraculous—
I do not retract the term—capacity of grammars to generate
counter-factuals, 'if'-propositions and, above all, future tenses,
which have empowered our species to hope, to reach far beyond
the extinction of the individual. We endure, we endure creatively
due to our imperative ability to say 'No' to reality, to build fictions
of alterity, of dreamt or willed or awaited 'otherness' for our
consciousness to inhabit. It is in this precise sense that the utopian
and the messianic are figures of syntax.

Each human language maps the world differently. There is life-
giving compensation in the extreme grammatical complication of
those languages (for example, among Australian Aboriginals or in
the Kalahari) whose speakers dwell in material and social contexts
of deprivation and barrenness. Each tongue—and there are no
'small' or lesser languages—construes a set of possible worlds
and geographies of remembrance. It is the past tenses, in their
bewildering variousness, which constitute history. Thus there is,
at the level of human psychic resources and survivance, an immensely positive, 'Darwinian' logic in the otherwise baffling and
negative excess of languages spoken on the globe. When a language
dies, a possible world dies with it. There is here no survival of the
fittest. Even where it is spoken by a handful, by the harried
remnants of destroyed communities, a language contains within
itself the boundless potential of discovery, of re-compositions of
reality, of articulate dreams, which are known to us as myths, as
poetry, as metaphysical conjecture and the discourse of the law.
Inherent in *After Babel* is the accelerating disappearance of
languages across our earth, the detergent sovereignty of so-called
major languages whose dynamic efficacy springs from the planetary spread of mass-marketing, technocracy, and the media.

Paradoxically, a comparable force for uniformity characterizes
the claims of transformational generative grammars. Paradoxically,
because the politics of Noam Chomsky have been anti-imperialist
in the extreme. The axiom of universal deep structures, innate in
the brain (though in ways never defined and, indeed, ruled as
beyond rational investigation), entails inevitably a relegation to

accident, to superficiality, of the facts of linguistic multiplicity and difference. The dissent from transformational generative claims throughout *After Babel* bears on the failure of such grammars to produce substantive examples of 'universals' in natural languages, and on the cardinal irrelevance of the Chomskyan project to poetics and hermeneutics. Today, generative grammars have retreated into almost total formality, into a degree of analytic and meta-mathematical algorithmic abstraction so great as to have hardly any bearing on the matter of actual 'speech-worlds' and of the creative differences between them. What has, arrestingly, replaced generative 'unitarianism' is the Nostratian theory, with its search for a single *Ur-* or primal language from which all others derive. There may or may not be evidence for such a common font: what engages the poet, what fascinates and perplexes the student of understanding is, on the contrary, what William Blake called 'the holiness of the minute particular'.

The fault may indeed be mine. So far as I am able to judge, the 'Darwinian' view of the psychic indispensability of the prodigality of diverse languages among mankind has not been grasped or argued. It is central to *After Babel*.

Between this book and its adoption into the current academic and high-journalistic canon (where the two are often so damagingly identical) lies the matter of 'theory'. There are theories in the exact and in the applied sciences. They have predictive obligation, can be crucially tested, and are falsifiable. A theory with a higher demonstrable yield of insight and application will replace its predecessor. Not one of these criteria obtains in the humanities. No configuration or classification of philosophic or aesthetic materials has any predictive force. There is no conceivable experimental validation or refutation of an aesthetic or philosophic judgement. In the disciplines of intuition and energized responses of sensibility, in the craft of apprehension and answerability which make up the humanities, no paradigm or school of judgement cancels out any other. Winckelmann does not erase or replace Aristotle; Coleridge does not render Dr Johnson obsolete; T. S. Eliot on Shelley cannot invalidate Matthew Arnold.

In consequence, I take the present-day, ubiquitous use of the term and rubric 'theory' in poetics, in hermeneutics, in aesthetics (also, I suspect in the social sciences) as spurious. It has no substantive status and radically obscures the subjective, imaginatively transcendental (in Kant's sense) tenor of all arguments, proposals, and findings in literature and the arts (there *are*, unquestionably, authentic theoretical, which is to say 'formalizable', elements in the analysis of music). There are no 'theories of literature', there is no 'theory of criticism'. Such tags are arrogant bluff, or a borrowing, transparent in its pathos, from the enviable fortunes and forward motion of science and technology. There are, most assuredly, and *pace* our current masters in Byzantium, no 'theories of translation'. What we do have are reasoned descriptions of processes. At very best, we find and seek, in turn, to articulate, narrations of felt experience, heuristic or exemplary notations of work in progress. These have no 'scientific' status. Our instruments of perception are not theories or working hypotheses in any scientific, which means falsifiable, sense, but what I call 'working metaphors'. At its finest, translation has nothing to gain from the (mathematically) puerile diagrams and flow-charts put forward by would-be theoreticians. It is, it always will be, what Wittgenstein called 'an exact art'.

The four-beat model of the hermeneutic motion in the act of translation argued in *After Babel*—'initiative trust–aggression–incorporation–reciprocity or restitution'—makes no claim to 'theory'. It is a narrative of process. What strength it has derives from the actual practice of translators, from the records, still too scant or unavailable, from their workshop. The concept of 'restitution', of the restoration of equilibrium between the original text and its translation, an equilibrium made vulnerable by translation itself, raises ethical questions of extreme complexity. There have been attempts to develop this elucidation since *After Babel* first appeared. But, like my own sketch, they remain inadequate. If I was to rewrite the book now, it is this question of the morality of appropriation via translation and of what I designate as 'transfiguration'—where the intrinsic weight and radiance of the translation eclipses that of the source—which I would want to hammer out at greater length. The dilemma seems to me of central importance precisely in an age in which deconstructive criticism and self-advertising scholarship dismiss texts as 'pre-texts' for their own scavenging.

At the time when *After Babel* was in progress, the increasing domination of an Anglo-American Esperanto across the planet looked to be obvious and possibly irreversible. To a large extent this is still the case. Science, technology, commerce, and world-finance speak a more or less American English. The collapse of centres of Marxism in the face of a triumphalist late capitalism and mass-distribution ideal has, if anything, underlined the linguistic hegemony of 'American' speech. Throughout most of the under-developed world, this speech is the only foreseeable escalator to economic–social emancipation. What matters more, the 'languages' of computers, the meta-linguistic codes and algorithms of electronic communication which are revolutionizing almost every facet of knowledge and production, of information and projection, are founded on a sub-text, on a linguistic 'pre-history', which is fundamentally Anglo-American (in the ways in which we may say that Catholicism and its history had a foundational Latinity). Computers and data-banks chatter in 'dialects' of an Anglo-American mother tongue.

None the less, the picture now strikes me as somewhat less clear than it was. Fierce ethnic and regional atavisms are re-surgent. Determinant of, determined by tribal, regional, and national passions for identity, languages are proving more resistant to rationalization, and the benefits of homogeneity and technical formalization, than one might have expected. Strenuous efforts towards uniformity, for example in India or South-East Asia, have, so far, proved abortive. If anything, the dislocation of the Soviet and East European power-blocs is bringing with it an almost fanatical wish for *apartheid*, for self-authenticating autochthony between neighbouring tongues (in the Ukraine, in the Caucasus, throughout the Balkans). Both Spanish and Chinese, moreover, are displaying energies of territorial and demographic acquisition which may come to challenge the American-English predomin-ance. The question, and with it the future functions of interlingual translation, remains open.

No less than the first edition, this amended *After Babel* aspires to reach philosophers of language, historians of ideas, scholars of poetics and the arts and of music, linguists, and, most obviously, translators. But it solicits the interest and pleasure of the general reader, of all those who love language, who experience language as formative of their humanity. Above all, it addresses itself, in hope of response, to poets. Which is to say to anyone who makes the language live and who knows that the affair at Babel was both a disaster and—this being the etymology of the word 'disaster'—a rain of stars upon man.

Geneva/Cambridge, July 1991

CONTENTS

ACKNOWLEDGEMENTS   vii

PREFACE TO THE SECOND
    EDITION   ix

One   Understanding as Translation   1

Two   Language and Gnosis   51

Three   Word against Object   115

Four   The Claims of Theory   248

Five   The Hermeneutic Motion   312

Six   Topologies of Culture   436

AFTERWORD   496

SELECT BIBLIOGRAPHY   500

INDEX OF PROPER NAMES
    AND TITLES   517

Der Mensch gebärdet sich, als sei er Bildner und Meister der Sprache, während doch sie die Herrin des Menschen bleibt. Wenn dieses Herrschaftsverhältnis sich umkehrt, dann verfällt der Mensch auf seltsame Machenschaften. Die Sprache wird zum Mittel des Ausdrucks. Als Ausdruck kann die Sprache zum blossen Druckmittel herabsinken. Dass man auch bei solcher Benutzung der Sprache noch auf die Sorgfalt des Sprechens hält, ist gut. Dies allein hilft uns jedoch nie aus der Verkehrung des wahren Herrschaftsverhältnisses zwischen der Sprache und dem Menschen. Denn eigentlich spricht die Sprache. Der Mensch spricht erst und nur, insofern er der Sprache entspricht, indem er auf ihren Zuspruch hört. Unter allen Zusprüchen, die wir Menschen von uns her nie zum Sprechen bringen dürfen, ist die Sprache der höchste und der überall erste.

MARTIN HEIDEGGER, '... Dichterisch Wohnet der Mensch...' 1954

Ningún problema tan consustancial con las letras y con su modesto misterio como el que propone una traducción.

J.-L. BORGES, 'Las versiones Homéricas', *Discusión*, 1957

La théorie de la traduction n'est donc pas une linguistique appliquée. Elle est un champ nouveau dans la théorie et la pratique de la littérature. Son importance épistémologique consiste dans sa contribution à une pratique théorique de l'homogénéité entre signifiant et signifié propre à cette pratique sociale qu'est l'écriture.

HENRI MESCHONNIC, *Pour la poétique II*, 1973

Man acts as if he were the shaper and master of language, while it is language which remains mistress of man. When this relation of dominance is inverted, man succumbs to strange contrivances. Language then becomes a means of expression. Where it is expression, language can degenerate to mere impression (to mere print). Even where the use of language is no more than this, it is good that one should still be careful in one's speech. But this alone can never extricate us from the reversal, from the confusion of the true relation of dominance as between language and man. For in fact it is language that speaks. Man begins speaking and man only speaks to the extent that he responds to, that he corresponds with language, and only in so far as he hears language addressing, concurring with him. Language is the highest and everywhere the foremost of those assents which we human beings can never articulate solely out of our own means.

No problem is as completely concordant with literature and with the modest mystery of literature as is the problem posed by a translation.

The theory of translation is not, therefore, an applied linguistics. It is a new field in the theory and in the practice of literature. Its epistemological importance lies in its contribution to the 'theoretical practice' of the homogeneity, of the natural union between the signifier and the signified. This homogeneity is proper to that social enterprise which we call writing.

Chapter One

UNDERSTANDING AS TRANSLATION

I

ACT II of *Cymbeline* closes with a monologue by Posthumus. Convinced that Iachimo has indeed possessed Imogen, Posthumus rails bitterly at woman:

> Is there no way for man to be, but women
> Must be half-workers? We are all bastards,
> And that most venerable man, which I
> Did call my father, was I know not where
> When I was stamp'd. Some coiner with his tools
> Made me a counterfeit: yet my mother seem'd
> The Dian of that time: so doth my wife
> The nonpareil of this. O vengeance, vengeance!
> Me of my lawful pleasure she restrain'd,
> And pray'd me oft forbearance: did it with
> A pudency so rosy, the sweet view on't
> Might well have warm'd old Saturn; that I thought her
> As chaste as unsunn'd snow. O, all the devils!
> This yellow Iachimo, in an hour, was't not?
> Or less; at first? Perchance he spoke not, but
> Like a full-acorn'd boar, a German one,
> Cried 'O!' and mounted; found no opposition
> But what he look'd for should oppose and she
> Should from encounter guard. Could I find out
> That woman's part in me—for there's no motion
> That tends to vice in man, but I affirm
> It is the woman's part: be it lying, note it,
> The woman's: flattering, hers; deceiving, hers:
> Lust, and rank thoughts, hers, hers: revenges, hers:

Ambitions, covetings, change of prides, disdain,
Nice longing, slanders, mutability;
All faults that name, nay, that hell knows, why, hers
In part, or all: but rather all. For even to vice
They are not constant, but are changing still;
One vice, but of a minute old, for one
Not half so old as that. I'll write against them,
Detest them, curse them: yet 'tis greater skill
In a true hate, to pray they have their will:
The very devils cannot plague them better.

This, of course, is only in part a realization of what Shakespeare wrote. *Cymbeline* was first printed in the Folio of 1623 and the distance between Shakespeare's 'manuscript' and the earliest printed texts continues to exercise scholars. But I am not, in fact, transcribing the Folio text. I am quoting from the Arden edition of the play by J. M. Nosworthy. His version of Posthumus's speech embodies a sum of personal judgement, textual probability, and scholarly and editorial precedent. It is a recension which seeks to gauge the needs and resources of the educated general reader of the mid-twentieth century. It differs from the Folio in punctuation, line-divisions, spelling, and capitalization. The visual effect is markedly different from that achieved in 1623. At one point, the editor substitutes for what he takes to be a corrupt reading what he, and previous scholars, assume to be the most likely emendation. The editor's task here is, in the full sense, interpretative and creative.

The direction of spirit and main rhetorical gestures of Posthumus's outburst are unmistakable. But only close reading will exhibit the details and manifold energies at work. A first step would deal with the meaning of salient words—with what that meaning may have been in 1611, the probable date of the play. Already this is a difficult step, because current meaning may not have been, or have been only in part, Shakespeare's. In short how many of Shakespeare's contemporaries fully understood his text? An individual and a historical context are both germane.

One might begin with the expressive grouping of *stamp'd*, *coiner*, *tools*, and *counterfeit*. Several currents of meaning and implication are interwoven. They invoke the sexual and the monetary and the strong, often subterranean links between these two areas of human will. The counterfeit coiner stamps false coin. One of the meanings of *counterfeit* is 'to pretend to be another' which is apposite to Iachimo. The *O.E.D.* cites a usage in 1577 in which *counterfeit* signifies 'to adulterate'. The meshing of *adulteration* with *adultery* would be characteristic of Shakespeare's total responsiveness to the field of relevant force and intimation in which words conduct their complex lives. *Tools* has a gross sexual resonance; is there, conceivably, an undertone of a sense of the verb *stamp*, admittedly rare, for which the *O.E.D.* finds an example in 1598: 'a blow with the pestle in pounding'? Certainly pertinent are such senses of the word as 'to imprint paper' (Italian: *stampare*), missives true and false playing so important a role in *Cymbeline*, and the meaning 'to stigmatize'. The latter is of especial interest: the *O.E.D.* and Shakespeare glossaries here direct us to *Much Ado About Nothing*. It soon becomes evident that Claudio's damnation of women in Act IV, Scene i foreshadows the rage of Posthumus.

Pudency is so unusual a word that the *O.E.D.* gives *Cymbeline* as authority for its undoubted general meaning: 'susceptibility to shame'. A 'rosy pudency' is one that blushes; but the erotic associations are insistent and part of a certain strain of febrile bawdy in this play. *Pudenda*, recorded as early as 1398, but not in common usage until the 1630s, cannot be ruled out. Both 'shame' and the 'sexual occasion of shame' are operative in *pudic*, which Caxton takes over from the French in 1490 as meaning 'chaste'. Shakespeare uses *chaste* three lines later with the striking image of *unsunn'd snow*. This touch of unrelenting cold may have been poised in his mind once reference was made to old Saturn, god of sterile winter. *Yellow* Iachimo is arresting. The aura of nastiness is distinct. But what is being inferred? Though 'green' is the more usual appurtenance of jealousy, Middleton in 1602 uses *yellow* to mean 'affected with jealousy'. Shakespeare does likewise in *The Winter's Tale*, a play contemporary with *Cymbeline*, and in *The Merry Wives of Windsor* (1. iii) 'yellowness' stands for 'jealousy' (could there be a false etymology somewhere in the background, associating the two words?). Iachimo *is* jealous, of Posthumus's nobility, of Posthumus's good fortune in enjoying the love and fidelity of Imogen. But does Posthumus know this, or does the dramatic strength of the epithet lie precisely in the fact that it exceeds Posthumus's conscious insight? Much later, and with American overtones, *yellow* will come to express both cowardice and mendacity—the 'yellow press'. Though these two nuances are beautifully apposite to Iachimo, neither was, so far as we can tell, available to Shakespeare. What latent undertones in the word and colour give rise to subsequent, negative usage? Shakespeare at times seems to 'hear' inside a word or phrase the history of its future echoes.

Encounter as 'erotic accosting' (cf. *Two Gentlemen of Verona*, II. vii) is easier to place; in the present context, the use of the term in *Much Ado About Nothing* (III. iii) is particularly relevant. Elizabethan bawdy suggests the proximity of a bitter pun. *Motion*, on the other hand, would require extensive treatment. Here it plainly signifies 'impulse'. But the development of the word, as it grows towards modern 'emotion', is a history of successive models of consciousness and volition. *Change of prides* has busied editors. The surface meaning is vivid and compact. Ought we to derive its suggestive force from an association of *prides* with 'ornate attire'? In *Doctor Faustus* that association is made explicit. Capitalized as they are in the Folio, Prides, Disdaine, Slanders, Mutability, and Vice direct us back to the personified, emblematic idiom of Tudor morality plays and allegoric pageants in which Marlowe and Shakespeare were at home and many of whose conventions recur, though in an intellectualized, reflective form, in Shakespeare's late tragi-comedies. By setting these nouns in lower-case, a modern text sacrifices a specific pictorial-sensory effect. The Folio prints *Nice-longing*. This may either be Shakespearean coinage or a printer's reading. In Posthumus's use of *nice*, Shakespeare exploits a certain instability in the word, a duplicity of ambience. The term can move either way, towards notions of delicacy, of educated finesse, or towards a faintly corrupt, hedonistic indulgence. Here, perhaps through a finely judged placing of vowel sounds, *nice* has a distinct unpleasantness. 'Wanton' and 'lascivious' are close at hand. Like 'motion', *mutability* would require extensive treatment. From Chaucer's *Troilus* to the unfinished seventh book of the *Faerie Queene*, the concept has a fascinating history. It embodies philosophic, perhaps astrologically-tinged notions of universal inconstancy, of an anarchic variable in the sum of human fortunes. But as early as Chaucer, and in Lydgate's *Troy Book* (1412–20), the word is strongly linked with the alleged infidelity of woman: 'They say that chaunge and mutabylyte / Apropred ben to femynyte.' *Mutability* climaxes and conjoins Posthumus's catalogue of reproach. If Imogen has yielded to Iachimo, all trust has ebbed from life and Hell is near.

Such a glossary, even if its lexical, historical elements aimed to be exhaustive, is only a preliminary move. A comprehensive reading would turn next to syntactic aspects of the passage. The study of Shakespeare's grammar is itself a wide field. In the late plays, he seems to develop a syntactic shorthand; the normal sentence structure is under intense dramatic stress. Often argument and feeling crowd ahead of ordinary grammatical connections or subordinations. The effects—*Coriolanus* is especially rich in examples—are theatrical in the valid sense. We hear discourse in a condition of heightened action. The words 'ache at us' with an immediacy, with an internalized coherence which come before the attenuated, often wasteful conventions of 'proper' public speech. But that coherence is not the same as that of common grammar. At two points in Posthumus's diatribe (lines 19 and 28) ordinary sequences and relations seem to break down. Thus some editors would read 'All faults that may be named, that hell knows'. Others prefer to keep the Folio text, judging Posthumus's lapses into incoherence to be a deliberate dramatic means. So nauseating is the image of Iachimo's easy sexual triumph, that Posthumus loses the thread of his discourse; in his enraged mind as in his syntax, Iachimo and Imogen are momentarily entangled.

Sustained grammatical analysis is necessary and cuts deep. But glossary and syntax are only instruments. The main task for the 'complete reader' is to establish, so far as he is able, the full intentional quality of Posthumus's monologue, first within the play, secondly in what is known of Shakespearean and Elizabethan dramatic conventions, and, most difficult of all, within the large context of early seventeenth-century speech-habits. What is involved here is the heart of the interpretative process. In seeking to apprehend Posthumus's meaning, and his own relations to such meaning, we attempt to determine the relevant 'tone-values' or 'valuations'. I use these terms for lack of a more rigorous designation of total operative context. I hope their definition will emerge in the course of this book.

Does Posthumus 'mean it' (itself a colloquialism charged with linguistic and psychological suppositions)? Does he believe what he is saying, or only in some measure? At what level of credence are we to respond? In part, the answers lie in our 'reading' of Posthumus's character. But that character is a semantic construct, an aggregate of verbal and gestural indicators. He is quick to anger and to despair. Perhaps we are to detect in his rhetoric a bent towards excess, towards articulation beyond the facts. What weight has this tirade in the immediate stage-setting? Granville-Barker supposed that it is delivered from the inner stage, after which Posthumus again comes forward. Iachimo and Philario would remain within earshot. In that case, we are dealing with a partial soliloquy only, with a statement at least some of which is intended as communication outward, here to Iachimo. Would this account for the grammatical compression, for the apparent ambiguity of focus at mid-point in the monologue? Or is Posthumus in fact alone and using the convention of the address to oneself which is intended to be 'overheard' by the entire audience?

Looking at the speech we are, I think, struck by certain elements of style and cadence subversive of any final gravity. The note of comic fury expressive of Claudio's myopia in *Much Ado* is not altogether absent from *Cymbeline*. The bulk of Posthumus's indictment has an undeniable seriousness and disgust; but the repeated 'hers', the naïve cumulation of vehemence produce a delicate counter-movement. 'I'll write against them' is near-comedy. Indeed, such is the effect of levity and doggerel at the close of the passage, that various editors regard the last line as a spurious addendum. Might it be that at some level immediately below articulate intent, Posthumus does not, cannot wholly believe

Iachimo's lies? If he did believe them without any reservation of consciousness, would he deserve reunion with Imogen (it is of the essence of tragi-comedy that self-destructive blindness be, wherever possible, qualified)? Moreover, as scholars point out, Posthumus's philippic is, at almost every stage, conventional; his vision of corrupt woman is a *locus communis*. Close parallels to it may be found in Harrington's translation of Ariosto's *Orlando Furioso* (XXVII), in Book X of *Paradise Lost*, in Marston's *Fawn*, and in numerous Jacobean satirists and moralists. This stylized fabric again alerts us to a certain distance between Posthumus's true self and the fury of his statement. The nausea of Othello, moving from sexual shock to a vision of universal chaos, and the infirm hysteria of Leontes in *The Winter's Tale* have a very different pitch.

The determination of tone-values, of the complete semantic event brought about by Posthumus's words, the attempt to grasp the full reach of those words both inward and in respect of other personages and the audience, moves in concentric and ever-widening circles. From Posthumus Leonatus at the close of Act II, we proceed to *Cymbeline* as a whole, then to the body of Shakespearean drama and to the context of cultural reference and literature on which it draws. But beyond these, large and complex as they are, lies the informing sphere of sensibility. This is, in certain respects, the most vital and the least explored. We know little of internal history, of the changing proceedings of consciousness in a civilization. How do different cultures and historical epochs use language, how do they conventionalize or enact the manifold possible relations between word and object, between stated meaning and literal performance? What were the semantics of an Elizabethan discourse, and what evidence could we cite towards an answer? The distance between 'speech signals' and reality in, say, Biblical Hebrew or Japanese court poetry is not the same as in Jacobean English. But can we, with any confidence, chart these vital differences, or are our readings of Posthumus's invective, however scrupulous our lexical studies and editorial discriminations, bound to remain creative conjecture?

And where are the confines of relevance? No text earlier than or contemporaneous with Shakespeare can, *a priori*, be ruled out as having no conceivable bearing. No aspect of Elizabethan and European culture is formally irrelevant to the complete context of a Shakespearean passage. Explorations of semantic structure very soon raise the problem of infinite series. Wittgenstein asked where, when, and by what rationally established criterion the process of free yet potentially linked and significant association in psychoanalysis could be said to have a stop. An exercise in 'total reading' is also potentially unending. We will want to come back to this odd truism. It touches on the nature of language itself, on the absence of any satisfactory or generally accredited answer to the question 'what is language?'

Jane Austen's *Sense and Sensibility* appeared in 1813, two centuries after *Cymbeline*. Consider Elinor Dashwood's reflections when hearing news of Edward Ferrars's engagement, in Chapter 1 of volume II:

The youthful infatuation of nineteen would naturally blind him to everything but her beauty and good nature; but the four succeeding years—years, which if rationally spent, give such improvement to the understanding, must have opened his eyes to her defects of education, while the same period of time, spent on her side in inferior society and more frivolous pursuits, had perhaps robbed her of that simplicity, which might once have given an interesting character to her beauty.

If in the supposition of his seeking to marry herself, his difficulties from his mother had seemed great, how much greater were they now likely to be, when the object of his engagement was undoubtedly inferior in connections, and probably inferior in fortune to herself. These difficulties, indeed, with an heart so alienated from Lucy, might not press very hard upon his patience; but melancholy was the state of the person, by whom the expectation of family opposition and unkindness, could be felt as relief!

This seems far easier to grasp confidently than a piece of dramatic poetry in Shakespeare's late manner. Indeed at the surface, Jane Austen's prose is habitually unresistant to close reading; it has a lucid 'openness'. Are we not making difficulties for ourselves? I think not, though the generation of obstacles may be one of the elements which keep a 'classic' vital. Arguably, moreover, these unobtrusive paragraphs, chosen almost at random, are more difficult to locate completely, to paraphrase fully, than is Posthumus's rhetoric.

The urbanity of Miss Austen's diction is deceptive. No less than Henry James, she uses style to establish and delimit a coherent, powerfully appropriated terrain. The world of an Austen novel is radically linguistic: all reality is 'encoded' in a distinctive idiom. What lies outside the code lies outside Jane Austen's criteria of admissible imaginings or, to be more precise, outside the legitimate bounds of what she regarded as 'life in fiction'. Hence the exclusive functions of her vocabulary and grammar. Entire spheres of human existence—political, social, erotic, subconscious—are absent. At the height of political and industrial revolution, in a decade of formidable philosophic activity, Miss Austen composes novels almost extraterritorial to history. Yet their inference of time and locale is beautifully established. The world of *Sense and Sensibility* and of *Pride and Prejudice* is an astute 'version of pastoral', a mid- and late-eighteenth-century construct complicated, shifted slightly out of focus by a Regency point of view. No fictional landscape has ever been more strategic, more expressive, in a constant if undeclared mode, of a moral case. What is left out is, by that mere omission, acutely judged. From this derives the distinctive pressure on Jane Austen's language of the unspoken.

Elinor Dashwood's agitated musings about Edward and the 'illiterate, artful, and selfish' Lucy Steele appear to require no glossary. The sentence structure in the second paragraph, on the other hand, attracts notice. There are two sentences, both unwieldy to a degree. By contrast, the preceding paragraph, though made up, remarkably enough, of only one long sentence, moves forward with a deliberately alternant, gliding cadence. The initial clause of paragraph two, 'If in the supposition of his seeking to marry herself...' is awkward. The repetition of 'herself' at the end of the sentence augments our impression of involution and discomfort. Both segments of the next sentence are ponderous and not immediately easy to construe. One wonders whether the exclamation mark is intended to introduce a certain simplification and renewal of narrative pace. The purpose of this grammatical opaqueness is evident. These gouty sentences seek to contain, to ravel up a rawness and disorder of feeling which Elinor herself would find inadmissible. She is endeavouring to give reasoned form to her turbulent, startled response. At the same time, she is so plainly involved in the situation that her pretence to considered, mundane judgement is transparent. The Augustan propriety of the passage, the profusion of abstract terms, the 'Chinese box' effect of dependent and conditional phrases, make for subtle comedy. The novelist's stance towards this little flutter of bruised sentiments and vanities is unmistakably arch. In the following paragraph ('As these considerations occurred to her in painful succession, she wept for him more than for herself . . .') the hint of whimsy shades into gentle irony.

But in this text, as so often in Jane Austen, even a detailed syntactical elucidation does not resolve the main difficulty. The crux lies in tonality, in the cumulative effect of key words and turns of phrase which may have behind them and, as it were, immediately beneath their own surface, a complex field of semantic and ethical values. A thorough gloss on Miss Dashwood's thoughts would engage not only problems of contemporary diction, but an awareness of the manifold ways in which Jane Austen enlists two previous bodies of linguistic convention: that of Restoration comedy, and that of post-Richardsonian sentimental fiction. The task is the more difficult because many of the decisive words have a 'timeless', immediately accessible mien. In fact, they are firmly localized in a transitional, partially artificial code of consciousness.

What precise intonations, what 'stress marks' ought we to put on 'good nature', on time 'rationally spent'? *Nature, reason,* and *understanding* are terms both of current speech and of the philosophic vocabulary. Their interrelations, implicit throughout the sentence, argue a particular model of personality and right conduct. The concision of Miss Austen's treatment, its assumption that the 'counters' of abstract meaning are understood and shared between herself, her characters, and her readers, have behind them a considerable weight of classic Christian terminology and a current of Lockeian psychology. By 1813 that conjunction is neither self-evident nor universally held. Jane Austen's refusal to underline what *ought* to be commonplace, at a time when it no longer is, makes for a covert, but forceful didacticism. 'Defects of education', 'inferior society', and 'frivolous pursuits' pose traps of a different order. No modern equivalent is immediately available. The exact note of derogation depends on a specific scale of social and heuristic nuances. Only by steeping oneself in Miss Austen's novels can one gauge the extent of Lucy Steele's imperfections. Used by a disappointed rival, moreover, these phrases may have an exaggerated, purely circumstantial edge. What results is objectively difficult, as difficult as anything met with in the excerpt from *Cymbeline*. Dealing with the problem of necessary and sufficient context, with the amount of prior material required to understand a given message-unit, some linguists have put forward the term 'pre-information'. How much pre-information do we need to parse accurately the notions of *simplicity* and of *interesting character*, and to visualize their relationship to Lucy Steele's *beauty*? The classic cadence of the sentence, its somewhat strained mundanity, direct us towards the possibility of mild satire. Elinor's supposition is couched in the modish idiom of sentimental fiction and reflects the domestic formalities of moral discourse after Addison and Goldsmith. It shows here a faintly dated, provincial coloration. At the same time, the aggrieved sharpness of Elinor's sentiments is unmistakable. If *simplicity* signifies 'freedom from artifice'—as in a handsome quote from Wesley in 1771 given in the *O.E.D.*—it also carries a charge of 'rusticity' and 'uncouthness'. The juxtaposition of 'illiterate' with 'artful' in the preceding sentence suggests a certain duplicity in Elinor's comment. How, next, are we to read 'an interesting character to her beauty'? In a usage which the utilitarian and pragmatic vocabularies of Malthus and Ricardo exactly invert, *interest* can mean 'that which excites pathos', 'that which attracts amorous, benevolent sympathies'. Sterne's *Sentimental Journey* of 1778, a work whose diction, though transposed, often underlies effects in Jane Austen, shows the narrator drawn to a countenance more *interesting* than handsome, the 'interest' betokening comeliness of spirit. The word *heart* in the common late-eighteenth-century locution 'she was a young woman of heart' (*elle avait du cœur*) would be cognate. Only in some such collocation can *simplicity* be said to give beauty an *interesting character*; and only by noting the stilted, eroded tenor of Elinor's parlance can we measure its cattiness, its betraying effort at self-control. But certain aspects of 'period flavour' (present, as well, in *alienated* and *melancholy* in paragraph two), and of the inferred body of idiomatic shorthand, remain elusive.

The obstacles to assured reading posed by Dante Gabriel Rossetti's sonnet on ' "Angelica Rescued by the Sea-Monster", by Ingres; in the Luxembourg', are of a very different sort:

> A remote sky, prolonged to the sea's brim:
>   One rock-point standing buffetted alone,
>   Vexed at its base with a foul beast unknown,
> Hell-spurge of geomaunt and teraphim:
> A knight, and a winged creature bearing him,
>   Reared at the rock: a woman fettered there,
>   Leaning into the hollow with loose hair
> And throat let back and heartsick trail of limb.
> The sky is harsh, and the sea shrewd and salt.
>   Under his lord, the griffin-horse ramps blind
>   With rigid wings and tail. The spear's lithe stem
> Thrills in the roaring of those jaws: behind,
>   The evil length of body chafes at fault.
>   She does not hear nor see—she knows of them.

Rossetti's 'Sonnets for Pictures' appeared in *The Germ* in 1850. The rubric is unclear. Are these poems acts of homage to Flemish, Italian, and French masters, notations of awed or exultant response? Are they transcriptions, representations in language of canvases which the poet has seen at Bruges and in Paris? Do they assume visual reference to the paintings? Most likely, these several schemes of relationship are active.

The verbs are in the 'immediate present', strongly suggesting that the speaker has Ingres's *Angelica* before his very eyes (in this arrangement *reared* makes for an awkward, momentarily ambiguous move). The reading eye—it 'reads' poem and painting simultaneously—is meant to travel from the horizon to the wild churn of waters, then back to the nude Angelica, a figure influenced by the pose of Leonardo's Leda, on whom Ingres focuses the storm-light. The actual painting is precise; it articulates dim, turbulent motion through firm contours. It draws on classical and Renaissance iconography to set out an elegant, somewhat predictable statement of sensuality and chivalric promise. What is going on in Rossetti's reproduction? What, except a search for rhyme, informs 'The evil length of body chafes at fault'? In what way does Ingres's nude, so firmly rounded in pictorial treatment, so neoclassically modelled, 'trail' her limbs? *Hell-spurge* is odd. Applied to a common genus of plants, the word may, figuratively, stand for any kind of 'shoot' or 'sprout'. One suspects that the present instance resulted from a tonal-visual overlap with *surge*. In the 1870 edition of the *Poems*, the phrase becomes *Hell-birth*. *Geomaunt* and *teraphim* make a bizarre pair. The *O.E.D.* gives Rossetti's sonnet as reference for 'geomant' or 'geomaunt', one skilled in 'geomancy', the art of divining the future by observing terrestrial shapes or the ciphers drawn when handfuls of earth are scattered (geomancy occurs in Büchner's *Wozzeck* when the tormented Wozzeck sees a hideous future writ in the shapes of moss and fungi). Rossetti's source for this occult term may well have been its appearance in Dante:

> quando i geomanti lor maggior fortuna
> veggiono in oriente, innanzi all' alba,
> surger per via che poco le sta bruna. . . .
> *(Purgatorio*, XIX. 4–6)

The occurrence of *surger* so close to *geomanti* makes it likely that a remembrance of Dante in fact underlies this part of Rossetti's sonnet and may be more immediate to it than Ingres's painting. *Teraphim* is, of course, Hebrew and figures as such in the Authorized Version. It signifies both 'small idols' and such idols used as means of divination. It has a markedly heathen ring and Milton used the word with solemn reprobation in his *Prelatical Episcopacy* of 1641. What does either noun have to do with a sea-monster, especially with the rather pathetic marine beast at the bottom right of Ingres's composition? If anything, these sonorous rarities are 'of the earth, earthy'. Nor is it easy to accord *the spear's lithe stem* with Ingres's unyielding, almost brutally emphatic

diagonal. It is as if some blurred recollection of Uccello's treatment of Saint George had intervened between Rossetti and the
actual *Roger Délivrant Angélique* with which Ingres in 1819 sought
to illustrate a famous episode from Canto X of the *Orlando Furioso*.

But surely these are the wrong questions to ask.

Ingres's composition is the merest occasion for Rossetti's exercise. The existence of the painting is essential, but paradoxically
so. It relieves the sonnet of the burden of genuine cogency. In a
way typical of Pre-Raphaelite verse, the linguistic proposition is
validated by another medium (music, painting, textile, the decorative arts). Freed from autonomy, Rossetti's evocative caption can
go through its motions. What do these amount to? No firm doctrine
of correspondence is operative: the sonnet makes no attempt to
simulate the style and visual planes of the picture. It embodies a
momentary ricochet: griffin, armoured paladin, the boiling sea, a
swooning figure on a phallic rock trigger a volley of 'poetic'
gestures. The life of the sonnet, so far as any is observable, derives
from the use of formulaic tags (*heartsick trail of limb, sea shrewd and
salt, ramps blind*). I mean by 'formulaic' ready bits of loftiness and
sonority whose focus is not internal to the poem but is underwritten by exterior, modish conventions—in the Pre-Raphaelite
case, an identification of the 'poetic' with a pseudo-medieval,
Keatsian idiom. The impertinent grandeur of 'Hell-spurge of
geomaunt and teraphim' only aggravates the offence of nullity.
'Vexed at its base', with the exact, Latinate control of the verb, is
the one redeeming item. Indeed, the whole of line three foreshadows the Pre-Raphaelite strain in Yeats.

This Baedeker sonnet is not worth belabouring. But the
dilemma of just response which it poses is, I think, representative.
By mid-twentieth-century standards of poetic reality, 'Angelica
Rescued' scarcely exists. Its opportunistic relation to Ingres's
painting is one we are hardly prepared to recognize as a motive for
poetry. Nothing is actually being said in these fourteen lines; no
expressive needs are being served. At various points a portentous
musicality is meant to fill empty space. To our current way of
feeling, Rossetti's poem is a hollow bauble. In short, at this stage
in the history of feeling and verbal perception, it is difficult to

'read at all' the 'Sonnets for Pictures'. Their words are on the page; scholar and textual critic can give us whatever lexical and syntactic help is needed. But to most of us, the only available mode of apprehension will be an artifice—a suspension of natural reflexes in the interest of some didactic, polemical, or antiquarian aim.

We are, in the main, 'word-blind' to Pre-Raphaelite and Decadent verse. This blindness results from a major change in habits of sensibility. Our contemporary sense of the poetic, our often unexamined presumptions about valid or spurious uses of figurative speech have developed from a conscious negation of *fin de siècle* ideals. It was precisely with the rejection, by the Modernist movement, of Victorian and post-Victorian aesthetics, that the new astringency and insistence on verifiable structure came into force. We have for a time disqualified ourselves from reading comprehensively (a word which has in it the root for 'understanding') not only a good deal of Rossetti, but the poetry and prose of Swinburne, William Morris, Aubrey Beardsley, Ernest Dowson, Lionel Johnson, and Richard Le Gallienne. Dowson's 'Cynara' poem or Arthur Symons's 'Javanese Dancers' provide what comes near to being a test-case. Even in the cool light of the early 1990s, the intimation of real poetry is undeniable. Something vital and with an authority of its own is taking place just out of reach. Much more is involved here than a change of fashion, than the acceptance by journalism and the academy of a canon of English poetry chosen by Pound and Eliot. This canon is already being challenged; the primacy of Donne may be over, Browning and Tennyson are visibly in the ascendant. A design of literature which finds little worth commending between Dryden and Hopkins is obviously myopic. But the problem of how to read the Pre-Raphaelites and the poets of the nineties cuts deeper. What conceivable revolution of spirit would redirect us to a land of clear colours and stories

> In a region of shadowless hours,
> Where earth has a garment of glories
> And a murmur of musical flowers . . . ?

It is, literally, as if a language had been lost or the key to a cipher mislaid.

No tone-values are more difficult to determine than those of a seemingly 'neutral' text, of a diction which gives no initial purchase to lexicographer or grammarian. What dates a well-known passage-at-arms out of Nöel Coward's *Private Lives*?

Amanda. And India, the burning Ghars, or Ghats, or whatever they are, and the Taj Mahal. How was the Taj Mahal?

Elyot. Unbelievable, a sort of dream.

Amanda. That was the moonlight I expect, you must have seen it in the moonlight.

Elyot. Yes, moonlight is cruelly deceptive.

Amanda. And it didn't look like a biscuit box did it? I've always felt that it might.

Elyot. Darling, darling, I love you so.

Amanda. And I do hope you met a sacred Elephant. They're lint white I believe, and very, very sweet.

Elyot. I've never loved anyone else for an instant.

Amanda. No, no, you musn't—Elyot—stop.

Elyot. You love me, too, don't you? There's no doubt about it anywhere, is there?

Amanda. No, no doubt anywhere.

Elyot. You're looking very lovely you know, in this damned moonlight. Your skin is clear and cool, and your eyes are shining, and you're growing lovelier and lovelier every second as I look at you. You don't hold any mystery for me, darling, do you mind? There isn't a particle of you that I don't know, remember, and want.

Amanda. I'm glad, my sweet.

Elyot. More than any desire anywhere, deep down in my deepest heart I want you back again—please—

Amanda. Don't say any more, you're making me cry so dreadfully.

The dialogue is a brittle wonder, as perfect within its trivial bounds as comparable scenes in Congreve and Marivaux. And as irretrievably 'period'. Not a touch but affirms 1930.

Yet to show this is extremely difficult. There are, of course, datable props: that 'biscuit box' and, more elusively, 'lint white'. Somehow it would be surprising if that particular shade, however, clearly and immediately one can visualize it, came to mind, casually, in 1992. 'Damned moonlight' is *passé*, though again it is difficult to say why. The term 'particle' has, since the late forties, acquired a more specialized, ominous intonation. 'You're making me cry so dreadfully' has a faintly remote, lavender flavour; we would not, I think, use the adverb in quite this way or put on it Amanda's stress. Other indices are subtler. The location of feeling is different from ours: 'anywhere' contains much of the poignant mock seriousness of the passage. 'More than any desire anywhere' is beautifully clear, yet defies paraphrase; both its precision and lilting generality derive from habits of speech which are no longer wholly ours. 'Cruelly deceptive' is, once again, immediately significant and banal. But the combination of words in regard to moonlight is, from the point of view of the 1990s slightly out of focus, like the blur in an old photograph.

But the sense of period lies, principally, in Nöel Coward's speech-rhythms. Being both actor and song-writer, Coward treats language with explicit musicality; pitch and cadence are minutely marked. The uses of 'and' in this scene are as distinctive as they are in the prose of Hemingway, Nöel Coward's contemporary. Sometimes the word acts as a bar division; in Elyot's declaration of love, it helps produce an effect of breathless, fragile impulse. Commas are placed to extraordinary effect: by current measure, the dialogue is over-punctuated, but each 'silence' or absence of a pause (after 'deepest heart') is dramatically pointed. The *presto* and the *andante* in *Private Lives* are as time-bound as the fox-trot. A wholly different metronome beats in our present phrasing. Moreover, such is the specificity of Coward's *métier* that one makes out a particular accent behind the words. Even in cold print they compel the inflections, the acuity of certain vowels, the falling strain of fashionable speech at the end of the jazz age. One would guess at the mannerisms of Gertrude Lawrence and Nöel Coward himself, even if one had never heard them in this *pas de deux*. Our current feelings move in another key.

2

These examples are meant to argue a simple point. Any thorough reading of a text out of the past of one's own language and literature is a manifold act of interpretation. In the great majority of cases, this act is hardly performed or even consciously recognized. At best, the common reader will rely on what instant crutches footnotes or a glossary provide. When reading any piece of English prose after about 1800 and most verse, the general reader assumes that the words on the page, with a few 'difficult' or whimsical exceptions, mean what they would in his own idiom. In the case of 'classics' such as Defoe and Swift that assumption may be extended back to the early eighteenth century. It almost reaches Dryden, but it is, of course, a fiction.

Language is in perpetual change. Writing about Clough in 1869, Henry Sidgwick remarked: 'His point of view and habit of mind are less singular in England in the year 1869 than they were in 1859, and much less than they were in 1849. We are growing year by year more introspective and self-conscious: the current philosophy leads us to a close, patient and impartial observation and analysis of our mental processes: we more and more say and write what we actually do think and feel, and not what we intend to think or should desire to feel.' Generalized, Sidgwick's comment applies to every decade of the history of English speech and consciousness of which we have adequate record. At many points a graph of linguistic change would have to plot points far closer in time than a decade. Language—and this is one of the crucial propositions in certain schools of modern semantics—is the most salient model of Heraclitean flux. It alters at every moment in perceived time. The sum of linguistic events is not only increased but qualified by each new event. If they occur in temporal sequence, no two statements are perfectly identical. Though homologous, they interact. When we think about language, the object of our reflection alters in the process (thus specialized or metalanguages may have considerable influence on the vulgate). In short: so far as we experience and 'realize' them in linear progression, time and language are intimately related: they move and the arrow is never in the same place.

As we shall see, there are instances of arrested or sharply diminished mobility: certain sacred and magical tongues can be preserved in a condition of artificial stasis. But ordinary language is, literally at every moment, subject to mutation. This takes many forms. New words enter as old words lapse. Grammatical conventions are changed under pressure of idiomatic use or by cultural ordinance. The spectrum of permissible expression as against that which is taboo shifts perpetually. At a deeper level, the relative dimensions and intensities of the spoken and the unspoken alter. This is an absolutely central but little-understood topic. Different civilizations, different epochs do not necessarily produce the same 'speech mass'; certain cultures speak less than others; some modes of sensibility prize taciturnity and elision, others reward prolixity and semantic ornamentation. Inward discourse has its complex, probably unrecapturable history: both in amount and significant content, the divisions between what we say to ourselves and what we communicate to others have not been the same in all cultures or stages of linguistic development. With the intensifying definition of the subconscious which marks post-Renaissance habits of feeling in the West, this 'redistribution' of linguistic mass—public speech being only the tip of the iceberg—has certainly been drastic. The verbal charge and polarity of dreams is a historic variable. So far as language is mirror or counterstatement to the world, or most plausibly an interpenetration of the reflective with the creative along an 'interface' of which we have no adequate formal model, it changes as rapidly and in as many ways as human experience itself.

What is the rate of linguistic change? A whole branch of study, 'lexico-statistics', has grown up around this question. But no general answer is known, nor is there any reason to suppose that universal rules apply. In *Language*, Bloomfield asserted that 'linguistic change is far more rapid than biological change, but probably slower than the changes in other human institutions'. I wonder, and is it in fact possible to separate language from those institutions which it largely informs and whose change is itself so often identified by linguistic description? What evidence we have is local and so diverse as to resist all but the most tentative conjectures. Totally different rates of transformation are at work throughout the history of any single language or language group.[1] To cite a textbook example: the Indo-European paradigm of singular, dual, plural, which may go back to the beginnings of Indo-European linguistic history, survives to this day in the English usage *better of two* but *best of three or more*. Yet the English of King Alfred's day, most of whose features are chronologically far more recent, is practically unintelligible. At certain moments, languages change at an extraordinary pace; they are acquisitive of lexical and grammatical innovation, they discard eroded units with conscious speed. This is true, so far as literature is a reliable witness, of English between the 1560s and the turn of the century. A comparable rate of change, though in a restrictive, normative direction, marks the history of literate French from the 1570s to the advent of Malherbe and Guez de Balzac. Less than a generation separates Herder from Kleist, but the German of the 1820s is, in many respects, a different language, a different vehicle of conscious being, from that of the 1770s and early 1780s. So far as films, humour, journalistic style, and fiction allow one to judge, American English is, at the moment, in a state of acquisitive brilliance but also of instability whereas 'English English' may be losing resilience. Words and values shift at bewildering speed.

At other moments, languages are strongly conservative. Such is

---

[1] Lexico-statistics and 'glottochronology' propose the following formula for the calculation of the time t that has elapsed since related languages split away from a common ancestral stem:

$$t = \frac{\log c}{2 \log r}$$

in which c stands for the percentage of cognates and r for the percentage of cognates retained after a millennium of separation (t is tabulated in millennia). This approach, largely associated with the work of the late M. Swadesh, remains controversial. Cf. R. B. Lees, 'The Basic of Glottochronology' (*Language*, XXIX, 1953), and M. Lionel Bender: 'Linguistic Indeterminacy: Why you cannot reconstruct "Proto-Human"' (*Language Sciences*, 26, 1973).

the prescriptive weight of post-Cartesian syntax that the French Romantics, while proclaiming themselves rebels and pioneers, cast their plays in traditional alexandrines and hardly modified the armature of French prose. During the 1760s English prose seems to have reached a confident, urbane plateau. Resistant to innovation, it extended its authority over much of poetic practice; late Augustan verse has a characteristic linguistic complacency. The conservatism, indeed the deliberate retention of the archaic, which marks several epochs in the history of Chinese has often been noted. Post-war Italian, despite the pressures of *verismo* and the conscious modernism of other media, such as film, has been curiously inert; Gadda's omnivorous demotic stands out as an exceptional, challenging case. No facile connection between political and linguistic change will serve. Both the French and the Bolshevik revolutions were linguistically conservative, almost academic in their rhetoric. The Second Empire, on the other hand, sees one of the principal movements of stress and exploration in the poetics and habits of sensibility of the French language. At most stages in the history of a language, moreover, innovative and conservative tendencies coexist. Milton, Andrew Marvell, and Dryden were contemporaries. In his 'old-fashionedness' Robert Frost drew on currents of speech as vital as those enlisted, or newly tapped, by Allen Ginsberg. The facts of language are as crowded with contrasting impulse as Leonardo's drawings of the braids and spirals of live water.

Even more difficult questions arise when we ask whether the notion of entropy applies to language. Do languages wane, do their powers of shaping response atrophy? Are there linguistic reflexes which have slowed and lost vital exactitude? The danger in putting the question this way is obvious: to think of the life and death of language in organic, temporal terms may be an animist fiction. Languages are wholly arbitrary sets of signals and conventionalized counters. Though the great master Tartakower thought otherwise, we do not ascribe feelings or some mystery of autonomous being to chess pieces. Yet the intimation of life-force and the concomitant notion of linguistic decay are difficult to discard. Some who have thought hardest about the nature of language and about the interactions of speech and society—De Maistre, Karl Kraus, Walter Benjamin, George Orwell—have, consciously or not, argued from a vitalist metaphor. In certain civilizations there come epochs in which syntax stiffens, in which the available resources of live perception and restatement wither. Words seem to go dead under the weight of sanctified usage; the frequence and sclerotic force of clichés, of unexamined similes, of worn tropes increases. Instead of acting as a living membrane, grammar and vocabulary become a barrier to new feeling. A civilization is imprisoned in a linguistic contour which no longer matches, or matches only at certain ritual, arbitrary points, the changing landscape of fact. There are aspects of paralysis, of language used to formalize rather than quicken the means of human response, in the Greek of the Byzantine liturgy. Is there some linguistic factor in the riddle of the collapse of Mayan culture? Did the language, with its presumably high proportion of immutable, hieratic phraseology, no longer provide a usable, generative model of reality? 'Words, those guardians of meaning, are not immortal, they are not invulnerable,' wrote Adamov in his notebook for 1938; 'some may survive, others are incurable'. When war came, he added: 'Worn, threadbare, filed down, words have become the carcass of words, phantom words; everyone drearily chews and regurgitates the sound of them between their jaws.'

The reverse may also be true. Historical relativism infers that there are no beginnings, that each human act has precedent. This could be spurious hindsight. The quality of genius in the Greek and Hebraic statement of human possibility, the fact that no subsequent articulation of felt life in the Western tradition has been either as complete or formally inventive, are undeniable. The totality of Homer, the capacity of the *Iliad* and *Odyssey* to serve as repertoire for most of the principal postures of Western consciousness—we are petulant as Achilles and old as Nestor, our homecomings are those of Odysseus—point to a moment of singular linguistic energy. (My own view is that the collation of the *Iliad* and the composition of the *Odyssey* coincide with the 'new immortality' of writing, with the specific transition from oral to written literature.) Aeschylus may not only have been the greatest of tragedians but the creator of the genre, the first to locate in dialogue the supreme intensities of human conflict. The grammar of the Prophets in Isaiah enacts a profound metaphysical scandal— the enforcement of the future tense, the extension of language over time. A reverse discovery animates Thucydides; his was the explicit realization that the past is a language-construct, that the past tense of the verb is the sole guarantor of history. The formidable gaiety of the Platonic dialogues, the use of the dialectic as a method of intellectual chase, stems from the discovery that words, stringently tested, allowed to clash as in combat or manoeuvre as in a dance, will produce new shapes of understanding. Who was the first man to tell a joke, to strike laughter out of speech?

In all these cases, language was 'new'; or, more accurately, the poet, the chronicler, the philosopher gave to human behaviour and to the current of mental experience an unprecedented 'second life'—a life they soon found to be more enduring, more exhaustive of meaning, than either biological or social existence. This insight, which is both exultant and tragic (the poet knows that the fictional personage he has created will outlive him), declares itself over and over in Homer and Pindar. It is difficult to suppose that the *Oresteia* was composed very long after the dramatist's first awareness of the paradoxical relations between himself, his personages, and the fact of personal death. The classic is the only total revolutionary: he is the first to burst not into that silent sea— language being rigorously coterminous with man—but into the *terra incognita* of symbolic expression, of analogy, of allusion, of simile and ironic counterpoint. We have histories of massacre and deception, but none of metaphor. We cannot accurately conceive what it must have been like to be the first to compare the colour of the sea with the dark of wine or to see autumn in a man's face. Such figures are new mappings of the world, they reorganize our habitation in reality. When the pop song moans that there is no new way of saying that I am in love or that her eyes are full of stars, it touches one of the main nerves in Western literature. Such was the acquisitive reach of Hellenic and Hebraic articulation, that genuine additions and new finds have been rare. No desolation has gone deeper than Job's, no dissent from mundanity

has been more trenchant than Antigone's. The fire-light in the
domestic hearth at close of day was seen by Horace; Catullus
came near to making an inventory of sexual desire. A great part of
Western art and literature is a set of variations on definitive
themes. Hence the anarchic bitterness of the late-comer and the
impeccable logic of Dada when it proclaims that no new impulses
of feeling or recognition will arise until language is demolished.
'Make all things new' cries the revolutionary, in words as old as
the Song of Deborah or the fragments of Heraclitus.

Why did certain languages effect a lasting grip on reality? Did
Hebrew, Greek, and Chinese (in a way that may also relate to the
history of writing) have distinctive resources? Or are we, in fact,
asking about the history of particular civilizations, a history reflected in and energized by language in ways so diverse and
interdependent that we cannot give a credible answer? I suspect
that the receptivity of a given language to metaphor is a crucial
factor. That receptivity varies widely: ethno-linguists tell us, for
example, that Tarascan, a Mexican tongue, is inhospitable to new
metaphors, whereas Cuna, a Panamanian language, is avid for
them. An Attic delight in words, in the play of rhetoric, was
noticed and often mocked throughout the Mediterranean world.
Qiryat Sepher, the 'City of the Letter' in Palestine, and the Syrian
Byblos, the 'Town of the Book', are designations with no true
parallel anywhere else in the ancient world. By contrast other
civilizations seem 'speechless', or at least, as may have been the
case in ancient Egypt, not entirely cognizant of the creative and
transformational powers of language. In numerous cultures blindness is a supreme infirmity and abdication from life; in Greek
mythology the poet and the seer are blind so that they may, by the
antennae of speech, see further.

One thing is clear: every language-act has a temporal determinant. No semantic form is timeless. When using a word we wake
into resonance, as it were, its entire previous history. A text is
embedded in specific historical time; it has what linguists call a
diachronic structure. To read fully is to restore all that one can of
the immediacies of value and intent in which speech actually
occurs.

There are tools for the job. A true reader is a dictionary addict. He knows that English is particularly well served, from Bosworth's *Anglo-Saxon Dictionary*, through Kurath and Kuhn's *Middle English Dictionary* to the almost incomparable resources of the *O.E.D.* (both Grimm's *Wörterbuch* and the Littré are invaluable but neither French nor German have found their history and specific genius as completely argued and crystallized in a single lexicon). Rossetti's *geomaunt* will lead to Shipley's *Dictionary of Early English* and the reassurance that 'the topic is capped with *moromancy*, foolish divination, a 17th century term that covers them all'. Skeat's *Etymological Dictionary* and *Principles of English Etymology* are an indispensable first step towards grasping the life of words. But each period has its specialized topography. Skeat and Mayhew's *Glossary of Tudor and Stuart Words* necessarily accompanies one's reading of English literature from Skelton to Marvell. No one will get to the heart of the Kipling world, or indeed clear up certain cruces in Gilbert and Sullivan without Sir H. Yule and A. C. Burnell's *Hobson-Jobson*. Dictionaries of proverbs and place-names are essential. Behind the façade of public discourse extends the complex, shifting terrain of slang and taboo speech. Without such quarries as Champion's *L'Argot ancien* and Eric Partridge's lexica of underworld usage, important elements of Western literature, from Villon to Genet is only partly legible.

Beyond such major taxonomies lie areas of relevant specialization. A demanding reader of mid-eighteenth-century verse will often find himself referring to the Royal Horticultural Society's *Dictionary of Gardening*. The old *Drapers' Dictionary* of S. William Beck clears up more than one erotic conundrum in Restoration comedy. Fox-Davies's *Armorial Families* and other registers of heraldry are as helpful at the opening of *The Merry Wives of Windsor* as they are in elucidating passages in the poetry of Sir Walter Scott. A true Shakespeare library is, of itself, very nearly a summation of human enterprise. It would include manuals of falconry and navigation, of law and of medicine, of venery and the occult. A central image in *Hamlet* depends on the vocabulary of wool-dyeing (wool greased or *enseamed* with hog's lard *over the nasty sty*); from *The Taming of the Shrew* to *The Tempest*, there is scarcely a Shakespearean play which does not use the extensive glossary of Elizabethan musical terms to make vital statements about human motive or conduct. Several episodes in Jane Austen can only be made out if one has knowledge, not easily come by, of a Regency escritoire and of how letters were sent. Being so physically cumulative in effect, so scenic in structure, the Dickens world draws on a great range of technicality. There is a thesaurus of Victorian legal practice and finance in *Bleak House* and *Dombey and Son*. The Admiralty's *Dictionary of Naval Equivalents* and a manual of Victorian steam-turbine construction have helped clear up the meaning of one of the most vivid yet hermetic similes in 'The Wreck of the Deutschland'.

But these are externals. The complete penetrative grasp of a text, the complete discovery and recreative apprehension of its life-forms (*prise de conscience*), is an act whose realization can be precisely felt but is nearly impossible to paraphrase or systematize. It is a matter of what Coleridge, in whom the capacity for vital comprehension was striking, called 'speculative instruments'. An informed, avid awareness of the history of the relevant language, of the transforming energies of feeling which make of syntax a record of social being, is indispensable. One must master the temporal and local setting of one's text, the moorings which attach even the most idiosyncratic of poetic expressions to the surrounding idiom. Familiarity with an author, the kind of restive intimacy which demands knowledge of all his work, of the best and the botched, of *juvenilia* and *opus posthumum*, will facilitate understanding at any given point. To read Shakespeare and Hölderlin is, literally, to prepare to read them. But neither erudition nor industry make up the sum of insight, the intuitive thrust to the centre. 'To read attentively, think correctly, omit no relevant consideration, and repress self-will, are no ordinary accomplishments,' remarked A. E. Housman in his London inaugural, yet more is needed: 'just literary perception, congenial intimacy with the author, experience which must have been won by study, and mother wit which he must have brought from his mother's womb'. Dr. Johnson, when editing Shakespeare, went further: conjectural criticism, by which he meant that final interaction with a text which allows a reader to emend his author, 'demands more than humanity possesses'.

Where the most thorough possible interpretation occurs, where our sensibility appropriates its object while, in this appropriation, guarding, quickening that object's autonomous life, the process is one of 'original repetition'. We re-enact, in the bounds of our own secondary but momentarily heightened, educated consciousness, the creation by the artist. We retrace, both in the image of a man drawing and of one following an uncertain path, the coming into form of the poem. Ultimate connoisseurship is a kind of finite *mimesis*: through it the painting or the literary text is made new— though obviously in that reflected, dependent sense which Plato gave to the concept of 'imitation'. The degree of re-creative immediacy varies. It is most radically life-giving in the case of musical performance. Every musical realization is a new *poiesis*. It differs from all other performances of the same composition. Its ontological relationship to the original score and to all previous renditions is twofold: it is at the same time reproductive and innovatory. In what sense does unperformed music exist? But what is the measure of the composer's verifiable intent after successive performances? The picture-restorer would come at the lesser end of the scale: for all its probing tact, the job is essentially conservative. It aims to arrest the naturally changing life of the work of art in a fiction of unique, static authenticity. But in either case a metaphor of love is not far distant. There is a strain of femininity in the great interpreter, a submission, made active by intensity of response, to the creative presence. Like the poet, the master executant or critic can say *Je est un autre*. As we shall see, two principal movements of spirit conjoin: the achievement of 'inscape' (*Einfühlung*) is both a linguistic and an emotive act.

In their use of 'speculative instruments', critic, editor, actor, and reader are on common ground. Through their diversely accentuated but cognate needs, written language achieves a continuation of life. It is they, in Ezra Pound's phrase, who see to it that literature is news that stays news. The function of the actor is particularly graphic. Each time *Cymbeline* is staged, Posthumus's monologue becomes the object of manifold 'edition'. An actor can choose to deliver the words of the Folio in what is thought to have been the pronunciation of Elizabethan English. He can adopt a neutral, though in fact basically nineteenth-century solemn register and *vibrato* (the equivalent of a Victorian prize calf binding). He may by control of caesura and vowel-pitch convey an impression of modernity. His—the producer's—choice of costume is an act of practical criticism. A Roman Posthumus represents a correction of Elizabethan habits of anachronism or symbolic contemporaneity—themselves a convention of feeling which we may not fully grasp. A Jacobean costume points to the location of the play in a unique corpus: it declares of *Cymbeline* that Shakespeare's authorship is the dominant fact. Modern dress production argues a trope of 'eternal relevance'; whatever the singularities of Jacobean idiom, the 'meaning' of Posthumus's outburst is to be enforced here and now. But there can also be, indeed there have been, presentations of *Cymbeline* in Augustan, Byronic, or Edwardian garb. Each embodies a specific commentary on the text, each realizes a particular mode of animation. A poem can also be recast. Make a collage of, say, Hieronymus Bosch motifs, Victorian erotica, and Dali squiggles—and place Rossetti's sonnet in the middle. It will take on a sudden queer vehemence. The blaze of life will be spurious. But only great art both solicits and withstands exhaustive or wilful interpretation.

'Interpretation' as that which gives language life beyond the moment and place of immediate utterance or transcription, is what I am concerned with. The French word *interprète* concentrates all the relevant values. An actor is *interprète* of Racine; a pianist gives *une interprétation* of a Beethoven sonata. Through engagement of his own identity, a critic becomes *un interprète*—a life-giving performer—of Montaigne or Mallarmé. As it does not include the world of the actor, and includes that of the musician only by analogy, the English term *interpreter* is less strong. But it is congruent with French when reaching out in another crucial direction. *Interprète/interpreter* are commonly used to mean *translator*.

This, I believe, is the vital starting point.

When we read or hear any language-statement from the past, be it Leviticus or last year's best-seller, we translate. Reader, actor, editor are translators of language out of time. The schematic model of translation is one in which a message from a source-language passes into a receptor-language via a transformational process. The barrier is the obvious fact that one language differs from the other, that an interpretative transfer, sometimes, albeit misleadingly, described as encoding and decoding, must occur so that the message 'gets through'. Exactly the same model—and this is what is rarely stressed—is operative within a single language. But here the barrier or distance between source and receptor is time. As we have seen, the tools employed in both operations are correlate: both the 'external' and 'internal' translator/*interprète* have recourse to lexica, historical grammars, glossaries of particular periods, professions, or social milieux, dictionaries of argot, manuals of technical terminology. In either case the means of penetration are a complex aggregate of knowledge, familiarity, and re-creative intuition. In either case also, as we shall see, there are characteristic penumbras and margins of failure. Certain elements will elude complete comprehension or revival. The time-barrier may be more intractable than that of linguistic difference. Any bilingual translator is acquainted with the phenomenon of 'false friends'—homonyms such as French *habit* and English *habit* which on occasion might, but almost never do, have the same meaning, or mutually untranslatable cognates such as English *home* and German *Heim*. The 'translator within' has to cope with subtler treasons. Words rarely show any outward mark of altered meaning, they body forth their history only in a fully established context. Where a passage is historically remote, say in Chaucer, the business of internal translation tends towards being a bilingual process: eye and ear are kept alert to the necessity of decipherment. The more seemingly standardized the language—the outward cast of the modern comes in with great speed after Dryden—the more covert are indices of semantic dating. We read as if time has had a stop. Thus a good deal of our theatre and the mass of our current literacy are founded on lazy translation. The received message is thinned and distorted. But so it is, more often than not, in a transfer between languages.

The process of diachronic translation inside one's own native tongue is so constant, we perform it so unawares, that we rarely pause either to note its formal intricacy or the decisive part it plays in the very existence of civilization. By far the greatest mass of the past as we experience it is a verbal construct. History is a speech-act, a selective use of the past tense. Even substantive remains such as buildings and historical sites must be 'read', i.e. located in a context of verbal recognition and placement, before they assume real presence. What material reality has history outside language, outside our interpretative belief in essentially linguistic records (silence knows no history)? Where worms, fires of London, or totalitarian régimes obliterate such records, our consciousness of past being comes on a blank space. We have no total history, no history which could be defined as objectively real because it contained the literal sum of past life. To remember everything is a condition of madness. We remember culturally, as we do individually, by conventions of emphasis, foreshortening, and omission. The landscape composed by the past tense, the semantic organization of remembrance, is stylized and differently coded by different cultures. A Chinese painting of figures in a garden differs from one by Poussin. Successive constructs of the past form a many-stranded helix, with imaginary chronologies spiralling around the neutral stem of 'actual' biological time. The Middle Ages experienced by Walter Scott were not those mimed by the Pre-Raphaelites. The Augustan paradigm of Rome was, like that of Ben Jonson and the Elizabethan Senecans, an active fiction, a 'reading into life'. But the two models were very different. From Marsilio Ficino to Freud, the image of Greece, the verbal icon made up of successive translations of Greek literature, history, and philosophy, has oriented certain fundamental movements in Western feeling. But each reading, each translation differs, each is undertaken from a distinctive angle of vision. The Platonism of the Renaissance is not that of Shelley, Hölderlin's Oedipus is not the Everyman of Freud or the limping shaman of Lévi-Strauss.

As every generation retranslates the classics, out of a vital compulsion for immediacy and precise echo, so every generation uses language to build its own resonant past. At moments of historical stress, mythologies of the 'true past' follow on each other at such speed that entirely different perspectives coexist and blur at the edges. There is, today, a 1914–19 *figura* for those in their seventies; to a man of forty, 1914 is the vague forerunner of realities which only gather meaning in the crises of the late 1930s; to the 'bomb-generation', history is an experience that dates to 1945; what lies before is an allegory of antique illusions. In the recent revolts of the very young, a surrealistic syntax, anticipated by Artaud and Jarry, is at work: the past tense is to be excluded from the grammar of politics and private consciousness. Being inevitably 'programmed' and selective in values, history is an instrument of the ruling caste. The present tense is allowed because it vaults, at once, into the confirming future. To remember is to risk despair; the past tense of the verb *to be* must infer the reality of death.

This metaphysic of the instant, this slamming of the door on the long galleries of historical consciousness, is understandable. It has a fierce innocence. It embodies yet another surge towards Eden, towards that pastoral before time (there could be no autumn before the apple was off the branch, no fall before the Fall) which the eighteenth century sought in the allegedly static cultures of the south Pacific. But it is an innocence as destructive of civilization as it is, by concomitant logic, destructive of literate speech. Without the true fiction of history, without the unbroken animation of a chosen past, we become flat shadows. Literature, whose genius stems from what Éluard called *le dur désir de durer*, has no chance of life outside constant translation within its own language. Art dies when we lose or ignore the conventions by which it can be read, by which its semantic statement can be carried over into our own idiom—those who have taught us how to reread the Baroque, for example, have extended the backward reach of our senses. In the absence of interpretation, in the manifold but generically unified meaning of the term, there could be no culture, only an inchoate silence at our backs. In short, the existence of art and literature, the reality of felt history in a community, depend on a never-ending, though very often unconscious, act of internal translation. It is no overstatement to say that we possess civilization because we have learnt to translate out of time.

3

Since Saussure, linguists distinguish between a diachronic (vertical) and synchronic (horizontal) structure of language. This distinction applies also to internal translation. If culture depends on the transmission of meaning across time—German *übertragen* carries the exact connotations of translation and of handing down through narrative—it depends also on the transfer of meaning in space.

There is a centrifugal impulse in language. Languages that extend over a large physical terrain will engender regional modes and dialects. Before the erosive standardizations of radio and television became effective, it was a phonetician's parlour-trick to locate, often to within a few dozen miles, the place of origin of an American from the border states or a north-country Englishman. The French spoken by a Norman is not that of the Touraine or the Camargue. *Hoch-* and *Plattdeutsch* are strongly differentiated. Indeed, in many important languages, differences of dialect have polarized to the degree that we are almost dealing with distinct tongues. The mutual incomprehensibility of diverse branches of Chinese such as Cantonese and Mandarin are notorious. A Milanese has difficulty in understanding the Italian spoken in neighbouring Bergamo. In all these cases comprehension demands translation along lines closer and closer to those of inter-lingual transfer. There are dictionaries and grammars of Venetian, Neapolitan, and Bergamasque.

Regional, dialectal disparities are the easiest to identify. Any body of language, spoken at the same time in a complex community, is in fact rifted by much subtler differentiations. These relate to social status, ideology, profession, age, and sex.

Different castes, different strata of society use a different idiom. Eighteenth-century Mongolia provides a famous case. The religious language was Tibetan; the language of government was Manchu; merchants spoke Chinese; classical Mongol was the literary idiom; and the vernacular was the Khalka dialect of Mongol. In very many cases, such as the sacred speech of the Zuni Indians, such differences have been rigorously formalized. Priests and initiates use a vocabulary and formulaic repertoire distinct from everyday language.[1] But special languages—hieratic, masonic, Ubuesque, mandarin, the semi-occult speech of the regimental mess or fraternity initiation—pose no essential difficulty. The need for translation is self-evident. Far more important and diffuse are the uses of inflection, grammatical structure, and word-choice by different social classes and ethnic groups to affirm their respective identities and to affront one another. It may be that the agonistic functions of speech inside an economically and socially divided community outweigh the functions of genuine communication. As we shall see throughout this study, languages conceal and internalize more, perhaps, than they convey outwardly. Social classes, racial ghettos speak at rather than to each other.

Upper-class English diction, with its sharpened vowels, elisions, and modish slurs, is both a code for mutual recognition—accent is worn like a coat of arms—and an instrument of ironic exclusion. It communicates from above, enmeshing the actual unit of information, often imperative or conventionally benevolent, in a network of superfluous linguistic matter. But this redundancy is itself functional: one speaks most completely to one's inferiors—the speech-act is most expressive of status, innuendo, and power—when a peer is in earshot. The ornamental irrelevancies and elided insinuations are not addressed so much to the tradesman or visitor as to one's fellow-officer or clubman who will recognize in them signals of complicity. Thackeray and Wodehouse are masters at conveying this dual focus of aristocratic semantics. As analysed by Proust, the discourse of Charlus is a light-beam pin-pointed, obscured, prismatically scattered as by a Japanese fan beating before a speaker's face in ceremonious motion. To the lower classes, speech is no less a weapon and a vengeance. Words may

---

[1] For a classic study of secret speech forms, cf. Michel Leiris, *La Langue secrète des Dogons de Sanga* (*Soudan Français*) (Paris, 1948). In this case, the special, occult language arises both from reasons of mythical initiation and from the differentiation between men and women. Cf. also M. Delafosse, 'Langage secret et langage conventionnel dans l'Afrique noire' (*L'Anthropologie*, XXXII, 1922). Though obviously dated, A. Van Gennep's 'Essai d'une théorie des langues spéciales' (*Revue des études ethnographiques et sociologiques*, I, 1908) remains of interest.

be appropriated and suborned, either by being given a clandestine significance or by being mocked through false intonation (in tribal warfare a captured fetish will be turned against its former owners). The pedantic decorum of 'menial' parlance in Molière, in Jeeves, is a stratagem of parody. Where there is no true kinship of interests, where power relations determine the conditions of meeting, linguistic exchange becomes a duel. Very often the seeming inarticulateness of the labourer, the thick twilight of Cockney speech, or the obeisant drag of Negro response are a well-judged feint. The illiteracy of the trooper or the navvy were porcupine quills, calculated to guard some coherence of inner life while wounding outward. The patronized and the oppressed have endured behind their silences, behind the partial incommunicado of their obscenities and clotted monosyllables.[1]

This, I suspect, makes for one of the radical differences between upper- and lower-class language habits. The privileged speak to the world at large as they do to themselves, in a conspicuous consumption of syllables, clauses, prepositions, concomitant with their economic resources and the spacious quarters they inhabit. Men and women of the lower class do not speak to their masters and enemies as they do to one another, hoarding what expressive wealth they have for internal use. For an upper- or middle-class listener, the authentic play of speech below stairs or in the proletarian home is more difficult to penetrate than any club. White and black trade words as do front-line soldiers lobbing back an undetonated grenade. Watch the motions of feigned responsiveness, menace, and non-information in a landlord's dialogue with his tenant or in the morning banter of tally-clerk and lorry-driver.

---

[1] Cf. the following for examples of the social stratification and social-strategic uses of speech: Felix M. and Marie M. Keesing, *Elite Communication in Samoa* (Stanford University Press, 1956); J. J. Gumperz and Charles A. Ferguson (eds.), *Linguistic Diversity in South Asia* (University of Indiana Press, 1960); Clifford Geertz, *The Religion of Java* (Illinois, 1960); Basil Bernstein, 'Social Class, Linguistic Codes and Grammatical Elements' (*Language and Speech*, V, 1962); William Labor, Paul Cohen, and Clarence Robbins, *A Preliminary Study of English Used by Negro and Puerto Rican Speakers in New York City* (New York, 1965); Robbins Burling, *Man's Many Voices: Language in its Cultural Context* (New York, 1970); Peter Trudgill, *The Social Differentiation of English in Norwich* (Cambridge University Press, 1974).

Observe the murderous undertones of apparently urbane, shared speech between mistress and maids in Genet's *Les Bonnes*. So little is being said, so much is 'being meant', thus posing almost intractable problems for the translator.

Polysemy, the capacity of the same word to mean different things, such difference ranging from nuance to antithesis, characterizes the language of ideology. Machiavelli noted that meaning could be dislocated in common speech so as to produce political confusion. Competing ideologies rarely create new terminologies. As Kenneth Burke and George Orwell have shown in regard to the vocabulary of Nazism and Stalinism, they pilfer and decompose the vulgate. In the idiom of fascism and communism, 'peace', 'freedom', 'progress', 'popular will' are as prominent as in the language of representative democracy. But they have their fiercely disparate meanings. The words of the adversary are appropriated and hurled against him. When antithetical meanings are forced upon the same word (Orwell's Newspeak), when the conceptual reach and valuation of a word can be altered by political decree, language loses credibility. Translation in the ordinary sense becomes impossible. To translate a Stalinist text on peace or on freedom under proletarian dictatorship into a non-Stalinist idiom, using the same time-honoured words, is to produce a polemic gloss, a counter-statement of values. At the moment, the speech of politics, of social dissent, of journalism is full of loud ghost-words, being shouted back and forth, signifying contraries or nothing. It is only in the underground of political humour that these shibboleths regain significance. When the entry of foreign tanks into a free city is glossed as 'a spontaneous, ardently welcomed defence of popular freedom' (*Izvestia*, 27 August 1968), the word 'freedom' will preserve its common meaning only in the clandestine dictionary of laughter.

That dictionary, one supposes, plays a large role in the language of children. Here diachronic and synchronic structures overlap. At any given time in a community and in the history of the language, speech modulates across generations. Or as psycho-linguists put it, there are 'phenomena of age grading' in all known languages. The matter of child-speech is a deep and fascinating one. Again, there are numerous languages in which such speech is formally set apart. Japanese children employ a separate vocabulary for everything they have and use up to a certain age. More common, indeed universal, is the case in which children carve their own language-world out of the total lexical and syntactic resources of adult society. So far as children are an exploited and mutinous class, they will, like the proletariat or ethnic minorities, pilfer and make risible the rhetoric, the taboo words, the normative idioms of their oppressors. The scatological doggerels of the nursery and the alley-way may have a sociological rather than a psychoanalytic motive. The sexual slang of childhood, so often based on mythical readings of actual sexual reality rather than on any physiological grasp, represents a night-raid on adult territory. The fracture of words, the maltreatment of grammatical norms which, as the Opies have shown, constitute a vital part of the lore, mnemonics, and secret parlance of childhood, have a rebellious aim: by refusing, for a time, to accept the rules of grown-up speech, the child seeks to keep the world open to his own, seemingly unprecedented needs. In the event of autism, the speech-battle between child and master can reach a grim finality. Surrounded by incomprehensible or hostile reality, the autistic child breaks off verbal contact. He seems to choose silence to shield his identity but even more, perhaps, to destroy his imagined enemy. Like murderous Cordelia, children know that silence can destroy another human being. Or like Kafka they remember that several have survived the song of the Sirens, but none their silence.

The anthropology or, as it would now be called, ethno-linguistics of child-speech is still at a rudimentary stage. We know far more of the languages of the Amazon. Adults tend to regard the language of children as an embryonic, inferior version of their own. Children, in turn, guard their preserve. Among early explorers were the novelists of the second half of the nineteenth century. Behind them lay certain tenacious eighteenth-century notions. Diderot had referred to 'l'enfant, ce petit sauvage', joining under one rubric the nursery and the natives of the South Seas. The sense of a dubious Eden, with its implications of a lost linguistic innocence and immediacy, colours our entire image of the child: we speak still of the *jardin d'enfants*, the *Kindergarten*. The passage from the transitional into the exploratory model is visible in Lewis Carroll. *Alice in Wonderland* relates to voyages into the language-world and special logic of the child as Gulliver relates to the travel literature of the Enlightenment. Both are subversive considerations of the general venture, and statements of limitation: they inform the voyager that he will, inevitably, find what he has brought with him and that there are blanks on the map beyond the reach of his survey.

Henry James was one of the true pioneers. He made an acute study of the frontier zones in which the speech of children meets that of grown-ups. *The Pupil* dramatizes the contrasting truth-functions in adult idiom and the syntax of a child. Children, too, have their conventions of falsehood, but they differ from ours. In *The Turn of the Screw*, whose venue is itself so suggestive of an infected Eden, irreconcilable semantic systems destroy human contact and make it impossible to locate reality. This cruel fable moves on at least four levels of language: there is the provisional key of the narrator, initiating all possibilities but stabilizing none, there is the fluency of the governess, with its curious gusts of theatrical *bravura*, and the speech of the servants so avaricious of insight. These three modes envelop, qualify, and obscure that of the children. Soon incomplete sentences, filched letters, snatches of overheard but misconstrued speech, produce a nightmare of untranslatability. 'I said things,' confesses Miles when pressed to the limit of endurance. That tautology is all his luminous, incomprehensible idiom can yield. The governess seizes upon 'an exquisite pathos of contradiction'. Death is the only plain statement left. Both *The Awkward Age* and *What Maisie Knew* focus on children at the border, on the brusque revelations and bursts of static which mark the communication between adolescents and those adults whose language-territory they are about to enter.

The speech of children and adolescents fascinated Dostoevsky. Its ferocious innocence, the tactical equivocations of the maturing child, are reproduced in *The Brothers Karamazov*. St Francis's ability to parley with birds is closely echoed in Alyosha's understanding of Kolya and the boys. But for all their lively truth, children in the novels of James and Dostoevsky remain, in large measure, miniature adults. They exhibit the uncanny percipience of the 'aged' infant Christ in Flemish art. Mark Twain's transcriptions of the secret and public idiom of childhood penetrate much further. A genius for receptive insight animates the rendition of Huck Finn and Tom Sawyer. The artfulness of their language, its ceremonies of insult and kinship, its tricks of understatement are as complex as any in adult rhetoric. But they are unfailingly re-creative of a child's way. The discrimination is made even more exact by the neighbouring but again very different 'childishness' of Negro speech. For the first time in Western literature, the linguistic terrain of childhood was mapped without being laid waste. After Mark Twain, child psychology and Piaget could proceed.

When speaking to a young boy or girl we use simple words and a simplified grammar; often we reply by using the child's own vocabulary; we bend forward. For their part, children will use different phrasings, intonations, and gestures when addressing a grown-up from those used when speaking to themselves (the iceberg mass of child language) or to other children. All these are devices for translation. J. D. Salinger catches us in the act:

> Sybil released her foot. 'Did you read "Little Black Sambo"?' she said.
> 'It's very funny you ask me that,' he said. 'It so happens I just finished reading it last night.' He reached down and took back Sybil's hand. 'What did you think of it?' he asked her.
> 'Did the tigers run all around that tree?'
> 'I thought they'd never stop. I never saw so many tigers.'
> 'There were only six,' Sybil said.
> '*Only* six!' said the young man. 'Do you call that *only*?'
> 'Do you like wax?' Sybil asked.
> 'Do I like what?' asked the young man.
> 'Wax.'
> 'Very much. Don't you?'
> Sybil nodded. 'Do you like olives?' she asked.
> 'Olives—yes. Olives and wax. I never go anyplace without 'em.'
>
> . . .
>
> Sybil was silent.
> 'I like to chew candles,' she said finally.
> 'Who doesn't?' said the young man, getting his feet wet.

This is the '*perfect* day for bananafish', the swift passage from Pentecost to silence. Being so near death, Seymour, the hero of the story, translates flawlessly. Usually, the task is more difficult. There is so much we do not know. Even more than the illiterate and the oppressed, children have been kept in the margin of history. Their multitudinous existence has left comparatively few archives. How, for instance, do class-lines cut across age gradients? Is it true that the current revolution in the language of sex is entirely a middle-class phenomenon, that sex-talk of the most anatomical and disenchanted kind has always been in use among children of the working-class? One thing is clear. The entry of the child into complete adult notice, a heightened awareness of its uniquely vulnerable and creative condition, are among the principal gains of the recent past. The stifled voices of children that haunt Blake's poetry are no longer a general fact. No previous society has taken as much trouble as ours to hear the actual language of the child, to receive and interpret its signals without distorting them.

In most societies and throughout history, the status of women has been akin to that of children. Both groups are maintained in a condition of privileged inferiority. Both suffer obvious modes of exploitation—sexual, legal, economic—while benefiting from a mythology of special regard. Thus Victorian sentimentalization of the moral eminence of women and young children was concurrent with brutal forms of erotic and economic subjection. Under sociological and psychological pressure, both minorities have developed internal codes of communication and defence (women and children constitute a symbolic, self-defining minority even when, owing to war or special circumstance, they outnumber the adult males in the community). There is a language-world of women as there is of children.

We touch here on one of the most important yet least understood areas of biological and social existence. Eros and language mesh at every point. Intercourse and discourse, copula and copulation, are sub-classes of the dominant fact of communication. They arise from the life-need of the ego to reach out and comprehend, in the two vital senses of 'understanding' and 'containment', another human being. Sex is a profoundly semantic act. Like language, it is subject to the shaping force of social convention, rules of proceeding, and accumulated precedent. To speak and to make love is to enact a distinctive twofold universality: both forms of communication are universals of human physiology as well as of social evolution. It is likely that human sexuality and speech developed in close-knit reciprocity. Together they generate the history of self-consciousness, the process, presumably millenary and marked by innumerable regressions, whereby we have hammered out the notion of self and otherness. Hence the argument of modern anthropology that the incest taboo, which appears to be primal to the organization of communal life, is inseparable from linguistic evolution. We can only prohibit that which we can name. Kinship systems, which are the coding and classification of sex for purposes of social survival, are analogous with syntax. The seminal and the semantic functions determine the genetic and social structure of human experience. Together they construe the grammar of being.

The interactions of the sexual and the linguistic accompany our whole lives. But again, much of this central area remains unexplored. If coition can be schematized as dialogue, masturbation seems to be correlative with the pulse of monologue or of internalized address. There is evidence that the sexual discharge in male onanism is greater than it is in intercourse. I suspect that the determining factor is articulateness, the ability to conceptualize with especial vividness. In the highly articulate individual, the current of verbal–psychic energy flows inward. The multiple, intricate relations between speech defects and infirmities in the nervous and glandular mechanisms which control sexual and excretory functions have long been known, at least at the level of popular wit and scatological lore. Ejaculation is at once a physiological and a linguistic concept. Impotence and speech-blocks, premature emission and stuttering, involuntary ejaculation and the word-river of dreams are phenomena whose interrelations seem to lead back to the central knot of our humanity. Semen, excreta, and words are communicative products. They are transmissions from the self inside the skin to reality outside. At the far root, their symbolic significance, the rites, taboos, and fantasies which they evoke, and certain of the social controls on their use, are inextricably inter-woven. We know all this but hardly grasp its implications.

In what measure are sexual perversions analogues of incorrect speech? Are there affinities between pathological erotic compul-sions and the search, obsessive in certain poets and logicians, for a 'private language', for a linguistic system unique to the needs and perceptions of the user? Might there be elements of homosexuality in the modern theory of language (particularly in the early Witt-genstein), in the concept of communication as an arbitrary mirror-ing? It may be that the significance of Sade lies in his terrible loquacity, in his forced outpouring of millions of words. In part, the genesis of sadism could be linguistic. The sadist makes an abstraction of the human being he tortures; he verbalizes life to an extreme degree by carrying out on living beings the totality of his articulate fantasies. Did Sade's uncontrollable fluency, like the garrulousness often imputed to the old, represent a psycho-physiological surrogate for diminished sexuality (pornography seeking to replace sex by language)?

Questions crowd upon one. No sphere of the *science de l'homme* is more compelling or nearer the core. But how much have we added to firm knowledge since Plato's myth of a lost, androgynous unity?

The difference between the speech of men and of women is one aspect, though crucial, of the interactions of language and eros. Ethno-linguists report a number of languages in which men and women use different grammatical forms and partially distinct vo-cabularies. A study has been made of men's and women's speech in Koasati, a Muskogean language of south-western Louisiana.[1] The differences observed are mainly grammatical. As they bring up male children, women know men's speech. Men, in turn, have been heard using women's forms when quoting a female speaker in a story. In a few instances, and this is an extraordinarily sugges-tive point, the speech of women is somewhat more archaic than

---

[1] Cf. Mary R. Haas, 'Men's and Women's Speech in Koasati' (*Language*, XX, 1944).

that of men. The same obtains in Hitchiti, another Creek Indian tongue. The formal duality of men's and women's speech has been recorded also in Eskimo languages, in Carib, a South American Indian language, and in Thai. I suspect that such division is a feature of almost all languages at some stage in their evolution and that numerous spoors of sexually determined lexical and syntactical differences are as yet unnoticed. But again, as in the case of Japanese or Cherokee 'child-speech', formal discriminations are easy to locate and describe. The far more important, indeed universal phenomenon, is the differential use by men and women of identical words and grammatical constructs.

No man or woman but has felt, during a lifetime, the strong subtle barriers which sexual identity interposes in communication. At the heart of intimacy, there above all perhaps, differences of linguistic reflex intervene. The semantic contour, the total of expressive means used by men and women differ. The view they take of the output and consumption of words is not the same. As it passes through verb tenses, time is bent into distinctive shapes and fictions. At a rough guess, women's speech is richer than men's in those shadings of desire and futurity known in Greek and Sanskrit as optative; women seem to verbalize a wider range of qualified resolve and masked promise. Feminine uses of the subjunctive in European languages give to material facts and relations a characteristic *vibrato*. I do not say they lie about the obtuse, resistant fabric of the world: they multiply the facets of reality, they strengthen the adjective to allow it an alternative nominal status, in a way which men often find unnerving. There is a strain of ultimatum, a separatist stance, in the masculine intonation of the first-person pronoun; the 'I' of women intimates a more patient bearing, or did until Women's Liberation. The two language models follow on Robert Graves's dictum that men do but women are.

In regard to speech habits, the headings of mutual reproach are immemorial. In every known culture, men have accused women of being garrulous, of wasting words with lunatic prodigality. The chattering, ranting, gossipping female, the tattle, the scold, the toothless crone her mouth wind-full of speech, is older than fairy-tales. Juvenal, in his Sixth Satire, makes a nightmare of woman's verbosity:

> cedunt grammatici, vincuntur rhetores, omnis
> turba tacet, nec causidicus nec praeco loquetur,
> altera nec mulier; verborum tanta cadit vis,
> tot pariter pelves ac tintinnabula dicas
> pulsari, iam nemo tubas, nemo aera fatiget:
> una laboranti poterit succurrere Lunae.

(The grammarians yield to her; the rhetoricians succumb; the whole crowd is silenced. No lawyer, no auctioneer will get a word in, no, nor any other woman. Her speech pours out in such a torrent that you would think that pots and bells were being banged together. Let no one more blow a trumpet or clash a cymbal: one woman alone will make noise enough to rescue the labouring moon [from eclipse].)

Are women, in fact, more spendthrift of language? Men's conviction on this point goes beyond statistical evidence. It seems to relate to very ancient perceptions of sexual contrast. It may be that the charge of loquacity conceals resentment about the role of women in 'expending' the food and raw material brought in by men. But Juvenal's allusion to the moon points inward, to the malaise which distances men from crucial aspects of feminine sexuality. The alleged outpouring of women's speech, the rank flow of words, may be a symbolic restatement of men's apprehensive, often ignorant awareness of the menstrual cycle. In masculine satire, the obscure currents and secretions of woman's physiology are an obsessive theme. Ben Jonson unifies the two motifs of linguistic and sexual incontinence in *The Silent Woman*. 'She is like a conduit-pipe', says Morose of his spurious bride, 'that will gush out with more force when she opens again.' 'Conduit-pipe', with its connotations of ordure and evacuation, is appallingly brutal. So is the whole play. The climax of the play again equates feminine verbosity with lewdness: 'O my heart! wilt thou break? wilt thou break? this is worst of all worst worsts that hell could have devised! Marry a whore, and so much noise!'

The converse are men's professions of delight in women's voices when their register is sweet and low. 'Comely speech' is, as the Song of Solomon affirms, an ornament to woman. Of an even greater and more concordant beauty is silence. The motif of the woman or maiden who says very little, in whom silence is a symbolic counterpart to chasteness and sacrificial grace, lends a unique pathos to the Antigone of *Oedipus at Colonus* or Euripides' Alcestis. A male god has cruelly possessed Cassandra and the speech that pours out of her is his; she seems almost remote from it, broken. Though addressed to an inanimate form, Keats's 'un-ravish'd bride of quietness' precisely renders the antique association of feminine quality with sparseness of speech. These values crystallize in Coriolanus' salute to Virgilia: 'My gracious silence, hail!' The line is magical in its music and suggestion, but also in its dramatic shrewdness. Shakespeare precisely conveys the idiom of a man, of a personage brimful with overweening masculinity. No woman would so greet her beloved.

Not that women have been slow to answer. Elvira's

> Non lo lasciar più dir;
> il labbro è mentitor . . .

has rung down through history. Men are deceivers ever. They use speech to conceal the true, sexually aggressive function of their lips and tongues. Women know the change in a man's voice, the crowding of cadence, the heightened fluency triggered off by sexual excitement. They have also heard, perennially, how a man's speech flattens, how its intonations dull after orgasm. In feminine speech-mythology, man is not only an erotic liar; he is an incorrigible braggart. Women's lore and secret mock record him as an eternal *miles gloriosus*, a self-trumpeter who uses language to cover up his sexual or professional fiascos, his infantile needs, his inability to withstand physical pain.

Before the Fall, man and woman may have spoken the same tongue, comprehending each other's meaning perfectly. Immediately after, speech divided them. Milton identifies the moment and its unending sequence:

> Thus they in mutual accusation spent
> The fruitless hours, but neither self-condemning:
> And of their vain contest appear'd no end.

The grounds of differentiation are, of course, largely economic and social. Sexual speech variations evolve because the division of labour, the fabric of obligation and leisure within the same community is different for men and for women. In many cases, such as the exclusively male use of whistle speech among the Mazateco Indians of Oaxaca, men mark their sociological and physical 'superiority' by reserving to themselves certain forms of communication. *Taceat mulier in ecclesia* is prescriptive in Judaic, Christian, and Muslim culture. But certain linguistic differences do point towards a physiological basis or, to be exact, towards the intermediary zone between the biological and the social. This is the area in which the problem of the relations of linguistic conventions to cognitive processes is most difficult. Are there biologically determined apprehensions of sense data which precede and generate linguistically programmed conceptualizations? This is a question we shall come back to. E. H. Lenneberg states: 'I have data on sex difference, and some colors are unanimously called by girls something and by men something else.' Using anthropological material, F. G. Lounsbury comments: 'I feel sure that a woman's color vocabulary is quite a bit greater than a man's.'[1] Both observations must have a social as well as a psychophysiological foundation. The sum of difference in the language habits of men and of women makes for two ways of fitting speech to the world: 'When all's done,' says Lady Macbeth to negate the fierce reality of Macbeth's vision of Banquo, 'You look but on a stool.'

Whatever the underlying causes, the resultant task of translation is constant and unfulfilled. Men and women communicate through never-ending modulation. Like breathing, the technique is unconscious; like breathing also, it is subject to obstruction and homicidal breakdown. Under stress of hatred, of boredom, of sudden panic, great gaps open. It is as if a man and a woman then heard each other for the first time and knew, with sickening conviction, that they share no common language, that their previous understanding had been based on a trivial pidgin which had left the heart of meaning untouched. Abruptly the wires are down and the

---

[1] H. Hoijer (ed.), *Language in Culture* (University of Chicago Press, 1954), p. 267.

nervous pulse under the skin is laid bare in mutual incomprehension. Strindberg is master of such moments of fission. Harold Pinter's plays locate the pools of silence that follow.

By far the greater proportion of art and historical record has been left by men. The process of 'sexual translation' or of the breakdown of linguistic exchange is seen, almost invariably, from a male focus. The relevant anthropology—itself a term charged with masculine presumptions—distorts evidence as does the white traveller's edge of power over his native informant. Few artists, though they are among the greatest, have rendered the genius of women's speech and seen the crisis of imperfect or abandoned translation from both sides. Much of the concentrated richness of the art of Racine lies in his 'ear' for the contrasting pressures of sexual identity on discourse. In every one of his major plays there is a crisis of translation: under extreme stress, men and women declare their absolute being to each other, only to discover that their respective experience of eros and language has set them desperately apart. Like no other playwright, Racine communicates not only the essential beat of women's diction but makes us feel what there is in the idiom of men which Andromaque, Phèdre, or Iphigénie can only grasp as falsehood or menace. Hence the equivocation, central in his work, on the twofold sense of *entendre*: these virtuosos of statement hear each other perfectly, but do not, cannot apprehend. I do not believe there is a more complete drama in literature, a work more exhaustive of the possibilities of human conflict than Racine's *Bérénice*. It is a play about the fatality of the coexistence of man and woman, and it is dominated, necessarily, by speech-terms (*parole, dire, mot, entendre*). Mozart possessed something of this same rare duality (so different from the characterizing, polarizing drive of Shakespeare). Elvira, Donna Anna, and Zerlina have an intensely shared femininity, but the music exactly defines their individual range or pitch of being. The same delicacy of tone-discrimination is established between the Countess and Susanna in *The Marriage of Figaro*. In this instance, the discrimination is made even more precise and more dramatically different from that which characterizes male voices by the 'bisexual' role of Cherubino. The Count's page is a graphic example of Lévi-Strauss's contention that women and words are analogous media of exchange in the grammar of social life. Stendhal was a careful student of Mozart's operas. That study is borne out in the depth and fairness of his treatment of the speech-worlds of men and women in Fabrice and la Sanseverina in *The Charterhouse of Parma*. Today, when there is sexual frankness as never before, such fairness is, paradoxically, rarer. It is not as 'translators' that women novelists and poets excel, but as declaimers of their own, long-stifled tongue.

I have been putting forward a truism, but one whose great importance and consequences usually go unexamined.

Any model of communication is at the same time a model of trans-lation, of a vertical or horizontal transfer of significance. No two historical epochs, no two social classes, no two localities use words and syntax to signify exactly the same things, to send identical signals of valuation and inference. Neither do two human beings. Each living person draws, deliberately or in immediate habit, on two sources of linguistic supply: the current vulgate corresponding to his level of literacy, and a private thesaurus. The latter is inextricably a part of his subconscious, of his memories so far as they may be verbalized, and of the singular, irreducibly specific ensemble of his somatic and psychological identity. Part of the answer to the notorious logical conundrum as to whether or not there can be 'private language' is that aspects of every language-act are unique and individual. They form what linguists call an 'idiolect'. Each communicatory gesture has a private residue. The 'personal lexicon' in every one of us inevitably qualifies the definitions, connotations, semantic moves current in public discourse. The concept of a normal or standard idiom is a statistically-based fiction (though it may, as we shall see, have real existence in machine-translation). The language of a community, however uniform its social contour, is an inexhaustibly multiple aggregate of speech-atoms, of finally irreducible personal meanings.

The element of privacy in language makes possible a crucial, though little understood, linguistic function. Its importance relates a study of translation to a theory of language as such. Obviously, we speak to communicate. But also to conceal, to leave unspoken.

The ability of human beings to misinform modulates through every wavelength from outright lying to silence. This ability is based on the dual structure of discourse: our outward speech has 'behind it' a concurrent flow of articulate consciousness. 'Al conversar vivimos en sociedad,' wrote Ortega y Gasset, 'al pensar nos quedamos solos.' In the majority of conventional, social exchanges, the relation between these two speech currents is only partially congruent. There is duplicity. The 'aside' as it is used in drama is a naïve representation of scission: the speaker communicates to himself (thus to his audience) all that his overt statement to another character leaves unsaid. As we grow intimate with other men or women, we often 'hear' in the slightly altered cadence, speed, or intonation of whatever they are saying to us the true movement of articulate but unvoiced intent. Shakespeare's awareness of this twofold motion is unfailing. Desdemona asks of Othello, in the very first, scarcely realized instant of shaken trust, 'Why is your speech so faint?'

Thus a human being performs an act of translation, in the full sense of the word, when receiving a speech-message from any other human being. Time, distance, disparities in outlook or assumed reference, make this act more or less difficult. Where the difficulty is great enough, the process passes from reflex to conscious technique. Intimacy, on the other hand, be it of hatred or of love, can be defined as confident, quasi-immediate translation. Having kept the same word-signals bounding and rebounding between them like jugglers' weights, year after year, from horizon to horizon, Beckett's vagrants and knit couples understand one another almost osmotically. With intimacy, the external vulgate and the private mass of language grow more and more concordant. Soon the private dimension penetrates and takes over the customary forms of public exchange. The stuffed-animal and baby-speech of adult lovers reflects this take-over. In old age the impulse towards translation wanes and the pointers of reference turn inward. The old listen less or principally to themselves. Their dictionary is, increasingly, one of private remembrance.

I have been trying to state a rudimentary but decisive point: interlingual translation is the main concern of this book, but it is also a way in, an access to an inquiry into language itself. 'Translation', properly understood, is a special case of the arc of communication which every successful speech-act closes within a given language. On the inter-lingual level, translation will pose concentrated, visibly intractable problems; but these same problems abound, at a more covert or conventionally neglected level, intra-lingually. The model 'sender to receiver' which represents any semiological and semantic process is ontologically equivalent to the model 'source-language to receptor-language' used in the theory of translation. In both schemes there is 'in the middle' an operation of interpretative decipherment, an encoding–decoding function or synapse. Where two or more languages are in articulate interconnection, the barriers in the middle will obviously be more salient, and the enterprise of intelligibility more conscious. But the 'motions of spirit', to use Dante's phrase, are rigorously analogous. So, as we shall see, are the most frequent causes of misunderstanding or, what is the same, of failure to translate correctly. In short: *inside or between languages, human communication equals translation*. A study of translation is a study of language.

The fact that tens of thousands of different, mutually incomprehensible languages have been or are being spoken on our small planet is a graphic expression of the deeper-lying enigma of human individuality, of the bio-genetic and bio-social evidence that no two human beings are totally identical. The affair at Babel confirmed and externalized the never-ending task of the translator—it did not initiate it. Logically considered, there was no guarantee that human beings would understand one another, that idiolects would fuse into the partial consensus of shared speech-forms. In terms of survival and social coherence such fusion may have proved to be an early and dramatic adaptive advantage. But, as William James observed, 'natural selection for efficient communication' may have been achieved at a considerable cost. This would have included not only the ideal of a totally personal voice, of a unique 'fit' between an individual's expressive means and his world-image, pursued by the poets. It meant also that the 'bright buzz' of non-verbal articulate codes, the sensory modes of smell, gesture, and pure tone developed by animals, and perhaps extra-sensory forms of communication (these are specifically adduced by James) all but vanished from the human repertoire. Speech would be an immensely profitable but also reductive, partially narrowing evolutionary selection from a wider spectrum of semiotic possibilities. Once it was 'chosen' translation became inevitable.

Thus any light I may be able to throw on the nature and poetics of translation between tongues has concomitant bearing on the study of language as a whole. The subject is difficult and ill-defined. Regarding the possible transfer into English of Chinese philosophic concepts, I. A. Richards remarks: 'We have here indeed what may very probably be the most complex type of event yet produced in the evolution of the cosmos.'[1] He may be right. But the complexity and range of implication were already present in the first moment of human speech.

[1] I. A. Richards, 'Towards a Theory of Translating' in Arthur F. Wright (ed.), *Studies in Chinese Thought* (University of Chicago Press, 1953), p. 250.

Chapter Two

LANGUAGE AND GNOSIS

TRANSLATION exists because men speak different languages. This truism is, in fact, founded on a situation which can be regarded as enigmatic and as posing problems of extreme psychological and socio-historical difficulty. Why should human beings speak thousands of different, mutually incomprehensible tongues? We live in this pluralist framework, have done so since the inception of recorded history, and take the ensuing farrago for granted. It is only when we reflect on it, when we lift the facts from the misleading context of the obvious, that the possible strangeness, the possible 'unnaturalness' of the human linguistic order strikes us. Conceivably there is here one of the more central questions in the study of man's cerebral and social evolution. Yet even the pertinent queries, the statements of astonishment which would put the facts into relief, are formulated only sporadically. Divisions between formal 'hard-edged' linguistics on the one hand and contrastive, anthropological investigations of actual language on the other, have further relegated the issue into the shadow of futile, metaphysical speculation.

We ought not, perhaps, to regard as either formally or substantively coherent, as responsible to verification or falsification, any model of verbal behaviour, any theory of how language is generated and acquired, which does not recognize as crucial the matter of the bewildering multiplicity and variousness of languages spoken on this crowded planet. In his foreword to Morris Swadesh's posthumous *The Origin and Diversification of Language*, Dell Hymes states: 'The diversity of languages, as they have developed and been adapted, is a patent fact of life that cries out for theoretical attention. It becomes increasingly difficult for theorists of language to persist in confounding potential equivalence with actual diversity.' This should have been a commonplace and respectable exigence among linguists long before 1972. Theories of semantics, constructs of universal and transformational grammar that have nothing of substance to say about the prodigality of the language atlas—more than a thousand different languages are spoken in New Guinea—could well be deceptive. It is here, rather than in the problem of the invention and understanding of melody (though the two issues may be congruent), that I would place what Lévi-Strauss calls *le mystère suprême* of anthropology.

Why does *homo sapiens*, whose digestive tract has evolved and functions in precisely the same complicated ways the world over, whose biochemical fabric and genetic potential are, orthodox science assures us, essentially common, the delicate runnels of whose cortex are wholly akin in all peoples and at every stage of social evolution—why does this unified, though individually unique mammalian species not use *one* common language? It inhales, for its life processes, one chemical element and dies if deprived of it. It makes do with the same number of teeth and vertebrae. To grasp how notable the situation is, we must make a modest leap of imagination, asking, as it were, from outside. In the light of anatomical and neurophysiological universals, a unitary language solution would be readily understandable. Indeed, if we lived inside one common language-skin, any other situation would appear very odd. It would have the status of a recondite fantasy, like the anaerobic or anti-gravitational creatures in science-fiction. But there is also another 'natural' model. A deaf, non-literate observer approaching the planet from outside and reporting on crucial aspects of human appearance and physiological behaviour, would conclude with some confidence that men speak a small number of different, though probably related, tongues. He would guess at a figure of the order of half a dozen with perhaps a cluster of dependent but plainly recognizable dialects. This number would be persuasively concordant with other major parameters of human diversity. Depending on which classification they adopt, ethnographers divide the human species into four or seven races (though the term is, of course, an unsatisfactory shorthand). The comparative anatomy of bone structures and sizes leads to the use of three main typologies. The analysis of human blood-types, itself a topic of great intricacy and historical consequence, suggests that there are approximately half a dozen varieties. Such would seem to be the cardinal numbers of salient differentiation within the species though the individual, obviously, is genetically unique. The development on earth of five or six major languages, together with a spectrum of derivative, intermediary dialects and pidgins, analogous to the gamut and blendings of skin-colour, would strike our imaginary observer as a profoundly natural, indeed inevitable pattern. If we lived within this pattern, we should experience it as inherently logical and take for granted the supporting or at least powerfully analogous evidence of comparative anatomy, physiology, and the classification of races. Under pressure of time and historical circumstance, the half dozen principal languages might well have bent quite far apart. Speakers would nevertheless be conscious of underlying uniformities and would expect to find that degree of mutual comprehension shared, for example, within the Romance-language family.

The actual situation is, of course, totally different.

We do not speak one language, nor half a dozen, nor twenty or thirty. Four to five thousand languages are thought to be in current use. This figure is almost certainly on the low side. We have, until now, no language atlas which can claim to be anywhere near exhaustive. Furthermore, the four to five thousand living languages are themselves the remnant of a much larger number spoken in the past. Each year so-called rare languages, tongues spoken by isolated or moribund ethnic communities, become extinct. Today entire families of language survive only in the halting remembrance of aged, individual informants (who, by virtue of their singularity are difficult to cross-check) or in the limbo of tape-recordings. Almost at every moment in time, notably in the sphere of American Indian speech, some ancient and rich expression of articulate being is lapsing into irretrievable silence. One can only guess at the extent of lost languages. It seems reasonable to assert that the human species developed and made use of at least twice the number we can record today. A genuine philosophy of language and socio-psychology of verbal acts must grapple with the phenomenon and rationale of the human 'invention' and retention of anywhere between five and ten thousand distinct tongues. However difficult and generalizing the detour, a study of translation ought to put forward some view of the evolutionary, psychic needs or opportunities which have made translation necessary. To speak seriously of translation one must first consider the possible meanings of Babel, their inherence in language and mind.

Even a cursory look at Meillet's standard compendium[1] or at more recent listings in progress under the direction of Professor Thomas Sebeok of Indiana University, shows a situation of utter intricacy and division. In many parts of the earth, the language-map is a mosaic each of whose stones, some of them minuscule, is entirely or partially distinct from all others in colour and texture. Despite decades of comparative philological study and taxonomy, no linguist is certain of the language atlas of the Caucasus, stretching from Bžedux in the north-west to Ruťul and Küri in the Tatar regions of Azerbeidjan. Dido, Xwarši, and Qapuči, three languages spoken between the Andi and the Koissou rivers, have been tentatively identified and distinguished, but are scarcely known to any but native users. Arči, a language with a distinctive phonetic and morphological structure, was, in the 1970s, spoken by only one village of approximately 850 inhabitants. Oubykh, once a flourishing tongue on the shores of the Black Sea, survives today in a handful of Turkish localities near Ada Pazar. A comparable multiplicity and diversity marks the so-called Palaeosiberian language families. Eroded by Russian during the nineteenth century, Kamtchadal, a language of undeniable resource and antiquity, survives in only eight hamlets in the maritime province of Koriak. In 1909, one old man was still conversant in the eastern branch of Kamtchadal. In 1845, a traveller came across five speakers of Kot (or Kotu). Today no living trace can be found. The history of Palaeosiberian cultures and migrations before the Russian conquest is largely obscure. But evidence of great linguis-

---

[1] A. Meillet and M. Cohen, *Les Langues du monde* (Paris, 1952).

tic variance and sophistication is unmistakable. With regard to nuances of action—possibility, probability, confirmation, necessity—Palaeosiberian languages possess a grammar of obvious precision. But we know little of the genesis of these tongues and of their affinities, if any, with other major linguistic groupings.

The Black Sea region and even Russian Siberia are well known; both have been involved in recorded history and in the spread of technology. By comparison, the language-map extending from the south-western United States to Tierra del Fuego is full of blanks and mere guesses. The fundamental divisions are uncertain: what, for instance, are the relations between the enormously ramified Uto-Aztec tree of languages and the great Mayan cluster? For Mexico and Central America alone, current listings reckon 190 distinct tongues. But the roll is incomplete, and entire language groups are designated as unclassified, as possibly extinct, or as identifiable only through hearsay and through their intrusions, in the guise of quotations and borrowings, into other idioms. The mind must be complacent to regard this situation without a radical sense of perplexity.

Tubatulabal was spoken by something like a thousand Indians at the southern spur of the Sierra Nevada as recently as the 1770s. All we know today is that this language was strikingly different from all neighbouring tongues. Kupeño survived into the late eighteenth century, but already then it was dwindling to a small patch of territory at the sources of the San Luis Rey. What may have been its wider past? What models of human similitude and cultural determination will account for the fact that Huite (or Yecarome), still spoken on the Rio Fuerte in the sixteenth century, should have been sharply different from the Cahita languages, themselves a branch of the Hopi family, which literally surrounded it? Mid-sixteenth-century travellers reported the currency of Matagalpa throughout north-west Nicaragua and in parts of present-day Honduras. Now only a handful of families living near the modern towns of Matagalpa and Esteli are thought to know the speech. In northern Mexico and along the Pacific coast, Nawa and then Spanish submerged a score of ancient, separate human tongues. Tomateka, Kakoma, Kučarete—these are now ghost names. Again, an intimation of enigmatic needs and energies crowds upon one.

Blank spaces and question marks cover immense tracts of the linguistic geography of the Amazon basin and the savannah. At latest count, ethno-linguists discriminate between 109 families, many with multiple sub-classes. But scores of Indian tongues remain unidentified or resist inclusion in any agreed category. Thus a recently discovered tongue spoken by Brazilian Indians of the Itapucuru river territory seems to be related to no previously defined set. Puelče, Guenoa, Atakama, and a dozen others are names designating languages and dialects spoken, perhaps over millions of square miles, by migrant and vanishing peoples. Their history and morphological structure are barely charted. Many will dim into oblivion before rudimentary grammars or word-lists can be salvaged. Each takes with it a storehouse of consciousness.

The language catalogue begins with Aba, an Altaic idiom spoken by Tatars, and ends with Zyriene, a Finno-Ugric speech in use between the Urals and the Arctic shore. It conveys an image of man as a language animal of implausible variety and waste. By comparison, the classification of different types of stars, planets, and asteroids runs to a mere handful.

What can possibly explain this crazy quilt? How are we to rationalize the fact that human beings of identical ethnic provenance, living on the same terrain, under equal climatic and ecological conditions, often organized in the same types of communal structure, sharing kinship systems and beliefs, speak entirely different languages? What sense can be read into a situation in which villages a few miles apart or valleys divided by low, long-eroded hills use tongues incomprehensible to each other and morphologically unrelated? I put the question repetitively because, for a long time, obviousness has disguised its extreme importance and difficulty.

A Darwinian scheme of gradual evolution and ramification, of adaptive variation and selective survival, may look credible. Consciously or not, many linguists seem to have worked with some such analogy. But it only masks the problem. Though many details of the actual evolutionary process remain obscure, the strength of

Darwin's argument lies in the demonstrable economy and specificity of the adaptive mechanism; living forms mutate with seemingly random profusion, but their survival depends on adjustment to natural circumstance. It can be shown, over a wide range of species, that extinction does relate to a failure or inexactitude of vital response. The language manifold offers no genuine counterpart to these visible, verifiable criteria. We have no standards (or only the most conjectural) by which to assert that any human language is intrinsically superior to any other, that it survives because it meshes more efficiently than any other with the demands of sensibility and physical existence. We have no sound basis on which to argue that extinct languages failed their speakers, that only the most comprehensive or those with the greatest wealth of grammatical means have endured. On the contrary: a number of dead languages are among the obvious splendours of human intelligence. Many a linguistic mastodon is a more finely articulated, more 'advanced' piece of life than its descendants. There appears to be no correlation, moreover, between linguistic wealth and other resources of a community. Idioms of fantastic elaboration and refinement coexist with utterly primitive, economically harsh modes of subsistence. Often, cultures seem to expend on their vocabulary and syntax acquisitive energies and ostentations entirely lacking in their material lives. Linguistic riches seem to act as a compensatory mechanism. Starving bands of Amazonian Indians may lavish on their condition more verb tenses than could Plato.

The Darwinian parallel also breaks down on the crucial point of large numbers. The multiplicity of fauna and flora does not represent randomness or waste. It is an immediate factor of the dynamics of evolutionary breeding, cross-fertilization, and competitive selection which Darwin set out. Given the range of ecological possibilities, the multiplication of species is, quite conceivably, economical. No language is demonstrably adaptive in this sense. None is concordant with any particular geophysical environment. With the simple addition of neologisms and borrowed words, any language can be used fairly efficiently anywhere; Eskimo syntax is appropriate to the Sahara. Far from being economic and demonstrably advantageous, the immense number and variety of human idioms, together with the fact of mutual incomprehensibility, is a powerful obstacle to the material and social progress of the species. We will come back to the key question of whether or not linguistic differentiations may provide certain psychic, poetic benefits. But the many ways in which they have impeded human progress are clear to see. No conceivable gain can have accrued to the crowded, economically harried Philippine islands from their division by the Bikol, Chabokano, Ermitano, Tagalog, and Wray-waray languages (to name only the most prominent of some thirty tongues), or from the related fact that for four of these five idioms the United States Employment Service can list only one qualified translator. Numerous cultures and communities have passed out of history as linguistic 'drop-outs'. Not because their own particular speech was in any way inadequate, but because it prevented communication with the principal currents of intellectual and political force. Countless tribal societies have withered inward, isolated by language barriers even from their near neighbours. Time and again, linguistic differences and the profoundly exasperating inability of human beings to understand each other have bred hatred and reciprocal contempt. To the baffled ear, the incomprehensible parley of neighbouring peoples is gibberish or suspected insult. Linguistically atomized, large areas of Africa, India, and South America have never gathered their common energies either against foreign predators or economic stagnation. Though sometimes sharing a lingua franca, such as Swahili, their consciousness of kinship and common need has remained artificial. The deeper springs of action stay rooted in linguistic separateness. Robbed of their own language by conquerors and modern civilization, many underdeveloped cultures have never recovered a vital identity. In short: languages have been, throughout human history, zones of silence to other men and razor-edges of division.

Why this destructive prodigality?

Few modern linguists, with the exception of Swadesh and Pei, have shown the curiosity which this situation ought to arouse. Where an answer is given at all, it is put in casually evolutionary terms: there are many different tongues because, over long stretches of time, societies and cultures split apart and, through accretion of particular experience, evolved their own local speech habits. The facile nature of such an explanation is worrying: it fails to engage precisely those central philosophical and logical dilemmas which spring from the admitted uniformities of human mental structures and from the economically and historically negative, often drastically damaging, role of linguistic isolation. Turn the argument around: let reasons be given why the adoption by the human race of a single language or a small number of related languages would have been natural and beneficial. It appears at once that *post hoc* justifications for the facts as we know them are wholly unconvincing. The problem lies deeper. And few linguists since Wilhelm von Humboldt, in the early decades of the nineteenth century, have thought about it at the required level of psychological insistence and historical sensibility. It was before Humboldt that the mystery of many tongues on which a view of translation hinges fascinated the religious and philosophic imagination.

No civilization but has its version of Babel, its mythology of the primal scattering of languages.[1] There are two main conjectures, two great attempts at solving the riddle via metaphor. Some awful error was committed, an accidental release of linguistic chaos, in the mode of Pandora's box. Or, more commonly, man's language condition, the incommunicados that so absurdly divide him are a punishment. A lunatic tower was launched at the stars; Titans savaged one another and of their broken bones came the splinters of isolated speech; eavesdropping, like Tantalus, on the gossip of the gods, mortal man was struck moronic and lost all remembrance of his native, universal parlance. This corpus of myth, springing from a very ancient, obstinate bewilderment, modulates gradually into philosophic and hermetic speculation. The history of such speculation, of the endeavours of philosophers, logicians, and *illuminati* to explain the confusion of human idioms, is itself a compelling chapter in the annals of the imagination. Much of it is turgid stuff. The argument is shot through with fantastications and

---

[1] The great work on this subject, and one of the most fascinating of intellectual histories, is Arno Borst, *Der Turmbau von Babel: Geschichte der Meinungen über Ursprung und Vielfalt der Sprachen und Völker* (Stuttgart, 1957–63).

baroque torsions. Stemming, as it must, from a meditation on its
own shell of being, words focused on the mirror and echoing surface of words, the metaphoric and esoteric tradition of philology
often loses touch with common sense. But via arcane images,
Kabbalistic and emblematic constructs, through occult etymologies
and bizarre decodings, the argument on Babel will feel its way—as
did the partially astrological, Pythagorean hypotheses of celestial
motion in Copernicus and Kepler—towards cardinal insights.
More justly amazed than modern linguistics at the whole business
of man's estrangement from the speech of his fellow man, the
tradition of language mysticism and philosophic grammar reaches
out to intuitions, to deeps of inquiry, which are, I think, often
lacking from current debate. Today we move on drier but shallower
ground.

Key images and lines of conjecture recur in the philosophy of
language from the Pythagoreans to Leibniz and J. G. Hamann. We
are told that the substance of man is bound up with language; the
mystery of speech characterizes his being, his mediate place in the
sequence leading from the inanimate to the transcendent order of
creation. Language is assuredly material in that it requires the play
of muscle and vocal cords; but it is also impalpable and, by virtue
of inscription and remembrance, free of time, though moving in
temporal flow. These antinomies or dialectical relations, which I
want to look at systematically in the next chapter, confirm the dual
mode of human existence, the interactions of physical with spiritual agencies. The occult tradition holds that a single primal
language, an *Ur-Sprache* lies behind our present discord, behind
the abrupt tumult of warring tongues which followed on the collapse of Nimrod's ziggurat. This Adamic vernacular not only
enabled all men to understand one another, to communicate with
perfect ease. It bodied forth, to a greater or lesser degree, the
original Logos, the act of immediate calling into being whereby
God had literally 'spoken the world'. The vulgate of Eden contained, though perhaps in a muted key, a divine syntax—powers of
statement and designation analogous to God's own diction, in
which the mere naming of a thing was the necessary and sufficient
cause of its leap into reality. Each time man spoke he re-enacted,

he mimed, the nominalist mechanism of creation. Hence the allegoric significance of Adam's naming of all living forms: 'and whatsoever Adam called every living creature, that *was* the name thereof.' Hence also the ability of all men to understand God's language and to give it intelligible answer.

Being of direct divine etymology, moreover, the *Ur-Sprache* had a congruence with reality such as no tongue has had after Babel or the dismemberment of the great, enfolding serpent of the world as it is recounted in the mythology of the Carib Indians. Words and objects dovetailed perfectly. As the modern epistemologist might put it, there was a complete, point-to-point mapping of language onto the true substance and shape of things. Each name, each proposition was an equation, with uniquely and perfectly defined roots, between human perception and the facts of the case. Our speech interposes itself between apprehension and truth like a dusty pane or warped mirror. The tongue of Eden was like a flawless glass; a light of total understanding streamed through it. Thus Babel was a second Fall, in some regards as desolate as the first. Adam had been driven from the garden; now men were harried, like yelping dogs, out of the single family of man. And they were exiled from the assurance of being able to grasp and communicate reality.

Theologians and metaphysicians of language strove to attenuate this second banishment. Had there not been a partial redemption at Pentecost, when the gift of tongues descended on the Apostles? Was not the whole of man's linguistic history, as certain Kabbalists supposed, a laborious swing of the pendulum between Babel and a return to unison in some messianic moment of restored understanding? Above all, what of the *Ur-Sprache* itself: had it been irretrievably lost? Here speculation hinged on the question of the veritable nature of Adam's tongue. Had it been Hebrew or some even earlier version of Chaldaean whose far lineaments could be made out in the names of stars and fabled rivers? Jewish gnostics argued that the Hebrew of the Torah was God's undoubted idiom, though man no longer understood its full, esoteric meaning. Other inquirers, from Paracelsus to the seventeenth-century Pietists, were prepared to view Hebrew as a uniquely privileged language, but itself corrupted by the Fall and only obscurely revelatory of the Divine presence. Almost all linguistic mythologies, from Brahmin wisdom to Celtic and North African lore, concurred in believing that original speech had shivered into seventy-two shards, or into a number which was a simple multiple of seventy-two.[1] Which were the primal fragments? Surely if these could be identified, diligent search would discover in them lexical and syntactic traces of the lost language of Paradise, remnants equitably scattered by an incensed God and whose reconstruction, like that of a broken mosaic, would lead men back to the universal grammar of Adam. If they did indeed exist, these clues would be deep-hidden. They ought to be ferreted out, as Kabbalists and adepts of Hermes Trismegistus sought to do, by scrutinizing the hidden configurations of letters and syllables, by inverting words and applying to ancient names, particularly to the diverse nominations of the Creator, a calculus as intricate as that of chiromancers and astrologers. The stakes were very high. If man could break down the prison walls of scattered and polluted speech (the rubble of the smashed tower), he would again have access to the inner penetralia of reality. He would know the truth as he spoke it. Moreover, his alienation from other peoples, his ostracism into gibberish and ambiguity, would be over. The name of Esperanto has in it, undisguised, the root for an ancient and compelling *hope*.

Starting with Genesis 11:11 and continuing to Wittgenstein's *Investigations* or Noam Chomsky's earliest, unpublished paper on morphophonemics in Hebrew, Jewish thought has played a pronounced role in linguistic mystique, scholarship, and philosophy. To both Jew and gentile, the text of the Books of Moses had a revealed character unlike that of any later body of language. Thus Hebrew has served time and again as the diamond edge of the cutter's tool. In Jewish tradition we find those rubrics that will largely organize the main directions of Western argument about the essence and enigmatic dismemberment of human tongues.

[1] Despite Arno Borst's exhaustive inquiries, the origins of this particular number remain obscure. The 6×12 component suggests an astronomical or seasonal correlation.

Each element of the received text has generated its own traditions of study in Jewish mysticism and rabbinical scholarship.[1] There is a philology and gnosis of the individual Hebrew letter as there is of the word and grammatical unit. In Merkabah mysticism, each written character may be regarded as embodying a fragment of the universal design of creation; all human experience, no less than all human discourse unto the end of time, is graphically latent in the letters of the alphabet. Those numinous letters whose combinations make up the seventy-two names of God may, if they are probed to the hidden core of meaning, reveal the cipher, the configurations of the cosmos. Accordingly, prophetic Kabbalism developed its 'science of the combination of letters'. Through self-hypnotic meditation on groupings of individual characters, groupings which need not in themselves be meaningful, the initiate may come to glimpse the great Name of God, manifest throughout the lineaments of nature, but enveloped, as it were, in the muffling layers of vulgate speech. But although Hebrew may have a privileged immediacy, the Kabbalist knows that all languages are a mystery and ultimately related to the holy tongue.

In German Hasidism, it is the word rather than the alphabetic sign whose hidden sense and unaltered preservation are of extreme importance. To mutilate a single word in the Torah, to set it in the wrong order, might be to imperil the tenuous links between fallen man and the Divine presence. Already the Talmud had said: 'the omission or the addition of one letter might mean the destruction of the whole world.' Certain *illuminati* went so far as to suppose that it was some error of transcription, however minute, made by the scribe to whom God had dictated holy writ, that brought on the darkness and turbulence of the world. Theosophy, as expressed in the *Zohar* and in the commentaries which followed, made use of mystical puns and word-games to prove some of its crucial doctrines. *Elohim*, the name of God, unites *Mi*, the hidden subject, with *Eloha*, the hidden object. The dissociation of

---

[1] Here, of course, I am drawing heavily on Gershom Scholem, *Major Trends in Jewish Mysticism* (Jerusalem, 1941 and New York, 1946).

subject from object is the very infirmity of the temporal world. Only in His name do we discern the promise of ultimate unity, the assurance of man's release from the dialectic of history. In brief: God's actual speech, the idiom of immediacy known to Adam and common to men until Babel, can still be decoded, partially at least, in the inner layers of Hebrew and, perhaps, in other languages of the original scattering.

The habits of feeling shown in these occult semantics are remote and often bizarre. But at several points, linguistic gnosis touches on decisive issues of a rational theory of language and of translation. There is a deceptively modern ring to the discriminations between deep structures of meaning, structures buried by time or masked by colloquialism, and the surface structures of spoken idiom. There is an acute understanding, essential to any treatment of communication within and between languages, of the ways in which a text may conceal more than it conveys. There is, above all, a clear sense, persistent in Spinoza as it is in Wittgenstein, of the numinous as well as problematic nature of man's life in language.

Numerous elements of gnostic speculation, often with reference to Hebrew, are evident in the great tradition of European linguistic philosophy. This sequence of visionary belief and conjecture extends unbroken from Meister Eckhart in the early fourteenth century to the teachings of Angelus Silesius during the 1660s and 1670s. Here also we find a stubborn wonder about the multiplicity and splintering of vernaculars. For Paracelsus, writing in the 1530s, there is little doubt that Divine providence shall one day restore the unity of human tongues. His contemporary, the Kabbalist Agrippa of Nettesheim, spun an arcane web around the figure seventy-two; in Hebrew, and particularly in Exodus with its seventy-two designations of the Divine name, magic forces were compacted. One day other languages would return to this fount of being. In the meantime, the very need for translation was like the mark of Cain, a witness to man's exile from *harmonia mundi*. There was, as Coleridge knew, no deeper dreamer on language, no sensibility more haunted by the alchemy of speech, than Jakob

Böhme (1575–1624).[1] Like Nicholas of Cusa long before him, Böhme supposed that the primal tongue had not been Hebrew, but an idiom brushed from men's lips in the instant of the catastrophe at Babel and now irretrievably disjected among all living speech (Nettesheim had, at one point, argued that Adam's true vernacular was Aramaic). Being erratic blocs, all languages share in a common myopia; none can articulate the whole truth of God or give its speakers a key to the meaning of existence. Translators are men groping towards each other in a common mist. Religious wars and the persecution of supposed heresies arise inevitably from the babel of tongues: men misconstrue and pervert each other's meanings. But there is a way out of darkness: what Böhme calls 'sensualistic speech'—the speech of instinctual, untutored immediacy, the language of Nature and of natural man as it was bestowed on the Apostles, themselves humble folk, at Pentecost. God's grammar sounds through echoing Nature, if only we will listen.

Kepler agreed that primal speech lay scattered. But it was not in the rough parlance of the primitive and uneducated that the sparks of Divine significance could be found. It was in the immaculate logic of mathematics and in the harmonics, also mathematical in essence, of instrumental and celestial music. The music of the spheres and of Pythagorean accords proclaimed, as they will in the Prologue to Goethe's *Faust*, the hidden architecture of Divine speech. In the visionary musings of Angelus Silesius (Johann Scheffler), Böhme's intimations are carried to extremes. Reaching back to the mysticism of Eckhart, Angelus Silesius asserts that God has, from the beginning of time, uttered only a single word. In that single utterance all reality is contained. The cosmic Word cannot be found in any known tongue; language after Babel cannot lead back to it. The bruit of human voices, so mysteriously diverse and mutually baffling, shuts out the sound of the Logos. There is no access except silence. Thus, for Silesius, the deaf and dumb are nearest of all living men to the lost vulgate of Eden.

[1] Cf. Alexandre Koyré, *La Philosophie de Jacob Boehme* (2nd edition, Paris, 1971), pp. 456–62.

In the climate of the eighteenth century these gnostic reveries faded. But we find them again, changed into model and metaphor, in the work of three modern writers. It is these writers who seem to tell us most of the inward springs of language and translation.

Walter Benjamin's 'Die Aufgabe des Übersetzers' dates from 1923.[1] Though influenced by Goethe's comments on translation in his famous notes to the *Divan*, and by Hölderlin's treatment of Sophocles, Benjamin's essay derives from the gnostic tradition. Benjamin posits, as he will throughout his extraordinarily refined, recreative work as exegetist, as 'secret sharer' of the poet's intent, that those who 'understand' a text have largely missed its essential significance. Bad translations communicate too much. Their seeming accuracy is limited to what is non-essential in the fabric of the original. Benjamin's approach to the question of translatability—can the work be translated at all? if so, for whom?—is Kabbalistic:

one might speak of a life or a moment as 'unforgettable' even if all men had forgotten it. If its essence required that it not be forgotten, then that assertion would not be false: it would only point to a requirement not satisfied by man and, simultaneously, to a realm in which it could be satisfied: the memory of God. By the same token, the question of the translatability of certain works would remain open even if they were untranslatable for man. And indeed, given an exacting concept of translation, should this not be the case to some extent? It is in the light of such an analysis that one can ask whether a given work of literature *requires* translation. The relevant proposition is this: if translation is a form, then the condition of translatability must be ontologically necessary to certain works.

Echoing Mallarmé, but in terms obviously derived from the Kabbalistic and gnostic tradition, Benjamin founds his metaphysic of translation on the concept of 'universal language'. Translation is both possible and impossible—a dialectical antinomy characteristic of esoteric argument. This antinomy arises from the fact that all known tongues are fragments, whose roots, in a sense which

[1] An English translation of this essay, by James Hynd and E. M. Valk, may be found in *Delos, A Journal on and of Translation*, 2 (1968).

is both algebraic and etymological, can only be found in and validated by 'die reine Sprache'. This 'pure language'—at other points in his work Benjamin will refer to it as the Logos which makes speech meaningful but which is contained in no single spoken idiom—is like a hidden spring seeking to force its way through the silted channels of our differing tongues. At the 'messianic end of their history' (again a Kabbalistic or Hasidic formulation), all separate languages will return to their source of common life. In the interim, translation has a task of profound philosophic, ethical, and magical import.

A translation from language A into language B will make tangible the implication of a third, active presence. It will show the lineaments of that 'pure speech' which precedes and underlies both languages. A genuine translation evokes the shadowy yet unmistakable contours of the coherent design from which, after Babel, the jagged fragments of human speech broke off. Certain of Luther's versions of the Psalms, Hölderlin's recasting of Pindar's Third Pythian Ode, point by their strangeness of evocatory inference to the reality of an *Ur-Sprache* in which German and Hebrew or German and ancient Greek are somehow fused. That such fusion can exist, that it must, is proved by the fact that human beings *mean* the same things, that the human voice springs from the same hopes and fears, though different words are *said*. Or to put it another way: a poor translation is full of apparently similar saying, but misses the bond of meaning. Philo-logy is love of the Logos before it is a science of differing stems. Luther and Hölderlin move German some distance 'back' towards its universal origin. But to accomplish this alchemy, a translation must, in regard to its own language, retain a vital strangeness and 'otherness'. Very little in Hölderlin's *Antigone* is 'like' ordinary German; Marianne Moore's readings of La Fontaine are thorn-hedges apart from colloquial American English. The translator enriches his tongue by allowing the source language to penetrate and modify it. But he does far more: he extends his native idiom towards the hidden absolute of meaning. 'If there is a language of truth, in which the final secrets that draw the effort of all thinking are held in silent repose, then this language of truth is—true language. And it is precisely this language—to glimpse or describe it is the only perfection the philosopher can hope for—that is concealed, intensively, in translations.' As the Kabbalist seeks the forms of God's occult design in the groupings of letters and words, so the philosopher of language will seek in translations—in what they omit as much as in their content—the far light of original meaning. Walter Benjamin's summation derives directly from the mystic tradition: 'For in some degree, all great writings, but the Scriptures in the highest degree, contain between the lines their virtual translation. The interlinear version of the Scriptures is the archetype or ideal of all translation.'

His loyalties divided between Czech and German, his sensibility drawn as it was, at moments, to Hebrew and to Yiddish, Kafka developed an obsessive awareness of the opaqueness of language. His work can be construed as a continuous parable on the impossibility of genuine human communication, or, as he put it to Max Brod in 1921, on 'the impossibility of not writing, the impossibility of writing in German, the impossibility of writing differently. One could almost add a fourth impossibility: the impossibility of writing'. Kafka often extended the latter to include the illusions of speech. 'Is it her singing that enchants us,' asks the narrator in 'Josephine the Singer, or the Mouse Folk', 'or is it not rather the solemn stillness enclosing her frail little voice?' And 'In the Penal Colony', perhaps the most desperate of his metaphoric reflections on the ultimately inhuman nature of the written word, Kafka makes of the printing press an instrument of torture. The theme of Babel haunted him: there are references to it in almost every one of his major tales. Twice he offered specific commentaries, in a style modelled on that of Hasidic and Talmudic exegesis.

The first occurs in his allegory on the building of the Great Wall of China, written in the spring of 1917. The narrative relates the two structures, though 'according to human reckoning' the purposes of the Wall were the very contrary to those of the insolent Tower. A scholar has written a strange book asserting that the destruction of Babel did not result from the causes generally alleged. Nimrod's edifice had fallen simply because its foundations had been defective. The sage argues that the Great Wall shall, itself, serve as plinth for a new Tower. The narrator confesses that he is bewildered. How can the Wall, being at most a semicircle, become a foundation for a Tower? Yet there must be some truth to the bizarre suggestion: architectural drawings for the Tower, albeit shadowy, are included among those for the Wall. And there are detailed proposals regarding the required labour force and gathering of nations. That gathering figures in 'Das Stadtwappen' ('The City Arms'), a brief parable which Kafka wrote in the autumn of 1920. This is among his most riddling texts. The first sentence refers to the presence of interpreters (*Dolmetscher*) on the building site. As no generation of men can hope to complete the high edifice, as engineering skills are constantly growing, there is time to spare. More and more energies are diverted to the erection and embellishment of the workers' housing. Fierce broils occur between different nations assembled on the site. 'Added to which was the fact that already the second or third generation recognized the meaninglessness, the futility (*die Sinnlosigkeit*) of building a Tower unto Heaven—but all had become too involved with each other to quit the city.' Legends and ballads have come down to us telling of a fierce longing for a predestined day on which a gigantic fist will smash the builders' city with five blows. 'That is why the city has a fist in its coat of arms.'

It would be fatuous to propose any single decoding or equivalence of meaning for Kafka's uses of Babel. That is not how his method of anagogic and allegoric anecdote works. The Talmud, which is often Kafka's archetype, refers to the forty-nine levels of meaning which must be discerned in a revealed text. But it is evident that Kafka saw in the Tower and its ruin a dramatic shorthand through which to convey certain exact, though not wholly articulate, intimations about man's linguistic condition and the relations of that condition to God. The Tower is a necessary move: it arises from some undeniable surge of human will and intelligence. The word *Himmelsturmbau* embodies a puzzling duality: the Tower is, as Genesis proclaims, an assault on Heaven (*Sturm*), but it is also a vast Jacob's ladder of stone (*Turm*) on which man would ascend towards his Creator. Rebellion and worship are inextricably mixed, as are the impulses of speech to lead towards and away from the truth. The foundations of the Tower preoccupy Kafka even more than the edifice itself. 'The Burrow', his last story, and an unmistakable comment on the relation of the writer to language and reality, shows how the Tower may be seen from its interior, spiralling galleries. Hence the uncanny remark in one of Kafka's notebooks: 'We are digging the pit of Babel.' But what are the concordances between the Tower and the Great Wall, which is usually in Kafka a symbol of the Mosaic Law? What are we to make of the precise shift in verb tenses in the final lines of 'Das Stadtwappen': sagas 'came from the city', presumably long ago, but 'the city has a fist in its coat of arms'? That of Prague happens not to have a fist but two towers. In all these allusions the menace of language and the mystery of its divided state are present. Another notebook entry may come nearest to being a summary of the range of paradox and tragic dialectic which Kafka concentrated in the emblem of the Tower: 'Had it been possible to build the Tower of Babel without ascending it, that would have been allowed.' If man could use language without pursuing meaning to the forbidden edge of the absolute, he might still be speaking a veritable and undivided tongue. Yet to use language without translations, without seeking out the hidden springs of the Law is also impossible, and perhaps prohibited. In Kafka speech is the paradoxical circumstance of man's incomprehension. He moves in it as in an inner labyrinth.

Labyrinths, circular ruins, galleries, Babel (or Babylon) are constants in the art of our third modern Kabbalist.[1] We can locate in the poetry and fictions of Borges every motif present in the language mystique of Kabbalists and gnostics: the image of the world as a concatenation of secret syllables, the notion of an absolute idiom or cosmic letter—alpha and aleph—which underlies the rent fabric of human tongues, the supposition that the entirety of knowledge and experience is prefigured in a final tome containing all conceivable permutations of the alphabet. Borges advances the occult belief that the structure of ordinary sensate

---

[1] See ch. 1 (pp. 3–54) of Jaime Alazraki's *Borges and the Kabbala, and Other Essays on his Fiction and Poetry* (Cambridge, 1988).

time and space interpenetrates with alternative cosmologies, with consistent, manifold realities born of our speech and of the fathomless free energies of thought. The logic of his fables turns on a refusal of normal causality. Gnostic and Manichaean speculation (the word has in it an action of mirrors)[1] provide Borges with the crucial trope of a 'counter-world'. Contrary streams of time and relation blow like high, silent winds through our unstable, itself perhaps conjectural, habitat. No poet has imaged with more density of life the possibility that our existence is being 'dreamt elsewhere', that we are the mere figure of another's speech, hurtling towards the close of that single, inconceivably vast utterance in which Jakob Böhme heard the sound of the Logos. As Borges writes in 'Compass':

> All things are words of some strange tongue, in thrall
> To Someone, Something, who both day and night
> Proceeds in endless gibberish to write
> The history of the world. In that dark scrawl
>
> Rome is set down, and Carthage, I, you, all,
> And this my being which escapes me quite,
> My anguished life that's cryptic, recondite,
> And garbled as the tongues of Babel's fall.
> (Richard Wilbur's translation)

There were times when Kafka felt the multiplicity of languages to be a gag in his throat. Borges moves with a cat's sinewy confidence and fun between Spanish, ancestral Portuguese, English, French, and German. He has a poet's grip on the fibre of each. He has rendered a Northumbrian bard's farewell to Saxon English, 'a language of the dawn'. The 'harsh and arduous words' of *Beowulf* were his before he 'became a Borges'. 'Deutsches Requiem' is not only as near as we get to a metamorphic realization of the murderous need which bound Nazi to Jew; in voice and narrative gist the story is also as German as those black woods. Though Borges's Spanish is often private and Argentine, he is

[1] Borges's 'The Mirror of Enigmas' (in *Labyrinths*, New York, 1962) argues the specific interactions of gnostic philosophy and the *speculum in aenigmate*.

possessed of the specific grain of the language, of the invariants which relate his own poetry to 'Seneca's black Latin'. But keen as is Borges's sense of the irreducible quality of each particular tongue, his linguistic experience is essentially simultaneous and, to use a Coleridgean notion, reticulative. Half a dozen languages and literatures interweave. Borges uses citations and literary–historical references, often invented, to establish the key, the singular locale of his verse and fables. Close-woven, these diverse idioms and legacies—the Kabbala, the Anglo-Saxon epic, Cervantes, the French symbolists, the dreams of Blake and De Quincey—constitute a mapping, a landscape of recognitions unique to Borges but also, somehow, familiar as sleep. Quick with interchange and mutation, Borges's several languages move towards a unified, occult truth (the Aleph glimpsed on the nineteenth step in the cellar of Carlos Argentino's house) as do the individual letters of the alphabet in the 'cosmic library' of one of the most secret of his *ficciones*.

'The Library of Babel' dates from 1941. Every element in the fantasia has its sources in the 'literalism' of the Kabbala and in gnostic and Rosicrucian images, familiar also to Mallarmé, of the world as a single, immense tome. 'The universe (which others call the Library) is composed of an indefinite, perhaps an infinite number of hexagonal galleries.' It is a beehive out of Piranesi but also, as the title indicates, an interior view of the Tower. 'The Library is total and . . . its shelves contain all the possible combinations of the twenty-odd orthographic symbols (whose number, though vast, is not infinite); that is, everything which can be expressed, in all languages. Everything is there: the minute history of the future, the autobiographies of the archangels, the faithful catalogue of the Library, thousands and thousands of false catalogues, a demonstration of the falsehood of the true catalogue, the Gnostic gospel of Basilides, the commentary on this gospel, the commentary on the commentary of this gospel, the veridical account of your death, a version of each book in all languages, the interpolation of every book in all books.' Any conceivable combination of letters has already been foreseen in the Library and is certain to 'encompass some terrible meaning' in one of its secret languages. No act of speech is without meaning: 'No one can articulate a syllable which is not full of tenderness and fear, and which is not, in one of those languages, the powerful name of some god.' Inside the burrow or circular ruins men jabber in mutual bewilderment; yet all their myriad words are tautologies making up, in a manner unknown to the speakers, the lost cosmic syllable or Name of God. This is the formally boundless unity that underlies the fragmentation of tongues.

Arguably, 'Pierre Menard, Author of the *Quixote*' (1939) is the most acute, most concentrated commentary anyone has offered on the business of translation. What studies of translation there are, including this book, could, in Borges's style, be termed a commentary on his commentary. This concise fiction has been widely recognized for the device of genius which it obviously is. But— and again one sounds like a pastiche of Borges's fastidious pedantry—certain details have been missed. Menard's bibliography is arresting: the monographs on 'a poetic vocabulary of concepts' and on 'connections or affinities' between the thought of Descartes, Leibniz, and John Wilkins point towards the labours of the seventeenth century to construe an *ars signorum*, a universal ideogrammatic language system. Leibniz's *Characteristica universalis*, to which Menard addresses himself, is one such design; Bishop Wilkins's *Essay towards a real character and a philosophical language* of 1668 another. Both are attempts to reverse the disaster at Babel. Menard's 'work sheets of a monograph on George Boole's symbolic logic' show his (and Borges's) awareness of the connections between the seventeenth century pursuit of an inter-lingua for philosophic discourse and the 'universalism' of modern symbolic and mathematical logic. Menard's transposition of the decasyllables of Valéry's *Le Cimetière marin* into alexandrines is a powerful, if eccentric, extension of the concept of translation. And *pace* the suave authority of the memorialist, I incline to believe that 'a literal translation of Quevedo's literal translation' of Saint François de Sales was, indeed, to be found among Menard's papers.

The latter's masterpiece, of course, was to consist 'of the ninth and thirty-eighth chapters of the first part of *Don Quixote* and a fragment of chapter twenty-two'. (How many readers of Borges have observed that Chapter IX turns on a translation from Arabic into Castilian, that there is a labyrinth in XXXVIII, and that Chapter XXII contains a literalist equivocation, in the purest Kabbalistic vein, on the fact that the word *no* has the same number of letters as the word *sí*?) Menard did not want to compose another *Quixote* 'which is easy—but *the Quixote itself*. Needless to say, he never contemplated a mechanical transcription of the original; he did not propose to copy it. His admirable intention was to produce a few pages which would coincide—word for word and line for line—with those of Miguel de Cervantes.' (So in James E. Irby's version. Anthony Bonner reads 'which would be so easy' and omits 'a few' before 'pages', striking what is surely a false note of prolixity.)[1]

Pierre Menard's first approach to the task of total translation or, one might more rigorously say, transubstantiation, was one of utter mimesis. But to *become* Cervantes by merely fighting Moors, recovering the Catholic faith, and forgetting the history of Europe between 1602 and 1918 was really too facile a métier. Far more interesting was 'to go on being Pierre Menard and reach the *Quixote* through the experiences of Pierre Menard', i.e. to put oneself so deeply in tune with Cervantes's being, with his onto-logical form, as to re-enact, inevitably, the exact sum of his realiz-ations and statements. The arduousness of the game is dizzying. Menard assumes 'the mysterious duty'—Bonner, rightly I feel, invokes the notion of 'contract'—of recreating deliberately and explicitly what was in Cervantes a spontaneous process. But al-though Cervantes composed freely, the shape and substance of the *Quixote* had a local 'naturalness' and, indeed, necessity now dissi-pated. Hence a second fierce difficulty for Menard: to write 'the *Quixote* at the beginning of the seventeenth century was a reason-able undertaking, necessary and perhaps even unavoidable; at the beginning of the twentieth, it is almost impossible. It is not in vain that three hundred years have gone by, filled with exceedingly

[1] Cf. 'Pierre Menard, Author of Don Quixote', translated by Anthony Bonner in *Fictions* (New York, 1962) with James E. Irby's version of the same story in *Labyrinths*.

complex events. Amongst them, to mention only one, is the *Quixote* itself' (Bonner's 'that same *Don Quixote*' both complicates and flattens Borges's intimation). In other words, any genuine act of translation is, in one regard at least, a transparent absurdity, an endeavour to go backwards up the escalator of time and to re-enact voluntarily what was a contingent motion of spirit. Yet Menard's fragmentary *Quixote* 'is more subtle than Cervantes's'. How wondrous is Menard's ability to articulate feelings, thoughts, counsels so eccentric to his own time, to find uniquely appropriate words for sentiments notoriously at variance with those he usually held:

Cervantes' text and Menard's are verbally identical, but the second is almost infinitely richer. (More ambiguous, his detractors will say, but ambiguity is richness.)

It is a revelation to compare Menard's *Don Quixote* with Cervantes'. The latter, for example, wrote (part one, chapter nine):

... truth, whose mother is history, rival of time, depository of deeds, witness of the past, exemplar and adviser to the present, and the future's counsellor.

Written in the seventeenth century, written by the 'lay genius' Cervantes, this enumeration is a mere rhetorical praise of history. Menard, on the other hand, writes:

... truth, whose mother is history, rival of time, depository of deeds, witness of the past, exemplar and adviser to the present, and the future's counsellor.

History, the *mother* of truth: the idea is astounding. Menard, a contemporary of William James, does not define history as an inquiry into reality, but as its origin. Historical truth, for him, is not what has happened; it is what we judge to have happened. The final phrases—*exemplar and adviser to the present, and the future's counsellor*—are brazenly pragmatic.

The contrast in style is also vivid. The archaic style of Menard—quite foreign, after all—suffers from a certain affectation. Not so that of his forerunner, who handles with ease the current Spanish of his time.

Menard's labours were Herculean. 'He dedicated his scruples and his sleepless nights to repeating an already extant book in an alien tongue. He multiplied draft upon draft, revised tenaciously and tore up thousands of manuscript pages.' To repeat an already extant book in an alien tongue is the translator's 'mysterious duty' and job of work. It cannot and must be done. 'Repetition' is, as Kierkegaard argued, a notion so puzzling that it puts in doubt causality and the stream of time. To produce a text verbally identical with the original (to make of translation a perfect transcription), is difficult past human imagining. When the translator, negator of time and rebuilder at Babel, comes near succeeding, he passes into that state of mirrors which is described in 'Borges and I'. The translator too 'must live on in Borges'—or in any other author he chooses—'not in myself—if indeed I am anyone—though I recognize myself less in his books than in many others, or than in the laborious strumming of a guitar.' A true translator knows that his labour belongs 'to oblivion' (inevitably, each generation retranslates), or 'to the other one', his occasion, begetter, and precedent shadow. He does *not* know 'which of us two is writing this page'. In that 'transubstantial ignorance'—I find no simpler, less unwieldy term—lies the misery of this whole business of translation, but also what repair we can make of the broken Tower.

We shall return to the Kabbalistic motifs and diverse models of translation inferred in the memoir written on the late Pierre Menard of Nîmes by his erudite friend. Irby qualifies the bonfire in which Menard burned his papers as 'merry'; Bonner as 'gay'. There are two psychologies here, two Christmases, two visions of heresy and of the phoenix.

2

It is via Leibniz and J. G. Hamann that language mysticism enters the current of modern, rational linguistic study. Both men were in active contact with Kabbalistic and Pietist thought.

Linguistic theory bears decisively on the question of whether or not translation, particularly between different languages, is in fact possible. In the philosophy of language two radically opposed points of view can be, and have been asserted. The one declares that the underlying structure of language is universal and common to all men. Dissimilarities between human tongues are essentially of the surface. Translation is realizable precisely because those deep-seated universals, genetic, historical, social, from which all grammars derive can be located and recognized as operative in every human idiom, however singular or bizarre its superficial forms. To translate is to descend beneath the exterior disparities of two languages in order to bring into vital play their analogous and, at the final depths, common principles of being. Here the universalist position touches closely on the mystical intuition of a lost primal or paradigmatic speech.

The contrary view can be termed 'monadist'. It holds that universal deep structures are either fathomless to logical and psychological investigation or of an order so abstract, so generalized as to be well-nigh trivial. That all men known to man use language in some form, that all languages of which we have apprehension are able to name perceived objects or to signify action—these are undoubted truths. But being of the class 'all members of the species require oxygen to sustain life', they do not illuminate, except in the most abstract, formal sense, the actual workings of human speech. These workings are so diverse, they manifest so bewilderingly complicated a history of centrifugal development, they pose such stubborn questions as to economic and social function, that universalist models are at best irrelevant and at worst misleading. The extreme 'monadist' position—we shall find great poets holding it—leads logically to the belief that real translation is impossible. What passes for translation is a convention of approximate analogies, a rough-cast similitude, just tolerable when the two relevant languages or cultures are cognate, but altogether spurious when remote tongues and far-removed sensibilities are in question.

Between these two poles of argument, there can be numerous intermediary and qualified attitudes. Neither position is maintained often with absolute rigour. There are relativist shadings in the universalist grammars of Roger Bacon, and the grammarians of Port Royal, and even in the transformational generative grammar of Chomsky. Nabokov, who regards all but the most rudimentary of interlinear translations as a fraud, as a facile evasion of radical impossibilities, is himself a master mover between languages. In their modern guise, moreover, both lines of argument can be traced to a common source.

In 1697, in his tract on the amelioration and correction of German, Leibniz put forward the all-important suggestion that language is not the vehicle of thought but its determining medium. Thought is language internalized, and we think and feel as our particular language impels and allows us to do. But tongues differ as profoundly as do nations. They too are monads, 'perpetual living mirrors of the universe' each of which reflects or, as we would now put it, structures experience according to its own particular sight-lines and habits of cognition. Yet at the same time, Leibniz had universalist ideals and hopes. Like George Dalgarno, whose *Ars Signorum* appeared in 1661, and Bishop Wilkins, who published his remarkable *Essay towards a real character and a philosophical language* in 1668, Leibniz was profoundly interested in the possibilities of a universal semantic system, immediately legible to all men. Such a system would be analogous to mathematical symbolism, so efficacious precisely because the conventions of mathematical operation seem to be grounded in the very architecture of human reason and appear independent of all local variation. It would be analogous also to Chinese ideograms. Once a lexicon of ideograms had been agreed to, all messages could be read instantaneously, whatever the language of the recipient, and the disaster at Babel would, on the graphic level at least, be mended. As we shall see, mathematical symbolism and Chinese writing are, to this day, implied models in almost all discussions of universal grammar and translation.

In Vico's 'philology', as in Leibniz's, universalist and 'monadist' strains coexist. Philology is the quintessential historical science, the key to the *Scienza nuova*, because the study of the evolution of language is the study of the evolution of the human mind itself. Vico knows, this is one of his great clairvoyances, that man enters into active possession of consciousness, into active cognizance of reality, through the ordering, shaping powers of language. All men do so, and in that sense language, and metaphor in particular, are a universal fact and a universal mode of being. In the genesis of the human spirit, all nations traverse the same stages of linguistic usage, from the immediate and sensory to the abstract. Simultaneously, however, Vico's opposition to Descartes and to the extensions of Aristotelian logic in Cartesian rationalism made of him the first true 'linguistic historicist' or relativist. He was acutely perceptive of the autonomous genius and historical coloration of different languages. All primitive men sought expression through 'imaginative universals' (*generi fantastici*), but in diverse tongues these universals rapidly acquired very different configurations. 'Almost infinite particulars' constitute both the syntactic and lexical corpus of different languages. These particulars both engender and reflect the differing world-views of races and cultures. The degree of 'infinite particularity' reaches so deep, that a universal logic of language, on the Aristotelian or Cartesian–mathematical model, is falsely reductionist. It is only by means of a scrupulous, essentially poetic recreation or translation of a given language-world, such as that of Homeric Greek and of Biblical Hebrew, that the 'new science' of myth and history can hope to retrace the growth of consciousness (and growth*s* would be more accurate).[1]

That Goethe, in a remark dated March 1787, compared Hamann to Vico is well known, as is the fact that Hamann had, ten years before, obtained a copy of the *Scienza nuova*. It remains unlikely, nevertheless, that there was any direct influence. Hamann's theories on language and culture go back to the very early 1760s. They spring both from the pregnant muddle of his extraordinary intellect and from his intimacy with theosophic and Kabbalistic speculations. Hamann's notions are usually fragmentary; they are veiled in a diction as 'radiantly dark' as was Blake's. But the originality and foresight of his conjectures on language are, particularly today, uncanny.

From the 1750s onward, the problem of 'l'influence réciproque du langage sur les opinions et des opinions sur le langage' was very much in vogue. Hamann addressed himself to the theme in his *Versuch über eine akademische Frage* (1760). He affirms that there is a determining concordance between the directions of

---

[1] Cf. Stuart Hampshire, 'Vico and the Contemporary Philosophy of Language', in G. Tagliacozzo (ed.), *Giambattista Vico, An International Symposium* (Baltimore, 1969).

thought and feeling in a community and 'the lineaments of its speech'. Nature has provided different races with different pigmentation and shapes of the eye. Similarly, it has caused in men imperceptible but decisive variations in the formation of lip, tongue, and palate. These variations are the source of the proliferation and diversity of languages. (This physiological hypothesis was not new, and Hamann himself draws on the English anatomist Thomas Willis.) Languages are as figurative of the particular nature of a civilization as are its garb and social rites. Each language is an 'epiphany' or articulate revelation of a specific historical–cultural landscape. Hebrew verb forms are inseparable from the niceties and strict punctualities that mark Jewish ritual. But that which a language reveals as being the specific genius of a community, the language itself has shaped and determined. The process is dialectical, with the formative energies of language moving both inward and outward in a civilization.

In 1761, Hamann applied these views to a comparative examination of the grammatical and lexical resources of French and German. Turgid, erratic as they are, the *Vermischte Anmerkungen* contain premonitions of genius. Though referring itself to Leibniz, Hamann's opening statement about the close kinship of linguistic and monetary exchanges, and his confident dictum that theories of language and of economics will prove mutually explanatory, are not only strikingly original but set out *in nuce* much of Lévi-Strauss's structural anthropology. Hamann is able to argue in this fashion because he is already working towards a general theory of significant signs, towards a semiology in the modern sense. Mystical exegesis underwrote Hamann's and Leibniz's belief that a nerve fabric of secret meanings and revelations lies below the surface structure of all languages. To read is to decipher. To speak 'is to translate (metapherein)'. Both skills constitute the decoding of the signs or vital hieroglyphs through which life acts on consciousness. In a usage which anticipates the whole of Kenneth Burke's 'grammar of motives', Hamann identifies 'action' (*Handlung*) with 'dynamic linguistic posture or structure' (*Sprachgestaltung*). Hamann opposes Kantian categories of universal, mental *a priori* in the name of those local, determinant energies inherent in a given language. Out of diverse tongues men will necessarily construe diverse mental and even sensory frameworks. Language generates specific cognition. Despite their rhapsodic, mystical format, the *Philologische Einfälle und Zweifel* of 1772 repay serious attention. Hamann throws out suggestions which anticipate the linguistic relativism of Sapir and Whorf. He seems to be saying that it is different languages that cause the different selections made by men from among that 'ocean of sensations' which tides, indiscriminately, through human sensibility. Hamann is arguing that neither Cartesian co-ordinates of general, deductive reasoning, nor Kantian mentalism will serve to account for the creative, irrational, and manifold proceedings through which language—unique to the species but so varied among nations—shapes reality and is, in turn, acted upon by local human experience.

It is one of the achievements of Romanticism to have sharpened the sense of locale, to have given specific density to our grasp of geographical and historical particularity. Herder was possessed of a sense of place. His '*Sprachphilosophie*' marks a translation from the inspired fantastications of Hamann to the development of genuine comparative linguistics in the early nineteenth century. Herder's quality can, I think, be overrated. He never shook himself free of the enigma of the natural or divine origin of language as he posed it in his famous essay of 1772. All the evidence seemed to point to an instinctual and evolutionary genesis of human speech, exactly as Lucretius and Vico had supposed. Yet the gap between spontaneous, mimetic speech-sounds and the wonder of mature language seemed too great. Thus the theory of a divine act of special bestowal was never far from Herder's thoughts. Like Leibniz, Herder had a vivid realization of the atomic quality of human experience, each culture, each idiom being a particular crystal reflecting the world in a particular way. The new nationalism and vocabulary of race provided Herder with a ready focus. He called for 'a general physiognomy of the nations from their languages'. He was convinced of the irreducible spiritual individuality of each language, and particularly of German, whose antique expressive strengths had lain dormant but were now armed for the light of a new age and for the creation of a literature of world rank. National character is 'imprinted on language' and, reciprocally, bears the stamp of language. Hence the supreme importance of the health of language to that of a people; where language is corrupted or bastardized, there will be a corresponding decline in the character and fortunes of the body politic. Herder carried this belief to curious lengths. He stated in the *Fragmente* that a language would derive great benefits by guarding 'itself from all translations'. The notion is very similar to that of mystical grammarians seeking to protect the holy text from traduction. An untranslated language, urges Herder, will retain its vital inno-cence, it will not suffer the debilitating admixture of alien blood. To keep the *Original- und Nationalsprache* unsullied and alive is the eminent task of the poet.

The short years between Herder's writings and those of Wilhelm von Humboldt were among the most productive in the history of linguistic thought. Sir William Jones's celebrated *Third Anniversary Discourse on the Hindus* of 1786 had, as Friedrich von Schlegel put it, 'first brought light into the knowledge of language through the relationship and derivation he demonstrated of Roman, Greek, Germanic and Persian from Indic, and through this into the ancient history of peoples, where previously every-thing had been dark and confused'. Schlegel's own *Ueber die Sprache und Weisheit der Indier* of 1808, which contains this tribute to Jones, itself contributed largely to the foundations of modern linguistics. It is with Schlegel that the notion of 'comparative grammar' takes on clear definition and currency. Not much read today, Mme de Staël's *De L'Allemagne* (1813) exercised tremen-dous influence. In her impressionistic but often acutely intelligent portrayal of a waking nation, Mme de Staël argued that there were crucial reciprocities between the German language and the charac-ter and history of the German people. Expanding on suggestions already made by Hamann, she sought to correlate the metaphysical ambience, internal divisions, and lyric bias of the German national spirit with the gnarled weave and 'suspensions of action' in German syntax. She saw Napoleonic French as antithetical to German, and found its systematic directness and rhetoric clearly expressive of the virtues and vices of the French nation.

All these lines of debate and conjecture anticipate Humboldt's work. But to enter on that work is to enter on an entirely different order of intellectual achievement. The play of intelligence, the delicacy of particular notation, the great front of argument which Humboldt exhibits, give his writings on language, incomplete though they are, a unique stature. Humboldt is one of the very short list of writers and thinkers on language—it would include Plato, Vico, Coleridge, Saussure, Roman Jakobson—who have said anything that is new and comprehensive.

Humboldt was fortunate. An extraordinary linguistic and psychological process was occurring all around him: a major literature was being created. It brought to bear on language and national sensibility a concentration of individual genius together with a common vision for which there are few parallels in history. Goethe, Schiller, Wieland, Voss, Hölderlin, and a score of others were doing more than composing, editing, translating masterpieces. With a high degree of policy and proclaimed intent, they were making of the German language an exemplar, a deliberate inventory of new possibilities of personal and social life. *Werther*, *Don Carlos*, *Faust* are supreme works of the individual imagination, but also intensely pragmatic forms. In them, through them, the hitherto divided provinces and principalities of the German-speaking lands could test a new common identity. Goethe and Schiller's theatre at Weimar, Wieland's gathering of German ballads and folk poetry, the historical narratives and plays of Kleist set out to create in the German mind and in the language a shared echo. As Vico had imagined it would, a body of poetry gave a bond of remembrance (partially fictive) to a new national community. As he studied the relations of language and society, Humboldt could witness how a literature, produced largely by men whom he knew personally, was able to give Germany a living past, and how it could project into the future great shadow-forms of idealism and ambition.

During his working years, Indo-European linguistics and the comparative study of classical, Hebraic, and Celtic antiquities, according to new criteria of philological and textual rigour, were laying the foundations for a genuine science of language. That such a science would have to enlist history, psychology, poetics, ethnography, and even various branches of biology, was clear to Humboldt. Like Goethe, he held the individual fact to be, as it were, shone through by the constant energies of universal, organic unity. It is the great weave and pulse of life itself that gives to each isolated phenomenon (isolated only because we may not yet have perceived the surrounding field of force) its meaning. To Humboldt and his brother, this intimation of universality was no empty metaphor. The Humboldts were among the last Europeans of whom it may be said with fair confidence that they had direct professional or imaginative notions of very nearly the whole of extant knowledge. Ethnographers, anthropologists, linguists, statesmen, educators, the two brothers were a nerve-centre for humanistic and scientific inquiry. Their active interests, like Leibniz's, ranged with authority and passionate curiosity from mineralogy to metaphysics, from the study of Amerindian antiquities to modern technology. When he posited language as the centre of man, Wilhelm von Humboldt was in a position to feel what such a pivot must inform and relate. Yet being in natural touch with the later eighteenth century, Humboldt still possessed a certain receptivity to those traditions of occult linguistic speculation which, as we have seen, led back unbroken to Nicholas of Cusa and Paracelsus. Both the very old and the newest were active in Humboldt's great enterprise.

That enterprise has come down to us in an incomplete, edited form.[1] It includes the lecture 'Ueber das Entstehen der grammatischen Formen und ihren Einfluss auf die Ideenentwicklung' (the title is itself a manifesto) of January 1822, and the *magnum opus* on which Humboldt was engaged from the 1820s until his death in 1835, and which was posthumously put together and published: *Ueber die Verschiedenheit des menschlichen Sprachbaues und ihren Einfluss auf die geistige Entwicklung des Menschengeschlechts.* Even translated, that title retains its proud scope: *On the Differentiation of the Structure of Human Language, and its Influence on the Spiritual Evolution of the Human Race.* Humboldt aims at nothing

---

[1] Edited by H. Steinthal (Berlin, 1883).

less than an analytic correlation of language and human experience. He would lay bare the concordance between the *Weltanschauung* of a given language and the history and culture of those who speak it. Essential to this analysis is the belief that language is the true or the only verifiable *a priori* framework of cognition. Perception is organized by the imposition of that framework on the total flux of sensations. 'Die Sprache ist das bildende Organ des Gedankens,' says Humboldt, using *bildend* in its forceful, twofold connotation of 'image' (*Bild*) and 'culture' (*Bildung*). Different linguistic frameworks will divide and channel the sensory flux differently: 'Jede Sprache ist eine Form und trägt ein Form-Princip in sich. Jede hat eine Einheit als Folge eines in ihr waltenden Princips.' This organic evolutionism goes well beyond and, indeed, against Kant. In so doing, Humboldt arrives at a key notion: language is a 'third universe' midway between the phenomenal reality of the 'empirical world' and the internalized structures of consciousness. It is this median quality, this material and spiritual simultaneity, that makes of language the defining pivot of man and the determinant of his place in reality. Seen thus, language is a universal. But so far as each human tongue differs from every other, the resulting shape of the world is subtly or drastically altered. In this way, Humboldt conjoins the environmentalism of Montesquieu and the nationalism of Herder with an essentially post-Kantian model of human consciousness as the active and diverse shaper of the perceived world.

The shaping agencies of intellect, Coleridge called them 'esemplastic powers', do not, as it were, perform via language. They are inherent in language. Speech is *poiesis* and human linguistic articulation is centrally creative. It may be that Humboldt derived from Schiller his emphasis on language as being itself the most comprehensive work of art. His own contribution is to insist, in a way that strikes a very modern note, on language as a total generative process. Language does not convey a pre-established or separately extant content, as a cable conveys telegraphic messages. The content is created in and through the dynamics of statement. The entelechy, the purposeful flow of speech—we find in Humboldt a kind of romantic Aristotelianism—is the communica-

tion of ordered, perceived experience. But experience only assumes order and cognizance in the language-matrix. Ultimately,
but inexplicably, language, *die Sprache*, is identical with 'the ideal
totality of spirit' or *Geist*. As we shall see, the fact that this radical
identity cannot be explained will undermine Humboldt's actual
linguistic analyses.

Under pressure of his extraordinary vision and emotional awareness of the life-giving, life-determining powers of language,
Humboldt advances the idea that language can be adverse to man.
So far as I am aware, no one before him had seen this point, and
even now we have hardly grasped its implications. Humboldt's
statement is arresting: 'Denn so innerlich auch die Sprache
durchaus ist, so hat sie dennoch zugleich ein unabhängiges,
äusseres, gegen den Menschen selbst Gewalt ausübendes Dasein'
('Albeit language is wholly inward, it nevertheless possesses at the
same time an autonomous, external identity and being which does
violence to man himself'). Language makes man at home in the
world, 'but it also has the power to alienate'. Informed by energies
proper to itself, more comprehensive and timeless than any who
make use of it, human speech can raise barriers between man and
nature. It can bend the mirrors of consciousness and of dreams.
There is a phenomenon of linguistic *Entfremdung* inseparable from
the creative genius of the word. The term is Humboldt's, and the
insight it expresses is of vital relevance to a theory of translation.

Ueber die Verschiedenheit des menschlichen Sprachbaues (particularly
sections 19 and 20) is crowded with linguistic conjectures of
prophetic brilliance. Man walks erect not because of some ancestral reaching out towards fruit or branch, but because discourse,
die Rede, 'would not be muffled and made dumb by the ground'.
More than a century before the modern structuralists, Humboldt
notes the distinctive binary character of the linguistic process: it
shares, it mediates between, the crucial antinomies of inner and
outer, subjective and objective, past and future, private and public.
Language is far more than communication between speakers. It is
dynamic mediation between those poles of cognition which give
human experience its underlying dual and dialectical form. Here
Humboldt clearly anticipates both C. K. Ogden's theory of opposition and the binary structuralism of Lévi-Strauss.

From this wide range of argument, I want to select those points which are immediate to our theme: the multiplicity of human tongues and the relations between *Weltansicht* and *Wort*.

'The bringing forth of language is an inner necessity for mankind.' It is, moreover, in the nature of 'spirit' to seek to realize, to energize into conscious being, all modes of possible experience. This is the true cause of the immense variety of speech forms. Each is a foray into the total potentiality of the world. 'Jede Sprache', writes Humboldt, 'ist ein Versuch.' It is a trial, an assay. It generates a complex structure of human understanding and response and tests the vitality, the discriminatory range, the inventive resources of that structure against the limitless potential of being. Even the noblest language is only *ein Versuch* and will remain ontologically incomplete. On the other hand, no language however primitive will fail to actualize, up to a point, the inner needs of a community. Humboldt is convinced that different tongues provide very different intensities of response to life; he is certain that different languages penetrate to different depths. He takes over Schlegel's classification of 'higher' and 'lower' grammars. Inflection is far superior to agglutination. The latter is the more rudimentary mode, a *Naturlaut*. Inflection allows and compels a far subtler, more dynamic treatment of action. It makes qualitative perception more acute and conduces necessarily to a more developed articulation (i.e. realization) of abstract relations. To pass from an agglutinate to an inflected tongue is to translate experience 'upward'.

Humboldt now sets out to perform the crucial experiment. He applies his theory of the reciprocal determinations of language and world-view to specific cases. He seeks to show how Greek and Latin respectively determine particular ethnic, national aggregates of feeling. He would demonstrate that these two great idioms produced contrasting structures of civilization and social reflex. The argument is intelligently set out and gives proof of Humboldt's at-homeness in classical philology and literature. But it falls unquestionably short of its theoretic aim and promise.

The Greek tone is light, delicate, *nuancé*. Attic civilization is incomparably inventive of intellectual and plastic forms. These virtues are engendered by and reflected in the precisions and shadings of Greek grammar. Few other languages have cast so finely-woven a net over the currents of life. At the same time, there is that in Greek syntax which helps explain the divisive quality of Greek politics, the excessive trust in rhetoric, the virtuosities of falsehood which sophisticate and corrode the affairs of the *polis*. Latin offers a grave contrast. The stern, masculine, laconic tenor of Roman culture is exactly correlate with the Latin language, with its sobriety, even paucity, of syntactic invention and *Lautformung*. The lettering of a Latin inscription is perfectly expressive of the linear, monumental weight of the language. Both are the active mould of the Roman way of life.

Humboldt's argument is circular. Civilization is uniquely and specifically informed by its language; the language is the unique and specific matrix of its civilization. The one proposition is used to demonstrate the other and vice versa. Knowing the Greeks to have been one thing and the Romans another, we argue back to linguistic differences. In what way do aorist and optative help or fail to account for the indiscriminate bluntness of Spartan life? Can we discern modulations in the ablative absolute as Rome passes from Republican to Augustan Latin? *Post hoc* and *propter hoc* are inevitably blurred. Humboldt's summarizing statement is eloquent, but also self-betraying in its lofty indistinction. Different languages engender different spiritual constructs of reality: 'der dadurch hervorgebrachte verschiedene Geist schwebt, wie ein leiser Hauch, über dem Ganzen' ('the differing Spirit thus produced hovers, like a silent breath, over the whole'). Having identified *Sprache* with *Geist* (Hegel's vocabulary is exactly contemporary with his own), Humboldt must conclude in this way. But having stated, at the outset, that this identification is, in the final analysis, inexplicable, he cannot use it to enforce demonstrable proof. His conviction remains fundamentally intuitive. For all its philosophic reach and sensibility to linguistic values, moreover, Humboldt's position is not fully worked out. The essential argument is 'monadist' or relativist, but a universalist tendency can also be found. Hence the lack of final incisiveness in Humboldt's key terms, 'structure of language' and 'structures determined by a particular language'. There is no doubt that these terms infer a wide range of example and historical evidence. But pressed home, they turn into metaphors, into shorthand formulations of the romantic criterion of organic life, rather than into verifiable concepts. Given the mystery at the core of the relations between 'Language' and 'Spirit', it could hardly be otherwise.

It has been said that the line from Herder and Humboldt to Benjamin Lee Whorf is unbroken.[1] Intellectually this is so. The actual history of linguistic relativity leads via the work of Steinthal (the editor of Humboldt's fragmentary texts) to the anthropology of Franz Boas. From there it reaches the ethno-linguistics of Sapir and Whorf. One can summarize that history as being an attempt to provide Humboldt's intuitions with a solid basis of semantic and anthropological fact. Much of the argument is developed in Germany. Nor is this surprising. The first true Germany was that of Luther's vernacular. Gradually the German language created those modes of shared sensibility from which the nation-state could evolve. When that state entered modern history, a late arrival burdened with myths and surrounded by an alien, partially hostile Europe, it carried with it a sharpened, defensive sense of unique perspective. To the German temper, its own *Weltansicht* seemed a special vision, whose foundations and expressive genius lay in the language. Reflecting on the drastic extremes of German history, on the apparently fatal attempts of the German nation to break out of the ring of more urbane or, in the east, more primitive and menacing cultures, German philosophers of history thought of their language as a peculiarly isolating yet also numinous factor. Other nations could not feel their way into its arcane depths. But great springs of renewal and metaphysical discovery would surge from what Schiller called *die verborgenen Tiefen*.

Cassirer's *Philosophy of Symbolic Forms* gave fresh impetus to Humboldt's ideas. Cassirer was in agreement with the theory that the different conceptual categories into which different languages place the same sensory phenomena must reflect linguistically de-

[1] Cf. R. L. Brown, *Wilhelm von Humboldt's Conception of Linguistic Relativity* (The Hague, 1967) and Robert L. Miller, *The Linguistic Relativity Principle and Humboldtian Ethno-linguistics* (The Hague, 1968).

termined differences of perception. The stimuli are demonstrably identical; the responses are often strikingly disparate. Between the 'physiological universal' of consciousness and the specific cultural–conventional process of identification and response lies the membrane of a particular language or, as Cassirer put it, the unique 'inner form' which distinguishes it from all other languages. In a series of books ranging from *Muttersprache und Geistesbildung* (1929) to *Vom Weltbild der Deutschen Sprache* in 1950, Leo Weisgerber sought to apply the 'monadic' or relativity principle to the actual, detailed features of German syntax and, correspondingly, to the history of German attitudes. It was his central affirmation that 'our understanding is under the spell of the language which it utilizes'. A very similar formulation was put forward by the linguist Jost Trier. Every language structures and organizes reality in its own manner and thereby determines the components of reality that are peculiar to this given language. This determination consti- tutes what Trier, in the early 1930s, called *das sprachliche Feld*. Thus, in a distinctly Leibnizian way, each tongue or language- monad constructs and operates within a total conceptual field (the imagistic correlation with quantum physics is obvious). This field may be understood as a *Gestalt*. Being linguistically diverse, dif- ferent cultures impose a different *Gestalt* on the same raw material and total aggregate of experience. In each case, the linguistic 'feedback' from experience is a particular one. Speakers of dif- ferent languages therefore inhabit different 'mediary worlds' (*Zwischenwelten*). The linguistic world-view of a given community shapes and gives life to the entire landscape of psychological and communal behaviour. It is language which decides how different conceptual groupings and contours are to be 'read' and related within the whole. Often a language will 'filter out' from the field of potential recognition even more information than it includes in that field. The gauchos of the Argentine know some 200 expres- sions for the colours of horses' hides, and such discrimination is obviously vital to their economy. But their normal speech finds room for only four plant names.

In American linguistics, relativism drew both on the legacy of Humboldt and on anthropological field-work. Though treated with reservations, Levy-Bruhl's concept of a 'primitive mind', in which the ethnographer could observe pre-rational or non-Cartesian linguistic—logical processes, had its influence. Anthropological study of American Indian cultures seemed to bear out Humboldt's conjectures on linguistic determinism and Trier's notion of the 'semantic field'. The whole approach is summarized by Edward Sapir in an article dated 1929:[1]

The fact of the matter is that the 'real world' is to a large extent unconsciously built up on the language habits of the group. No two languages are ever sufficiently similar to be considered as representing the same social reality. The worlds in which different societies live are distinct worlds, not merely the same world with different labels attached.

The emphasis on 'group' is worth noting. The 'semantic field' of a given culture is a dynamic, socially motivated construct. The particular 'language and reality game' played by the community depends, in a way very similar to that argued by Wittgenstein in the *Philosophical Investigations*, on the actions, on the historically evolved and agreed-to customs of the particular society. What we find here is a 'dynamic mentalism': language organizes experience, but that organization is constantly acted upon by the collective behaviour of the particular group of speakers. Thus there occurs a cumulative dialectic of differentiation: languages generate different social modes, different social modes further divide languages.

The 'monadist' case has philosophic origins of great distinction in the work of Leibniz and of Humboldt. Its crowning statement is also of great intellectual fascination. The 'metalinguistics' of Whorf have for some time been under severe attack by both linguists and ethnographers. It looks as if a good deal of his work cannot be verified. But the papers gathered in *Language, Thought and Reality* (1956) constitute a model which has extraordinary intellectual elegance and philosophic tact. They are a statement of vital possibility, an exploration of consciousness relevant not only to the linguist but also to the poet and, decisively, to the translator. Whorf was an outsider. He brought to ethno-linguistics a sense of

[1] In D. Mandelbaum (ed.), *Selected Writings in Language, Culture and Personality by Edward Sapir* (Berkeley and Los Angeles, 1949).

the larger issues, of the poetic and metaphysical implications of language study such as is rare among professionals. He had something of Vico's philosophic curiosity, but was a chemical engineer with a distinctively modern awareness of scientific detail. The years in which Roman Jakobson, I. A. Richards and Benjamin Lee Whorf were active simultaneously must count among decisive moments in the history of the investigation of the human mind.

Whorf's theses are well known. Linguistic patterns determine what the individual perceives in his world and how he thinks about it. Since these patterns—observable in the syntax and lexical means of the language—vary widely, the modes of perception, thought, and response in human groups using different language systems will be very different. World-views that are basically unlike will result. Whorf designates these as 'thought worlds'. They make up the 'microcosm that each man carries about within himself, by which he measures and understands what he can of the macrocosm'. There is, so far as human consciousness goes, no such entity as a universally objective physical reality. 'We dissect nature along lines laid down by our native language.' Or to be more exact: there is a fundamental duality in the exercise of human perception (Whorf is drawing on *Gestalt* psychology). There is a universal but also rudimentary neuro-physiological apprehension of space that may have preceded language in the evolution of the species and that may still precede articulate speech in the growth of the infant. But once a particular language is used, a particular conceptualization of space follows (Whorf is not altogether clear as to whether language determines that conceptualization or only conditions it). Spatialization, and the space–time matrix in which we locate our lives, are made manifest in and by every element of grammar. There is a distinctive Indo-European time-sense and a corresponding system of verb tenses. Different 'semantic fields' exhibit different techniques of numeration, different treatments of nouns denoting physical quantity. They divide the total spectrum of colours, sounds, and scents in very diverse ways. Again, Wittgenstein's use of 'mapping' offers an instructive parallel: different linguistic communities literally inhabit and traverse different landscapes of conscious being. In one of his very last papers, Whorf summarized his entire vision:[1]

Actually, thinking is most mysterious, and by far the greatest light upon it that we have is thrown by the study of language. This study shows that the forms of a person's thoughts are controlled by inexorable laws of pattern of which he is unconscious. These patterns are the unperceived intricate systematizations of his own language—shown readily enough by a candid comparison and contrast with other languages, especially those of a different linguistic family. His thinking itself is in a language—in English, in Sanskrit, in Chinese. And every language is a vast pattern-system, different from others, in which are culturally ordained the forms and categories by which the personality not only communicates, but also analyses nature, notices or neglects types of relationship and phenomena, channels his reasoning, and builds the house of his consciousness.

To show that this doctrine 'stands on unimpeachable evidence', Whorf was prepared to apply comparative semantic analyses to a wide range of languages: Latin, Greek, Hebrew (there are important links between his own work and the eccentric Kabbalism of Fabre d'Olivet), Kota, Aztec, Shawnee, Russian, Chinese and Japanese. Unlike many universalists, Whorf had an obvious linguistic ear. But it is his work on the languages of the Hopis of Arizona that carries the weight of evidence. It is here that the notion of distinct 'pattern-systems' of life and consciousness is argued by force of specific example. The key papers on 'an American Indian model of the universe' date from *circa* 1936 to 1939, at which point Whorf extended his analyses to the Shawnee language.

Examining the punctual and segmentative aspects of verbs in Hopi, Whorf concludes that the language maps a certain terrain 'of what might be termed primitive physics'. As it happens, Hopi is better equipped to deal with wave processes and vibrations than is modern English. 'According to the conception of modern physics, the contrast of particle and field of vibrations is more fundamental in the world of nature than such contrasts as space and time, or past, present, and future, which are the sort of contrasts our

---

[1] *Language, Thought, and Reality: Selected Writings of Benjamin Lee Whorf*, ed. John B. Carroll (Cambridge, Mass., 1956), p. 252.

own language imposes upon us. The Hopi aspect-contrast...
being obligatory upon their verb forms, practically forces the
Hopi to notice and observe vibratory phenomena, and furthermore encourages them to find names for and to classify such
phenomena.' Whorf finds that the Hopi language contains no
words, grammatical forms or idiomatic constructions referring
directly to what we call 'time', or to the vectors of time and motion
as we use them. The 'metaphysics underlying our own language,
thinking, and modern culture' necessarily imposes a static three-
dimensional infinite space, but also a perpetual time-flow. These
two 'cosmic co-ordinates' could be harmoniously conjoined in the
physics of Newton and the physics and psychology of Kant. They
confront us with profound internal contradictions in the world
of quantum mechanics and four-dimensional relativity. The
metaphysical framework which informs Hopi syntax is, according
to Whorf, far better suited to the world-picture of modern science.
Hopi verb tenses and phrasings articulate the existence of events
'in a dynamic state, yet not a state of motion'. The semantic
organization of 'eventuating and manifesting' phenomena allows—
indeed enforces—precisely those modulations from subjective
perceptions or 'ideal mappings' of events to objective status, which
Indo-European grammar finds it so difficult to accommodate or
must express wholly in mathematical terms.

In translating into English, the Hopi will say that these entities in process
of causation 'will come' or that they—the Hopi—'will come to' them, but
in their own language, there are no verbs corresponding to our 'come' and
'go' that mean simple and abstract motion, our purely kinematic concept.
The words in this case translated 'come' refer to the process of eventuating without calling it motion—they are 'eventuates to here' (*pew'i*) or
'eventuates from it' (*angqö*) or 'arrived' (*pitu*, pl. *öki*) which refers only to
the terminal manifestation, the actual arrival at a given point, not to any
motion preceding it.[1]

Thus the entire Hopi treatment of happenings, inferential reasoning, and distant events is delicate and susceptible of provisional
postures in just the way so often required by twentieth-century

[1] *Language, Thought, and Reality*, p. 60.

astrophysics or wave-particle theory. The shaping influence of the observer on the process observed, the statistics of indeterminacy, are inherent in Hopi as they are not, or only by virtue of explanatory metaphor, in English.

Crucial to Whorfian semantics is the notion of the *cryptotype*. He defines it 'as a submerged, subtle, and elusive meaning, corresponding to no actual word, yet shown by linguistic analysis to be functionally important in the grammar'. It is these 'cryptotypes' or 'categories of semantic organization'—dispersion without boundaries, oscillation without agitation, impact without duration, directed motion—which translate the underlying metaphysics of a language into its overt or surface grammar. It is the study of such 'cryptotypes' in different languages, urges Whorf, that will lead anthropology and psychology to an understanding of those deep-seated dynamics of meaning, of chosen and significant form, that make up a culture. It is, no doubt, exceedingly difficult for an outsider, operating inevitably within the world-frame of his own tongue, to penetrate to the active symbolic deeps of a foreign tongue. We reach for the bottom and stir up further darkness. 'Cryptotypes', moreover, are 'so nearly at or below the threshold of conscious thinking' that even the native speaker cannot put them into adequate words. Patently, they elude translation (we shall return to this point). Yet careful, philosophically and poetically disciplined observation does allow the linguist and anthropologist to enter, in some degree at least, into the 'pattern-system' of an alien tongue. Particularly if he acts on the principles of ironic self-awareness which underlie a genuine relativist view.

Whorf was tireless in emphasizing the built-in bias, the axiomatic arrogance of traditional and universalist philology, with its scarcely veiled presumption that Sanskrit and Latin constitute the natural, optimal model of all human speech or, at the least, a model manifestly preferable to all others. Whorf's revaluation of 'thinking in primitive communities' coincides in date and spirit with Lévi-Strauss's early studies of the genius of *La Pensée sauvage*. Lévi-Strauss would fully endorse Whorf's assertion that 'many American Indian and African languages abound in finely wrought, beautifully logical discriminations about causation, action, result, dynamic or energic quality, directness of experience, etc., all matters of the function of thinking, indeed the quintessence of the rational. In this respect they far outdistance the European languages.' Whorf offers telling instances: the four forms of the pronoun in the Algonkian languages allow compact notations of intricate social situations; the distinction between a tense for past events with present result or influence, and for those with none, in Chichewa, 'a language related to Zulu, spoken by a tribe of unlettered Negroes in East Africa'; the three causal verb forms in the Cœur d'Alène language, spoken by a small Indian tribe in Idaho. Here again, Whorf finds the paradox that the 'semantic field' of numerous so-called primitive communities segments experience into a phenomenology which is closer than that of the Indo-European language family to the data of twentieth-century physics and *Gestalt* psychology. Equally fascinating are Whorf's hints—any theory of translation will want to explore and extend them—that different languages show different degrees of accord between phonetics (which must, in some measure, be universal) and the 'inner music of meaning'. German *zart*, meaning 'tender', calls up tonal associations of bright hardness. English *deep* ought to go with such sounds of quick, sharp lightness as 'peep'. Meaning in a given tongue may go against the grain of apparently universal aural associations. This clash between 'mental' and 'psychic' codes of recognition may be crucial to the evolution of a particular language and will assume very different forms in different tongues.

A picture of language, mind, and reality based almost exclusively on Cartesian–Kantian logic and on the 'semantic field' of Standard Average European (SAE) is a hubristic simplification. The close of 'Science and Linguistics', a paper published in 1940, is worth quoting in full—especially at a time when the study of language is so largely dominated by a theory of dogmatic generality and mathematical aspect:

A fair realization of the incredible degree of the diversity of linguistic system that ranges over the globe leaves one with an inescapable feeling that the human spirit is inconceivably old; that the few thousand years of history covered by our written records are no more than the thickness of a pencil mark on the scale that measures our past experience on this planet; that the events of these recent millenniums spell nothing in any evolutionary wise, that the race has taken no sudden spurt, achieved no commanding synthesis during recent millenniums, but has only played a little with a few of the linguistic formulations and views of nature bequeathed from an inexpressibly longer past. Yet neither this feeling nor the sense of precarious dependence of all we know upon linguistic tools which themselves are largely unknown need be discouraging to science but should, rather, foster that humility which accompanies the true scientific spirit, and thus forbid that arrogance of the mind which hinders real scientific curiosity and detachment.

Whatever may be the future status of Whorf's theories of language and mind, this text will stand.

3

Such are the distinction and consequence of Whorf's metalinguistics, that, even of themselves, critiques of Whorf constitute a fair statement of the universalist case. These critiques bear on the circularity of Whorf's evidence. Seeing a dripping spring, an Apache will describe it as 'whiteness moving downward'. The verbal formulation is clearly different from that in current English. But what direct insight does it afford into Apache *thinking*? It is tautological to argue that a native speaker perceives experience differently from us because he talks about it differently, and then infer differences of cognition from those of speech. Behind such inference lies a rudimentary, untested scheme of mental action. In 'A Note on Cassirer's Philosophy of Language', E. H. Lenneberg summarizes a whole range of philosophic doubts: 'There is no cogent reason to assume that the grammarian's articulation of the stream of speech is coterminous with an articulation of knowledge or the intellect.' Words are not the embodiments of invariant mental operations and fixed meanings. The idea that conventional syntactic patterns incorporate uniquely determined and determinant acts of perception is itself the reflection of a primitive dualism. It corresponds to the mind–body image of early psychology. Any operational model of the linguistic process, e.g. Wittgenstein's proposal that 'the meaning of a word is its use in the language', will refute Whorf's deterministic parallelism of thought and speech.

Moreover, if the Humboldt–Sapir–Whorf hypothesis were right, if languages were monads with essentially discordant mappings of reality, how then could we communicate interlingually? How could we acquire a second tongue or traverse into another language-world by means of translation? Yet, manifestly, these transfers do occur continually.

The empirical conviction that the human mind actually does communicate across linguistic barriers is the pivot of universalism. To the twelfth-century relativism of Pierre Hélie, with his belief that the disaster at Babel had generated as many kinds of irreconcilable grammar as there are languages, Roger Bacon opposed his famous axiom of unity: 'Grammatica una et eadem est secundum substantiam in omnibus linguis, licet accidentaliter varietur.' Without a *grammatica universalis*, there could be no hope of genuine discourse among men, nor any rational science of language. The accidental, historically moulded differences between tongues are, no doubt, formidable. But underlying these there are principles of unity, of invariance, of organized form, which determine the specific genius of human speech. Amid immense diversities of exterior shape, all languages are 'cut from the same pattern'.

We have met this intuitive certitude in Leibniz and even among the relativistic arguments of Humboldt. The successes obtained by nineteenth-century Indo-European philology in formalizing, in giving a normative and predictive account of the great mass of discrete phonological and grammatical facts, strengthened the universalist bias. Today, the working vision of a universal grammar is shared by almost all linguists. Indeed, it is because it deals with phenomena of a universal, deep-seated character, with the general ground rules of human cognitive processes, that current linguistic theory advances claims to philosophic and psychological authority. 'The main task of linguistic theory must be to develop an account of linguistic universals that, on the one hand, will not be falsified by the actual diversity of languages and, on the other, will be sufficiently rich and explicit to account for the rapidity and uniformity of language learning, and the remarkable complexity and range of the generative grammars that are the product of language learning.'[1]

The axiom of universality and the aim of comprehensive description are clear. What remains of great difficulty is the question of levels (it already perplexed such universalists of the late eighteenth century as James Beattie). At what level of the structure of language can 'universals' be accurately located and described? How deep must we go below the live, obstinately diverse layers of linguistic usage? During the past forty years, the direction of universalist argument has been one of ever-deepening formalization and abstraction. In turn, each level of proposed universality has been found to be contingent or subverted by anomalies. Singularities have cropped up in what looked like the most general of assumptions. Instead of being rigorous and exhaustive, the description of 'universal linguistic traits' has often proved to be no more than an open-ended catalogue.

There are three obvious planes of language on which to seek out universals: the phonological, the grammatical, and the semantic.

All human beings possess the same neurophysiological equipment with which to emit and receive sounds. There are notes whose pitch lies outside the range of the human ear; there are tones which our vocal cords cannot produce. All languages, therefore, fall within certain definable material bounds. All are combinations of a limited set of physical phenomena. It is an obvious move to seek to identify and enumerate the physiological or phonological universals of which each and every spoken tongue is a selective aggregate. One of the most influential of such enumerations is N. S. Trubetskoy's *Grundzüge der Phonologie* published in Prague in 1939. Comparing some 200 phonological systems, Trubetskoy set out those acoustic structures without which there cannot be a language and which all languages exhibit. Roman Jakobson's theory of 'distinctive features' is a refinement

[1] N. Chomsky, *Aspects of the Theory of Syntax* (Cambridge, Mass., 1965), pp. 27–8.

of Trubetskoy's universals. Jakobson identifies some twenty universal phonetic elements, each of which can be rigorously characterized according to articulatory and acoustic criteria (e.g. every language must contain at least one vowel). In different combinations, these features make up the phonology, the physical presence and transmission of all languages. Using these crucial markers, a science-fiction writer or computer could devise a new tongue, and one could affirm in advance that it would fall within the set limits of human expressive potentiality. A signal-system lacking these 'distinctive universals' would lie, literally, outside the human octave.

In practice, the analysis of phonological universals turns out to be a rather simple-minded and blunt enterprise. A good many conclusions are, again, of the order of unsurprising generality implicit in the statement that all human beings require oxygen. Where the argument becomes prescriptive, problems of rigorous description arise. It seems safe enough to assert that all languages on this earth have a vowel system. In fact, the proposition is true only if we take it to include segmented phonemes which occur as syllabic peaks—and even in that case, at least one known tongue, Wishram, poses problems. There is a Bushman dialect called Kung, spoken by a few thousand natives of the Kalahari. It belongs to the Khoisan group of languages, but is made up of a series of clicking and breathing sounds which, so far as is known, occur nowhere else, and which have only recently been transcribed. Obviously, these sounds lie within the physiological bounds of human possibility. But why should this anomaly have developed at all, or why, if efficacious, should it be found in no other phonological system? A primary nasal consonant 'is a phoneme of which the most characteristic allophone is a voiced nasal stop, that is, a sound produced by a complete oral stoppage (e.g. apical, labial), velic opening, and vibration of the vocal cords'.[1] Having thus defined a PNC, phonologists can identify the conditions under which it occurs in all languages and the determined ways in which it affects the position and stress of other

[1] Charles A. Ferguson, 'Assumptions about Nasals: A Sample Study in Phonological Universals', in J. H. Greenberg (ed.), *Universals of Language* (Cambridge, Mass., 1963), p. 56.

phonemes. But the plain statement that every human tongue has at least one primary nasal consonant in its inventory requires modification. Hockett's *Manual of Phonology* (1955) reports a complete absence of nasal consonants from Quileute and two neighbouring Salishan languages. Whether such nasals once existed and have, in the course of history, become voiced stops, or whether, through some arresting eccentricity, Salishan speech never included nasal phonemes at all, remains undecided. Such examples can be multiplied.

Consequently, the universalist case proceeds beyond the somewhat rudimentary and 'soft-edged' material of phonology to that of grammar. If all languages are indeed cut from the same pattern, a comparative analysis of syntactic systems will reveal those elements that truly constitute a *grammatica universalis*.

The pursuit of such a 'fundamental grammar' is itself a fascinating chapter in the history of analytic thought. A considerable distance has been covered since Humboldt's hope that a generalized treatment of syntactic forms would be devised to include all languages, 'from the rawest' to the most accomplished. The notion that certain fixed syntactic categories—noun, verb, gender—can be found in every tongue, and that all languages share certain primary rules of relation, became well established in nineteenth-century philology. That 'same basic mould' in which all languages are cast came to be understood quite precisely: as a set of grammatical units, of markers which themselves denote nothing but make a difference in composite forms, and of rules of combination.

Some of these rules are of very great generality. No language has been found to lack a first- and second-person singular pronoun. The distinctions between 'I', 'thou', and 'he' and the associated network of relations (so vital to kinship terms) exist in every human idiom. Every language in use among men has a class of proper names. No language has a vocabulary that is grammatically entirely homogeneous. A type of clause in which a 'subject' is talked about or modified in some manner, is observable in every linguistic system. All speech operates with subject–verb–object combinations. Among these, the sequences 'verb–object–subject', 'object–subject–verb', and 'object–verb–subject' are exceedingly rare. So rare, as to suggest an almost deliberate violation of a deep-rooted ordering of perception. Other 'grammatical universals' are points of detail: for example 'when the adjective follows the noun, the adjective expresses all the inflectional categories of the noun. In such cases the noun may lack overt expression of one or all of these categories.' The most ambitious list of syntactic universals to have been established 'on the basis of the empirical linguistic evidence' was that of J. H. Greenberg.[1] It enumerates forty-five fundamental grammatical relations, and leads to the conclusion that 'the order of elements in language parallels that in physical experience or the order of knowledge'. The underlying grammar of all human speech forms is a mapping of the world. It emphasizes those features of the landscape and of bio-social experience which are common to all men. Differences of stress, organized sequence, relations of hierarchy as between the general and the particular or the sum and the part, these are the counters of reason from which all languages develop. If a language 'has the category of gender, it always has the category of number'. Otherwise, there would be human aggregates trapped in eccentric chaos.

Again, the scheme looks more impressive than it actually is. Compared to the total of languages in current use, the number whose grammar has been formalized and thoroughly examined is absurdly small (Greenberg's empirical evidence is drawn almost exclusively from thirty languages). In syntax, moreover, no less than in phonology, tenacious singularities occur. One would expect all languages with a distinction of gender in the second-person singular to show this distinction in the third person as well. In nearly every known instance, this holds. But not in a very small cluster of tongues spoken in central Nigeria. The Nootka language provides an often-cited example of a grammatical system in which it is very difficult to draw any normal distinction between noun and verb. The alignment of genitive constructions looks like a primal typological marker according to which all languages can be classified into a small number of major groups. Araucanian, an

---

[1] Joseph H. Greenberg, 'Some Universals of Grammar with Particular Reference to the Order of Meaningful Elements' in op. cit., pp. 73–113.

Indian tongue spoken in Chile, and some Daghestan languages of the Caucasus do not fit the scheme. Such anomalies cannot be dismissed as mere curios. A single genuine exception, in any language whether living or dead, can invalidate the whole concept of a grammatical universal. Indeed, this whole approach has since been largely abandoned.

It is, in part, because the statistical, ethno-linguistic approach to syntactic universals has proved unsatisfactory or merely descriptive, that generative transformational grammars propose to argue at much greater phenomenological depths. In doing so, they have sought to drive the very notion of grammar inward, to a specifically linguistic innate faculty of human consciousness.

Chomskian grammar is emphatically universalist (but what other theory of grammar—structural, stratificational, tagmemic, comparative—has not been so?). No theory of mental life since that of Descartes and the seventeenth-century grammarians of Port Royal has drawn more explicitly on a generalized and unified picture of innate human capacities, though Chomsky and Descartes mean very different things by 'innateness'. In Descartes this 'innateness', specifically underwritten by a transcendental wager on God, on the congruence between word and world, entails a social context. It anticipates some of the very 'stimulus and response' configurations which Chomsky will rebuke. Chomsky's starting-point was the rejection of behaviourism. No simple pattern of stimulus and mimetic response could account for the extreme rapidity and complexity of the way in which human beings acquire language. All human beings. Any language. A child will be able to construct and understand utterances which are new and which are, at the same time, acceptable sentences in his language. At every moment of our lives we formulate and understand a host of sentences different from any that we have heard before. These abilities indicate that there must be fundamental processes at work quite independently of 'feedback from the environment'.[1] Such processes are innate to all men: 'human

---

[1] These and the immediately following quotations are taken from N. Chomsky's review of B. F. Skinner's *Verbal Behavior*. First published in *Language*, 35 (1959), the article is reprinted in John P. De Cecco (ed.), *The Psychology of Language, Thought, and Instruction* (New York and London, 1967).

beings are somehow specially designed to do this, with data-handling or "hypothesis-formulating" ability of unknown character and complexity.' Each individual on earth has somehow and in some form internalized a grammar from which his, but also any other language is generated. ('Generation' translates Humboldt's *erzeugen*. Here, as in the shared axiom that language 'makes infinite use of finite means', Chomskian universalism is congruent with the relativism of Humboldt.)

Differences between languages represent differences of 'surface structure' only. They are accidents of terrain which impress the eye but tell us scarcely anything of the underlying 'deep structure'. Via a set of rules, of which 'rewriting rules' are fundamental, 'deep structures' generate, i.e. bring to the phonetic surface, the sentences we actually use and hear. We are then able to work back from the actual physical sentence, together with the derivation tree or 'phrase marker' constructed for it, to obtain some insight into the underlying 'deep structure'. More complex sentences are, in turn, generated by a second class of rules, the 'rules of trans-formation'. These rules—for which the theory of recursive functions offers the best analogy—must be applied in an ordered sequence. Some of them are not 'context-free'; their correct application depends on the surrounding linguistic material. It is at this point, presumably, that a universal system modulates into a particular language or family of languages. But any 'real progress in linguistics consists in the discovery that certain features of given languages can be reduced to universal properties of language, and explained in terms of these deeper aspects of linguistic form'.[1]

Chomsky contends that a search for universals at the phono-logical or ordinary syntactic level is wholly inadequate. The shaping centres of language lie much deeper. In fact, surface analogies of the kind cited by Greenberg may be entirely mis-leading: it is probable that the deep structures for which uni-versality is claimed are quite distinct from the surface structure of sentences as they actually appear. The geological strata are not reflected in the local landscape.

---

[1] N. Chomsky, *Aspects of the Theory of Syntax*, p. 35.

But what are these 'universal deep structures' like?

It turns out that it is exceedingly difficult to say anything about them. In the vocabulary of Wittgenstein, the transition from 'surface grammar' to 'depth grammar' is a step towards clarity, towards a resolution of those philosophic muddles which spring from a confusion of linguistic planes. Chomskian 'deep structures', on the other hand, are located 'far beyond the level of actual or even potential consciousness'. We may think of them as relational patterns or strings of an order of abstraction far greater than even the simplest of grammatical rules. Even this is too concrete a representation. 'Deep structures' are those innate components of the human mind that enable it to carry out 'certain formal kinds of operations on strings'. These operations have no *a priori* justification. They are of the category of essential arbitrariness inherent in the fact that the world exists. Thus 'there is no reason to expect that reliable operational criteria for the deeper and more important theoretical notions of linguistics ... will ever be forthcoming.' Try to draw up the creature from the deeps of the sea, and it will disintegrate or change form grotesquely.

Yet 'only descriptions concerned with deep structure will have serious import for proposals concerning linguistic universals'. Since descriptions of this sort are rare, rather like cores from the great marine trenches, 'any such proposals are hazardous, but are clearly no less interesting or important for being hazardous'. Chomsky then proceeds to offer one example of a genuine formal universal. It concerns the rules which govern the operations and legitimacy of deletion in the underlying structure of sentences of the type 'I know several more successful lawyers than Bill'. These rules or 'erasure transformations' may be proposed 'for consideration as a linguistic universal, admittedly on rather slender evidence'.[1]

Some grammarians would go even 'deeper' than Chomsky in locating the universal base of all languages. The sequential order of rules of transformation may itself lie near the surface and be specific to different languages. The whole notion of sequence may

[1] Ibid., pp. 180 ff.

have to be modified when it is applied to 'the rules of a universal base'. Professor Emmon Bach suggests that 'deep structures are much more abstract than had been thought'.[1] It may be erroneous to think of them, even by analogy, as linguistic units or 'atomic facts' of grammatical relation. At this final level of mental organization, we may be dealing with 'abstract kinds of pro-verbs which receive only indirect phonological representation' (I take 'pro-verbs' to signify potentialities of meaning 'anterior to' even the most rudimentary verbal units). At one level such a scheme of 'universal base rules' resembles the logical systems of Carnap and Reichenbach. At another level, most probably metaphoric, it suggests the actual patterning of the cortex, with its immensely ramified yet, at the same time, bounded or 'programmed' network of electro-chemical and neurophysiological channels. A system of variables, the set of all names, 'general predicates', and certain rules of constraint and relation between these, would, as it were, be imprinted on the fabric of human consciousness.

This imprint may never be susceptible of direct observation. But the 'selectional constraints and transformational possibilities' which we can discern at the surface of language give undeniable proof of its existence, efficacy, and universality. 'Such a system expresses directly the idea that it is possible to convey any conceptual content in any language, even though the particular lexical items available will vary widely from one language to another—a direct denial of the Humboldt–Sapir–Whorf hypothesis in its strongest form.'[2]

Whether it is indeed 'possible to convey any conceptual content in any language' is what I seek to investigate.

Granted the extreme difficulty of defining universals of grammar, many linguists feel that it is far too early to identify any

---

[1] E. Bach and R. T. Harms (eds.), *Universals in Linguistic Theory* (New York, 1968), p. 121.

[2] *Aspects of the Theory of Syntax*, pp. 121–2. In *Problems of Knowledge and Freedom* (New York, 1971), Chomsky puts forward a more cautious view: 'It is reasonable to formulate the hypothesis that such principles are language universals. Quite probably the hypothesis will have to be qualified as research into the variety of languages continues.'

'semantic universals'. Nevertheless, such identifications have been proposed, certainly since Vico's suggestion that all languages contain key anthropomorphic metaphors. One of these, the comparison of the pupil of the eye to a small child (*pupilla*), has been traced in all Indo-European languages, but also in Hebrew (*bath-'ayin*), Swahili, Lapp, Chinese, and Samoan.[1] Every language contains both 'opaque' and 'transparent' words, i.e. words in which the relation between sound and sense is purely arbitrary (German *Enkel*) and those in which it is obviously figurative (French *petit-fils*). The existence and statistical distribution of these two types of words 'is in all probability a semantic universal'.[2] The presence in every known tongue of certain *taboo* words, of expressions circumscribed by a zone of prohibition or sacred power, may well be a universal though also context-bound semantic feature. The thought that onomatopoeic patterns, sibilants, lateral consonants, may be rooted in specific modes of human perception—that there are universal ways of 'sounding the world'—is very ancient. It underlies a number of Plato's conjectural etymologies. And indeed, *i* carries values of smallness in almost every Indo-European and Finno-Ugrian language. But English *big* and Russian *velikij* suffice to show that we are not dealing with anything like a universal semantic reflex. Lévi-Strauss and several psycho-linguists agree in finding 'universal binomials' or contrast-pairs which tend to divide reality for us, and whose polarization is reflected in metaphors and stress patterns throughout all languages (white/black, straight/crooked, rising/falling, sweet/sour). The white/black dichotomy is of particular interest, as it appears to convey a positive/negative valuation in all cultures, regardless of skin-colour. It is as if all men, since the beginning of speech, had set the light above the dark.

Chomsky puts forward a number of semantic universals of a very broad but suggestive type: 'proper names in any language,

---

[1] Cf. C. Tagliavini, 'Di alcune denominazioni della pupilla' in *Annali dell' Istituto Universitario di Napoli* (1949).

[2] Stephen Ullmann, 'Semantic Universals', in J. H. Greenberg (ed.), *Universals of Language*, p. 221.

must designate objects meeting a condition of spatiotemporal contiguity, and that the same is true of other terms designating objects; or the condition that color words of any language must subdivide the color spectrum into continuous segments; or the condition that artifacts are defined in terms of certain human goals, needs, and functions instead of solely in terms of physical qualities.'[1] Again, the problem is one of the degree of precision which can be attached to such generalizations. All languages do subdivide the colour spectrum into continuous segments (though 'continuous' begs difficult issues in the neurophysiology and psychology of perception), but, as R. W. Brown and E. H. Lenneberg have shown, they go about their segmentation in ways which can be startlingly different. Indeed, basic questions about the relations between physical perception and linguistic coding remain far more open than Chomsky's statement suggests.

The history of Chomsky's own thought and of the modulation from transformational generative grammar to generative grammar has been one of the erosion of semantics. Chomsky's *Remarks on Nominalization* emphasize that 'meaning' is not at issue. The connections between generative grammatical investigations and semantic studies are all but severed. The true issue is now one of 'grammaticality'. The question asked is this: 'How would a natural-language speaker know that what he is enunciating is part of an organized grammar, how would he know, when encountering a new sentence, whether it is grammatically plausible or not?' A formal universal is defined as the operational rule whereby grammars generate sentences. The exploration does not touch on stylistics or constructs of meaning in the semantic range. What are to be elicited and formalized are rule-constrained natural usages. When we say 'He said John laughed', we are obeying and enacting the rule whereby 'he' and 'John' cannot co-refer. Generative grammarians are today not studying words and their manifold, possibly indeterminate, semantic fields. Nor are they mapping the neuro-physiological foundations of human speech. They are positing 'abstract formalizations in human psychology' (where

[1] N. Chomsky, *Aspects*, p. 29.

'psychology' continues to evade closer definition). The results obtained are not those which Chomsky announced in his earlier works. They are formal, meta-mathematical paradigms of the generation of grammaticality in speech.

To summarize: the evidence for the universality of those linguistic structures of which there is phenomenal evidence is, until now, provisional and putative. It oscillates between postulated levels of extreme formal abstraction at which the language-model becomes meta-mathematical and is divorced more or less completely from the phonetic fact, and levels which are crudely statistical (for example, Charles Osgood's proposal that the ratio of the number of phonemes to the number of distinctive features in any and every language will vary around an efficiency value of 50 per cent). The guarded conclusion of at least one linguist opposed to facile universalism may prove justified: 'Linguistic structures do differ, very widely indeed, among all the attested languages of the earth, and so do the semantic relationships which are associated with linguistic structures. The search for linguistic universals ... has recently come to the fore again, but it is still premature to expect that we can make any except the most elementary observations concerning linguistic universals and expect them to be permanently valid. Our knowledge of two-thirds or more of the world's languages is still too scanty (or, in many instances, non-existent).'[1] It may be that too many linguists have assumed that the

---

[1] Robert A. Hall, Jr., *An Essay on Language* (Philadelphia, 1968), pp. 53–4. For a sanely balanced discussion of the respective, ultimately collaborative claims and merits of Whorfian and universalist linguistics, cf. Helmut Gipper: 'Der Beitrag der inhaltlich orientierten Sprachwissenschaft zur Kritik der historischen Vernunft', in *Das Problem der Sprache*, ed. Hans-Georg Gadamer (Munich, 1967), pp. 420–5; also, in the same symposium, Wilhelm Luther, 'Sprachphilosophie und geistige Grundlagenbildung', pp. 528–31. Johannes Lohmann's *Philosophie und Sprachwissenschaft* (Berlin, 1965) contains a fascinating but idiosyncratic argument for a division of world languages into six fundamental structural types, each correlated with certain ways of experiencing the world, and each corresponding to certain phonetic and alphabetic features. A careful survey of the evidence, and further bibliography, may be found in Helmut Gipper, *Bausteine zur Sprachinhaltsforschung* (Düsseldorf, 1963), pp. 215 ff. Cf. also the important debate on the linguistic determination of Greek philosophic terms between E. Benveniste in *Problèmes de linguistique générale* (Paris, 1966), pp. 63 ff., and P. Auberique, 'Aristote et le langage, note annexe sur les catégories d'Aristote. A propos d'un

'deep structures' of all languages are identical because they have equated universal criteria of constraint and possibility with what could be in truth aspects only of the grammar of their own tongue or language group.

None the less, the belief that 'all languages are cut to the same pattern' is, currently, widespread. Few grammarians would hold with Osgood that eleven-twelfths of any language consist of universals and only one-twelfth of specific, arbitrary conventions, but the majority would agree that the bulk and organizing principles of the iceberg belong to the subsurface category of universals. To most professional linguists today the question is less *whether* there are 'formal and substantive universals of language' but precisely *what* they are, and to what extent the depths at which they lie will ever be accessible to either philosophic or neurophysiological investigation.

The postulate of linguistic universals or, to be exact, of substantive universals, should lead by direct inference to a working theory of interlingual translation. Proof that mutual transfer between languages is possible should follow immediately on the principle of substantive universality. Translation ought, in effect, to supply that principle with its most palpable evidence. The very possibility of motion of meaning between languages would seem to be firmly rooted in the underlying templet or common architecture of all human speech. But how is one to distinguish substantive from formal universals? How, except by theoretical *fiat* at one end or local intuition at the other, can one determine whether perfect translation should be possible because formal universals underlie all speech, or whether actual untranslatabilities persist because universals are only rarely or obscurely substantive? The discrimination is cogent in theory but has not been shown to be so in practice. It shares implicit ambiguities with the related distinction between 'deep' and 'surface' structures. Formal universals can be postulated at remote depths beyond concrete investigation or

article de M. Benveniste' (*Annales de la faculté des lettres d'Aix*, 43 (1965)). This debate and its implications are in turn reviewed by Jacques Derrida in *Marges de la philosophie* (Paris, 1972), pp. 214–46.

possible paraphrase. Substantive universals will, inevitably, overlap with the pragmatic, obstinately particularized realities of natural language. Translation is, plainly, the acid test. But the uncertainties of relation between formal and substantive universality have an obscuring effect on the relations between translation and universality as such. Only if we bear this in mind can we understand a decisive hiatus or shift in terms of reference in Chomsky's *Aspects of the Theory of Syntax*:

The existence of deep-seated formal universals ... implies that all languages are cut to the same pattern, but does not imply that there is any point by point correspondence between particular languages. It does not, for example, imply that there must be some reasonable procedure for translating between languages.

A footnote reinforces the sense of a fundamental uncertainty or *non sequitur*: 'The possibility of a reasonable procedure for translation between arbitrary languages depends on the sufficiency of substantive universals. In fact, although there is much reason to believe that languages are to a significant extent cast in the same mould, there is little reason to suppose that reasonable procedures of translation are in general possible.'[1]

How can the two suppositions be separated? 'Point by point' merely obscures the logical and substantive issue. The 'topological mapping' in which linguistic universals can be transferred from language to language—note the curious evasion in the phrase 'between arbitrary languages'—may lie very deep, but if it exists at all, a 'point by point correspondence' must be demonstrable. If translation can be achieved, is it not precisely because of the underlying 'sufficiency of substantive universals'? If, on the contrary, there is little reason to suppose that reasonable procedures of translation are 'in general' possible (and what does 'in general' really signify?), what genuine evidence have we of a universal structure? Are we not back in a Whorfian hypothesis of autonomous language-monads? Could Hall be right when he polemicizes against the whole notion of 'deep structures', calling

[1] N. Chomsky, *Aspects*, p. 30, and the relevant footnote on pp. 201–2.

them 'nothing but a paraphrase of a given construction, concocted *ad hoc* to enable the grammarian to derive the latter from the former by one kind of manipulation or another'?[1] Might it be that the transformational generative method is forcing all languages into the mould of English, as much seventeenth-century grammar endeavoured to enclose all speech within the framework of classical Latin?

Once more, the problem of the nature of translation appears to be central to that of language itself. The lacuna between a system of 'universal deep structures' and an adequate model of translation suggests that the ancient controversy between relativist and universalist philosophies of language is not yet over. It also suggests that the theory whereby transformational rules map semantically interpreted 'deep structures' into phonetically interpreted 'surface structures' may be a meta-mathematical ideal of considerable intellectual elegance, but not a true picture of natural language. 'No set of rules, however complete, is sufficient to describe ... the utterances possible in any living language.'[2] By placing the active nodes of linguistic life so 'deep' as to defy all sensory observation and pragmatic depiction, generative grammar, certainly in its 'strong' pristine versions, may have put the ghost out of all reach of the machine.

There is room, I submit, for an approach whose bias of interest focuses on languages rather than Language; whose evidence will derive from semantics (with all the implicit stress on meaning) rather than from 'pure syntax'; and which will begin with words, difficult as these are to define, rather than with imaginary strings or 'pro-verbs' of which there can never be any direct presentation. Investigation has shown that even the most formal rules of grammar must take into account those aspects of semantics and performance which Chomsky would exclude. Even individual sounds are concept-bound and act in a particular semantic field. It is doubtful, as well, whether a real grammar can start from and allow pre- or ungrammatical sentences as transformational

---

[1] Robert A. Hall, Jr., *An Essay*, p. 53.
[2] Ibid., p. 77.

generative grammar must. 'Grammaticality is, in any case, not a phenomenon that can be measured in terms of simple binary opposition, declaring any linguistic phenomenon to be either grammatical or ungrammatical. There is an infinite gradation between something which every member of a speech-community would use and recognize unhesitatingly as completely normal, to the opposite extreme of something that every speaker would declare was never used ... new formations resulting from analogy or blending are taking place all the time, and are being recognized and understood without difficulty.'[1]

Or to put it in summary fashion: a meta-mathematical view of language, working principally with pre- or pseudo-linguistic atomic units, will fail to account for the nature and possibility of relations between languages as they actually exist and differ.[2]

Noam Chomsky's *Theory of Government and Binding*, his *The Science of Language*, show a sharpening distance from the universalist innocence and sovereign claims of earlier theories. The promises initially offered to non-technical readers of literature, of philosophy, of translation, have been relinquished. What few attempts have been made at systematic applications of generative grammatical methods to matters and texts in the 'poetical' or 'rhetorical' modes, to actualities of speech and writing other than rudimentarily formalized, have departed widely from any strict Chomskyan position.

Hence the need of looking in directions which are, I fully admit, more impressionistic and far less amenable to formal codification. But language itself is 'open-ended' and charged with energies of the utmost diversity and intricacy. 'The really deep results of transformational grammar', writes George Lakoff, 'are, in my opinion, the negative ones, the hosts of cases where transformational grammar fell apart for a deep reason: it tried to study the structure of language without taking into account the fact that

---

[1] Ibid., p. 72.
[2] The case is put succinctly by I. A. Richards in 'Why Generative Grammar does not Help' (*English Language Teaching*, 22, i and ii (1967–8)). An expanded version of this critique forms Chapter IV of Richards's *So Much Nearer: Essays Towards a World English* (New York, 1970).

language is used by human beings to communicate in a social context.'[1] Time moves through every feature of language as a shaping force. No true understanding can arise from synchronic abstraction. Even more than the linguists, and long before them, poets and translators have worked inside the time-shaped skin of human speech and sought to elucidate its deepest springs of being. Men and women who have in fact grown up in a multilingual condition will have something to contribute towards the problem of a universal base and a specific world-image. Translators have left not only a great legacy of empirical evidence, but a good deal of philosophic and psychological reflection on whether or not authentic transfers of meaning between languages can take place.

Much of current linguistics would have things neater than they are. Before conceding that the deeper, more important proceedings of language lie far beyond the level of actual or potential consciousness (Chomsky's postulate), we must look to the vital disorders of literature in which that consciousness is most incisively at work. To know more of language and of translation, we must pass from the 'deep structures' of transformational grammar to the deeper structures of the poet. 'Man weiss nicht, von wannen er kommt und braust', wrote Schiller of the surge of language from the depths to the light. No man knows from whence it comes:

> Wie der Quell aus verborgenen Tiefen,
> So des Sängers Lied aus dem Innern schallt
> Und wecket der dunkeln Gefühle Gewalt,
> Die im Herzen wunderbar schliefen.

---

[1] *New York Review of Books* (8 February 1973), p. 34.

Chapter Three

WORD AGAINST OBJECT

I

WHAT follows is personal and, as I have said, partly impressionistic. This may not be entirely a defect. Whether there is
a genuine 'science of language' is a moot point. An extended,
often unexamined analogy underlies the whole concept of scientific linguistics. We borrow the idiom and posture of sensibility of
an exact science—in this case mathematics, clinical psychology,
mathematical logic—and transfer them to a body of perception, to
a phenomenology, which lie essentially outside the natural limits
of scientific hypotheses and verification. The claims made for
a scientific linguistics derive their substance from an assumed
parallelism with formal logic and with the kinds of experimental
psychological and statistical investigation which are, in fact,
susceptible of precise, quantifiable treatment. It may well be that
human speech is not of this order. The problems posed by the
indissoluble bond of the examining process with the examined, the
dynamics of instability which result from the need to use language
in order to study language—these are very probably resistant to
rigorous, let alone exhaustive, construction. This dilemma is at the
root of epistemology. It is not of a technical or conventional
nature. There is an inescapable ontological autism, a proceeding
inside a circle of mirrors, in any conscious reflection on (reflection
of) language.

Mediate thought about language is an attempt to step outside
one's own skin of consciousness, a vital cover more intimately
enfolding, more close-woven to human identity than is the skin of
our body. To declare that the idiom of modern linguistics is
a 'metalanguage' is to say little. Once again, a loan image is
operative: that of mathematical logic in relation to mathematics.

Though tricked out with logical symbols and markers from the theory of recursive functions, the metalanguage of scientific linguistics is compelled to draw on common syntax and current words. It has no extraterritorial immunity. It does not carry out its investigations from an exterior, neutral zone. It remains inalienably a member of the language or language family which it seeks to analyse. 'Was sich in der Sprache spiegelt,' wrote Wittgenstein in his diary for 1915, 'kann ich nicht mit ihr ausdrücken.' The inter-actions of observer and observed are of extreme methodological and psychological opaqueness. This is a crucial point, about which there is much confusion. The elementary or tree-structures arrived at by the application of transformational rules to an English sentence are not an X-ray. There has been no empirically verifiable probe from surface to depth. Roentgen rays stem from a demonstrably external, objective source and reveal that which cannot be otherwise seen and that which may totally contravene theoretic postulates or expectations. A transformational analysis however abstract, however suggestive of the formal moves of pure logic, is itself a language-act, a procedure which interpenetrates at every stage with the object of its analysis. The linguist no more steps out of the mobile fabric of actual language—his own language, the very few languages he knows—than does a man out of the reach of his shadow. Or as Merleau-Ponty puts it: 'Il nous faut penser la conscience *dans* les hasards du langage et impossible sans son contraire.'[1] These 'hasards' are the cognitive substance of our being. The sole mediate, truly external view of them conceivable is that of a total leap out of language, which is death.

Formal schemata and metalanguages are of undoubted utility. They produce fictions of isolation whereby we can study one or another element of phonology, of grammar, or of semantics. Used with the definitional awareness found, for instance, in Chomsky's classic paper on 'The Structure of Language and its Mathematical Aspects' (1961), they can lead to the projection of strong models. What needs careful note is the nature of such models. A model will comprehend a more or less extensive and significant range of

[1] M. Merleau-Ponty, *La Prose du monde* (Paris, 1969), p. 26.

linguistic phenomena. For reasons that are philosophic and not merely statistical, it can never include them all. If it could, the model would be the world. It can give to that which it includes a more or less coherent, economic, intellectually persuasive pattern of interrelation. But to assert that any given pattern is uniquely concordant with 'underlying reality' and therefore normative and predictive, is to take a very large, philosophically dubious step. It is just at this point that the implied analogy with mathematics is decisive and spurious. The revelatory, 'forward-moving' nature of mathematical argument and proof is itself a very difficult, disputed topic (what is 'moving forward', what is being 'discovered'?). But the difficulty as well as the explanations offered are based on the arbitrary, internally consistent, possibly tautological quality of the mathematical fact. It is this quality which makes the mathematical model *verifiable*. The facts of language are otherwise. No momentary cut, no amount of tissue excised from the entirety of the linguistic process can represent or guarantee a determination of all future forms and inherent possibilities. A language-model is no more than a model. It is an idealized mapping, not a living whole.

Merleau-Ponty rightly identifies the psychological source of the current tendency to confuse formal linguistic models with the phenomenal totality of actual language: 'L'algorithme, le projet d'une langue universelle, c'est la révolution contre le langage donné.'[1] That 'revolt', I repeat, has great analytic and heuristic merits. It prevents the submersion of linguistics under a tide of

[1] Ibid., p. 10. The literature which deals with the theory of linguistic models and with the discriminations to be made between formal and natural languages is large. Cf. I. I. Revzin, *Models of Language* (London, 1966), pp. 4–14; Y. Bar-Hillel, 'Communication and Argumentation in Pragmatic Languages' in *Linguaggi nella società e nella tecnica* (Milan, 1970) and S. K. Šaumjan on 'Linguistic Models as Artificial Languages simulating Natural Languages' in the same volume. As Šaumjan states (p. 285): 'a linguistic model is nothing more than an artificial system of symbols, an artificial language which simulates a natural language'. He concludes: 'A natural language is an immensely involved system which is a mixture of the rational and the irrational, and this system defies direct mathematical description. Now, if a natural language cannot be considered a well-defined object in a mathematical sense ... we cannot construct a device which will generate the sentences of a natural language' (pp. 287–8). For a practical exemplification of this fact (with its drastic consequences for the Chomskyan approach), see Richard B. Noss, 'The Ungrounded Transformer' (*Language Sciences*, XXIII, 1972).

inchoate particularity. It makes salient and, as it were, visible anomalies of language as well as profound economies and resources. It shows 'how things might in fact work'. Or how they would work optimally, given the kind of frictionless, homogeneous, perfectly measurable reality in which the laws of physics, such as we learn them in school-books, are said to operate. But it is the *langage donné* in which we conduct our lives, whether as ordinary human beings or as linguists. We have no other. And the danger is that formal linguistic models, in their loosely argued analogy with the axiomatic structure of the mathematical sciences, may block perception. The marginalia, the anarchic singularities and inefficiencies which generative transformational grammars leave to one side or attempt to cover with *ad hoc* rules, may in fact be among the nerve-centres of linguistic change, as the turbulent dust-clouds and 'black spots' in the galaxy are, on present evidence, the intricate locale of the formation of stars. It is quite conceivable that, in language, continuous induction from simple, elemental units to more complex, realistic forms is not justified. The extent and formal 'undecidability' of context—and every linguistic particle above the level of the phoneme is context-bound—may make it impossible, except in the most abstract, meta-linguistic sense, to pass from 'pro-verbs', 'kernels', or 'deep deep structures' to actual speech. The simple assertion that surface features need not in any way 'be like' their underlying deep structures does not meet the central philosophic difficulty. Once more, the seductive precedent of Euclidean geometry or classic algebraic demonstration, as each proceeds from axiomatic simplicities to high complexity, must not be invoked uncritically. The 'elements' of language are not elementary in the mathematical sense. We do not come to them new, from outside, or by postulate. Behind the very concept of the elementary in language lie pragmatic manoeuvres of problematic and changing authority. I shall return to this point.

It may be that today's formal linguistics and construction of transformational models are a prelude to a genuine science of language, that the ground is being cleared in a way which is, inevitably, a reductive simplification. One can even specify the substantive basis of a future science. It would lie in a neuro-chemical or neurophysiological location of the mental structures or 'imprinting' through which human beings internalize a grammar and the necessary rules of transformation. Arguably, a more penetrating neurochemistry or electrophysiology of the brain will throw unequivocal light on these innate settings of human linguistic competence. Chomsky himself, here at his least Cartesian, does not allow such expectations: 'molecular biology, ethology, the theory of evolution, and so on, have absolutely nothing to say about this matter, beyond the most trivial observations. And on this issue . . . linguistics has nothing to say either.'[1] Other linguists and psychologists of language would disagree sharply. Some would hold that dynamic singularities in brain action, when properly elucidated, will prove to be the physiological correlates of precisely those preferential or consistent linguistic patterns which transformational grammars regard as innate and universal. Work by Lorenz and Piaget suggests that logico-mathematical structures and the kind of relational strings that underlie the generation of sentences have their biological roots in the structure and function of the nervous system. If this is so, neurophysiology and molecular biology will have relevance to an analysis of human behaviour at the conscious symbolic–linguistic level.[2] The long-established study of speech defects, moreover, of aphasia and so-called speech blocks, provides ample evidence of direct and often highly specific relations between physiology and language. Nevertheless, the prospects of a 'physically grounded' theory of the evolution and generation of human speech remain uncertain. Today, and for the foreseeable future, linguistics must proceed with the aid of partially arbitrary metalanguages and within a framework of formal conjecture and analytic models which are, only in a wide or metaphoric sense, scientific. The application of the concept of exact science to the study of language is an idealized simile.

This is not a negative judgement. It is only an attempt to state the criteria of exactitude, of predictive force, and of proof with

[1] Private communication of 18 November 1969.
[2] Cf. Arthur Koestler and J. R. Smythies (eds.), *Beyond Reductionism, New Perspectives in the Life Sciences* (New York, 1970), p. 302.

which linguistics and a study of translation can reasonably operate. The sixteenth and seventeenth centuries had their 'science of rhetoric'. The 'science of aesthetics' plays a major part in the analytic thought of the nineteenth century. In these instances, the use of the term 'science' is complex, being in some measure analogy, and in some measure expectation. Many humane disciplines have regarded themselves as 'sciences' during a particularly energetic phase of growth or internal debate. Linguistics is, currently, in this condition of heightened and confident life. This obscures the fact that many of its essential philosophic and phenomenological aspects are less akin to the exact or mathematical sciences than they are to the study of literature, of history, and of the arts. The counters of linguistics, where it is most saliently a 'meta-science', are generalized and abstract in the extreme. I am arguing that such generality and abstraction go against the grain of other, perhaps equally important elements in the structure of language. To do so concretely, I must argue from within.

My father was born to the north of Prague and educated in Vienna. My mother's maiden name, Franzos, points to a possible Alsatian origin, but the nearer background was probably Galician. Karl Emil Franzos, the novelist and first editor of Büchner's *Wozzeck*, was a grand-uncle. I was born in Paris and grew up in Paris and New York.

I have no recollection whatever of a first language. So far as I am aware, I possess equal currency in English, French, and German. What I can speak, write, or read of other languages has come later and retains a 'feel' of conscious acquisition. But I experience my first three tongues as perfectly equivalent centres of myself. I speak and I write them with indistinguishable ease. Tests made of my ability to perform rapid routine calculations in them have shown no significant variations of speed or accuracy. I dream with equal verbal density and linguistic-symbolic provocation in all three. The only difference is that the idiom of the dream follows, more often than not, on the language I have been using during the day (but I have repeatedly had intense French- or English-language dreams while being in a German-speaking milieu, as well as the reverse). Attempts to locate a 'first language' under hypnosis have failed. The banal outcome was that I responded in the language of the hypnotist. In the course of a road accident, while my car was being hurled across oncoming traffic, I apparently shouted a phrase or sentence of some length. My wife does not remember in what language. But even such a shock-test of linguistic primacy may prove nothing. The hypothesis that extreme stress will trigger one's fundamental or bedrock speech assumes, in the multilingual case, that such a speech exists. The cry might have come, quite simply, in the language I happened to have used the instant before, or in English because that is the language I share with my wife.

My natural condition was polyglot, as is that of children in the Val d'Aosta, in the Basque country, in parts of Flanders, or among speakers of Guarani and Spanish in Paraguay. It was habitual, unnoticed practice for my mother to start a sentence in one language and finish it in another. At home, conversations were interlinguistic not only inside the same sentence or speech segment, but as between speakers. Only a sudden wedge of interruption or roused consciousness would make me realize that I was replying in French to a question put in German or English or vice versa. Even these three 'mother tongues' were only a part of the linguistic spectrum in my early life. Strong particles of Czech and Austrian-Yiddish continued active in my father's idiom. And beyond these, like a familiar echo of a voice just out of hearing, lay Hebrew.

This polyglot matrix was far more than a hazard of private condition. It organized, it imprinted on my grasp of personal identity, the formidably complex, resourceful cast of feeling of Central European and Judaic humanism. Speech was, tangibly, option, a choice between equally inherent yet alternate claims and pivots of self-consciousness. At the same time, the lack of a single native tongue entailed a certain apartness from other French schoolchildren, a certain extraterritoriality with regard to the surrounding social, historical community. To the many-centred, the very notion of 'milieu', of a singular or privileged rootedness, is suspect. No men inhabit a 'middle kingdom', all are each other's guests. The realization that the chestnut tree on the quai outside our house was no less a *marronnier* than a *Kastanienbaum* (the English tree, as it happens, carries a French 'flambeau'), and that these three configurations coexisted, though in the actual moment of utterance at varying distances of synonymy and real presence, was essential to my sense of a meshed world. From the earliest of memories, I proceeded within the unexamined cognition that *ein Pferd*, a horse, and *un cheval* were the same and/or very different, or at diverse points of a modulation which led from perfect equivalence to disparity. The idea that any one of these phonetic incarnations could have seniority or pride of depth over the other did not occur to me. I later came to feel almost the same, though not entirely, about *un cavallo* and *un albero castagno*.

When I began thinking about language—this vaulting across one's shadow and attempt to examine the skin of one's shadow from within and without being itself a peculiar action and one to which few cultures have been prone—obvious questions turned up. Questions implicit in my own circumstance, but also of a far wider theoretic interest.

Was there, despite my inability to 'feel the fact', a first language after all, a *Muttersprache* vertically deeper than the other two? Or was my sense of complete parity and simultaneity accurate? Either alternative led to problematic models. A vertical structure suggests an alignment of strata throughout. In that case, which language came second, which third? If, on the other hand, my three languages are equally native and primary, what manifold space contains their coexistence? Does one imagine them as a continuum on some kind of Moebius strip, intersecting itself yet preserving the integrity and distinctive mappings of its surface? Or ought one, rather, to picture the dynamic foldings and inter-penetration of geological strata in a terrain that has evolved under multiple stress? Do the languages I speak, after they diverge to separate identity from a common centre and upward thrust, combine and recombine in an interleaved set, each idiom being, as it were, in horizontal contact with the others, yet remaining itself continuous and unbroken? Such infolding would presumably be a constant mechanism. When speaking, thinking, dreaming French, I would selectively compress, selectively energize with currents of stored use and present feedback, the 'nearest' stratum or rift of the French component in my levels of subconsciousness and consciousness. This stratum would, under stress of generation and reciprocal stimulus (French coming in from outside), 'fold upward', and become the momentary surface, the visible contour of the mental terrain. When I reverted to German or to English, an analogous process would occur. But with each linguistic shift or 'new folding', the underlying stratification has, in some measure, altered. With each transfer of energy to the articulate surface, the most recently used plane of language must be traversed or enfolded and the most recent 'crust' broken.

And if there is a common centre, what geological or topological simile can provide a model? During the first eighteen to twenty-six months of my life, did French, English, and German constitute a semantic magma, a wholly undifferentiated agglomerate of linguistic competence? At some deep level of energized consciousness or, rather, pre-consciousness, do they still? Does the linguistic core, to continue the image, stay 'molten', and do the three relevant language streams intermingle completely, though 'nearer the surface' they crystallize into distinct formations? In my own case, such a magma would contain three elements. This is the case for every individual in the much-studied trilingualism (German, Friulian, Italian) of Sauris, a German linguistic enclave in the Carnian Alps of north-east Italy. Can there be more? Are there human beings wholly and unreflectively quadrilingual? Could there be anyone whose sense of primary speech-reflexes extends to five languages? At the stage of conscious, learnt mastery, of course, there is plenty of evidence that gifted individuals can truly possess anywhere up to a dozen tongues. Or is any native configuration above that of bilingualism suspect, so that, as some psycho-linguists seem to believe, even my own experience of an undivided triplet would, in some way I can give no account of, have derived from an even earlier split into only two language centres? And what of the original congeries itself? Is it radically individualized or, to stick to my own case, is the same dynamic core of compressed semantic material present in anyone who starts out with these three particular languages? Are all children who grow up totally bilingual in, say, Malay and English, carriers of the same generative centre (the matrix, as it were, of nascent linguistic competence), or are the elemental proportions of admixture somewhat different for each individual, even as no two steel ingots, cast from the same crucible and furnace in successive instants are, at the molecular level, identical?

Does a polyglot mentality operate differently from one that uses a single language or whose other languages have been acquired by subsequent learning? When a natively multilingual person speaks, do the languages not in momentary employ press upon the body of speech which he is actually articulating? Is there a discernible, perhaps measurable sense in which the options I exercise when uttering words and sentences in English are both enlarged and complicated by the 'surrounding presence or pressure' of French and German? If it truly exists, such tangential action might subvert my uses of English, making them in some degree unsteady, provisional, off-centre. This possibility may underlie the pseudo-scientific rumour that multilingual individuals or children reared simultaneously in 'too many' languages (is there a critical number?) are prone to schizophrenia and disorders of personality. Or might such 'interference' from other languages on the contrary render my use of any one language richer, more conscious of specificity and resource? Because alternative means lie so very near at hand, the speech forms used may be more animate with will and deliberate focus. In short: does that 'intertraffique of the minde', for which Samuel Daniel praised John Florio, the great translator, inhibit or augment the faculty of expressive utterance? That it must have marked influence is certain.

How does a multilingual sensibility internalize translation, the actual passage from one of its first languages to another? Certain experts in the field of simultaneous translation declare that a native bilingual speaker does not make for an outstanding interpreter. The best man will be one who has consciously gained fluency in his second tongue.[1] The bilingual person does not 'see the difficulties', the frontier between the two languages is not sharp enough in his mind. Or as Quine puts it, sceptically, in *Word and Object*, it may be 'that the bilingual has his own private semantic correlation—in effect his private implicit system of analytical hypotheses—and that this is somehow in his nerves'. If this is true, it suggests that a bi- or trilingual individual does not proceed laterally when translating. The polyglot mind undercuts the lines of division between languages by reaching inward, to the symbiotic core. In a genuinely multilingual matrix, the motion of spirit performed in the act of alternate choice—or translation—is parabolic rather than horizontal. Translation is inward-directed discourse, a descent, at least partial, down Montaigne's 'spiral staircase of the self'. What light does this process throw on the vital issue of the primal direction or target of human speech? Are the mechanisms of self-address, of interior dialogue between syntax and identity, different in a polyglot and in a single-language speaker? It may be—I will argue so—that communication outward is only a secondary, socially stimulated phase in the acquisition of language. Speaking to oneself would be the primary function (considered by L. S. Vygotsky in the early 1930s, this profoundly suggestive hypothesis has received little serious examination since). For a human being possessed of several native tongues and a sense of personal identity arrived at in the course of multilingual interior speech, the turn outward, the encounter of language with others and the world, would of necessity be very different, metaphysically, psychologically different, from that experienced by the user of a single mother-tongue. But can this difference be formulated and measured? Are there degrees of linguistic monism and of multiplicity or unhousedness that can be accurately described and tested?

In what language am *I*, suis-*je*, bin *ich*, when I am inmost? What is the tone of self?

[1] This point is discussed in the Proceedings of the Symposium of the International Congress of Translators held at Hamburg in 1965 and published in R. Italiander (ed.), *Uebersetzen* (Frankfurt, 1965).

One finds few answers to these questions in the literature.[1] Indeed, they are not often asked. Theoretic and psycho-linguistic investigations of a natural multilingual condition are still relatively rare. Most of the available research deals with the historical and anthropological features of bilingual territories. Even in this

[1] The technical literature is, of course, considerable and has expanded rapidly since the early 1960s under the impulse of ethno-linguistics and psycho-linguistics. V. Vildomeč, *Multilingualism* (Leiden, 1963) remains a standard survey and contains a large bibliography. Charles Ferguson's article 'Diglossia' (*Word*, XV, 1959) set out much of the vocabulary of subsequent study. The latter can be divided into two main branches: the theoretic discussion of multi- and plurilingualism in relation to a general understanding of human speech, and the study of actual cases of multilingual usage in polyglot communities. Cf. Uriel Weinreich, *Languages in Contact* (The Hague, 1962); Jean-Paul Vinay, 'Enseignement et apprentissage d'une langue seconde' in *Le Langage*, ed. A. Martinet (Paris, 1968); R. B. Le Page, 'Problems of Description in Multilingual Communities' (*Transactions of the Philological Society*, 1968); John J. Gumperz, 'Communication in Multilingual Communities' in S. Tyler (ed.), *Cognitive Anthropology* (New York, 1969); Neils Anderson (ed.), *Studies in Multilingualism* (Leiden, 1969); J. R. Rayfield, *The Languages of a Bilingual Community* (The Hague, 1970); Dell Hymes (ed.), *Pidginization and Creolization of Languages* (Cambridge University Press, 1971); Paul Pimsleur and Terence Quinn (eds.), *The Psychology of Second Language Learning* (Cambridge University Press, 1971); J. J. Gumperz and D. Hymes (eds.), *The Ethnography of Communication* (Wisconsin, 1964) contains important material on actual plurilingual societies. Cf. also Einar Hagen, *Language Conflict and Language Planning: The Case of Modern Norwegian* (Harvard, 1966), and P. David Seaman, *Modern Greek and American English in Contact* (The Hague, 1972). J. A. Fishman's article 'Who Speaks What Language to Whom and When?' (*Linguistique*, II, 1965) outlines an approach to multilingualism in terms of the 'pluralistic' levels of social usage, of contextually determined idiom, which occur crucially even when we speak only one language. This approach is illustrated in N. Denison, 'A Trilingual Community in Diatypic Perspective' (*Man*, III, 1968), and 'Sociolinguistics and Plurilingualism' (*Acts of the Xth International Congress of Linguistics*, 1969). Cf. also W. H. Whiteley (ed.), *Language Use and Social Change* (Oxford, 1971), and the papers assembled in Edwin Ardener (ed.), *Social Anthropology and Language* (London, 1971), notably: N. Denison, 'Some Observations on Language Variety and Plurilingualism'; Elizabeth Tonkin, 'Some Coastal Pidgins of West Africa'; W. H. Whiteley, 'A Note on Multilingualism'. There have also been attempts to devise statistical models and exact measurements of 'interference effects' in bilingual individuals and communities. Cf. A. R. Diebold, 'Incipient Bilingualism' (*Language*, XXXVII, 1961), W. F. Mackey, 'The Measurement of Bilingual Behavior' (*Canadian Psychologist*, VII, 1966); J. J. Gumperz, 'On the Linguistic Markers of Bilingual Communication' (*The Journal of Social Issues*, XXIII, 1967); Susan Kaldor and Ruth Snell, 'Decoding in a Second Language' (*Linguistics*, LXXXVIII, 1972). Until now, results are tentative. Leonard Forster's *The Poet's Tongues: Multilingualism in Literature* (Cambridge University Press, 1970), introduces a large, unexplored field.

domain, attention tends to focus on the relations between a local dialect and the national speech-forms. We have few, if any, detailed accounts of an individual's coming of age or realization of self-consciousness under natural polyglot circumstances. What records there are of a primary at-homeness in two or more languages may be found disseminated in the memoirs of poets, novelists, and refugees. They have never been seriously analysed. (Nabokov's *Speak Memory* and the material ironized and inwoven in *Ada* are of the first importance.)

There are reasons for this lack. If we except the Moscow and Prague language-circles, with their explicit association with contemporary poets and literature in progress, it can fairly be said that many modern analytic linguists are no great friends to language. Not many, and this applies particularly to the American school of 'mathematical linguistics', have inhabited the husk of more than one speech. Linguistic cross-reference, at any but the severest level of structural universality, recalls to them the discredited habits of nineteenth-century *vergleichende Philologie*. Even as there is in certain branches of modern literary criticism a covert distaste for literature, a search for 'objective' or verifiable criteria of poetic exegesis though such criteria are obstinately alien to the way in which literature acts, so there is in scientific linguistics a subtle but unmistakable displeasure at the mobile, perhaps anarchic prodigality of natural forms.

But there is also a more respectable reason. Multilingualism is a special case. Moreover, it is a case of obvious complication. At a time when strict phonological investigations and transformational grammars are, at last, establishing a truly autonomous and professional science of language, it would be absurd, we are told, to go beyond the analysis of the deep structures of one language or, as it were, of Language itself. It is only when such analyses have

But despite the extent of technical literature, very little is known of the psychological experience of the polyglot, and no substantive case has been put forward as to the type of mental lattice and multidimensional transpositions which may well be involved. For a preliminary view of the difficulties of the subject, cf. W. E. Lambert, 'Psychological Studies of the Interdependencies of the Bi-lingual's two Languages', in J. Puhvel (ed.), *Substance and Structure of Language* (University of California Press, 1969).

been pressed home, when it is possible to give an account (this account will have to be a total one in order to satisfy the prerequisites of a transformational grammar) of the strings, of the first- and second-order transformational rules, and of the surface mappings that correctly describe the competence of the 'idealized native speaker', that linguistics can proceed to the class of 'more than one mother tongue'. A sane man will start with simple equations, not with the topology of Banach spaces.

Leaving aside the question of whether the transformational generative model of human speech is an adequate one, of whether there can ever be a complete and/or verifiable description of the internalization of grammars in the human mind, the assumption that 'several languages' merely represents a more complex variant of 'one language' may be fallacious. To take it as proven is to beg the whole point. At levels above those of the most abstract, meta-mathematical idealization, primary multilingualism may be an integral state of affairs, a case radically on its own. If some species of bilingual or polyglot matrix does underlie the very earliest steps from innate linguistic competence to actual performance in an individual multilingual child or community, then these steps will differ from those taken by the 'idealized native speaker' of a single tongue. So far as all sentences are acts, utterances from within a particular speech-situation, the nature of that situation is bound to affect the early acquisition of language. It is, at the least, conceivable that multilingualism, where the individual has no recollection of any other personal state, constitutes a determinant situation.

Again, we touch on an absolutely central issue of reductionism, on the belief, axiomatic in modern scientific linguistics, certainly since Bloomfield and Harris, that formal analyses of postulated elementary strings will lead, by progressive inference, to an understanding of the complex structures found in natural language. As we have noted, the forceful analogue to this belief is the inductive process in the logical, mathematical, and physical sciences. There, indeed, movement proceeds characteristically from atomic facts or minimal designations to forms of increasing elaboration and 'reality'. But does this motion of analytic ascent apply to human speech?

2

The median nature of language is an epistemological common-place. So is the fact that every general statement worth making about language invites a counter-statement or antithesis. In its formal structure, as well as in its dual focus, internal and external, the discussion of language is unstable and dialectical. What we say about it is momentarily the case. In an idealized framework in which articulate energy would be totally conserved—Rabelais's fable that all speech utterances are preserved intact 'somewhere'—the sum of all preceding statements would, however minutely, be altered every time something new was said. Such alteration would, in turn, affect all possibilities of future speech. What is said, what conventions are observed by our latest uses of meaning and response, modify future forms. A user of language is like Cyrano's moon-voyager, throwing the magnet of his motion before him. I would argue, therefore, that general propositions about language can never be entirely validated. Their truth is a kind of momentary action, an assumption of equilibrium. Each statement, if it is of any serious interest at all, will be another way of asking. The kinds of thing said about death offer a grammatical and ontological parallel. Language and death may be conceived of as the two areas of meaning or cognitive constants in which grammar and ontology are mutually determinant. The ways in which we try to speak of them, or rather to speak them, are not satisfactory statements of substance, but are the only ways in which we can question, i.e. experience their reality. According to the medieval Kabbalah, God created Adam with the word '*emeth*, meaning 'truth', writ on his forehead. In that identification lay the vital uniqueness of the human species, its capacity to have speech with the Creator and itself. Erase the initial '*aleph* which, according to certain Kabbalists, contains the entire mystery of God's hidden Name and of the speech-act whereby He called the universe into being, and what is left is *meth*, 'he is dead'.[1] What we can say best of language, as of death, is, in a certain sense, a truth just out of reach.

It is knowledge older than Plato that language has both material and immaterial aspects, that there is a speech-system that is markedly physical and one that is not. Recent study underlines the specific finesse and adaptive resources of the human articulatory apparatus. It insists on the difference between that apparatus and that possessed by even the best-endowed of primates.[2] There could be no language as we know it without the complex but also unmistakable evolutionary advance of the human larynx and of the control of our vocal organs in the central nervous system. Anatomical and neurophysiological investigation of the engineering of vocal signals, of the muscular means whereby we set air in significant wave-motion, reveals an accord of extreme precision and discriminatory range between larynx, palate, tongue, and the facts of language. Speech depends on the long pharyngeal cavity unique to humans. One recalls Roman Jakobson's intriguing explanation why so many languages of the world have 'Mama' and 'Papa'. In terms of the position of the child's mouth and of the funnelling of sound, 'p' and 'm' are the optimum consonants and 'a' the optimum vowel. To any human organism seeking the plainest possible oppositional contrasts, these sounds are the natural starting-point.[3] Man's auditory equipment is similarly elaborate. Here, however, there is less instrumental specificity. The audition and vibratory transmission of incoming speech sounds is only one of the many functions of the ear. It performs

[1] Cf. Gershom Scholem, *On the Kabbalah and its Symbolism* (New York, 1965), p. 179.

[2] Cf. J. Bronowski and Ursula Bellugi, 'Language, Name and Concept', in T. G. Bever and W. Weksel (eds.), *The Structure and Psychology of Language* (New York, 1969), II, and the decisive paper by Philip Lieberman, Edmund S. Crelin, and Dennis H. Klatt, 'Phonetic Ability and Related Anatomy of the Newborn and Adult Human, Neanderthal Man, and the Chimpanzee' (*American Anthropologist*, LXXIV, 1972).

[3] Cf. Roman Jakobson, 'Why "mama" and "papa"?', in B. Kaplan and S. Wagner (eds.), *Perspectives in Psychological Theory* (New York, 1960). See also the full treatment of phonological determinants in R. Jakobson, *Child Language, Aphasia, and Phonological Universals* (The Hague, 1968).

others as well or better. Indeed, one suspects that the reception of meaning is as much, or even more, a process of internalized mimesis, of reconstructive decoding, as it is one of immediate hearing. What biologists and linguists are convinced of is that no other mode of sensory transmission and reception of sound known to us could have generated or would allow the tremendous range, diacritical exactitude, and flexibility of human speech. Thus there is a very important sense in which man's linguistic nature, with all that that entails in relation to the rest of organic life, is a matter of comparative anatomy and neurophysiological history.

Yet in another sense we have said almost nothing when we analyse the operations of the larynx or transcribe on to graph paper the extraordinarily intricate, rapid, and rigorous moves whereby tongue and palate collaborate to exteriorize speech sounds, many of them scarcely distinguishable but vitally different in purpose. Even as we speak, we feel that instrumentalities of an entirely different, much 'deeper' order are implicit. The lesion of our vocal organs may arrest audible speech; it can intensify the current of language which at all times seems to stream inward (mutes have recorded dreams that are full of voices). Again, no doubt, this deeper order has material aspects.

We know, since Paul Broca at least, that certain areas of the brain act as language centres and that there are specific correlations between certain speech defects and localized brain damage. A number of psychologists and psycho-linguists would go further. They argue that identifiable features of cerebral anatomy provide a basis for such primary linguistic devices as naming and the use of symbols. They postulate the existence, singular to man, of special circuits which allow the formation of cross-connections between 'non-limbic' or 'borderless' sensory impressions. It is these cortical hook-ups that relate the mechanism of sight or touch or taste, or any combination of these, to the sound by which we designate the relevant object. Work done with patients who have recovered eyesight after long periods of blindness or first acquire normal vision in mature age does suggest that we only see completely or accurately what we have touched. These very complicated sensory-motor interrelations may precede, or at least underlie, the acquisition and development of language.[1] Or to put it generally: evidence is accumulating that our ability to subsume different experiences of the same object under one name or symbol and to manipulate some of the primary procedures of logic and grammar which are based on relation, may depend on physical features of the architecture and 'wiring' of the cortex. The Platonic account of metaphor as the bringing into relation of areas of perception hitherto discrete may have its material analogue or mapping in the actual topology of the brain.

The emphasis has to be put on 'may'. It is, indeed, reasonable to suppose that progress in the anatomical and neurophysiological understanding of the human brain will throw light on the generation and ordering of language. It has been widely noticed that some of the most striking analogies and working models to emerge from recent discoveries in genetics and molecular biology have a distinctly 'linguistic' ring. Notions of coding, information storage, feedback, punctuation, and replication have their suggestive analogues in descriptions of language. To the extent that life itself is viewed as a dynamic transfer of information in which implicit coded signals trigger and release complex pre-set mechanisms, the study of neurophysiological processes at the molecular level and that of the foundations of language are bound to draw close. Quantitatively, the twenty-six-letter alphabet is richer than the genetic code with its 'three-letter words'. But the lettering analogy may, as one biologist has put it, be 'of intriguing pertinence'.[2] This is particularly so when it is extended to the fact that in both the genetic and the linguistic scheme an appropriate receptor or auditor is needed to complete the message. Without the concordant interface or surrounding field structure, the gene-sequence cannot 'communicate'.

[1] Cf. Jean Piaget and Bärbel Inhelder, 'The Gaps in Empiricism' in *Beyond Reductionism*, pp. 128–56. Of great interest also is the discussion of the relations between linguistic development and the formation of mathematical concepts in A. I. Wittenberg, *Vom Denken in Begriffen, Mathematik als Experiment des reinen Denkens* (Basel and Stuttgart, 1957). The whole question of the child's acquisition of linguistic and 'extra-linguistic' concepts, notably those of spatial relation, is related both to Kantian mentalism and to the experimental tradition in modern psychology.

[2] Paul A. Weiss, 'The Living System: Determinism Stratified' in *Beyond Reductionism*, p. 40.

But other scientists and linguists feel that such hopes of direct empirical penetration are illusory. What, in fact, is being looked for? What would constitute evidence of the molecular basis for the generation of symbolic functions? At the level of elementary logic there is the classic conundrum of machine-intelligence theory: 'given a set of input symbol-strings which have been presented to a finite automaton and the corresponding outputs, is there a possible way of determining the internal structure of the machine, and what would such a way be?' But we are not, of course, inquiring into a finite automaton. The belief is growing that the organizing principles of the human brain are of an order of hitherto undefined complication and autonomy. Add the bits together and there is a great deal 'else' left to account for. Not in any occult sense. But on a plane of systematic interaction between genetic, chemical, neurophysiological, electro-magnetic, and environmental factors for whose numerous relations and spatial contiguities we have, until now, no examinable analogue or inductive model. Such a model may not be forthcoming. The Vedantic precept that knowledge shall not, finally, know the knower points to a reasonable negative expectation; consciousness and the elucidation of consciousness as object may prove inseparable. The needed distance for reflexive cognition is lacking. Even, perhaps, at the physiological level. Hence Jacques Monod's speculations on the emergence of 'a new realm' within the biosphere. Language, proposes Monod, may have appeared in pre-humans with the help of 'new, and not in themselves especially complex interconnections'. But once it had come into even rudimentary existence, language was bound to confer an immensely increased selective value on the capacity for recording and for symbolic combination. 'In this hypothesis, language may have preceded, perhaps by some time, the emergence of a central nervous system particular to man and have contributed decisively to the selection of those variants aptest to utilize all its resources. In other words, language may have created man, rather than man language.'[1]

---

[1] Jacques Monod, *From Biology to Ethics* (San Diego, California, 1969), pp. 15–16.

This sense of 'another realm', which would be both central and diffuse, as are our perceptions of life processes, does attach to our awareness of language. At least in those moments in which we stop to isolate and externalize that awareness. The meridian of language seems to pass through concrete and abstract poles of reality. We cross it each time we speak or recollect speech. No one has given a satisfactory schematic picture of this duality, though C. D. Broad's suggestion, put forward in his study on *Scientific Thought* in 1923, of a cross-section of physical space-time and various mental space-times does have intuitive attractions. The notion of interface phenomena between 'brain space' and 'mind space' would meet some of the facts of language-experience. We do not know. What we are unquestionably aware of is a constant movement towards immateriality, a process of metamorphosis from the phonetic into the spiritual. Jean Paulhan, on whose practical poetics Merleau-Ponty often draws, describes this transmutation: 'métamorphose par quoi les mots cessent d'être accessibles à nos sens et perdent leur poids, leur bruit, et leurs lignes, leur espace (pour devenir pensées). Mais la pensée de son côté renonce (pour devenir mots) à sa rapidité ou sa lenteur, à sa surprise, à son invisibilité, à son temps, à la conscience intérieure que nous en prenions.'[1] This simultaneous transformation in contrary direction is, adds Merleau-Ponty, *le mystère du langage*.

Paulhan infers a reality of thought previous to or outside words. We all make this inference in numerous contexts. But what meaning has this concept of pre- or extra-linguistic thought? Is William James justified in maintaining that except in the case of newborn babes, the comatose or the drugged, there can be no *that* which is not yet a definite *what*, i.e. that can be named? In *Ordinary Language*, Ryle affirms that conceptual thought consists in 'operating with words'. The statement was made in 1953. Today, the picture is less clear. Work by Piaget and J. S. Bruner suggests that in the young child adaptive, generic, intelligent organization of behaviour precedes, by a considerable margin, the development of anything that can reasonably be termed

---

[1] Cit. in M. Merleau-Ponty, *La Prose du monde*, pp. 162–3.

language. In this early sensory-motor period there seem to occur the adaptations of the brain to logical and mathematical relations and procedures that are of fundamental importance. Do these 'pre-verbal' schemata continue active and independent when language develops its full resources? Are there, in the common locution, felt realities 'too deep for words'? The analogy of music and of the invention of melody—about which so very little is known—does allow the notion of forms of 'thought' or energized significance that are, in some highly *abstract* yet also *physical* mode, relations between levels or centres of internal tension. One can imagine psycho-physical consonances or dissonances of inner pitch creating a condition of unbalance, of 'overloading' or 'short-circuit', that can only be resolved through an expressive, performative act. Is there, as is felt in dreams and the penumbra of uncertain waking, a syntax of shape, colour, motion, spatial relations, that is somehow located in the mind but 'lies further' than words? Do we experience it when we 'grope for' a word?

We distort the question even when we merely ask it. We give it, inevitably, the flatness and coherence of normal speech. What is discoverable of the thought processes of infants or deaf-mutes, or rather how can the evidence be gathered except in forms already marked by a ready stamp of verbal convention? Only this is evident: that the hybrid nature of the language-experience, its material–immaterial, abstract–concrete, physical–mental dualism is a central *donnée* of consciousness. We cannot escape from the inherent *coincidentia oppositorum*. Each assertion based on either the neurophysiological or the transcendental model of speech utterances is defective to the extent that it does not comprehend its opposite. We are able to speak because we do not, except in the momentary artifice of philosophic doubt, speak of speech. ('Le langage ne reste énigmatique que pour qui continue de l'interroger, c'est-à-dire d'en parler.')[1]

A cognate duality marks the coexistence of language and of time. There is a sense, intuitively compelling, in which language occurs *in* time. Every speech act, whether it is an audible utterance

[1] Ibid., p. 165.

or only voiced innerly, 'takes time'—itself a suggestive phrase. It can be measured temporally. It shares with time the sensation of the irreversible, of that which streams away from us, 'backward', in the moment in which it is realized. As I think my thought, time passes; it passes again as I articulate it. The spoken word cannot be called back. Because language is expressive action in time, there can be no unsaying, only denial or contradiction, which are themselves forward motions. Hence the wish, so literal when it refers to menace, to curses, to taboo speech, 'if only I could call back my words'; but as Artemis reminds Theseus in the *Hippolytus*:

> ἀλλὰ θᾶσσον ἤ σ᾽ ἐχρῆν
> ἀρὰς ἐφῆκας παιδὶ καὶ κατέκτανες
>
> (with evil haste you loosed the murdering malediction on your son.)
>
> (1323–4)

But this occurrence of language in time is only one aspect of the relation, and the easier to grasp. Time, as we posit and experience it, can be seen as a function of language, as a system of location and referral whose main co-ordinates are linguistic. Language largely composes and segments time. I mean this in both a 'weak' and a 'strong' sense. The weak sense relates to the actual psychology of time-perception, to the ways in which the language-flow in and amid which we pass much of our conscious existence, helps determine our experience of temporality. Speech rhythms obviously punctuate our sensation of time-flow and may well have synchronic relations with other nervous and somatic beats. Speech which is deliberately metrical, and even the slackest prose has elements of syncopation, will play with or against this temporal matrix; it amplifies or interferes with the dominant frequency of language in and across time. Speech segments probably have an even more significant chronometric role in subconscious and unconscious psychological phenomena. It is likely that the current of language passing through the mind, either in voluntary self-address or in the perhaps random but almost certainly un-interrupted soliloquy of mental activity, contributes largely to the definition of 'interior time'. Here the sequence of speech signals or named images may well be the principal clock. Nevertheless, these are the 'weak' forms of the co-ordination of language with time. Other agencies do as much or more to structure and to alter our time-consciousness. Drugs, schizophrenic disturbances, exhaustion, hunger, common stress, and many other factors can bend, accelerate, inhibit, or simply blur our feeling and recording of time. The mind has as many chronometries as it has hopes and fears. During states of temporal distortion, linguistic operations may or may not exhibit a normal rhythm.[1]

The 'strong' sense of the time—language relation is grammatical. It is no Whorfian fantasy to say that our uses of time are mainly generated by the grammar of the verb. If evidence derived from ritual, myth, and anthropological language-studies is to be trusted, different cultures operate with and within different conceptualizations or, at least, different images of time. We know of constructs that are cyclical, spiralling, recursive, and, in some instances of hieratic representation, almost static. Whether language 'causes' these different architectonics, or whether a given grammar merely reflects and codifies a time-scheme elaborated 'outside language', is difficult to say. Most probably, linguistic and non-linguistic factors interact at stages of cultural evolution so rudimentary that we have no real evidence about them. But it is a commonplace to insist that much of the distinctive Western apprehension of time as linear sequence and vectorial motion is set out in and organized by the Indo-European verb system.[2] That system with, as Émile Benveniste emphasizes, its referral only to the subject and not to the object, and its supple classifications of conditions of state, makes up the locale, the 'time-space' of our cultural identity. An entire anthropology of sexual equality before and in time is implicit in the fact that our verbs, with the exception of Indo-European past participles, in distinction from those of Semitic tongues, do not indicate the gender of the agent. The past—

---

[1] There is an interesting but at times obscure discussion of these points in R. Wallis, *Quatrième dimension de l'esprit* (Paris, 1966).

[2] It is on this crucial point that Lévi-Strauss's account of the logic of 'primitive' time and of 'primitive' non-historicism is most acutely in conflict with the 'linear-universalism' of Hegelian—Marxism and of Sartre's *Raison dialectique*.

present–future axis is a feature of grammar which runs through our experience of self and of being like a palpable backbone. The modulations of inference, of provisionality of conjecture, of hope through which consciousness 'maps ahead' of itself, are facts of grammar.

Does the past have any existence outside grammar? The notorious logical teaser—'can it be shown that the world was not created an instant ago with a complete programme of memories?'—is, in fact, undecidable. No raw data from the past have absolute intrinsic authority. Their meaning is relational to the present and that relation is realized linguistically. Memory is articulated as a function of the past tense of the verb. It operates through a deep-seated, intuitively obvious, yet in large measure conventional application of past tenses to a scanning of 'stored material' whose own stacking, if there is such, may not be time-bound in any sense we can conceive of. The violation of natural order in the proposition that 'it happened tomorrow' is immediately sensible, but awkward to analyse. Given a relativistic structure or one of a number of only partially congruent n-dimensional 'time-spaces', the required picture could be devised. If a characteristic discomfort arises over such a phrase (there can be a curious 'nausea of the illogical' which is not the same as the imitation caused by a syntactic impossibility such as 'one men'), if the instantaneous metamorphosis of present into past attaches to our every word and act, the reason is that the inflection of verbs as we practise it has become our skin and natural topography. From it we construe our personal and cultural past, the immensely detailed but wholly impalpable landscape 'behind us'. Our conjugations of verb tenses have a literal and physical force, a pointer backward or forward along a plane which the speaker intersects as would a vertical, momentarily at rest yet conceiving of itself as in constant forward motion. When Petrarch, in his *Africa* of 1338, deliberately reverses the time-axis and bids the young 'walk back into the pure radiance of the past' because that classic past *is* the true future, the shock of the image is tangible:

> Poterunt discussis forte tenebris
> Ad purum priscumque iubar remeare nepotes.

Western historicism and that stress on the uniqueness of individual recollection which underwrites our notion of the integrity and privacy of the person, are inseparable from the wealth of 'pasts' available to our speech. French knows a *passé défini*, a *passé indéfini*, a *passé antérieur*, a *parfait* (more properly, *prétérit parfait*), and an *imparfait*, to name only the principal modes.[1] No philosophic grammar has until now provided an analysis of the diverse logics, tonal values, semantic properties of past tenses and of the modulations between them to rival that of *A la recherche du temps perdu*—a title which is itself a pun on grammar. Proust's minutely discriminated narrative pasts are reconnaissances of the 'language-distances' which we postulate and traverse when stating memories. Proust's control of grammar is so deeply felt, his collation of language with psychological stimuli so vital and examined, that he makes of the verb tense not only a precisely fixed location—at each moment of utterance we know where we were—but an investigation of the essentially linguistic, formally syntactic nature of the past. If the Abbé Sièyes could make of the laconic *j'ai vécu* a comprehensive reply to those who asked for an account of his life during the French Revolution, the reason is that the setting of the verb in the perfect preterite and the use of it without any prepositional adjunct, define a special 'pastness', an area of recall seemingly vague, yet made exact by inference of ironic judgement. A set of simple statements occurs towards the close of the preface to *La Vie de Rancé*, Chateaubriand's masterpiece: 'il tombait dans un silence consterné qui épouvantait ses amis. Il fut délivré de ses tourments par suite du changement des choses humaines. On passa du crime à la gloire. . . . ' No fewer than three co-ordinate systems interact in this short sequence. A narrative imperfect that is almost present modulates abruptly into

---

[1] Cf. the pioneering work on the 'semantics and grammars of time' in Gustave Guillaume, *Temps et verbe* (Paris, 1929) and *L'Architectonique du temps dans les langues classiques* (Copenhagen, 1946). Further discussion will be found in Jean Pouillon, *Temps et roman* (Paris, 1946); Alessandro Ronconi, *Interpretazioni grammaticali* (Padua, 1958); William E. Bull, *Time, Tense and the Verb* (Berkeley, California, 1960). For an illuminating study of narrative tenses in the French novel cf. Harald Weinrich, *Tempus: Besprochene und Erzählte Welt* (Stuttgart, 1964). The most complete treatment of the whole topic of time in language is to be found in André Jacob, *Temps et langage* (Paris, 1967). This work includes an extensive bibliography.

a definitiveness whose finality is accentuated by the passive voice (itself prepared for by the complications both positive and negative of *délivré*). After which a dynamic but also impersonally stylized 'simple past', *on passa*, enfolds the event and gives it a very subtle but unmistakable coloration, as of ironic pardon.

What is psychoanalysis if it is not an attempt to derive and give substantive authority to a verbal construct of the past? The past is to be re-called by present discourse, Orpheus walking to the light but with his eyes resolutely turned back. Free association and the provocative echo of the analyst are designed to make recollection or, more accurately, collection, spontaneous as well as significant. But whatever the methodology, the resurrection is verbal. A past is created as one is abolished when revolutionaries re-start time from *l'An I*. So far as it depends on identifying a 'true past' with what are, in fact, word-strings in the past tense, so far as it seeks to exhume reality through grammar, psychoanalysis remains a circular process. Each instant begets the one before. Whatever the tense used, all utterance is a present act. Remembrance is always now.[1]

Croce's dictum 'all history is contemporary history' points directly at the ontological paradox of the past tense. Historians are increasingly aware that the conventions of narrative and of implicit reality with which they work are philosophically vulnerable. The dilemma exists on at least two levels. The first is semantic. The bulk of the historian's material consists of utterances made in and about the past. Given the perpetual process of linguistic change, not only in vocabulary and syntax but in meaning, how is he to interpret, to translate, his sources? Frege, using what is essen-

---

[1] For an, at times, almost incomprehensibly opaque but widely influential attempt to deal with the validity of a 'past' which is, in fact, 'present speech', cf. Jacques Lacan, *Écrits* (Paris, 1966), and particularly his 'Fonction et champ de la parole et du langage en psychoanalyse'. In my opinion, Paul Ricœur's *De l'interprétation* (Paris, 1965) will remain the classic statement of the ontological 'fictions' in propositions about the past, and of the role of such 'fictions' in psychoanalysis. For a discussion of the logical issues involved, cf. G. E. M. Anscombe, 'The Reality of the Past', in M. Black (ed.), *Philosophical Analysis* (Cornell University Press, 1950), and Paul Weiss, 'The Past; Its Nature and Reality' (*Review of Metaphysics*, V, 1952).

tially a Platonic idiom, postulated that there must somehow be 'a third realm' beyond the flux of language in which meaning has a timeless status. More prudently, Carnap argued, in *Philosophy and Logical Syntax*, for the permanence of major 'emotional and volitional dispositions'. But even if such 'permanent units of meaning' do exist, how is the historian to elicit them? Reading a historical document, collating the modes of narrative in previous written history, interpreting speech-acts performed in the distant or nearer past, 'he finds himself becoming more and more the translator in the technical sense'.[1]

I have tried to show, at the start of this book, what are some of the delicate manoeuvres and unexamined assumptions in such 'translation'. It can be urged, though I would reject the argument, that the case is more crucial in history than it is in literature. There is a sense in which successive misreadings or imitative re-enactments of a literary text constitute a new yet possibly valid 'meaning'. So far as dominant values of literature are metaphoric and/or non-discursive, later readings can be said to form a natural variation and guarantee of continued life. The truth-functions cannot, as it were, be nailed down. Hence J. L. Austin's revealing phrase about 'joking or writing poetry' being 'parasitic uses of language which are "not serious", not the "full normal use".[2] The historian must 'get it right'. He must determine not only *what* was said (which may prove exceedingly difficult given the state of documents and the conflicts of testimony), but what was *meant* to be said and at what diverse levels of understanding the saying was to be received. The scheme is Austin's when he identified an 'illocutionary force of utterance' co-ordinate with the meaning of the utterance itself, yet somehow 'additional' and essential to grasp. Whether this notion of 'illocutionary force' is sound (Austin himself voiced serious doubts),[3] or whether it adds much to the Ogden–Richards distinction between 'symbolic' and 'emotional' functions of meaning, need not concern us. The historian's prob-

---

[1] J. H. Hexter, 'The Loom of Language and the Fabric of Imperatives: The Case of *Il Principe* and *Utopia*' (*American Historical Review*, LXIX, 1964), p. 946.

[2] J. L. Austin, *How to do Things with Words* (Oxford, 1962), p. 104.

[3] Ibid., p. 148.

lem as to what he is talking about is a genuine one. He must not only 'explain' his verbal document, i.e. paraphrase, transcribe, gloss it at the lexical–grammatical level, but also 'understand' it, i.e. show '*how* what was said was meant and thus what *relations* there may have been between various different statements even within the same general context'.[1] And the meaning thus arrived at must be the 'true one'. By what metamorphic magic is the historian to proceed?

He 'must study all the various situations, which may change in complex ways, in which the given form of words can logically be used—all the functions the words can serve, all the various things that can be done with them'.[2] Looking at an oration by Pericles or an edict by Robespierre, he must determine 'the whole range of communications which could have been conventionally performed on the given occasion by the utterance of the given utterance'.[3] This is a handsome ideal, and it sharply illuminates the nature of the historian's dilemma. But the solution offered is linguistically and philosophically naïve. There can be no determination of *all* 'the functions words can serve' at any given time; 'the whole range of communications that could have been conventionally performed' can never be registered or analysed. The determination of the dimensions of pertinent context (what are all the factors that may have genuine bearing on the meanings of this statement?) is very nearly as subjective, as bordered by undecidability in the case of the historical document as in that of the poetic or dramatic passage. The meaning of a word or sentence uttered in the past is no single event or sharply defined network of events. It is a recreative selection made according to hunches or principles which are more or less informed, more or less astute and comprehensive. The illocutionary force of any past statement is diffused in a complex pragmatic field which surrounds the lexical core. Moreover, as I have already suggested, where is the evidence that the function of language itself, its place in the entirety of the semiological, cultural context has remained unaltered? Different

---

[1] Quentin Skinner, 'Meaning and Understanding in the History of Ideas' (*History and Theory*, VII, 1969), p. 47.

[2] Ibid., p. 37.   [3] Ibid., p. 49.

ages and civilizations work differently with words, with verbal taboos, with levels of vocabulary. They probably attach differing truth-values and postulates of reality to their designation of objects. Thucydides' valuation of the truth of the speeches which he 'reports', reportage being in this instance an intricate hybrid of typology and dramatic maximization, involves the whole question of Greek views on the authority of language over or 'toward' reality. How are we to legislate on these views, who know only conjecturally some of the lexical equivalents for the words used?[1] Thus the elucidation of what was meant, implied, concealed, inferentially omitted, equivocated on 'in these circumstances, to this audience, for these purposes and with these intentions' (Austin's defining rubric for the truth or falsity of an utterance), can never be reduced to a single, stringently verifiable method. It must remain a selective, highly intuitive proceeding, at the very best self-conscious of its restricted and, in certain regards, fictional status. It hinges, in Schleiermacher's phrase, on the 'art of hearing'.

But the dilemma is not only semantic. There can, as Rudolf Bultmann has shown in his study of the Gospels, be no 'presuppositionless readings' of the past. To all past events, as to all present intake, the observer brings a specific mental set. It is a set programmed for the present. 'A la vérité,' writes Marc Bloch, 'consciemment ou non, c'est toujours à nos expériences quotidiennes que, pour les nuancer, là ou il se doit de teintes nouvelles, nous empruntons en dernière analyse les éléments qui nous servent à reconstituer le passé: les noms mêmes dont nous usons afin de caractériser les états d'âmes disparus, les formes

[1] This is the central problem of hermeneutics. In *Wahrheit und Methode* (Tübingen, 1960), pp. 370–83, H.-G. Gadamer argues the problematic status of all historical documentation at a level which is, philosophically, a good deal deeper than that touched on by Skinner. His conclusion is lapidary, 'Der Begriff des ursprünglichen Lesers steckt voller undurchschauten Idealisierung' (p. 373). Oddly enough, Gadamer does not point out how drastically Heidegger—who is so clearly the source of the current hermeneutic movement—commits errors of arbitrary recreation in his definitions of the supposedly 'true, authentic' meaning of key terms in early Greek philosophy. Cf. in particular Heidegger's *Einführung in die Metaphysik* of 1935 and 1953. See Richard E. Palmer, *Hermeneutics* (Evanston, Illinois, 1969) for an admirable introduction to the literature.

sociales évanouies, quel sens auraient-ils pour nous si nous n'avions d'abord vu vivre des hommes?"[1] The historian's perception of past tenses, his own personal usage of them, are generated by a linguistic set 'in' and 'of' the present. Except in mathematics and, perhaps, in formal logic—the issue is controversial—there are no non-temporal truths. The articulation now of a supposed past fact involves an elaborate, mainly subconscious network of conventions about the 'reality-contents' of language, about the 'real presence' of past time in the symbolic practices of language, and about the accessibility of memory to grammatical coding. None of these conventions is susceptible of final logical analysis. When we use past tenses, when we remember, when the historian 'makes history' (for that is what he is actually doing), we rely on what I shall call from here on, and throughout the discussion of translation, *axiomatic fictions*.

These may well be indispensable to the exercise of rational thought, of speech, of shared remembrance, without which there can be no culture. But their justification is comparable to that of the foundations of Euclidean geometry whereby we operate, with habitual comfort, in a three-dimensional and mildly idealized space. They are axiomatic, but need not be either inevitable or absolute. Other spaces are possible. Other co-ordinate systems than that of the past–present–future axis are conceivable. And even where we work from and within our particular axiomatic fictions, border-areas of paradox, of significant singularity, will turn up. This likelihood is crucial to a study of language and of mind. Certain grammars do not entirely 'fit', and we are brought up sharp against local or arbitrary assumptions in what may have seemed until then to be 'natural' moves. The edge of paradox in our uses of the past tense, aptly rendered in Augustine's phrase *praesens de praeteritis* (the past is ever present) can never be wholly resolved. There is a level on which Hume's demonstration that 'our past experience presents no determinate object' (*Treatise*, II. xii), remains valid and persistently challenging. It directs us towards that duality of relation through which language happens in time but also, very largely, creates the time in which it happens.

[1] Marc Bloch, *Apologie pour l'histoire, ou métier d'historien* (Paris, 1961), p. 14.

It may be, to use Kierkegaard's distinction, that doubts about the past tense are 'aesthetic'. The status of the future of the verb is at the core of existence. It shapes the image we carry of the meaning of life, and of our personal place in that meaning. No single individual or even culture can produce a comprehensive statement of the notions of futurity. Each of the relevant branches—an ontology of the future, a metaphysic, a poetic and grammar of future tenses, a rhetoric of political, sociological, utopian futures, a modal logic of future consequence—is a major discipline *per se*. Several are in a rudimentary phase. I can do no more than point in certain directions.

Again, as in the matter of the prodigality of languages, the proper start is wonder, a tensed delight at the bare fact, that there *are* future forms of verbs, that human beings have developed rules of grammar which allow coherent utterances about tomorrow, about the last midnight of the century, about the position and luminosity of the star Vega half a billion years hence. Such supple immensity of linguistic projection, and the discriminations it allows between nuances of anticipation, doubt, provisionality, probabilistic induction, fear, conditionality, hope, may well be the major achievement of the neocortex, which is that part of the brain that distinguishes man from more primitive mammals. I recall the shock I experienced as a young child when I first realized that statements could be made about the far future, and that these were, in some sense, licit. I remember a moment by an open window when the thought that I was standing in an ordinary place and 'now' and could say sentences about the weather and those trees fifty years on, filled me with a sense of physical awe. Future tenses, future subjunctives in particular, seemed to me possessed of a literal magic force. That force can bring vertigo, as can very large numbers (scholars of Sanskrit suggest that the development of a grammatical system of futurity may have coincided with an interest in recursive series of very large numbers, possibly with astronomical reference, in the surrounding civilization). I found it difficult to believe that the *code civil* put no restriction whatever on uses of the future, that such occult agencies as the *futur actif*, the *futur composé*, the *futur antérieur* should be in indiscriminate employ. Only the *futur prochain*, which is the present bending forward a little, had a household mien. I nursed the belief that there must be republics more prudent than ours, more attentive to the cross-weave of language and life, in which our lavish consumption of predictive, hypothetical, counterfactual forms was prohibited. In such a culture uses of future predicates, of optatives, of future indefinites would be reserved for ceremonious occasions. They would be numinous as are taboo words which cannot figure in common speech but are included in certain religious rites. Manipulation of unknowns and of future time through language would be the business of an initiate caste or, at least, the number of such manipulations allowed to the vulgar would be carefully regulated (no man in the prudent city being entitled to make more than, say, a dozen statements about the future in a month). Such rationing is perfectly conceivable, as are the restrictions a community imposes on alchemy or the distillation of poisons. Stalinism has shown how a political system can outlaw the past, how it can determine exactly what memories are to be allowed to the living and what dose of oblivion to the dead. One can imagine a comparable prohibition of the future, the point being that tenses beyond the *futur prochain* necessarily entail the possibility of social change. What would existence be like in a total (totalitarian) present, in an idiom which limited projective utterances to the horizon of Monday next?

One writer has tried to visualize a body politic that is end-stopped. In *Die Befristeten* (1956), Elias Canetti postulates a city, long after the nuclear terrors and enigmas of our current condition, in which every inhabitant is named by a number. The number proclaims his life span. A child called 'Ten' will not be scolded; it has so little time. A man baptized 'Eighty' luxuriates his whole life long, be he ever so fatuous or incompetent. No one outlives his 'Moment' (*Augenblick*); no one dies before it is due. A perfect certitude has replaced the ancient, scarcely imaginable torments of unknowing. But it is a subtly tempered certitude. No citizen would reveal the exact date of his own birthday, nor gossip of anyone else's. The true date is contained in a sealed locket which everyone must carry around their necks. The Custodian of Lockets breaks the seal at the time of death—he alone is entitled to do so—and confirms that life span and baptismal number are indeed in perfect accord. Canetti's play tells of a rebel in the city, of a man haunted by the freedom of the future indefinite. Rebellion succeeds (the lockets are shown to be empty), but victory is ambiguous. At the open doors of the future tense, chaos and ancient panics wait.

Much of the interest of the fantasy lies in a flattening out of syntax. When lovers meet, when colleagues discuss work, they communicate in an extended but airless present. Vital stresses of doubt have been excised from the fabric of thought and speech. Hope trots on a short lead. Like Dostoevsky's 'Legend of the Grand Inquisitor', Canetti's fable points to the necessary kinship of freedom and uncertainty. The moral is plain. But the largesse with which we dispose of 'futures' in common life and language also has its haunting aspects. I wondered as a child whether the plethora of forward-flung utterances about tomorrow and tomorrow did not, as might a sorcerer's spell, pre-empt the open future? Did those many proud verbs of conjecture, expectation, intent, and promise not waste the available store of time? Were men always so prodigal, or were proto-grammars parsimonious, advancing only very gradually into the future tense, as we enter the water when it is morning and cold?

No one knows. The prehistory of languages, meaning primarily a theoretic construction of proto-languages through comparative analyses of existing phonetic and grammatical forms, hardly reaches to 4000 B.C.[1] The fact that young children begin by

---

[1] Cf. Mary R. Haas, *The Prehistory of Languages* (The Hague, 1969), pp. 13–34.

Today (1991), there is wide-ranging debate over what is called the 'Nostratic' hypothesis in comparative phonology. This hypothesis, championed most notably by comparative linguists and ethno-linguists in Russia and Israel (e.g. A. B. Dolgopolsky and V. Shevoroshkin), argues for a unitary source of all human tongues. Putting forward a new vision of pre-history, 'Nostratian' linguists draw on material from molecular biology, genetics, archaeology, and anthropology. They reclassify the languages known to have been spoken and being spoken on the globe into 'macro-families' which are interrelated and whose ultimate ancestry, before dispersal and local modification, would be a paradigmatic 'proto-language'. A forceful survey of the argument, together with a bibliography, is to be found in Colin Renfrew, 'Before Babel: Speculations on the Origins of Linguistic Diversity' (*Cambridge Archaeological Journal*, 1 (1991)).·

It is precisely the appositeness, the historical-psychological context of the

using verbs unmarked by tense may or may not tell us something regarding the genesis of language itself. Clearly, we have no history of the future tense.

Part of that history would be philosophic. It would comprise the views which metaphysicians, theologians, logicians have held regarding the grammatical and formal validity of future forms. It would be, at many points, a history of the problem of induction. Limiting itself purely to Western thought and to the most obvious names, such a record would include Aristotle, the Stoics, Augustine, Aquinas, Ockham, and Malebranche. It would study the argument on time in Leibniz, Hume, Kant, and Bergson. Presumably, it would review the discussions on the reality and logical structure of tense-propositions by C. S. Peirce, Eddington, McTaggart, Frege, and C. D. Broad. On each of these philosophic positions, and on the historical and formal relations between them, the literature is vast and often technical.[1]

There are few questions concerning the logic and substantive status of futures that are not already raised in Aristotle's *Physics*, in the *Metaphysics*, and in the famous ninth chapter of *De Interpretatione*. The Aristotelian investigation of cause, of motion, and of the entelechy or teleological intensionality of living forms, obviously involves a view of future propositions. The variousness of Aristotelian argument and the range of differing contexts in which the problem is set make it difficult to elicit any one doctrine. Greek allows Aristotle to speak of 'the nows' (τὰ νῦν) in a manner which seems to foreshadow modern manifolds. Elsewhere, however, he goes so far as to say that verbs in non-present tenses are

---

'Nostratian' model which induce caution. Unification, the search for the monistic 'ultimate' are very much in the air. This is true of 'Big Bang' cosmology, of the micro-biology of DNA, of evolutionary genetics, of particle physics. This quest may reflect deep-lying anguish in the face of seemingly intractable ethnic and cultural conflicts. A 'centre that can hold' is an intense *desideratum*. But whether evidence will come to substantiate a unitarian or a pluralistic understanding of the proto-history of human speech, and whether seductive 'reconstructions' of original phonological units are more than ingenious play, remains to be seen.

[1] A useful selection of articles and bibliography may be found in J. T. Fraser (ed.), *The Voices of Time* (New York, 1966), and Richard M. Gale (ed.), *The Philosophy of Time* (London, 1968).

not true verbs at all, but 'cases' similar to the oblique cases of nouns. Perhaps one comes nearest to the facts by saying that Aristotle's theory of time as cyclical but not precisely repetitive provides for a generalized rather than individually designative logic of future tenses. The entelechy of forms out of, as it were, a 'pre-setting' of potentialities necessitates a logic of future statements, though it is a logic which, when having to formalize such concepts as motion and duration, will run into anomalies.[1] It was some of these which Stoic logicians, notably Diodorus Cronos, seem to have fixed on.

In the early history of the Christian churches and their principal heresies, issues of predestination, of foreknowledge, and of the nature of Divine omniscience played a large part. These issues, together with the ontological and grammatical debates they provoked, have continued to mark the course of Western logic. Thus the treatment of linguistic and conceptual time-flow in Book XI of Saint Augustine's *Confessions* has lost nothing of its intense, probing interest.[2] 'Quid est ergo tempus? si nemo ex me quaerat, scio; si quaerenti explicare velim, nescio'. ('What then is time? If no one asks me, I know. If I want to explain it to a questioner, I do not know.') This experience of temporality as the most obvious yet inexplicable datum of consciousness underlies Augustine's argument. There was no time before Creation, there was no 'then'; *non enim erat tunc*. God's time is an eternal present, extra-territorial to the passage of past–present–future. Yet it is only 'in time' that we perceive human experience. It is only by virtue of temporal sequence that essential motions of spirit such as remorse, responsibility for consequent action, prayer, and resolution can assume meaning. What relations can there be between God's

---

[1] The literature on the Aristotelian treatment of time is large. I have found the following of particular value: J. L. Stocks, *Time, Cause and Eternity* (London, 1938); Hugh R. King, 'Aristotle and the Paradoxes of Zeno' (*Journal of Philosophy*, XLVII, 1949); Ernst Vollrath, 'Der Bezug von Logos und Zeit bei Aristoteles' in *Das Problem der Sprache*, (ed.) H. G. Gadamer (Munich, 1967). Cf. also Jean Guitton, *Le Temps et l'éternité chez Plotin et Saint Augustin* (Paris, 1969).

[2] For an interesting analysis of Augustine's argument in the light of modern philosophy cf. R. Suter, 'Augustine on Time with some Criticisms from Wittgenstein' (*Revue internationale de philosophie*, XVI, 1962).

timelessness and the tense-structure of man? Augustine answers by internalizing human time. He believes that 'a present time of past things', 'a present time of present things', and 'a present time of future things' are realities of the mind, related to the everlastingness of God as is human knowledge to omniscience. It was the latter concept—'in what sense does God's cognition include, i.e. pre-determine, all future events, and could God set Himself an insoluble problem?'—which generated the discussions of tense in Aquinas, in Ockham, and in the fifteenth-century debates on contingent futures.[1] Even today, there is a moving quality in the taut finesse, in the commitment to abstruse and transcendental worry, which animates these analytic texts. Here modal logic touches on the centre of man's relations to God and on those vital contingencies without which that relation would be an empty terror.

Undoubtedly, the scientific advances of the seventeenth century and the scepticism of the Enlightenment took the theological sting out of the argument. The coolness and frankly psychologized character of Hume's solution are well known. Utterances, judgements about the future are neither reports of experience nor logical consequences of it. They arise simply from an assumption of natural uniformity and from ineluctable grooves of mental and linguistic habit. Thus the notion crucial to induction that the future will resemble the past 'is not founded on arguments of any kind, but is derived entirely from habit' *(Enquiry*, I. ii). Problems of contingency, of possibility, of doubt, may best be treated as problems of differentiation between valid and invalid predictions. There is a logic of induction whose rules are grounded in the same fabric of customary association and propinquity that makes up all normal mental life. The sober force of Hume's model impressed itself on the main tradition of Western thought. Even where it reacts against it, the Kantian device of spatial–temporal categories, the assertion that time and our necessary experience of time as directed sequence are 'buried in the depths of the human

[1] The account of Aquinas's and Ockham's thought in Étienne Gilson, *La Philosophie au Moyen Age* (3rd edn., Paris, 1947) remains indispensable.

mind', can be seen as a deepening and 'centralization' of Hume's psychology. Kant's moralism, however, does carry further. His brief tract of 1794, *Das Ende aller Dinge*,[1] expresses the uncanny but innate compulsion of man to reflect on 'last things'. The concept is lofty and dark, but closely meshed with human understanding: 'Der Gedanke ... ist furchtbar erhaben; zum Theil wegen seiner Dunkelheit, in der die Einbildungskraft mächtiger, als beim hellen Lichte zu wirken pflegt. Endlich muss er doch mit der allgemeinen Menschenvernunft auf wundersame Weise verwebt sein. ...' The idea of 'an end of time', as it is foretold in Revelation 10, has 'mystical truth' but no intelligibility. Nevertheless, the urge of the mind to meditate on futurity and the logic of internal sequence that gives future forms to predicative statements have their great moral significance. The extension of causality to future consequence, together with the rational conceit—it may be no more than that—of a finality to human affairs is, says Kant, indispensable to right conduct. Futurity is a necessary condition of ethical being. Beyond that we need not speculate, 'denn die Vernunft', in Kant's haunting phrase, 'hat auch ihre Geheimnisse'.

Whether these 'secrets of reason' would comprise Bergson's *élan vital* is a moot point. What is certain is the extent to which modern logicians have reacted against the rhapsodic blur of Bergson's intuitive–vitalist theory of inner duration. When applied to the future, the laws of identity, excluded middle, and non-contradiction seemed to carry with them fatalist consequences. Bergson's evolutionary subjectivism, on the other hand, had once more focused attention on the pivotal role of time in mental operations. But it offered little solid ground for choosing between alternative schemata, some of them wholly solipsistic, of time-flow. The development of many-valued logics, allowing not only 'true' and 'false' markers but a whole range of indeterminate, neuter, and potential aspects, has been an attempt to clarify the issues. McTaggart's celebrated proof that time is unreal first

---

[1] I am grateful to Prof. Donald McKinnon of Cambridge University who drew my attention to this text, as to a number of others referred to in this section.

appeared in 1908; Bergson's *Évolution créatrice* a year later.
Refutations of McTaggart and critiques of Bergson are at the
source of the development of modern 'tense-logic'. The questions
asked are old. What logical validation can be found for statements
of future contingency? What is the status of 'always'? Is it possible
to devise a consistent logical system embodying the assertion that
time will have an end?[1] What is new is the rigour and formal
power of the calculus of tenses. For the first time the unstable
factor of futurity is formalized in a strict modal logic. I am not
competent to judge of the results—though some are of obvious
wit and poetic suggestion. All I would emphasize is the alertness of 'tense-logic' to the profoundly problematic nature of
language when it speaks about tomorrow. Even at its most meta-
mathematical, 'tense-logic' focuses unmistakably on the shaping
strangeness of man's ability to make statements concerning 'sea-
battles that will be'.

Far more difficult to establish than the history of the analytic,
formal treatments of futurity is the history of actual human
'futures' and optatives. As I noted before, we have no such history
and only problematic notions of what its documentation and evidence would be like. Yet the probability that substantive changes
have taken place in the psychological and social conventions
governing the future tense, in the ways in which different cultures
have articulated inductive or premonitory speech-acts, is very
strong. It declares itself in literary texts, in ritual, in a comparative study of idiomatic forms. We neither experience nor phrase
anticipatory, stochastic, projective conditions of statement as did
the Ionians of the sixth century B.C. But how, even by the most
scrupulous reference to philology, is one to recapture a 'past
future', given the fact that concepts of futurity are determined by
and determinant of numerous social, historical, religious variables
in the relevant speech community? Again there is the dilemma of

---

[1] For an examination of McTaggart's 'proof' cf. G. Schlesinger, 'The Structure
of McTaggart's Argument' (*Review of Metaphysics*, XXIV, 1971). The best history of
'tense-logic' and the most thorough investigation of the issues involved are to be
found in the two books by A. N. Prior, *Past, Present, and Future* (Oxford, 1967), and
Papers on Time and Tense (Oxford, 1968).

circularity, language being used to make explicit and translate earlier or deep-buried linguistic reflexes. All I would indicate are some of the obvious pivots and synapses to be looked for by a putative historian of future forms in certain Western grammars (that qualification being itself severely restrictive).[1]

Futures play a major role in the 'tenseless' syntax of the Hebrew Old Testament, where 'tenseless' simply refers to a grammatical system in which past, present, and future are not distinguished and demarcated as they are in non-Semitic grammars such as Greek and Latin. Timeless, but enunciated *in time* (a paradox which Augustine will explore in Book XII of his *Confessions*), the words of God mesh closely but also 'strangely' with the understanding, with the historical self-identification of a people itself committed to a special, eschatological temporality. Early on, a critical distinction seems to have been drawn between two orders of foresight, of projection into the future. None, prescribes Deuteronomy 18: 10, is to employ divination or augury. The Jew, repeats Leviticus 19: 26, is not 'to observe the times' as do the idolatrous magi and astrologers in surrounding faiths. As the parable of Balaam makes emphatic, it is because the Law prohibits soothsaying that 'there is no enchantment against Jacob, neither is there any divination against Israel'. The necromancer, the witch at Endor claim to decipher God's hidden purpose instead of reading His manifest will. The relation of the genuine prophet (*nabi*) to the future is, in the classic period of Hebrew feeling, unique and complex. It is one of 'evitable' certitude. In as much as he merely transmits the word of God, the prophet cannot err. His uses of the future of the verb are tautological. The future is entirely present to him in the literal presentness of his speech-act. But at the same

---

[1] Ideally, a history of 'past futures' would begin with prehistory. Neanderthal burial practices and the probable evolution of the incest taboo point to an early, evident concern with actual and symbolic projection into future time. The whole question of the accuracy and sophistication of the time-sense of prehistoric cultures is currently under discussion. Some evidence seems to indicate a formidable degree of mathematical and symbolic prevision. Cf. A. Thom, *Megalithic Lunar Observatories* (Oxford, 1971). Such prevision could have far-reaching linguistic consequences. But as in the case of certain possibilities raised by Mayan hieroglyphs the evidence remains conjectural and probably beyond rigorous assessment.

moment, and this is decisive, his enunciation of the future makes
that future alterable. If man repents and changes his conduct, God
can bend the arc of time out of foreseen shape. There is no
immutability except His being. The force, the axiomatic certainty
of the prophet's prediction lies precisely in the possibility that the
prediction will go unfulfilled. From Amos to Isaiah, the true
prophet 'does not announce an immutable decree. He speaks into
the power of decision lying in the moment, and in such a way that
his message of disaster just touches this power'.[1] The abrupt,
time-retracting motion of argument in Chapter 5 of Amos is
characteristic. Israel shall rise no more, 'there is none to raise her
up'. But simultaneously, on a plane of total potentiality which
intersects human time, the prophet speaks the Lord's promise:
'Seek ye me, and ye shall live.' Thus 'behind every prediction
of disaster there stands a concealed alternative.'[2] It is from the
inspired duplicity of the prophet's task that the tale of Jonah
derives its intellectual comedy.

A deep shift begins with Isaiah and the use of the word te'udah
meaning 'testimony'. It is in Isaiah 11 that the Messianic prophecy
'which hitherto stood in the full reality of the present hour and
all its potentialities, becomes "eschatology"'.[3] Henceforth the
optative, future indefinite character of the Messianic promise
is stressed. The Redeemer is latent in the historic choices of
man, he is the evolving consequence as much as the agent of
man's return to God. After the disaster at Megiddo in 609 B.C.,
God's will, says Buber, becomes an enigma. Jeremiah is a bachun
('watch tower') who seeks to resolve that enigma through moral
perception.[4] Now human grammar interacts directly, creatively
with the mystery of God's speech. The 'watchman's' call has a

[1] Martin Buber, *The Prophetic Faith* (New York, 1949), p. 103. Throughout this
section I have drawn also on Ernst Sellin, *Der alttestamentliche Prophetismus*
(Leipzig, 1912), C. A. Skinner, *Prophecy and Religion* (London, 1922), and Shalom
Spiegel, *The Last Trial* (New York, 1969).

[2] Buber, op. cit., p. 134.

[3] Ibid., p. 150.

[4] Buber is borrowing the actual word *bahun* from Isaiah and is using its
traditional signification. (Though often rendered as 'watch-tower', the word *bahun*
remains obscure.)

vital but also externalized function: Jeremiah 'has to *say* what God does'.[1] He does not foretell so much as he glosses. Hence Jeremiah's unprecedentedly 'equal', parallel dialogue with God. Ezekiel marks the close of the original prophetic tradition, he stands on the borderline between prophecy and apocalypse, between open message and hermetic code. The elements of riddle and image in his foresight are nearly Persian or Hellenic.

But in its initial forms the prophetic literature of the Old Testament expresses a unique apprehension of the relations of time and the word. Complete adherence to the Covenant, a rigorous observance of the Law, puts the house of Jacob in a state of concordance with the 'naturalness' of the unknown. Or to say it another way: the 'unknownness' of the future is made ontologically and ethically trivial. It only assumes a veritable quality, either illusory or menacing, through human failure, through departure from the Law. No threat, no lament voiced by the prophet is not already wholly contained in the act of transgression. As is also the divine promise of a future which can be recalled, held back. 'I will heal their backsliding,' proclaims God through the mouth of Hosea, 'for mine anger is turned away.' The dominant syntax, not strictly comparable with any other that we know, is one of 'future present', of anticipation that is also, at every historical moment, remembrance and tautology. In ancient Judaism man's freedom is inherent in a complex logical–grammatical category of reversibility. The prophecy is authentic: what is foretold *must* be. But it *need* not be, for God is at liberty to non-corroborate His declared truths. The eternal present of His relation to Israel both confirms and subverts tense. (Though he could assert that *sentimus nos aeternos esse*, Spinoza, no less than Jonah, found the paradox of unfulfilled necessity philosophically vexing.)

The conditional futures of Hebrew prophecy, by which I mean an ontological and psychological potential in the language, not a grammatical or morphological feature, contrast sharply with what one might call the ambiguous fatalities of a Greek oracle. The oracle, at least during the early stages of Greek history, is

---

[1] Buber, op. cit., p. 166.

never mistaken (during the Persian wars Delphi will prove to be erroneous and untrustworthy). Oracular uses of the future tense are severely deterministic. As in the grammar of malediction, the words cannot be called back or the fatality undone. But more often than not the phraseology of oracular pronouncements is susceptible of contrary interpretations. The language of the pythoness is forked as are the roads from Daulis. Frequently the questioner misreads the gnomic answer. Indeed the entire stance of those who consult the oracle is that of the unraveller. Such confrontation between deceptive message and code-breaker is characteristic of many aspects of Greek intellectual life. The augur is 'deciphering a cryptogram by means of a key'.[1] This is the origin of the ambivalent relations and, later on, of the conflicts between oracular foresight and scientific prediction.[2] As philosophic and scientific investigations develop, they seek to distinguish their own mechanisms of inference and syllogistic projection from the art of the diviner. The latter springs from archaic and pathological impulse. In the *Phaedrus* Plato discriminates between four species of divinely-occasioned madness. Just beneath the urbanities of divination lie more ancient modes of ecstatic prophecy. The Greeks knew that prophetic shamanism points back to a twilit zone between gods and men, a metamorphic time in which mantic agencies flowed unchecked into the open, perhaps incompletely defined consciousness of mortals. As Dodds points out, early Indo-European speech forms retain the association of prophecy with madness.[3]

From these currents of visionary possession and foresight through induction stems a distinctive free fatalism. Much of Greek drama and of the Greek theory of history is founded on the tensions which occur between realized necessity and meaningful action.[4] More vividly than any other cultural forms, Greek

---

[1] F. M. Cornford, *Principium Sapientiae: A Study of the Origins of Greek Philosophical Thought* (Cambridge, 1952), p. 73.

[2] Cf. Cornford, pp. 133–7.

[3] Cf. E. R. Dodds, *The Greeks and the Irrational* (University of California Press, 1951), Chapter III.

[4] Cf. William Chase Green, *Moira: Fate, Good, and Evil in Greek Thought* (Harvard, 1944). Chapter XI contains a well-documented account of the strain of fatalism in different forms and periods of Greek thought.

tragedy, Thucydidean history embody a coexistence, a dialectical reciprocity between that which is wholly foreseen and yet shatters the mind. We *know* what will happen to Agamemnon when he enters the house, each instant of the *agon* has been announced and prepared for. We *know* precisely what Oedipus will discover—in a crucial sense he too has known all along. Yet with each narration or performance of the fable our sense of shock is renewed. The tragic vision of Greek literature turns on this deep paradox: the event most expected, most consequent on the internal logic of action, is also the most surprising. Conceive of the strange, subtle nausea which would come over us if Agamemnon sprang back from the net, if Oedipus heeded Jocasta and stopped asking. Freedom—the will to launch the Sicilian expedition when every portent and pulse of instinctual clairvoyance spells disaster—is the correlate of necessity. The final exchanges between Eteocles and the Chorus in the *Seven Against Thebes* are a perfect instance of free fatalism. Eteocles' knowledge that death waits for him at the seventh gate does not void his action; it gives it the dignity of meaning. Men move, as it were, in the interstices, in the lacunae of misunderstanding left by the oracle; or in a space of necessity made coherent, made logical by foresight. It is an extraordinarily complex psychological and cultural framework. It may well be more consonant than any other we know with the actual grain of things.

From it derive Stoicism and a braced gaiety in the face of the unknown, of the inhuman. Anyone seeking to render key passages in Aeschylus or Heraclitus knows that the particular idiom of freedom within inevitability, of the optative interacting with the necessary, can be no more than approximated in any other speech. Cicero's version, in the *De Divinatione* and *De Fato* already lacks the tense paradoxality of the Greek source. Probably Yeats comes nearest, in 'Lapis Lazuli':

> They know that Hamlet and Lear are gay;
> Gaiety transfiguring all that dread.

Clearly early Christianity benefited from a widely diffused mood of eschatological and apocalyptic expectancy. At almost no place or level of Mediterranean and Near-Eastern society were there not strong currents of millenarian fantasy. Virgil's all too often invoked annunciation in the Fourth Eclogue, seems in fact to have expressed a widespread truth of feeling:

> ultima Cumaei venit iam carminis aetas;
> magnus ab integro saeclorum nascitur ordo.
> iam redit et Virgo, redeunt Saturnia regna;
> iam nova progenies caelo demittitur alto.

(Now, at last, the season of the song of the Sibyl of Cumae has come. Now the great cycle of the centuries begins anew. Now the Virgin returns, and the reign of Saturn. Now a new generation descends from the lofty heaven.)

'The world's great age begins anew'; through the resurrection of the god, through cleansing fire, through personal initiation into the mysteries of eternal life. How literal were these awaitings? What pressures did they bring to bear on actual social behaviour? We know something of extreme sectarian visions, of withdrawals from a world soon to end, of a making ready for the great noon by zealot communities and Mithraic cults. For a good many Jews and Christian Jews the destruction of the temple at Jerusalem marked a hinge of time. But almost from the outset, and notably in the Fourth Gospel and in Revelation, a symbolic eschatology overlies literal psychological, historical sentiment. We cannot recapture what may have been rapid or profound mutations in time-sense, in the grammars of temporal statement among the first Christians and initiates in the mystery religions. Evidence suggests that there was a relatively brief spell during which Christ's coming was regarded as imminent, as an event occurring in time but bringing time to a stop. As normal sunrise persisted, this anticipation shifted to a millenary calendar, to the numerological and cryptographic search for the true date of His return. Very gradually this sense of speculative but exact futurity altered, at least within orthodox teaching, to a preterite. The Redeemer's coming had happened already; that 'pastness' being replicated and made present in each true sacrament. Even the most lucid of modern Christologists can do little more than state the paradox: 'So it seems we must say that for the early Church the coming of Christ was both present and future, both at once.'[1] Such coterminous duality could fit no available syntax. The event, formidably concrete as it was held to have been, 'lies outside our system of time-reckoning'. The mystery of the transubstantiative rite, enacted in each mass, has its own tense-logic. It literally bodies forth, says Dodd, a 'coming of Christ which is past, present and future all in one'.[2]

These sovereign antinomies and suspensions of the common grammar of tense recur in fundamentalist and chiliastic movements throughout Western history. Repeatedly, conventicles, *illuminati*, messianic communities have proclaimed the imminent end of time and striven to act accordingly. The *paniques de l'an mille*, analysed by Henri Focillon, the Adamite visionaries of the late Middle Ages, the men of the Fifth Monarchy in seventeenth-century England, the 'doom churches' now proliferating in southern California, produce a similar idiom. There is no day after tomorrow. The promise of Revelation is at hand: 'there shall be time no longer'. From a sociolinguistic point of view, it would be of extreme interest to know the extent to which such convictions actually reshape speech habits. One would, for instance, want to know whether the faithful who accepted voluntary, ritual suicide at Jonestown avoided the future tense as their apocalypse neared. But hardly any evidence is available. The history of visionary sects is made up principally of the distorting testimony of their destroyers. Only tantalizing scraps remain. Reportedly, the Old Believers in Russia, seeking martyrdom and immediate ascent into the kingdom of God, used the future tense of verbs sparingly, if at all.[3]

---

[1] C. H. Dodd, *The Coming of Christ* (Cambridge, 1951), p. 8.

[2] Ibid. Cf. also Ernst von Dobschütz, 'Zeit und Raum im Denken des Urchristentums' (*Journal of Biblical Literature*, XLI, 1922), and two important articles by Henri-Charles Puech, 'La Gnose et le temps' (*Eranos-Jahrbuch*, XX, 1951) and 'Temps, histoire et mythe dans le christianisme des premiers siècles' (*Proceedings of the VIIth Congress for the History of Religion*, Amsterdam, 1951). A stimulating but highly compressed analysis of early Christian doctrines of time and future, with particular reference to St Irenaeus and the latter's influence on St Augustine, will be found in Mircea Eliade, *Le Mythe de l'éternel retour: archétypes et répétition* (Paris, 1949).

[3] I owe this arresting detail to a personal communication from Prof. James Billington, now Librarian of Congress.

There is an abundant literature concerning the new linearity and open-endedness of felt time brought on by Galilean and Newtonian physics.[1] Newton's religious scruples inhibited him from drawing temporal inferences clearly implicit in his celestial mechanics. But his successors, notably Buffon, did not flinch from the immensities of time allowed, indeed required by a mechanistic, evolutionary model of the earth and of the solar system. A palpable spaciousness animates late-seventeenth- and eighteenth-century natural philosophy, a confidence that there are in fact worlds enough and time for even the most forward-vaulting of sensibilities to draw deep breath. It is no longer the containment by the crystalline and concentric, still vivid in Kepler, nor a Pascalian terror of the void, which characterizes the new cosmography, but a logic of infinite sequence. We hear its bracing note as early as 1686, in the poetry of vast spaces, of ordered eternity, in Fontenelle's discourse *Sur la pluralité des mondes*. Kant's astronomical speculations, set down in the *Allgemeine Naturgeschichte und Theorie des Himmels* during the 1750s, conjoin divine determinism with the largesse of an unbounded future: 'The infinity and the future succession of time, by which Eternity is unexhausted, will entirely animate the whole range of Space to which God is present, and will gradually put it into that regular order which is conformable to the excellence of His plan.' In Newtonian–Kantian co-ordinates, time and number without end are a necessary derivation from the Creator's presence: in the word 'presence'—still more in *Gegenwart*—a temporal and spatial constancy are fused. Limit time and, as Newton plainly observed, you must limit the authority of natural law and God's initiatory omnipotence.

Yet, strictly considered, the belief in 'an infinity and future succession of time by which eternity is unexhausted' did not last long. For some inquiring spirits at least, it cannot have survived intact Sadi Carnot's *Réflexions sur la puissance motrice du feu et les moyens propres à la développer* of 1824. In a preliminary way (which

---

[1] Cf. in particular A. Koyré, *La Révolution astronomique* (Paris, 1961), and *Études newtoniennes* (Paris, 1968). For general background, cf. Stephen Toulmin and June Goodfield, *The Discovery of Time* (New York, 1965).

Clapeyron's *Mémoire* of 1834 was to make mathematically more rigorous) this monograph formulated the entropy principle. Here is set out, not in terms of apocalyptic speculation or metaphoric conjecture but with an almost elementary ease of algebraic–mechanical deduction, the first of a number of related theories of irreversibility in the flow of energy. The arrow of time is directional. The true condition of the universe is one of thermodynamic processes approaching equilibrium and, therefore, inertness. Past the zero point and the cessation of any energy-yield from the motion of particles there can be no 'time'. Given a statistical framework of sufficient comprehensiveness, it can be shown that the grammar of the future tense is end-stopped, that entropy reaches a maximum at which the future ends. Even if it is regarded as no more than a statistical and idealized paradigm, applicable only where the microscopically discontinuous nature of matter enters the picture, the Clausius–Carnot principle is, surely, one of the extraordinary leaps of the human mind. The ability to conceive of a calculable finish to the energy exchanges in one's own cosmos must draw on some of the subtlest, most proudly abstractive of cerebral centres. Few texts go further than Carnot's treatise, severely technical as it is, to instance the singular dignity and risks of human thought.

What effect had the statement of the Second Law of Thermodynamics on sensibility and speech at large?

The 'interior history' of the entropy concept and of its relations to contemporary philosophic and linguistic consciousness is difficult to make out.[1] The 1849 *Account of Carnot's Theory* by

---

[1] There is no adequate history of the philosophic and psychological implications of the formulation of the entropy principle. F. Auerbach's *Die Königin der Welt und ihre Schatten* (Jena, 1909) and B. Brunhes's *La Dégradation de l'énergie* (Paris, 1909) represent influential popularizations of the concept of universal heat death. Hans Reichenbach's *The Direction of Time* (University of California Press, 1956) contains acute insights into the logic of entropy. Volume II of J. T. Merz's *A History of European Thought in the Nineteenth Century* (Edinburgh and London, 1927) is still useful in regard to the general historical context of thermodynamic theory. Background material and a summary of the latest cosmological aspects of the Second Law may be found in Wilson L. Scott, *The Conflict Between Atomism and Conservation Theory 1644–1860* (London and New York, 1970), and in F. O. Koenig, 'The History of Science and the Second Law of Thermodynamics', in

W. Thomson (Lord Kelvin) did a good deal to disseminate the analytic treatment of irreversibility. The word 'entropy' however, and the extrapolation of the notion of thermal or heat death to include the whole universe, are due to a paper by Clausius in the *Annalen der Physik und Chemie* for 1865. This paper contains the famous sentence 'die Entropie der Welt strebt einem Maximum zu'. It is not clear at all whether the extension of the Second Law to the entire cosmos is mathematically or empirically valid. Boltzmann's refutations of Clausius, in his work on the theory of gases, has, in turn, been found inadequate. But one need look only at the strident rejections of entropy by Engels and of the concept of 'universal heat death' by Soviet textbooks on thermodynamics to realize that issues of the utmost political, philosophic force are involved.

My question is narrower. Has the notion of a thermal death of the universe, of 'our' universe at least, affected the psychological tenor and linguistic conventions of uses of the future tense? Are the uses of futures in Western speech after Carnot and Clausius in some degree terminal or 'full-stopped'? The common-sense rejoinder that the remote immensities of time envisaged in theoretic speculations on entropy cannot press on a sane imagination, that magnitudes and statistical generalities of this order have no felt meaning, is only partly convincing. Eschatological images of a comparable distance and abstraction did influence patterns of feeling and idiom at earlier points in history. There are moods in

H. M. Evans (ed.), *Men and Moments in the History of Science* (Seattle, 1959). The most complete, rigorous formulation of the Clausius–Carnot law and of its mechanical implications can be found in G. N. Hatsopoulos and J. H. Keenan, *Principles of General Thermodynamics* (New York, 1965). Whether all energy transformations will 'eventually come to an end', or whether, as Boltzmann argued, we live in a universe of 'different times' separated by immense spaces, obviously remains a moot point. Recent astrophysical considerations and Planck's principle that the evolution of any system can be shown to represent an increase of entropy if the system is incorporated into a more comprehensive system that is sufficiently large, strongly suggest that the whole will run down even if certain parts show a downgrade of entropy. 'Although this principle leads to the unwelcome consequence that someday our universe will be completely run down and offer no further possibilities of existence to such unequalized systems as living organisms, it at least supplies us with a direction of time: positive time is the direction toward higher entropy' (Reichenbach, op. cit., p. 54).

which indistinct immensity takes on a concrete insistence. I can recall the queer inner blow I experienced when learning, as a boy, that the future thermodynamics of the sun would inevitably consume neighbouring planets and the works of Shakespeare, Newton, and Beethoven with them. As in Canetti's parable, the crux is one of distinct perception. Events a billion years off are fully conceptual in mathematical calculation and in language, but lie outside any zone of imaged, sensorily analogical apprehension. What then of ten million years, of half a million, of five generations? The quality of grasp, of registered impression, will be specific to different cultures and professional milieux. The quotient of substantive association in an astrophysicist's or geologist's consciousness of great time spans is obviously larger than that normal to an insurance actuary. The temporal horizons of Mayan civilization seem to have exceeded by far, and by deliberate expansion those available to other Central American cultures. Studies of Indo-European philology and of early Indian arithmetic point to a particular fascination with immensely extended numerical series and time projections.[1] But whatever the degree of individual and cultural diversity, there is a time-point, a location of thermal death, at which the threat of maximal entropy *would* assume reality for the general run of consciousness. The uses of futures of verbs *would* alter or take on a stylized, propitiatory cast of fiction, as perhaps they ought to have done already after Carnot. Condemned men probably bring complex idiomatic attenuations to any discourse on the 'day after tomorrow'. From a psycho-linguistic and socio-linguistic point of view, as well as in the perspective of cultural history, it would be valuable to know a good deal more than we do about the 'cut-off points' in future imaginings for different societies and epochs. There is more than wit to Lévi-Strauss's proposal that the science of man is an *entropologie*.[2]

---

[1] Cf. Karl Menninger, *Number Words and Number Symbols* (Cambridge, Mass., and London, 1969), pp. 102–3 and 135–8.

[2] There have recently been fascinating conjunctions between entropy and language or, more exactly, between thermodynamics and information theory. The notion that information can be treated as 'negative entropy' originates in the work of Leo Szilard and Norbert Wiener. It has been developed since, notably by Léon Brillouin in *Science and Information Theory* (New York, 1962), and *Scientific*

Even these cursory examples should suggest that the shapes of time are entrenched in grammar. The use of projectable predicates on which the validity of induction depends 'is effected by the use of language and is not attributed to anything inevitable or immutable in the nature of human cognition'.[1] The coiled spring of cause and effect, of forward inference, of validation through recurrence, indispensable to the ordered motion of feeling, is inseparable from the fabric of speech, from a syntax of the world as the latter 'has been described and anticipated in words'.[2] On this issue poets, formal logicians, and casual common sense are at one.

The difficulty arises when we ask whether and to what degree actual linguistic practice determines or is determined by underlying time-schemes. Are logicians, such as Nelson Goodman, right in assuming that all languages embody time in the same way or, more exactly, that every natural language can accommodate any conceivable temporality? Or does evidence point rather towards the well-known image, put forward in the late 1860s by Friedrich Max Mueller, the orientalist and ethnolinguist, of 'petrified philosophies' and psychologies of time buried in and specific to different grammars? Is the chronological scale of human history sufficient to register, at anything deeper than levels of idiomatic fashion, genuine and differentiated changes in man's time sense?

Most empirical investigation (it remains meagre) has borne on Biblical Hebrew and classical Greek. C. von Orelli's *Die hebräischen*

Uncertainty and Information (New York, 1964). The attempt to refute the well-known paradox of Maxwell—a decrease in entropy brought about without any apparent input of work—by treating information or knowledge as a species of energy, is suggestive. But it remains exceedingly difficult to grasp, let alone quantify. The Einsteinian concept of the transformation of mass into energy is one thing; the analogous transformation of knowledge, of 'bits of information', into energy, is quite another.

[1] Nelson Goodman, *Fact, Fiction, and Forecast* (London, 1954), p. 96. Cf. the critique of Goodman by S. F. Barker and P. Achinstein, 'On the New Riddle of Induction' (*Philosophical Review*, LXIX, 1960), and Goodman's rejoinder in 'Positionality and Pictures' (*The Philosophy of Science*, ed. P. H. Nidditch, Oxford, 1968).

[2] Goodman, op. cit., p. 117. Cf. G. H. von Wright's discussion of alternative 'time-grammars' in *Time, Change and Contradiction* (Cambridge, 1969).

Synonyma der Zeit und Ewigkeit genetisch und sprachvergleichend dargestellt of 1871 marks the beginning of methodical attempts to relate grammatical possibilities and constraints to the development of such primary ontological concepts as time and eternity. It had long been established that the Indo-European framework of threefold temporality—past, present, future—has no counterpart in Semitic conventions of tense. The Hebrew verb views action as incomplete or perfected. Even archaic Greek has definite and subtly discriminatory verb forms with which to express the linear flow of time from past to future. No such modes developed in Hebrew. In Indo-European tongues 'the future is preponderantly thought to lie before us, while in Hebrew future events are always expressed as coming after us'.[1] But how, if at all, do these differences relate to the contrasting morphology and evolution of Greek and Hebrew thought, of the Biblical as against the Herodotean code of history? Is the convention that spoken facts are strictly contemporaneous with the presentness of the speaker—a convention which, as Kierkegaard saw, is crucial to Hebraic–Christian doctrines of revelation—a generator or a consequence of grammatical forms?

We do not know, because here also the evidence is circular. The linguistic structure articulates and seems to organize the ruling image and philosophic stance; but it is via the philosophic or ritual text that we determine the grammatical base. If, in Semitic languages, 'the notion of recurrence coincides with that of duration',[2] which came first: the lexical and grammatical rule or the mental picture, with its primordial but likely source in conjectures on the orbital motion of the stars?

---

[1] Thorlief Boman, *Hebrew Thought Compared with Greek* (London, 1960), p. 51. Boman's treatment of individual texts and etymologies is fascinating, but his thesis suffers from considerable anthropological and hermeneutic *naïveté*. The assumption that one can 'translate' the semantics of ancient Hebrew and Greek speech modes into our own, the proposition that the 'idiosyncracy of a nation or family of nations, a race, finds expression in the language peculiar to them', cannot be taken for granted. It is just these points that require demonstration. Cf. also the analysis of Hebrew 'temporalities' in John Marsh, *The Fullness of Time* (London, 1952).

[2] Boman, op. cit., p. 136.

It is banal but necessary to insist on a manifold reciprocity between grammar and concept, between speech form and cultural pressure. Intricate grooves of possibility and of limitation, neuro-physiological potentialities of many-branched but not unbounded realization, prepare, in ways we can only guess at schematically, for anything as complex as a grammar and system of symbolic reference. Presumably the dialectic of interaction is persistent, between linguistic 'spaces' and the trajectories of thought and feeling within them, between such trajectories and the unfolding or mapping of new spaces. Hebrew speech-consciousness informs and is informed by the sovereign tautology 'I am that I am' which alone and axiomatically defines an undefinable, inconceivable yet omnipresent God. It is this 'present absence' and 'self-erasing' tautology from which has sprung the current grammatology of deconstruction, itself a variant on Talmudic–Kabbalistic speech speculations.

The spectrum of Greek tenses occasions but is also realized in the genius of Thucydidean historicism. The pattern is one of reciprocal 'triggering' and actualization. If current biology is right, precisely the same reciprocity obtained between the origins of language itself and the enabling-responsive growth of the cortex. Pre-condition and consequences are aspects of a continuum. 'Il est impossible de ne pas supposer', writes Monod, 'qu'entre l'évolution privilégiée du système nerveux central de l'Homme et celle de la performance unique qui le caractérise, il n'y ait pas eu un couplage très étroit, qui aurait fait du langage non seulement le produit, mais l'une des conditions initiales de cette évolution.'[1]

What I would emphasize is the interdependence between that evolution and the availability of the future tense.

Whatever may be the proto-linguistic or meta-linguistic codes of other species, I would want to argue strongly that man alone has developed a grammar of futurity. Primates use rudimentary tools but, so far as has been observed, they do not store tools for future usage. There is a vital sense in which that grammar has 'developed

---

[1] Jacques Monod, *Le hasard et la nécessité: essai sur la philosophie naturelle de la biologie moderne* (Paris, 1970), p. 145. The entire section, pp. 144–51, is highly relevant to an understanding of the model of 'informing reciprocity'.

man', in which we can be defined as a mammal that uses the future of the verb 'to be'. Only he, as writes Paul Celan in *Atemwende*, can cast nets 'in rivers north of the future'. The syntactic development is inextricably inwoven with historical self-awareness. The 'axiomatic fictions' of forward inference and anticipation are far more than a specialized gain of human consciousness. They are, I believe, a survival factor of the utmost importance. The provision of concepts and speech acts embodying the future is as indispensable to the preservation and evolution of our specific humanity as is that of dreams to the economy of the brain. Cut off from futurity, reason would wither. Such is the posture of the doomed prophets in the *Inferno* (X):

> Però comprender puoi che tutta morta
> fia nostra conoscenza da quel punto,
> che del futuro fia chiusa la porta.

Close the door on the future and all perception, all knowledge is made inert.

There could be no personal, no social history as we know them, without the ever-renewed springs of life in future-tense propositions. These constitute what Ibsen called 'the Life-lie', the complex dynamism of projection, of will, of consoling illusion, on which our psychic and, conceivably, our biological perpetuation hinge. There can be spasms of despair in the individual and in the community, solicitations of 'neverness' and of that last great repose which haunted Freud in *Beyond the Pleasure Principle*. Suicide is a recurrent option, as are resolutions of communal extinction, by sacrificial violence or a refusal to bear children. But these nihilistic temptations remain fitful and, statistically considered, rare. The language fabric we inhabit, the conventions of forwardness so deeply entrenched in our syntax, make for a constant, sometimes involuntary, resilience. Drown as we may, the idiom of hope, so immediate to the mind, thrusts us to the surface. If this was not the case, if our system of tenses was more fragile, more esoteric and philosophically suspect at its open end, we might not endure. Through shared habits of articulate futurity the individual forgets, literally 'overlooks', the certainty and

absoluteness of his own extinction. Through his constant use of a
tense-logic and time-scale beyond that of personal being, private
man identifies, however abstractly, with the survival of his species.

Social psychologists such as Robert Lifton, in his study of
Revolutionary Immortality (1968), and philosophers such as Adorno
and Ernst Bloch, have investigated the collective, historical implications of futurity. The ability of the race to recover from local
or widespread disaster, the resolve to 'continue history' when
so much of it has been frustration and terror, seem to originate
in those centres of consciousness which 'imagine ahead', which
extrapolate but at the same time alter the model. Very probably,
the self-perpetuation of animals takes place in the matrix of a
constant present. Like the replication of molecular organisms, the
generation and nurture of offspring does not, of itself, instance
a concept of the future. The drive of human expectations or, as
Bloch calls it, 'das Prinzip Hoffnung', relates to those probabilistic,
partly Utopian reflexes which every human being displays each
time he expresses hope, desire, even fear. We move forward in the
slipstream of the statements we make about tomorrow morning,
about the millennium. Only because the relevant grammar is
available to us—the grammar which articulates the perception of
evolution and which evolution, in turn, must have generated—can
we grasp Nietzsche's definition of man as 'an animal not yet
determined, not yet wholly posited' ('ein noch nicht festgestelltes
Tier').

I hope to indicate shortly in what ways the capacity of language
to put forward propositions about the future and to map logical
and grammatical 'spaces' for such propositions, is a subclass of a
larger category. Future tenses are an example, though one of the
most important, of the more general framework of non- and
counter-factuality. They are a part of the capacity of language for
the fictional and illustrate the absolutely central power of the
human word to go beyond and against 'that which is the case'.

Our languages simultaneously structure and are structured by
time, by the syntax of past, present, and future. In Hell, that is to
say in a grammar without futures, 'we literally hear how the verbs
kill time' (Mandelstam's penetrating comment on Dante and on

linguistic form echoes his own asphyxia under political terror, in the absence of tomorrow). But 'at other times', itself an extraordinary locution, it is only through language and, perhaps through music, that man can make free of time, that he can overcome momentarily the presence and presentness of his own punctual death.

3

Language is in part physical, in part mental. Its grammar is temporal and also seems to create and inform our experience of time. A third polarity is that of private and public. It is worth looking at closely because it poses the question of translation in its purest form. In what ways can language, which is by operative definition a shared code of exchange, be regarded as private? To what degree is the verbal expression, the semiotic field in which an individual functions, a unique idiom or idiolect? How does this personal 'privacy' relate to the larger 'privacy of context' in the speech of a given community or national language? The paradoxical possibility of the existence of private language has widely exercised modern logic and linguistic philosophy. It may be that a muddle between 'idiolect' and 'privacy' has frustrated the whole debate. It may be also that only a close reading of actual cases of translation, particularly of poetry, will isolate and make concrete the elements of privacy within public utterance. But the philosophic discussion should be summarized first.

Currently, reference to 'private language' implies, almost inevitably, reference to Wittgenstein's treatment of the topic in the *Philosophical Investigations*. The canonic texts can be found in sections 203–315, with special emphasis on 206–7, 243–4, 256 and 258–9. These, together with N. Malcolm's well-known review of the *Investigations* in the *Philosophical Review* (LXIII, 1954), have given rise to a voluminous, often highly abstruse literature.[1] Obviously there are facets of the discussion which lie

---

[1] An extensive bibliography is to be found in K. T. Fann, *Wittgenstein's Conception of Philosophy* (Oxford, 1969). Much of the literature sprang directly from A. J. Ayer's 'Can There Be a Private Language?' and R. Rhees's rejoinder under the

outside the grasp of anyone not qualified in technical aspects of modern philosophy. Nevertheless, the material leaves one with the sense of an impasse, with the suspicion that a subject of intense interest to philosophy at large and to the theory of language has been unduly narrowed and, perhaps, muddled. In part, this is a matter of mandarin idiom, of the strong inclination of logicians to deal more with each other's previous papers and animadversions than with the intrinsic question. But it may well be that the trouble lies with Wittgenstein's own handling of the private-language argument. 'It seems impossible to state with complete assurance exactly what Wittgenstein took the private language argument to be or to show,' remarks one logician.[1] 'It is not clear at all what the Private-Language argument is supposed to come to or what its assumptions and its reasoning are,' concludes another.[2]

Wittgenstein's opaqueness at pivotal moments in the discussion may have its own intent. As so often in the *Investigations*, he is concerned with the most honest articulation possible of difficulties, with the instigation of heuristic malaise, not with the proposal of systematic answers. Moreover, and this again is characteristic, Wittgenstein seems to be directing attention to one problem while, in fact, sketching the contours of a larger, less immediately designated area of philosophic inquiry. The actual considerations on private language are pointers towards a wider questioning of sensations and sensation words (notably 'pain').[3] They are also involved with Wittgenstein's perennial aim to discriminate between

same title (both in *Proceedings of the Aristotelian Society,* Suppl. Vol. XXVIII, 1954). A number of the most important articles on the private-language argument have been reprinted in H. Morick (ed.), *Wittgenstein and the Problem of Other Minds* (New York, 1967), and O. K. Jones (ed.), *The Private Language Argument* (New York, 1969). The issues are summarized in Warren B. Smerud, *Can There Be a Private Language?* (The Hague, 1970).

[1] Michael A. G. Stocker, 'Memory and the Private Language Argument' (*Philosophical Quarterly*, XVI, 1966), p. 47.

[2] J. F. Thomson, 'Symposium on the Private Language Argument', in C. D. Rollins (ed.), *Knowledge and Experience* (University of Pittsburgh Press, 1964), p. 119.

[3] Cf. P. von Morstein, 'Wittgensteins Untersuchungen des Wortes, "Schmerz"' (*Archiv für Philosophie*, XIII, 1964), and L. C. Halborow, 'Wittgenstein's Kind of Behaviourism?' (*Philosophical Quarterly*, XVII, 1967).

empirical, analytic, and grammatical forms of statement and with the whole, more general controversies between phenomenalist and behaviourist views of human speech and action. The claim that Wittgenstein was not 'trying to show something about language but rather about sensations or mental phenomena'[1] goes too far. The issues were not separate for him. But it is fair to say that the focus of interest is not always declared and that the links between the private-language problem, strictly posed, and the inferred epistemological and psychological aspects, are at times ambiguous.

Baldly put, Wittgenstein's criteria for a private language are that it should be used by exactly one person, that it should be intelligible to him alone, and that it can refer to inner mental events. He then shows or, rather, suggests how one would demonstrate, that such a 'language' is neither a logical nor a practical possibility. The analysis is at once fragmentary and, as is often the case in the later Wittgenstein, of great delicacy. It hinges on the conviction that language is a social function which depends upon the possibility of correction by another person, and that there can be no objective check upon memory mistakes in a purely phenomenal language (whatever the latter oddity might be). The use of language is the use of a system of rules. These rules must be consistent if the propositions which they inform are to have meaning. If we check a rule privately we cannot distinguish between actually observing the rule, and merely thinking that we have done so. Given the fallibility of personal memory, the hermit cannot tell whether today's rules are the same as yesterday's. A community of speakers is required in order to provide a standard of correct usage. Meaning and public verification are reciprocal aspects of a genuine speech-act.

References to inner mental events—this is the crux of Wittgenstein's whole investigation—are in fact a social phenomenon. They depend for meaning on a network of recognitions and behavioural responses on the part of those to whom the reference is uttered. Wittgenstein insists that any sign which has a use

---

[1] V. C. Chappell, 'Symposium on the Private Language Argument' (op. cit.), p. 118.

cannot be simply associated with a personal sensation. In language utility and mutual intelligibility are indivisible. 'A privately referring-with-a-word person is not a referring-with-a-word person at all. A person who is privately referring with a word is not a logical possibility.'[1] Despite appearances, argues Wittgenstein, such words as 'pain' do not and cannot refer to 'private objects'. The latter, whose status is at best implausible, cannot be spoken of in a public language. But a linguistic proposition has meaning only in so far as it can be verified, and such verification is necessarily social. Hence language must be public.[2] Meaning is, in fact, a process, a consequence of exchange, correction, and reciprocity. For language to work 'there must be something more like an organization in which different people are, as we may put it, playing different roles. . . . Language is something that is spoken.'[3] It is something that can be translated.

Every filament in Wittgenstein's argument, an argument to which Malcolm's restatement gave more edge and sequence than the original may have intended, has been the object of minute elucidation and critique. Wittgenstein's case does not emerge intact. Following suggestions made by Ayer, a number of logicians have felt that a distinction must be drawn between a language which only one person *does* use and understand (the last member of a moribund community or speech-culture), and a language which only one person *can* use and understand. Not only could Robinson Crusoe develop a language of his own, but given 'a certain sort of language', he could also make solitary use of it.[4] Strictly speaking, Wittgenstein has done no more than demonstrate that 'if a language is to communicate, at least some of the

---

[1] Moreland Perkins, 'Two Arguments Against a Private Language', in H. Morick (ed.), *Wittgenstein and the Problem of Other Minds*, p. 109. Cf. also N. Garver, 'Wittgenstein on Private Language' (*Philosophy and Phenomenological Research*, XX, 1960) for a similar conclusion.

[2] Cf. N. Malcolm, *Knowledge and Certainty* (New York, 1964), and D. Locke, *Myself and Others: A Study in Our Knowledge of Minds* (Oxford, 1968), Chapter V, for thorough discussion of the issue of criteria of verification.

[3] R. Rhees, 'Can There Be a Private Language?' (*Proceedings of the Aristotelian Society*), p. 76.

[4] N. P. Tanburn, 'Private Languages Again' (*Mind*, LXXII, 1963), p. 90.

entities to which it refers must be publicly available'.[1] Acute criticisms have been made of Wittgenstein's treatment of memory in the argument. It has been asserted that the entire private-language denial in the *Investigations* is founded on 'the epistemically invidious distinction between private and public memory judgements'.[2] Ultimately, the criteria of verification applicable to public speech-acts are no more infallible than those which Wittgenstein denies to private utterance. Strict analysis, moreover, shows that 'there are at least some cases where there are independent criteria for discovering whether the rules of a private language have been obeyed'.[3] Wittgenstein's case conceals a *reductio ad absurdum*, for it can be made to demonstrate that no language at all is possible.

The matter of 'sensation words' has also been closely debated. Using Moritz Schlick's image of a world which we would perceive in different colours according to our changing and unpredictable moods, C. L. Hardin finds that there are words which can in fact be 'known only by a single individual if there are situations in which only he can decide whether or not the word can be properly applied'.[4] Accordingly, Wittgenstein would have failed to prove the *logical* impossibility of a purely phenomenalist language. Other critics go further. Persuaded that natural language does indeed refer to private data, and that such reference is both a valid and inevitable part of communication they detect in Wittgenstein a fairly naïve behaviourism.[5] Furthermore, the demonstration that another individual will not fully understand a 'personal sensation statement' does not prove that such statements are logically and causally impossible. In what is until now the most thorough dissent from Wittgenstein's whole position, C. W. K. Mundle, in *A Critique of Linguistic Philosophy* (1970), finds that there is in the *Investigations* a set of fundamental confusions. The rules governing the use of a word are confounded with the way in which it was learnt, and privacy of reference is confused with

[1] Ibid., p. 98.

[2] Michael A. G. Stocker, op. cit., p. 47.

[3] W. Todd, 'Private Languages' (*Philosophical Quarterly*, XII, 1962), p. 216.

[4] C. L. Hardin, 'Wittgenstein on Private Languages' (*Journal of Philosophy*, LVI, 1959), pp. 519–20.

[5] Cf. C. W. K. Mundle: '"Private Language" and Wittgenstein's Kind of Behaviourism' (*Philosophical Quarterly*, XVI, 1966).

incommunicability. Sometimes, argues Mundle, Wittgenstein uses 'private' to characterize language which refers to or describes private experiences. At other times, he means a language whose significance can be known only by its inventor. 'Wittgenstein and his followers oscillate at their own convenience between using "private language" in different senses.' Disturbed by the opaqueness and discontinuity of the entire argument, J. F. Thomson concludes: '(1) It is widely held that Wittgenstein showed something important about the notion of a private language. (2) When we look into the claim, it is not obvious that he did anything of the sort.'[1]

One need not endorse this finding. The points made in the *Investigations* and the large literature which has followed are of the most vivid interest to poetics and to the philosophy of language. What does strike the layman is the deceptive uniformity and idealization of the model. If there was such a thing as a private language, how could one tell that one was in fact hearing or reading it? What would distinguish it, beyond any conceivable doubt, from a 'lost' language of the past, from a language spoken to himself or in fever by the last speaker of an extinct tongue? Some of Wittgenstein's remarks seem to indicate that potential acquisition by a second person is a sufficient criterion to define a public language. Is the converse necessarily true? The question of memory is also troubling. Having suffered a spell of amnesia, or returning to his solitude after a lengthy absence, the hermit might well regard the entries in his old diary as being gibberish. In actual fact, it might simply be the case that he no longer knew how to decipher them. Would this prove anything, *either way*, about the status of the original sign-system? No. Suppose he did decipher these diary entries: could there be any logical proof that his decoding was the right one? Conversely, would the lack of such a proof be sufficient to show that he was not dealing with a genuine language in the first place? Seeking to grasp the force of Wittgenstein's criticism of 'private objects', one is made aware of the possibility that the obscurities, the indeterminacies in the logic

[1] J. F. Thomson, op. cit., p. 124.

of the case, stem from a refusal to distinguish between 'reference' and 'meaning'. 'The fact that a word has a private reference does not mean that it has to have a private meaning; there is no reason why a word should not refer to a private object and yet have a meaning that is publicly ascertainable and publicly checkable.'[1] The decision to reject this distinction dates back to the very beginnings of Wittgenstein's philosophy and to his quarrel with Frege's system. It is this rejection which may account for some of the enigmas and behaviourist naïvetés in the private-language argument.[2]

Running through the argument is the assumption that any 'secret' or personal language invented by an individual must be parasitic on previous languages. However ingenious, it will be no more than a translation inward from public grammars and conventions of speech. 'To use language "in isolation" is like playing a game of *solitaire*. The names of the cards and the rules of manipulation are publicly given and the latter enable the player to play without the participation of other players. So, in a very important sense, even in a game of *solitaire* others participate, namely those who had made up the rules of the game.'[3] Is this necessarily so, or ought the assumption of 'necessary transposition' from an extant language to be looked at more closely? Even at the most immediate level of plausibility, a problem is posed. An unknown game played by an individual in total solitude is, precisely, one we could know nothing about. Yet the contrivance of such a game, and even its perception by a hidden observer who might not make out that anything rule-governed and regular was being performed (he sees the game being played only once) are logically entirely conceivable though psychologically implausible. As we shall see, the perplexity is one of degree, of the distance of the singular phenomenon from a preceding, analogous norm of verification. Cryptography provides a crude model. The practice

[1] D. Locke, op. cit., p. 99.
[2] For the importance of Frege's distinction cf J. R. Searle in J. R. Searle (ed.), *The Philosophy of Language* (Oxford, 1971), pp. 2–3.
[3] Gershon Weiler, *Mauthner's Critique of Language* (Cambridge University Press, 1970), p. 107.

of encoding information in hidden characters, which can be transmitted either orally or in writing, is probably as ancient as human communication itself, and certainly older than the coded hieroglyphics incised in *c.*1900 B.C. in a nobleman's tomb at Menet Khufu. It seems to be an inference from the private-language argument that all codes are based on a known public speech-system and can, therefore, be broken (i.e. understood, learned by at least one person beyond the original encoder). I am not certain whether there is a logical proof of this contention, or indeed whether there can be. But factually this appears to be the case. If certain texts—the Indus Valley script, the pictographs found on Easter Island, Mayan glyphs—have, until now, remained undeciphered, the reasons are contingent. They lie in human error or the lack of a critical mass of samples. Yet even here there are suggestive border-cases, puzzles which make of contingency a complex matter of degree. The so-called Voynich manuscript first turned up in Prague in 1666 (a date with emphatic apocalyptic–numerological overtones). Its 204 pages comprise a putative code of twenty-nine symbols recurring in what appear to be ordered 'syllabic' units. The text gives every semblance of common non-alphabetic substitution. It has, up to the present time, resisted every technique of crypto-analysis including computer-simulation. We do not even know whether we are dealing with, as was formerly held, a thirteenth- or, as now seems probable, a late-sixteenth- or seventeenth-century device.[1] I have wondered whether we are, in fact, looking at an elaborate nonsense-structure, at an assemblage of systematic, recurrent, rule-governed characters signifying strictly nothing. Though immensely laborious and absurd, such an exercise is, logically, entirely possible. But could there be any proof of nullity of meaning now that the original contriver is long dead? Would the absence of any such proof be evidence, however tenuous, towards the privacy of the 'language' in question? And what of the 'one-time pad' codes instituted by the German diplomatic service in the early 1920s? By its use of random non-repeating keys, this system makes of every message a unique, non-

[1] Cf. David Kahn, *The Codebreakers* (London, 1966) for a detailed discussion of the Voynich manuscript.

repeatable event. Does this undecipherable singularity throw any light on the logical paradigm of a language spoken only once, of a diary, in Wittgenstein's model, whose rules of notation would apply only in and for the moment at which they were set down? It is the bizarre extremity of such cases which may help to point up, to elicit some of the untested assumptions in the private-language debate.

The most powerful of these assumptions is either anthropological or philosophical or both. The postulate that any language devised by man is finally reducible to known and public precedents, that the concept of 'linguistic privacy' is a logical and substantive muddle standing, at best, for individual variants on or translations from existing speech, can have a decisive evolutionary consequence. It could point to a common origin for all languages. The eroded metaphor of 'root' and 'stem' as applied to etymology evokes the abiding image of a common tree (the pictorial overlap is striking, for instance in Leibniz's argument on universality).[1]

The stronger hypothesis adduces a universal speech-potential and grammatical programme innate in the human mind. This is the conclusion put forward by generative linguistics. 'So far as evidence is available,' writes Chomsky, 'it seems that very heavy conditions on the form of grammar are universal. Deep structures seem to be very similar from language to language, and the rules that manipulate and interpret them also seem to be drawn from a very narrow class of conceivable formal operations.'[2] Despite their manifest diversity and mutual unintelligibility, all past, extant and *conceivable* languages satisfy the same fixed set of deep, invariant, highly restrictive principles. The 'wolf-child' imagined by natural philosophy or the hermit cut off by amnesia from all remembrance of former speech, will develop an idiom related to all other human tongues through a recognizable system of constraints and transformational rules. The human brain is so constructed that it cannot but do so. All grammars belong to a definable sub-class of

[1] Cf. Hans Aarsleff, 'The Study and Use of Etymology in Leibniz' (*Erkenntnislehre. Logik, Sprachphilosophie Editionsberichte*, Wiesbaden, 1969, III).

[2] N. Chomsky, 'Recent Contributions to the Theory of Innate Ideas' in J. R. Searle (ed.), *The Philosophy of Language*, p. 125.

the class of all transformational grammars, being the product of specific and structured elements of innateness in man. A creature speaking a 'language' not in this sub-class would, by definition, be non-human and we could not learn its 'Martian' speech. This impossibility accounts for the imaginary 'translation-machines' which play so characteristic a part in the equipment of science-fiction space-argonauts.

The two hypotheses can be taken as congruent and mutually reinforcing though logically they need not be. They tell us that there are no private speech-acts. Wherever speech occurs on the earth, it will evolve along universal grooves of grammatical possibility. All new languages, however secret or eccentric, will be parasitic on a public and preceding model. As it happens, there is as yet no strong evidence in anthropology to demonstrate either a single and diffusive or a multiple origin of human speech. The generative postulate of innateness remains highly controversial and is thought by many to be the weakest aspect of the new linguistics.[1] Nevertheless, the philosophic corollaries of the rejection of 'private language' and the bearing of the private-language argument on a theory of translation should be obvious.

But whether in Wittgenstein's critique or in controversies over the innateness and universality of grammatical constraints, it is clear that 'privacy' is being used in a formalized, sharply restrictive sense. There are other, more immediately significant ways in which an impulse towards privacy of intent and reference is one of the vital, problematic realities in human communication.

No two human beings share an identical associative context. Because such a context is made up of the totality of an individual existence, because it comprehends not only the sum of personal memory and experience but also the reservoir of the particular subconscious, it will differ from person to person. There are

[1] Cf. the vehement critiques of Chomsky's argument by Hilary Putnam and Nelson Goodman reprinted in *The Philosophy of Language*, pp. 130–44. The debate was resumed at the Ninth Annual Meeting of the New York University Institute of Philosophy in 1968. The proceedings generated a fair amount of acrimony but little fresh light. So long as Chomsky does not specify what kind of innate mechanism he is adducing, it is difficult to imagine what would constitute evidence for or against the innateness of deep structures and transformational procedures.

no facsimiles of sensibility, no twin psyches. All speech forms
and notations, therefore, entail a latent or realized element of individual specificity. They are in part an idiolect. Every counter of
communication carries with it a potential or externalized aspect of
personal content. The zone of private specification can extend to
minimal phonetic units. As children and poets bear witness, even
individual letters and the sound-unit which they vocalize, can
assume particular symbolic values and associations. To a literate
member of Western culture in the mid-twentieth century, the
capital letter K is nearly an ideogram, invoking the presence of
Kafka or of his eponymous doubles. 'I find the letter K offensive,
almost nauseating,' noted Kafka mordantly in his diary, 'and yet I
write it down, it must be characteristic of me.' Such vividness and
personal focus of associative content can colour even the most
abstract, formally neutral of expressive terms. Contrary to what
logicians have asserted, numerals do not necessarily satisfy the
condition of an identity and universality of associative content.
The erotic innuendo of 'sixty-nine' belongs to a particular cultural
and linguistic milieu. In French, *quatre-vingt-treize* and *soixante-
quinze* have carried a specific associative nimbus, in the one
case mainly historical–political (a time of revolutionary terror
and survival), in the other military (the famous field-gun). But
it is by no means necessary that the relevant numeral should
suggest a picture or be attached to a preceding verbal context.
Mathematicians will invest individual numbers with personal
values; particular primes or cardinals can take on a lively context
of association, a tonality wholly independent of any extraneous
non-mathematical reference. 'Every positive integer was one of his
personal friends,' said J. E. Littlewood in his recollections of his
colleague Ramanujan.

The associative mechanism has profound consequences for the
theory of language and of translation. The distinction between
phonetic and semantic constituents of a speech-act is, nearly
always, approximative. All phonetic elements above the level of
morphemes (perhaps even prior to that level) can become carriers
of semantic values. Because every speech form and symbolic code
is open to contingencies of memory and of new experience,

semantic values are necessarily affected by individual and/or historical–cultural factors.

As we observed, the associative content which contingencies import into letters, numbers, syllables, and words can be private or social or both. The associative contour lies along a spectrum which extends the whole way from the solipsism of the maniac to human generality (but being historical and cultural, this generality has nothing to do with the 'innate universality' postulated by generative theory). At one pole we find a 'pathology of Babel', autistic strategies which attach hermetic meanings to certain sounds or which deliberately invert the lexical, habitual usage of words. At the other extreme, we encounter the currency of banal idiom, the colloquial shorthand of daily chatter from which constant exchange has all but eroded any particular substance. Every conceivable modulation exists between these two extremes. Even the sanest among us will have recourse, as does the deranged solipsist, to words and numerals, to phrases or sound-clusters, whose resonance and talismanic invocation are deeply personal. The cornered child will loose such signals on a deaf world. Families have their own thesaurus often irritatingly opaque to the newest member or outsider. So do priesthoods, guilds, professions, mysteries. There are as many lexica and glossaries of shared association as there are constructs of kinship, of generation, of *métier*, of special inheritance in a society.

As concentric spheres of association move outward, they come to include the community, the province, the nation. There are innumerable near-identities or, more strictly speaking, overlaps of associative content which Englishmen share by virtue of historical or climatic experience but which an American, emitting the same speech-sounds, may have no inkling of. The French language, as self-consciously perhaps as any, is a palimpsest of historical, political undertones and overtones. To a remarkable degree, these embed even ordinary locutions in a 'chord' of associations which anyone acquiring the language from outside will never fully master. There is no dictionary that lists but a fraction of the historical, figurative, dialectic, argotic, technical planes of significance in such simple words as, say, *chaussée* or *faubourg*; nor could there be, as these planes are perpetually interactive and changing. Where experience is monotonized, on the other hand, the associative content grows progressively more transparent. There is, currently, a stylistic and emotional esperanto of airport lounges, a vulgate identically inexpressive from Archangel to Tierra del Fuego.

In short, whether consciously or unconsciously, every act of human communication is based on a complex, divided fabric which may, fairly, be compared to the image of a plant deeply and invisibly rooted or of an iceberg largely under water. Active inside the 'public' vocabulary and conventions of grammar are pressures of vital association, of latent or realized content. Much of this content is irreducibly individual and, in the common sense of the term, private. When we speak to others we speak 'at the surface' of ourselves. We normally use a shorthand beneath which there lies a wealth of subconscious, deliberately concealed or declared associations so extensive and intricate that they probably equal the sum and uniqueness of our status as an individual person. It was from this central fact of the dual or subsurface phenomenology of speech that Humboldt derived his well-known axiom: 'All understanding is at the same time a misunderstanding, all agreement in thought and feeling is also a parting of the ways.' Or as Fritz Mauthner put it, it was via language, with its common surface and private base, that men had 'made it impossible to get to know each other'.[1]

But this opaqueness, this part of illusion in all public speech-acts is probably essential to the equilibrium of the psyche. Articulated or internalized, language is the principal component and validation of our self-awareness. It is the constantly tested carapace of distinct identity. Yet at the phonological, grammatical, and, in significant measure, semantic levels it is also among the most ubiquitous and common of human properties. There is a sense in which our own skin belongs to every man. This apparent contradiction is resolved by the individuation of associative content. Without that individuation, in the absence of a decided

[1] Fritz Mauthner, *Beiträge zu einer Kritik der Sprache* (Leipzig, 1923), I, p. 56.

private component in all but the most perfunctory, unreflecting of our speech-acts, language would possess only a surface. Lacking roots in the irreducible singularity of personal remembrance, in the uniqueness of the 'association-net' of personal consciousness and subconsciousness, a purely public, common speech would severely impair our sense of self. Harold Pinter and Peter Handke have strung together inert clichés, tags of commercial, journalistic idiom, to produce discourse which would show no indeterminacy, no roughage of personal reference. These satiric exercises have a direct bearing on the theory of language. The ego, with its urgent but vulnerable claims to self-definition, withers among hollow, blank phrases. Dead speech creates a vacuum in the psyche.

Linguistic taboos illustrate the role of a 'non-public' associative content in the vital economy of individual and social feeling. Kept 'out of sight' certain words, formulas, combinations of letters, retain a numinous, life-giving energy. Because he can use them rarely, if at all, because such usage will take place in situations abstracted from the random banality of ordinary occurrence, the priest, the initiate, the private individual will surround his utterance with a field of special force. Often the edge of meaning will not have been entirely defined and the associative contour of power or sanctity will have been drawn by the tensed conjecture of the speaker. The semantics of sex provide an incisive example. At one end of their associative range, taboo words for sexual activities, for bodily parts and functions, were deliberately defused. Their menacing and comic implications were 'secularized' by their use in slang or were devalued by conspicuous waste (the unending epithets of army prose). At the opposite pole, however, many of these same terms were reserved for the most intense, private of erotic approaches. When spoken aloud for the first time to the beloved, when taught her—such 'teaching' being itself, perhaps, based on a myth of preceding innocence and purity—'obscene' words took on a fierce, almost ritual privacy. Repeated, echoed by the beloved, they marked the private heart of privacy, of that aloneness to which one other speaker or listener is indispensable.

I say 'marked' because this condition, which may have been

largely a middle-class phenomenon, has altered radically. Over the past forty years, the vocabulary of sex has been massively publicized. It has been all but neutralized by constant exploitation on the stage, in print and in emancipated colloquialism. The educated Western sensibility has been rapidly immunized against the ancient terrors and instigations of the 'private parts' of speech. Social psychologists welcome this change. They see it as a liberation from needless shadows. I wonder. The balance between subterranean argot and quintessential, exploratory privacy—lover to lover—must have been a mechanism of extreme complexity and emotional logic. The capacity of words to be at once devalued, loudly demeaned, and magical points to a dynamic poise between private and public aspects of language. These delicate strengths have been eroded. Moreover, the imaginative and expressive resources of most men and women are limited. The enrichments of intimacy, of evocative excitement, that came from the use of taboo words, the sense of a uniquely shared access to a new and secret place, were real. Being, today, so loud and public, the diction of eros is stale; the explorations past silence are fewer.

The issue is larger. A diffuse rationalism, the levelling impress of the mass media, the increasing monochrome of the technological milieu, are crowding on the private components of speech. Under stress of radio and television, it may be that even our dreams will be standardized and made synchronic with those of our neighbours. Religion, magic, regionalism, the relative isolation of communities and individuals, verbal taboos were the natural sources and custodians of the numinous aspects of language. Each of these agencies is now decaying. The effects on the vital stability of the speech-structure, on the complex verticality which relates the subconscious and the central privacies of language to the public surface, may be severe. Ballast is lacking.

There can hardly be an awakened human being who has not, at some moment, been exasperated by the 'publicity' of language, who has not experienced an almost bodily discomfort at the disparity between the uniqueness, the novelty of his own emotions and the worn coinage of words. It is almost intolerable that needs, affections, hatreds, introspections which we feel to be over-whelmingly our own, which shape our awareness of identity and the world, should have to be voiced—even and most absurdly when we speak to ourselves—in the vulgate. Intimate, un-precedented as is our thirst, the cup has long been on other lips. One can only conjecture as to the blow which this discovery must be to the child's psyche. What abandonments of autonomous, radical vision occur when the maturing sensibility apprehends that the deepest instrumentalities of personal being are cast in a ready public mould? The secret jargon of the adolescent coterie, the conspirator's pass-word, the nonsense-diction of lovers, teddy-bear talk are fitful, short-lived ripostes to the binding commonness and sclerosis of speech. In some individuals the original outrage persists, the shock of finding that words are stale and promiscuous (they belong to everyone) yet wholly empowered to speak for us either in the inexpressible newness of love or in the privacies of terror. It may be that the poet and philosopher are those in whom such outrage remains most acute and precisely remembered; witness Sartre's study of himself in *Les Mots* and his analysis of Flaubert's 'infantile' refusal to enter the matrix of authorized speech. 'O Wort, du Wort das mir fehlt!' cries Moses at the enigmatic climax of Schoenberg's *Moses und Aron*. No word is adequate to speak the present absence of God. None to articulate a child's discovery of his own unreplicable self. None to persuade the beloved that there has been neither longing nor trust like this in any other time or place and that reality has been made new. Those seas in our personal existence into which we are 'the first that ever burst' are never silent, but loud with commonplaces.

The concept of 'the lacking word' marks modern literature. The principal division in the history of Western literature occurs between the early 1870s and the turn of the century. It divides a literature essentially housed in language from one for which language has become a prison. Compared to this division all pre-ceding historical and stylistic rubrics or movements—Hellenism, the medieval, the Baroque, Neo-classicism, Romanticism—are only subgroups or variants. From the beginnings of Western literature until Rimbaud and Mallarmé (Hölderlin and Nerval are decisive but isolated forerunners), poetry and prose were in organic accord with language. Vocabulary and grammar could be expanded, distorted, driven to the limits of comprehension. There are deliberate obscurities and subversions of the logic of common discourse throughout Western poetry, in Pindar, in the medieval lyric, in European amorous and philosophic verse of the sixteenth and seventeenth centuries. But even where it is most explicit, the act of invention, of individuation in Dante's *stil nuovo*, in the semantic cosmography of Rabelais, moves with the grain of speech. The *métier* of Shakespeare lies in a realization, a bodying forth more exhaustive than any other writer's, more delicately manifold and internally ordered, of the potentialities of public word and syntax. Shakespeare's stance in language is a calm tenancy, an at-homeness in a sphere of expressive, executive means whose roots, traditional strengths, tonalities, as yet unexploited riches, he recognized as a man's hand will recognize the struts and cornices, the worn places and the new in his father's house. Where he widens and grafts, achieving reaches and interactions of language unmatched before him, Shakespeare works from within. The process is one of generation from a centre at once conventional (popular, historically based, current) and susceptible of augmented life. Hence the normative poise, the enfolding coherence which mark a Shakespearean text even at the limits of pathos or compactness. Violent, idiosyncratic as it may be, the statement is made from inside the transcendent generality of common speech. A classic literacy is defined by this 'housedness' in language, by the assumption that, used with requisite penetration and suppleness, available words and grammar will do the job. There is nothing in the Garden or, indeed, in himself, that Adam cannot name. The concord between poetry and the common tongue dates back at least to the formulaic elements in Homer. It is because it is so firmly grounded in traditional and communal speech, taught Milman Parry, that a Homeric simile retains its force. So far as the Western tradition goes, an underlying classicism, a pact negotiated between word and world, lasts until the second half of the nineteenth century. There it breaks down abruptly. Goethe and Victor Hugo were probably the last major poets to find that language was sufficient to their needs.[1]

Rimbaud's *lettres du voyant* were written in 1871. They do no less than proclaim a new programme for language and for literature: 'Trouver une langue;—Du reste, toute parole étant idée, le temps d'un langage universel viendra!' The first version of Mallarmé's 'Sonnet allégorique de lui-même' is dated 1868; the *Éventails* poems followed in the 1880s and 1891. With them Western literature and speech-consciousness enter a new phase. The poet no longer has or can confidently hope for tenure in a generalized authority of speech. The languages waiting for him as an individual born into history, into society, into the expressive conventions of his particular culture and milieu, are no longer a natural skin. Established language is the enemy. The poet finds it sordid with lies. Daily currency has made it stale. The ancient metaphors are inert and the numinous energies bone-dry. It is the writer's compelling task, as Mallarmé said of Poe, 'to purify the language of the tribe'. He will seek to resuscitate the magic of the word by dislocating traditional bonds of grammar and of ordered space (Mallarmé's 'Un Coup de dés jamais n'abolira le hasard'). He will endeavour to rescind or at least weaken the classic continuities of reason and syntax, of conscious direction and verbal form (Rimbaud's *Illuminations*). Because it has become

[1] The causes of this breakdown lie outside the scope of the argument. They are obviously multiple and complex. One would want to include consideration of the phenomenology of alienation as it emerges in the industrial revolution. The 'discovery' of the unconscious and subconscious strata of the individual personality may have eroded the generalized authority of speech. Conflicts between artist and middle class make the writer scornful of the prevailing idiom (this will be the theme of Mallarmé's homage to Poe). 'Entropy' effects could be important: the major European tongues, which are themselves offshoots from an Indo-European and Latin past, tire. Language bends under the sheer weight of the literature which it has produced. Where is the Italian poet to go after Dante, what untapped sources of life remain in English blank verse after Shakespeare? In 1902, Edmund Gosse will say of the Shakespearean tradition: 'It haunts us, it oppresses us, it destroys us.' But the whole question of the aetiology and timing of the language-crisis in Western culture remains extremely involved and only partly understood. I have tried to deal with certain political and linguistic aspects of the problem in *Language and Silence* (1967) and *Extraterritorial* (1971).

calcified, impermeable to new life, the public crust of language must be riven. Only then shall the subconscious and anarchic core of private man find voice. Since Homer, literature, the utterance of vision, had moved with the warp of language. After Mallarmé nearly all poetry which matters, and much of the prose that determines modernism, will move against the current of normal speech. The change is immense and we are only now beginning to grasp it.

One consequence is an entirely new, ontologically motivated, order of difficulty. The whole question of 'difficulty' is more startling, nearer the heart of a theory of language, than is ordinarily realized. What is meant by saying that a linguistic proposition, a speech-act—verse or prose, oral or written—is 'difficult'? Assuming the relevant language is known and the message plainly heard or transcribed, how can it be? Where does its 'difficulty' lie? As Mauthner's critique shows exhaustively, it is merely an evasion to affirm that the 'thought' or 'sentiment' in, behind the words is difficult. The words themselves, the linguistic fact, are the sole demonstrable locus of difficulty. Language articulates sense; it is intended to externalize and communicate meaning. In what ways can it fail to do so, and which of these ways can, possibly, be construed as intentional?[1] The topic is large and logically opaque. I want to touch here only on its historical–formal aspect, with special reference to the private language argument.

One is given to understand that there are 'difficult' passages in Shakespeare. Consider Aufidius's spasm of nettled pride in *Coriolanus* (I. x):

> My valour's poisoned
> With only suff'ring stain by him; for him
> Shall fly out of itself. Nor sleep nor sanctuary,
> Being naked, sick, nor fane nor Capitol,
> The prayers of priests nor times of sacrifice,
> Embarquements all of fury, shall lift up
> Their rotten privilege and custom 'gainst
> My hate to Marcius.

[1] Cf. G. Ryle, 'Systematically Misleading Expressions' (*Proceedings of the Aristotelian Society*, XXXII, 1932).

Or take Timon's soliloquy by the sea-shore in *Timon of Athens* (IV. iii):

> O blessed breeding sun, draw from the earth
> Rotten humidity; below thy sister's orb
> Infect the air. Twinned brothers of one womb,
> Whose procreation, residence, and birth,
> Scarce is dividant, touch them with several fortunes,
> The greater scorns the lesser. Not nature,
> To whom all sores lay siege, can bear great fortune
> But by contempt of nature.
> Raise me this beggar and deject that lord,
> The senator shall bear contempt hereditary,
> The beggar native honour.
> It is the pasture lards the wether's sides,
> The want that makes him lean.

In both passages the 'difficulty' is largely one of pace, of the sovereign haste of Shakespeare's late style. Transit and modulation fall away under the pressure of intensely compressed, close-knit dramatic advance. So far as we may reconstruct it, punctuation is at once decisive, as in the case of a musical interval, and provisional. It marks only imperfectly the underlying sequence, coil, and 'leaps of implication' in the speaker's mind. But with attention the gaps can be filled and a reasonable paraphrase offered. Complex, abbreviated as it is, the motion of meaning is beautifully consonant with that of visible grammar. A second source of 'difficulty' lies in the vocabulary: 'fane', 'embarquements all of fury', 'dividant', 'wether'. Here again there is no genuine obstacle. Our ignorance of a word is purely contingent and can be remedied by reference to a glossary. A third level of 'difficulty' arises out of Timon's usage of 'nature', 'contempt', and 'fortune'. The pertinent range of significance is not immediately transparent. One needs to experience the play as a living entity and to have some acquaintance with the ambient philosophic, emblematic idiom in order to gauge the weight of key terms. At this level, the 'difficulties' are a matter of reference. The language points to areas of knowledge, of special context and recognition which we may or may not possess. But, obviously, these can be acquired.

The theory of contagions and celestial motion invoked by Timon can be 'looked up'.

It remains the case that our own sensibilities, our capacity to hear the full tonal range of speech fall drastically short of Shakespeare's. As we re-read, we take in what we were too obtuse to grasp before. But such insufficiency is contingent. It is not a 'difficulty' logically inherent in the text.

Until the modernist crisis, by far the greater proportion of 'difficulty' in Western literature was referential. It could be resolved through recourse to the lexical and cultural context (an 'omniscient' reader or listener would have no feeling of difficulty, in the 'complete library' all answers may be found). There is an important sense, though I am not satisfied that I can delimit it, in which contextual difficulties are of the same order as those which face us in, say, a treatise on chemistry. A vocabulary, a body of rules and denotative conventions, an area of knowledge (of conceptual images) must be mastered before the message can be adequately delivered and received. But the elements of decipherment lie entirely in the public domain. There is neither indeterminacy nor intent of concealment. This is still true of *Ulysses*, which is in this cardinal respect a classic work, no less responsible to a public grid and tradition than were the works of Milton and of Goethe. The fissure opens with *Finnegans Wake*.

No 'difficulty' in Shakespeare, none in Browning's *Sordello*, reputedly the most obscure of romantic poems, is of the same nature, of the same semantic purpose and meaning, as are the difficulties in Mallarmé's

> Une dentelle s'abolit
> Dans le doute du Jeu suprême
> A n'entr'ouvrir comme un blasphème
> Qu'absence éternelle de lit.
>
> Cet unanime blanc conflit
> D'une guirlande avec la même
> Enfui contre la vitre blême
> Flotte plus qu'il n'ensevelit.
>
> Mais, chez qui du rêve se dore
> Tristement dort une mandore
> Au creux néant musicien

Telle que vers quelque fenêtre
Selon nul ventre que le sien,
Filial on aurait pu naître.

There are overlaps with the older, classic devices of difficulty: puns, exotic words, contractions of grammar. Explication and paraphrase will have some hold on the text.[1] But the energies of concealment are of an entirely new species. The poem presses against the confines of language. It works not in the mould of public speech but in spite of it (the visible logic of meaning derives mainly from the patterns of vowels and accents, in a very strong sense this is a poem *about* 'l'accent circonflexe' which, in a manner the sonnet demonstrates, embodies a conjunction, a poised tension between acute and grave). The wit and visionary exactitude of the exercise lie in the suggestion, constantly urged by Mallarmé, that alternative languages, purer, more rigorous, flourish at increasing distances from or below the surface of common discourse. The meanings of the statement are not directed outward to a context of allusion or lexical equivalence. They pivot inward and we follow as best we may. The process is, as Mallarmé, Khlebnikov, and Stefan George taught, one of calculated failure: characteristically, a modern poem is an active contemplation of the impossibilities or near-impossibilities of adequate 'coming into being'. The poetry of modernism is a matter of structured débris: from it we are made to envision, to hear the poem that might have been, the poem that will be if, when, the word is made new. This conceit of 'unfulfilment', of an adumbration which is almost archaeological— these are the spoors, the lineaments of suggestion left by the absent poem—is one of Rilke's principal themes:

Gesang, wie du ihn lehrst, ist nicht Begehr,
nicht Werbung um ein endlich noch Erreichtes....

Ineluctably, the stress of internalization, of a descent 'inward' from the norms of general syntax, leads to deepening difficulty.

---

[1] Cf. Octavio Paz's acute analysis of Mallarmé's 'Sonnet in "ix"' in *Delos*, IV, 1970.

We reach the 'darkling splinterecho' of Paul Celan, almost certainly the major European poet of the period after 1945:

> Das Gedunkelte Splitterecho,
> hirnstrom—
> hin,
>
> die Bühne über der Windung,
> auf die es zu stehn kommt,
>
> soviel Unverfenstertes dort,
> sieh nur,
>
> die Schütte
> müssiger Andacht,
> einen Kolbenschlag von
> den Gebetssilos weg,
>
> einen und keinen.

This is by no means the most gnomic of Celan's poems. But the point to be made is obvious. There had been almost no 'difficulties' of this nature in Western literature before the 1880s. The secrecy of the text stems from no esoteric knowledge, from no abstruseness of supporting philosophic argument. By themselves the words are nakedly simple. Yet they cannot be elucidated by public reference. Nor will the poem as a whole admit of a single paraphrase. It is not clear that Celan seeks 'to be understood', that our understanding has any bearing on the cause and necessity of his poem.[1] At best, the poem allows a kind of orbit or cluster of possible responses, tangential readings, and 'splintered echoes'. The meanings of Celan's verse are not ambiguous or hermetic in the sense in which these terms may be used of riddling *dizains* by Maurice Scève and a metaphysical conceit in Donne. Though they are incisive at any given moment of full response—when the echo is made whole—the meanings are also indeterminate, provisional, susceptible of constant reorganization (the crystal revolves to show a new ordering of living form). These

---

[1] For discussions of the 'difficulty' of Celan and of the hermeneutic issues which it raises cf. Alfred Kelletat, 'Accessus zu Celan's "Sprachgitter"'; Harald Weinrich, 'Kontraktionen'; Hans-Georg Gadamer, 'Wer bin ich und wer bist Du?' in Dietlind Meinecke (ed.), *Über Paul Celan* (Frankfurt, 1970).

subversions of linearity, of the logic of time and of cause so far as they are mirrored in grammar, of a significance which can, finally, be agreed upon and held steady, are far more than a poetic strategy. They embody a revolt of literature against language— comparable with, but perhaps more radical than any which has taken place in abstract art, in atonal and aleatory music. When literature seeks to break its public linguistic mould and become idiolect, when it seeks untranslatability, we have entered a new world of feeling.

In a short, uncannily dense lyric, Celan speaks of 'netting shadows written by stones'. Modern literature is driven by a need to search out this 'lithography' and *écriture d'ombres*. They lie outside the clarity and sequent stride of public speech. For the writer after Mallarmé language does violence to meaning, flattening, destroying it, as a living thing from the deeps is destroyed when drawn to the daylight and low pressures of the sea surface.

But hermeticism, as it develops from Mallarmé to Celan, is not the most drastic of moves counter to language in modern literature. Two other alternatives emerge. Paralysed by the vacuum of words, by the chasm which has opened between individual perception and the frozen generalities of speech, the writer falls silent. The tactic of silence derives from Hölderlin or, more accurately, from the myth and treatment of Hölderlin in subsequent literature (Heidegger's commentaries of 1936–44 are a representative instance). The fragmentary, often circumlocutionary tenor of Hölderlin's late poetry, the poet's personal collapse into mental apathy and muteness, could be read as exemplifying the limits of language, the necessary defeat of language by the privacy and radiance of the inexpressible. Rather silence than a betrayal of felt meaning. Or as Wittgenstein wrote of his *Tractatus*, in a letter to Ludwig Ficker dated, it is thought, late October or early November 1919: 'my work consists of two parts: the one presented here plus all that I have *not* written. And it is precisely this second part which is the important one.'

The classic statement of the paradox is Hofmannsthal's 'Letter of Lord Chandos' of 1902. The young Elizabethan nobleman has been fired by poetic and philosophic dreams, by the design of penetrating art and mythology to their hidden, Orphic centre. The whole of natural creation and of history have seemed to him an articulate cipher. But now he finds that he can scarcely speak and that the notion of writing is an absurdity. Vertigo assails him at the thought of the abyss which separates the complexity of human phenomena from the banal abstraction of words. Haunted by microscopic lucidity—he has come to experience reality as a mosaic of integral structures—Lord Chandos discovers that speech is a myopic shorthand. Looking at the most ordinary object with obsessive notice, Chandos finds himself entering into its intricate, autonomous specificity: he espouses the life-form of the wheelbarrow in the garden shed, of the water-bug paddling across the ocean of the pail. Language, as we know it, gives no access to this pure pulse of being. Hofmannsthal's rendition of this paralysing empathy is cunning:

Es ist mir dann, als geriete ich selber in Gärung, würfe Blasen auf, wallte und funkelte. Und das Ganze ist eine Art fieberisches Denken, aber Denken in einem Material, das unmittelbarer, flüssiger, glühender ist als Worte. Es sind gleichfalls Wirbel, aber solche, aber solche, die nicht wie die Wirbel der Sprache ins Bodenlose zu führen scheinen, sondern irgenwie in mich selber und in den tiefsten Schoss des Friedens.

We shall come back to this description of a matrix of thought more immediate, more fluid and intense than is that of language. Stemming from a writer who was steeped in music, the notion of introspective vortices, 'leading' to foundations deeper, more stable than those of syntax, is of great interest. Clearly, however, no earthly language can rival this vehemence of vision and repose. Chandos seeks a tongue 'of which not a single word is known to me, a tongue in which mute objects speak to me and in which I shall one day, perhaps, and in the grave, have to give account of myself before an unknown judge'. So far as the natural world goes, it is the language of total privacy or of silence.

The disasters of world war, the sober recognition that the finalities of lunacy and barbarism which occurred during 1914–18 and the Nazi holocaust could neither be adequately grasped nor described in words—what is there to *say* about Belsen?—

reinforced the temptations of silence. A good deal of what is representative in modern literature, from Kafka to Pinter, seems to work deliberately at the edge of quietness. It puts forward tentative or failed speech-moves expressive of the intimation that the larger, more worth-while statements cannot, ought not to be made (Hofmannsthal came to speak of the 'indecency of eloquence' after the lies and massacres of world war). An entry in Ionesco's diary summarizes the ironic, crippled posture of the writer when words fail him:

It is as if, through becoming involved in literature, I had used up all possible symbols without really penetrating their meaning. They no longer have any vital significance for me. Words have killed images or are concealing them. A civilization of words is a civilization distraught. Words create confusion. Words are not the word (*les mots ne sont pas la parole*). . . . The fact is that words say nothing, if I may put it that way. . . . There are no words for the deepest experience. The more I try to explain myself, the less I understand myself. Of course, not everything is unsayable in words, only the living truth.

No writer can arrive at a more desolate conclusion. Its philosophic implications, the 'negative creativity' which it has exercised in recent literature, are of great importance. An *Act Without Words*, Beckett's title, represents the logical extreme of the conflict between private meaning and public utterance. But so far as a model of language goes, silence is, palpably, a dead end.

There is a second alternative. So that 'words may again be the word' and the living truth said, a new language must be created. For meaning to find original untarnished expression, sensibility must shake off the dead hand of precedent as it is, ineradicably, entrenched in existing words and grammatical moulds. This was the programme set out by the Russian 'Kubofuturist', Alexei Kručenyx, in his *Declaration of the Word As Such* (1913): 'The worn-out, violated word "lily" is devoid of all expression. Therefore I call the lily *éuy*—and original purity is restored.' As we have seen, this notion of a language made pure and veritable again as the morning light has a theological provenance. But it springs also from a specific historical conjecture prevalent in the late eighteenth and nineteenth centuries. Considering the innocent finality of Hebrew poetry and of Greek literature, the paradox of freshness combined with ripeness of form, thinkers such as Winckelmann, Herder, Schiller, and Marx argued that Antiquity and the Greek genius in particular had been uniquely fortunate. The Homeric singer, Pindar, the Attic tragedians had been, literally, the first to find shaped expression for primary human impulses of love and hatred, of civic and religious feeling. To them metaphor and simile had been novel, perhaps bewildering suppositions. That a brave man should be like a lion or dawn wear a mantle of the colour of flame were not stale ornaments of speech but provisional, idiosyncratic mappings of reality. No Western idiom after the Psalms and Homer has found the world so new.

Presumably, the theory is spurious. Even the earliest literary texts known to us have a long history of language behind them.[1] What we notice of the formal building-blocks in even the most archaic of Ugarit poetical fragments and Biblical passages and what we understand of the formulaic composition of the *Iliad* and *Odyssey* point to a lengthy, gradual process of selection and conventionality. No techniques of anthropological or historical reconstruction will give us any insight into the conditions of consciousness and social response which may have generated the beginnings of metaphor and the origins of symbolic reference. It could be that there was a speaker of genius or manic longing who first compared the magnitude of his love to that of the sea. But we can observe nothing of that momentous occasion. Nevertheless, factitious as it is, the model of a lost *poiesis* has a powerful negative influence. It spurs on the intuition, widespread after the 1860s, that there can be no progress in letters, no embodiment of private and exploratory vision, if language itself is not made new.

---

[1] The most recent anthropological and linguistic hypotheses put at *c.*100,000 years ago the emergence of 'characteristically human speech'. The breakthrough would coincide with the last Ice Age and the manufacture of new types of elaborate stone and bone implements. Cf. Claire Russell and W. M. S. Russell, 'Language and Animal Signals', in N. Minnis (ed.), *Linguistics at Large* (London, 1971), pp. 184–7. Our earliest literatures are very late forms.

This making new can take three forms: it can be a process of dislocation, an amalgam of existing languages, or a search for self-consistent neologism. These three devices do not normally occur in isolation. What we find from the 1870s to the 1930s are numerous variants on the three modes, usually drawing on some element from each.

Nonsense poetry and prose, nonsense taxonomies, and nonsense alphabets of many sorts are an ancient genre often active just below the surface of nursery rhymes, limericks, magic spells, riddles, and mnemonic tags.[1] The art of Edward Lear and of Lewis Carroll, however, is probably cognate with the new self-consciousness about language and the logical investigations of semantic conventions which develop in the late nineteenth century. An obvious force and sophistication of psychological conjecture lie behind Lewis Carroll's disturbing assertion that nonsense languages, however esoteric, would be totally understandable to 'a perfectly balanced mind'. As Elizabeth Sewell points out, the dislocations of normal vocabulary and grammar in nonsense have a specific method. The world of nonsense poetry concentrates 'on the divisibility of its material into ones, units from which a universe can be built. This universe, however, must never be more than the sum of its parts, and must never fuse into some all-embracing whole which cannot be broken down again into the original ones. It must try to create with words a universe that consists of bits.'[2] None of these bits can be allowed to engender external references or accumulate towards a final manifold. In other words: nonsense-speech seeks to inhibit the constant polysemy and contextuality of natural language. The grammar of nonsense consists primarily of pseudo-series or alignments of discrete units which imitate and intermingle with arithmetic progressions (in Lewis Carroll these are usually familiar rows and factorizations of whole numbers).

[1] Throughout this section I am drawing on the great study by Alfred Liede, *Dichtung als Spiel: Studien zur Unsinnspoesie an den Grenzen der Sprache* (Berlin, 1963). The best analyses of the language of nonsense with special reference to English may be found in Emile Cammaerts, *The Poetry of Nonsense* (London, 1925), and Elizabeth Sewell, *The Field of Nonsense* (London, 1952).

[2] Elizabeth Sewell, *The Field of Nonsense*, pp. 53–4.

The idiom of Jabberwocky, says Miss Sewell, aims at 'making no direct connection for the mind with anything in experience'. On closer inspection, however, this does not turn out to be the case. Eric Partridge's witty gloss on the four new verbs, ten new adjectives, and eight new nouns in Jabberwocky shows how near these coinages lie to the resonance of familiar English, French, and Latin constituents.[1] It is not enough to adduce some 'half-conscious perception of verbal likeness'.[2] That perception is more often than not immediate and inescapable. Hence the fact that the feats of the Dong and of the Snark can be and have been brilliantly translated into other tongues.

> 'Twas brillig, and the slithy toves
>   Did gyre and gimble in the wabe:
> All mimsy were the borogroves,
>   And the mome raths outgrabe

haunts us by analogy. Thoroughly familiar phonetic associations and sequences from English ballads lie in instant, explicit reach. In Celan's terms, the echoes are not 'splintered' but knit in mildly unexpected ways.

From the point of view of the renewal of language, there lies the weakness of the whole undertaking. The material is too pliant, the translation too immediate. It draws too readily on counters of feeling and of imagery long-established in the sound-associations of English or any other public speech. The best of Lear, in particular, is Victorian, post-Blakeian verse delicately out of focus, as is a solid shape when the air beats about it, blurring it faintly, on a hot day.

'I said it in Hebrew—I said it in Dutch— / I said it in German and Greek—' proclaims Lewis Carroll in 'The Hunting of the Snark', 'But I wholly forgot (and it vexes me much) / That English is what you speak!' There has been poetry made of this oversight. Bilingual and multilingual poetry, i.e. a text in which lines or stanzas in different languages alternate, goes back at least to

---

[1] Cf. Eric Partridge, 'The Nonsense Words of Edward Lear and Lewis Carroll' in *Here, There and Everywhere: Essays upon Language* (London, 1950).
[2] Elizabeth Sewell, op. cit., p. 121.

the Middle Ages and to contrapunctal uses of Latin and the vulgate. The minnesinger Oswald von Wolkenstein composed a notorious *tour de force* incorporating six languages, and there are combinations of Provençal, Italian, French, Catalan, and Galician-Portuguese in troubadour verse. In his monograph on *The Poet's Tongues*, Professor Leonard Forster cites a delightful poem of the fifteenth century made up of alternating lines of English, Anglo-Norman, and Latin. A simpler, well-known example is provided by a German Christmas carol also of the fifteenth century:

> Ubi sunt gaudia?
>> Niendert mehr denn da,
>> Da die Engel singen
> Nova cantica
>> Und die Schellen klingen
> In Regis curia
> Eia wärn wir da!

The finest instance I am aware of, from both a literary and linguistic point of view, is modern. Meeting in Paris in April 1969, Octavio Paz, Jacques Roubaud, Edoardo Sanguineti, and Charles Tomlinson produced a *renga*. This is a collective poem or set of poems modelled on a Japanese form which may date back to the seventh or eighth century. But this *renga* is more than a collective act of composition: it is quadrilingual. Each poet wrote in his own tongue echoing, countering, transmuting through sound-play and masked translation the lines written immediately before him, in turn, by the three other authors. The resulting English–French–Italian–Spanish texts are of extreme imaginative density and raise issues of language and of translation to which I will return. Even one example, (II. i) will show something of the interactive energies released:

> *Aime criaient-ils aime gravité*
> *de très hautes branches tout bas pesait la*
> *Terre aime criaient-ils dans le haut*
> *(Cosí, mia sfera, cosí in me, sospesa, sogni: soffiavi, te-*
> *nera, un cielo: e in me cerco i tuoi poli, se la*

tua lingua è la mia ruota, Terra del Fuoco, Terra di Roubaud)
Naranja, poma, seno esfera al fin resuelta
en vacuidad de estupa. Tierra disuelta.
Ceres, Persephone, Eve, sphere
earth, bitter our apple, who at the last will hear
that love-cry?

A good measure of the prose in *Finnegans Wake* is polyglot. Consider the famous riverrounding sentence on page one: 'Sir Tristram, violer d'amores, fr'over the short sea, has passencore rearrived from North Armorica. . . .' Not only is there the emphatic obtrusion of French in *triste, violer, pas encore* and *Armoric* (ancient Brittany), but Italian is present in *viola d'amore* and, if Joyce is to be believed, in the tag from Vico, *ricorsi storici*, which lodges partly as an anagram, partly as a translation, in 'passencore rearrived'. Or take a characteristic example from Book II: 'in deesperation of deispiration at the diasporation of his diesparation'. In this peal a change is rung on four and, possibly, five languages: English 'despair', French *déesse*, Latin *dies* (perhaps the whole phrase *Dies irae* is inwoven), Greek *diaspora*, and Old French or Old Scottish *dais* or *deis* meaning a stately room and, later, a canopied platform for solemn show. In Joyce's 'nighttalk' banal monosyllables can knit more than one language. Thus 'seim' in 'the seim anew' near the close of 'Anna Livia Plurabelle' contains English 'same' and the river Seine in a deft welding not only of two tongues but of the dialectical poles of identity and flux.

Joyce represents a borderline case between synthesis and neologism. But even in *Finnegans Wake*, the multilingual combinations are intended towards a richer, more cunning public medium. They do not aim at creating a new language. Such invention may well be the most paradoxical, revolutionary step of which the human intellect is capable.

We have no real history of these enigmatic constructs. They turn up in the apocrypha of heresy trials, alchemy, and occultism. The inquisitor will report or the heretic profess the use of a secret, magical idiom impenetrable to the outsider. The orthodox investigators—Gottfried von Strassburg denouncing the great poet Wolfram von Eschenbach for his resort to *trobar clus*, the secret diction of the courts of love, the pursuers of Paracelsus—assign a Satanic origin to the hidden words. The initiate, such as the early prophets of the Mormon Church, on the other hand, claims angelic inspiration or a direct Pentecostal visitation by 'words robed in fire'.[1] In the nature of the case, the evidence is either puerile or lost.

The same is, on the whole, true of the new and private tongues invented by individuals for their own singular use. But it is probable that many writers, certainly since Rimbaud and Mallarmé, have at some point and, perhaps, to an intense degree, shared Stefan George's wish 'to express themselves in a language inaccessible to the profane multitude'. In George's own case, the thirst for hermeticism was compelling. He made an orphic exercise of his personal life and art so far as modern circumstance would allow. His language-artefacts include at least two poems in a *lingua romana* made up of transparent elements drawn from French, Spanish, and Italian.[2] Pursuing his search for untainted purity and originality of statement, George constructed an entirely secret speech. Reportedly, he translated Book I of the *Odyssey* into this 'neology'. If George's disciples are to be trusted,[3] the master had this translation destroyed before his death lest vulgar scholarship ransack its secrets. The tale is, very likely, a *canard*, but the theoretic design of deepening and renewing the authority of a classic text by 'translating it forward' into a language hitherto unknown and itself innocent of literature, is astute and suggestive. Two somewhat haunting verses of this alleged translation survive. They are embedded in 'Ursprünge', a poem which deals,

[1] For the theological and social problems posed by claims to direct instruction in Divine or angelic speech during, for example, the seventeenth century, cf. L. Kolakowski, *Chrétiens sans église* (Paris, 1969).

[2] For examinations of Stefan George's views on a synthesis of romance languages and classic German to renew the vitality of European poetry, cf. H. Arbogast, *Die Erneuerung der deutschen Dichtersprache in den Frühwerken Stefan Georges. Eine stilgeschichtliche Untersuchung* (Tübingen, 1961), and Gerd Michels, *Die Dante-Übertragungen Stefan Georges* (Munich, 1967).

[3] The story is told by both Ernst Morwitz and Friedrich Gundolf in their memoirs of George.

appropriately, with the persistence of antique, necromantic energies under the ascetic surface of early Christianity:

> Doch an dem flusse im schilfpalaste
> Trieb uns der wollust erhabenster schwall:
> In einem sange den keiner erfasste
> Waren wir heischer und herrscher vom All.
> Süss und befeuernd wie Attikas choros
> Ueber die hügel und inseln klang:
> CO BESOSO PASOJE PTOROS
> CO ES ON HAMA PASOJE BOAÑ.

'A song which none can grasp yet which makes us riddler and master of All.' I have seen something indistinctly like these syllables only once, on a Maltese inscription. It might be worth imagining just which two lines in *Odyssey* I George is 'translating'. The formulaic pattern is unmistakable.

By far the most interesting exercises in neologism in Western literature are those performed by Russian futurists and by Dada and the Surrealists and *lettristes* who derive from the Dada movement after 1923. This is not the place to go into the extensive, intricate literary aspects of Dada.[1] But it now seems probable that the entire modernist current, right to the present day, to minimalist art and the happening, to the 'freak-out' and aleatory music, is a footnote, often mediocre and second-hand, to Dada.

---

[1] The field has reached an extension and complexity such that there is nearly need for a 'bibliography of bibliographies'. The following are of particular use: R. Motherwell (ed.), *The Dada Painters and Poets* (New York, 1951); Willy Verkauf (ed.), *Dada. Monographie einer Bewegung* (Teufen, Switzerland, 1957); the catalogue on *Cubisme, Futurisme, Dada, Surréalisme* issued by the Librairie Nicaise in Paris in 1960; Hans Richter, *Dada—Kunst und Antikunst. Der Beitrag Dadas zur Kunst des 20. Jahrhunderts* (Cologne, 1964); Herbert S. Gershman, *A Bibliography of the Surrealist Revolution in France* (University of Michigan Press, 1969). Valuable material on Dada poetry is contained in G. E. Steinke, *The Life and Work of H. Ball, founder of Dadaism* (The Hague, 1967), and in Reinhard Döhl's authoritative monograph, *Das literarische Werk Hans Arps 1903–1930* (Stuttgart, 1967). But wherever possible, it is best to refer to the letters, documents and memoirs written by those actually involved in Dada. Hugo Ball's *Briefe 1911–1927* (Cologne, 1957), Ball's autobiographical novel *Flametti oder vom Dandysmus der Armen* first published in Berlin in 1918, and Otto Flak's *roman à clef, Nein und Ja. Roman des Jahres 1917* (Berlin, 1923), remain indispensable.

The verbal, theatrical, and artistic experiments conducted first in Zürich in 1915–17 and then extended to Cologne, Munich, Paris, Berlin, Hanover, and New York, constitute one of the few undoubted revolutions or fundamental 'cuts' in the history of the imagination. The genius of Dada lies less in what was accomplished (the very notion of 'finish' being in question) than in a purity of need and disinterestedness of creative and collaborative impulse. The slapstick and formal inventions of Hugo Ball, Hans Arp, Tristan Tzara, Richard Huelsenbeck, Max Ernst, Kurt Schwitters, Francis Picabia, and Marcel Duchamp have a zestful integrity, an ascetic logic notoriously absent from a good many of the profitable rebellions that followed.

Many instigations, themselves fascinating, lie behind the Dada language-routines as they erupt at the Cabaret Voltaire in 1915. It seems likely that Ball chose the name of the cabaret in order to relate Dada to the Café Voltaire in Paris at which Mallarmé and the Symbolists met during the late 1880s and 1890s. For it was Mallarmé's programme of linguistic purification and private expression which Ball and his associates sought to carry out.[1] The notion of automatic writing, of the generation of word groups freed from the constraints of will and public meaning, dates back at least to 1896 and Gertrude Stein's experiments at Harvard. These trials, in turn, were taken up by Italian Futurism and are echoed in Marinetti's call for *parole in libertà*. The crucial concept of 'randomness' (*Zufall*) applied to language referred itself not only to Mallarmé's *Igitur* but to the 'trance poetry' attempted by the Decadent movement of the 1890s. The techniques of *collage* in the plastic arts show a parallel development with Dada verse and had a direct influence on Arp's treatment of language. Sound-poetry and *poésie concrète* were very much in the air; witness Kandinsky's *Klänge* published in Munich in 1913. The Zürich milieu at the time was rootless and polyglot. German, French, Italian, Spanish, Rumanian, and Russian were current in and around the Dada circle. The idea of syncretism and of a personal *patois* lay close at hand.

[1] Cf. R. Döhl, op. cit., p. 36.

Yet these several strains would, I believe, have remained loose and modish but for the shock of world war. It was from that shock and its implications for the survival of human sanity that Dada derived its morality. The 'neologies' and silences of Ball, of Tristan Tzara, of Arp have affinities of despair and nihilistic logic with the exactly contemporaneous language-critiques of Karl Kraus and the early Wittgenstein. 'We were seeking an elemental art', recalls Hans Arp, 'which would cure man of the lunacy of the time.'[1] As Dada sprang up, 'madness and death were competing.... Those people not immediately involved in the hideous insanity of world war behaved as if they did not understand what was happening all around them.... Dada sought to rouse them from their piteous stupor.'[2] One of the instruments of awakening was the human voice (Giacometti running along the Limat and shouting into the houses of solid Zürich citizens). But the sounds uttered could not, as Hugo Ball urged, belong to languages corrupted to the marrow by the lies of politics and the rhetoric of slaughter. Hence the endeavour to create 'poetry without words'.

The most penetrating record of this attempt is contained in Ball's memoir, *Die Flucht aus der Zeit*, issued in 1927. The 'flight from the times' could only succeed if syntax, in which time is given binding force, could be broken. Ball's account is of extreme interest to both literature and linguistics:

I do not know whence came the inspiration for the cadence. But I began to chant my rows of vowels in the manner of a liturgical plain song and sought not only to maintain a serious mien but to enforce seriousness on myself. For a moment it seemed to me as if the pale, distraught face of a young boy had emerged from my cubist mask, the half-terrified, half-inquisitive face of a ten year-old hanging, tremulous and eager, on the lips of the priest during the requiem masses and high masses in his home parish.

Before speaking the lines, I had read out a few programmatic words. In this kind of 'sound-poetry' (*Klanggedichtung*) one relinquishes—lock, stock, and barrel—the language which journalism has polluted and made

[1] Hans Arp, *Unsern täglichen Traum. Erinnerungen, Dichtungen und Betrachtungen aus den Jahren 1914–1954* (Zürich, 1955), p. 51.

[2] Ibid., p. 20.

impossible. You withdraw into the inmost alchemy of the word. Then let the word be sacrificed as well, so as to preserve for poetry its last and holiest domain. Give up the creation of poetry at second-hand: namely the adoption of words (to say nothing of sentences) which are not immaculately new and invented for your own use.

A quotation from Ball's *Elefantenkarawane* gives some idea of the intended effect:

> jolifanto bambla ô falli bambla
> grossiga m'pfa habla horem
> égiga goramen
> higo bloika russula huju
> hollaka hollala
> blago bung
> blago bung
> bosso fataka
> ü üü ü
> schampa wulla wussa ólobo
> hej tatta gôrem
> eschige zunbada
> wulubu ssubudu uluw ssubudu . . .

What is here onomatopoeic foolery (*blago*) can, in the famous *Totenklage*, become enigmatic and strangely suffocating.

Ball's programme, like Khlebnikov's attempt to create a 'star-language', calls for absolute linguistic renovation.[1] They lead directly to the principles enunciated in the *lettrist* manifestos of the mid-1940s: 'elevation beyond the WORD', 'the use of letters to destroy words', 'the demonstration that letters have a destiny other than their incorporation in known speech'. Surrealism, *lettrisme* and 'concrete poetry' have gone forward to break the association not only between words and sense, but between semantic signs and that which can be spoken. Poetry has been produced solely for the reading eye. Take, for instance, Isidore Isou's

[1] For detailed discussions of Khlebnikov's 'star-language', see Ronald Vroon, *Velimir Xlebnikov's Shorter Poems: A Key to the Coinages* (Ann Arbor, 1983); and Raymond Cook, *Velimir Khlebnikov* (Cambridge, 1987).

Larmes de jeune fille
—poème clos—

M dngoun, m diahl Θhna îou
hsn îoun înhlianhl M pna iou
vgaîn set i ouf! saî iaf
fln plt i clouf! mglaî vaf
Λ o là îhî cnn vîi
snoubidi î pnn mîi
A gohà îhîhî gnn gî
klnbidi Δ blîglîhlî
H mami chou a sprl
scami Bgou cla ctrl
gue! el înhî nî K grîn
Khlogbidi Σ vî bîncî crîn
cncn ff vsch gln iééé . . .
gué rgn ss ouch clen dééé . . .
chaîg gna pca hi
Θ snca grd kr di.

The result is a disturbing sensation of possible events and densities (Heidegger's *Dichtung*) just below the visual surface. No signals, or very few apart from the title, are allowed to emerge and evoke a familiar tonal context. Yet there is no doubt in my mind that we are looking at a poem, and that it is, in some way, oddly moving. The wall is at the same time blank and expressive.

Whether such devices unlock 'the inmost alchemy of the word' or preserve the sanctum of poetry is a moot point. With Isou's confection we are at the limits of language and of semantic systems about which anything useful can be said. This latter restriction—the impossibility of cogent metaphrase—may not be as conclusive or condemnatory as it seems. There are other expressive modes which also defy useful comment.[1] Moreover,

[1] One of the most instructive border areas between 'normal' and 'private' linguistic practices is that of schizophrenia. As L. Binswanger and other psychiatrists have pointed out, the distinction between schizophrenic speech-patterns and certain forms of Dada, Surrealist, and *lettrist* literature lies mainly in the fact of historical and stylistic context. The inventions of the patient have no external aetiology and he cannot comment on them historically. Cf. David V. Forrest,

what occurs at the limits, in the region where linguistic signs, very much in Saussure's famous definition, shade into arbitrary 'non-significance', is not trivial. One need only recite Ball's *Klanggedichte* to a child to realize that a great deal of meaning, of presence—partly musical, partly kinetic, partly in the form of subliminal or incipient imagery—is being communicated. The problem consists in locating the point at which contingent, increasingly private signals cease emitting any coherent stimulus or any stimulus to which there could be a measure of agreed, repeatable response. Obviously, there is no general rule. In 'Larmes de jeune fille', some of the signs will convey to a mathematician possible specificities of intent, possible relevancies to the sound and theme of the poem which other readers may miss altogether. The self-defeating paradox in private language, be it the *trobar clus* of the Provençal poet or the *lettrisme* of Isou, lies in the simple fact that privacy diminishes with every unit of communication. Once utterance becomes address, let alone publications, privacy, in any strict sense, ceases.

But the 'frontier zone' need be neither one of the literary striving after personal style nor one of experimental strangeness. It is a constant of natural language. This is the overriding point. Private connotations, private habits of stress, of elision or periphrase make up a fundamental component of speech. Their weight and semantic field are essentially individual. Meaning is at all times the potential sum total of individual adaptations. There can be no definitive lexicon or logical grammar of ordinary language or even of parts of it because different human beings, even in simple cases of reference and 'naming', will always relate different associations to a given word. These differences are the

'The Patient's Sense of the Poem: Affinities and Ambiguities', in *Poetry Therapy* (Philadelphia, 1968). But as Augusto Ponzio shows in his essay, 'Ideologia della anormalità linguistica' (*Ideologie*, XV, 1971), the very definition and perception of speech-pathology are themselves a social and historical convention. Different periods, different societies draw different lines between permissible and 'private' linguistic forms. Cf. also B. Grassi, 'Un contributo allo studio della poesia schizofrenica' (*Rassegna neuropsichiatrica*, XV, 1961), David V. Forrest, 'Poiesis and the Language of Schizophrenia' (*Psychiatry*, XXVIII, 1965), and S. Piro, *Il linguaggio schizofrenico* (Milan, 1967).

life of normal speech. Few of us possess the genius needed to invent new words or to imprint on existing words, as the great poet or thinker does, a fresh value and contextual scope. We make do with the worn counters minted long since by our particular linguistic and social inheritance. But only up to a point. As personal memory ramifies, as the branches of feeling touch deeper and nearer the stem of the evolving, irreducible self, we crowd words and phrases with singular sense. Only their phonetics, if that, will remain wholly public. Below the lexical tip—a dictionary is an inventory of consensual, therefore eroded and often 'sub-significant' usages—the words we speak as individuals take on a specific gravity. Specific to the speaker alone, to the unique aggregate of association and preceding use generated by his total mental and physical history. When memory or occasion serve, we may externalize and make explicit certain levels of private content. In his self-analysis, *L'Âge d'homme*, Michel Leiris observes that the *s* in 'suicide' retains for him the precise shape and whistling sibilance of a *kris* (the serpentine dagger of the Malays). The *ui* sound stands for the hiss of flame; *cide* signifies 'acidity' and corrosive penetration. A picture of oriental immolation in a magazine had fixed and interwoven these associations in the child's mind. No dictionary could include them, no grammar formalize the process of collocation. Yet this is precisely the way in which all of us put meaning into meaning. The difference is that, more often than not, the active sources of connotation remain subconscious or outside the reach of memory.

Thus, in a general sense, though not in that of the Wittgenstein–Malcolm argument, there is 'private language' and an essential part of all natural language is private. This is why there will be in every complete speech-act a more or less prominent element of translation. All communication 'interprets' between privacies.

As we have seen in the first chapter, such mediation is at best uncertain. Though generically the same, the uncertainty is of course compounded and made visible where interpretation has to take place between languages. This dilemma of intra- and inter-linguistic 'privacies' has inspired a strong counter-current: the search for unambiguous and universal codes of communication. Because so much of natural language is private, there have been numerous attempts to strengthen the public sector.

There are several reasons why these attempts should have been particularly frequent and sustained during the seventeenth and early eighteenth centuries. The decline of Latin from general currency had created important gaps in mutual comprehension. These deepened with the rise of linguistic nationalism. At the same time, both intellectual and economic relations were developing on a scale that required ease and exactitude of communication. The constant ramifications of knowledge in the seventeenth century, moreover, led to a search for universal taxonomies, for a comprehensive, clearly articulated vocabulary and grammar for all science. Progress in mathematical analysis and logic, together with a sketchy but fascinated awareness of Chinese ideograms and of the part these played in allowing communication between different Far Eastern tongues, gave further impetus to the pursuit of a *lingua universalis* or 'Universal Character'.[1]

The concept of such an *interlingua* in fact comprises three principal aims. There was need of an international auxiliary language, such as Latin had been, to expedite and universalize scientific, political, and commercial exchanges. Secondly, a 'universal character' would generate a logistic treatment of science; ideally it would provide a simplified, rigorous set of symbols for the expression of all actual and possible knowledge. Finally—and this is the desideratum to which the educators and natural philosophers of the seventeenth century attached foremost importance—a true universal semantic would prove to be an instrument of discovery and verification.

These three goals are already implicit in Bacon's plea, in *The Advancement of Learning* (1605), for the establishment of a hierarchy of 'real characters' capable of giving precise expression to fundamental 'things and notions'. Some twenty years later

---

[1] L. Couturat and L. Léau, *Histoire de la langue universelle* (Paris, 1903), with its investigation of fifty-six artificial languages, remains the standard work. Cf. also the incisive, though selective article by Jonathan Cohen, 'On the Project of a Universal Character' (*Mind*, LXIII, 1954).

Descartes, in his correspondence with Mersenne, welcomed the project but doubted whether it could be executed before the elaboration of a complete analytic logic and 'true philosophy'. Comenius's *Janua linguarum reserata* and an English translation, *The Gate of Tongues Unlocked and Opened*, followed in 1633. Though intended mainly to facilitate and clarify the learning of Latin (along lines already pursued by the Jesuits of Salamanca), Comenius's treatise looks forward to the constitution of a universal idiom for the liberation and improvement of mankind. That ideal found expression in the famous *Orbis sensualium pictus* of 1658. The English title, *Comenius's Visible World, or a Picture and Nomenclature of All the Chief Things That Are in the World; and of Mens Employments Therein*, illustrates the encyclopaedic and taxonomic foundations of Comenius's grammar. There must be an unambiguous, universal concordance between words and things. *Pansophia* can be achieved only by means of *panglottia*. The imperfections and controversies which beset human knowledge and emotions are a direct consequence of the disorder within and between tongues. Beyond Latin lies the promise of a perfect philosophical language in which nothing false can be expressed and whose syntax will, necessarily, induce new knowledge.[1]

By the 1650s and early 1660s such hopes were being widely canvassed. Raymond Lully's *Ars Magna* of 1305-8, revised and developed by Athanasius Kircher, offered a remote but prestigious model of the use of symbolic notations and combinatory diagrams to classify and interrelate all intellectual disciplines. Here were the first hints towards a universal algebra able to initiate and systematize analytic processes in the human mind. Sir Thomas Urquhart's *Logopandecteision* of 1653 is a characteristic example of the universalist scheme. Urquhart was a notorious joker and one need not take very seriously the claim that a full-scale glossary of his new language had been destroyed at the Battle of Worcester in 1650. The bare outlines, as set out in his prospectus, are intriguing enough. The object is 'to appropriate the words of the

[1] The best account of Comenius's linguistic work is contained in H. Geissler, *Comenius und die Sprache* (Heidelberg, 1959). I am indebted also to a private communication from Prof. H. Aarsleff of Princeton University.

universal language with the things of the universe'. Only a 'Grammatical Arithmetician' (the term is itself prophetic) will bring about this indispensable accord. Urquhart's *interlingua* contains eleven genders and ten cases besides the nominative. Yet the entire edifice is built on 'but two hundred and fifty prime radices upon which all the rest are branches'. Its alphabet counts ten vowels, which also serve as digits, and twenty-five consonants; together these articulate all sounds of which the vocal organs of man are capable. This alphabet is a powerful means of arithmetical logic: 'What rational Logarithms do by writing, this language doth by heart; and by adding of letters, shall multiply numbers; which is a most exquisite secret.' The number of syllables in a word, moreover, is proportionate to the number of its significations. Urquhart kept his 'exquisite secret' but the anticipation of his claim on modern symbolic logic and computer languages is striking. As is Urquhart's assurance that the phonetic and syntactic rules of his 'universal character' have inherent mnemonic advantages. A child, he says, will acquire fluency in the new speech with little effort because the structure of the idiom in fact reproduces and reenacts the natural articulations of thought.

The 1660s produce a spate of linguistic blueprints. Some, such as J. J. Becher's *Character, pro notitia linguarum universali* (1661), and Kircher's own *Polygraphia Nova et Universalis* of 1663 are, as Cohen points out, no more than 'systems for ciphering a limited group of languages on a unitary pattern'. They are merely an *interglossa* and auxiliary shorthand for the sciences. But other schemes were of fundamental interest. Dalgarno's *Ars Signorum, vulgo Character Universalis et Lingua Philosophica* (1661) did not fulfil the promise of its title, but spurred John Wilkins to produce his *Essay towards a Real Character and a Philosophical Language* seven years later. Bishop Wilkins was a man of genius and his project foreshadows many elements in modern logistic theory.

Although Leibniz's *de Arte Combinatoria* dates back at least to 1666, and although Leibniz's early linguistic thought is probably more indebted to the German Pietists and to J. H. Bisterfeld than it is to any other source, Wilkins's influence on Leibniz's life-long search for a universal combinatorial grammar of communication and discovery is unmistakable.[1] That search, which is still discernible in the *Collectanea etymologica* of 1717, bore obvious fruit in Leibniz's epistemology and mathematics. It added to European awareness of Chinese. But it did not achieve that *mathesis* of unambiguous denotation and discovery which the seventeenth century and Leibniz himself had intended. 'It was clearly a mistake to think that the same language could serve adequately both as an unspecialized international auxiliary and also as a scientific terminology.'[2]

Modern universalists have sought to avoid this mistake. The artificial languages proposed since J.-M. Schleyer's Volapük (1879) and the Esperanto of L. L. Zamenhof (1887) are auxiliary *interlinguae* calculated to expedite economic and social intercourse and meant to counteract the threats of chauvinism or isolation in a tensely nationalist world. No less than their ancestor, the *Langue nouvelle* outlined by the *Encyclopédistes* in the 1760s, these synthetic constructs take their components from existing major tongues. This is entirely the case for Esperanto, Ido, Occidental, Novial, and a dozen others. Volapük and the *Latine sine flexione* on which the eminent Italian mathematician and mathematical logician Peano worked from 1903 to 1930, are more ambitious. Both embody elements of logistic formalization of the kind the seventeenth century strove for, and Peano's initial project refers explicitly to Wilkins and to Leibniz. Nevertheless, as Peano makes clear in his *Notitias super lingua internationale* (1906), the main purpose of his scheme is not analytic but social and psychological. Swift, agreed understanding between neighbouring nation states and ideologically divided communities is necessary to the survival

---

[1] L. Couturat's treatment of Leibnizian linguistics in *La Logique de Leibniz* (Paris, 1901) remains authoritative. Cf. also Hans Werner Arndt, 'Die Entwicklungsstufen von Leibniz's Begriff einer Lingua Universalis' in H.-G. Gadamer (ed.), *Das Problem der Sprache* (Heidelberg, 1966). A useful survey of the topic as Leibniz found it is contained in Paolo Rossi, *Clavis Universalis. Arti mnemoniche e logica combinatoria da Lullo a Leibniz* (Milan and Naples, 1960).

[2] J. Cohen, op. cit., p. 61.

of man.[1] Few of these confections have shown much vitality. Only Esperanto continues to lead a somewhat Utopian, vestigial existence.

The analytic current, on the other hand, has been among the most influential in modern philosophy. The attempt, initiated in the seventeenth century, to formalize mental operations and to systematize the rules of definition, inference, and proof, has been extensively pursued in modern symbolic logic, in the study of the foundation of mathematics and in such semantic theories of truth as those of Tarski and of Carnap. The connection between the *characteristica universalis* of Leibniz and the early logical investigations of Russell and of Russell and Whitehead has often been stressed. The attempt to develop a formally rigorous 'science of sciences', such as Wilkins envisaged, is of central importance to the later philosophy of Carnap. In computer languages traditional concepts of *mathesis*, of symbolic representation and of universality are implicit though in a special framework.[2]

Neither the 'interlingual' nor the logical–analytic approach has done very much to deepen our understanding or modify our uses of natural language. This is not to say that linguistic philosophy and formal logic from Frege and Wittgenstein to Prior and Quine have failed to produce results of extraordinary subtlety. But the focus, the purpose of relevant insight need careful definition. As we have seen, 'purifications' and idealizations of extreme stringency are being applied. The actual relations between the language-model investigated by the analytic logician and language 'at large' are themselves being tested. But the trial is often tacit or, as it were, 'left for later'. The consequence may be a kind of

[1] For a balanced discussion of modern artificial languages cf. Chapter VI of J. R. Firth, *The Tongues of Men* (London, 1937).

[2] There are numerous treatments of the logical and linguistic aspects of computer languages. Several important papers are gathered in T. B. Steel (ed.), *Formal Languages and Description Languages for Computer Programming* (Amsterdam, 1961), and in M. Minsky (ed.), *Semantic Information Processing* (M.I.T. Press, 1968). Cf. also B. Higman, *A Comparative Study of Programming Languages* (London and New York, 1967). A more general introduction to the whole field of modern linguistic logic may be found in L. Linsky (ed.), *Semantics and the Philosophy of Language* (University of Illinois Press, 1952).

depth which is isolated from the contaminations of real context. Authentic as it is, the penetration by the logician will breed its own 'meta-context' and autonomous problems. The difficulties encountered are genuine, but their reality is of a special, self-sustaining nature. The slippery, ambiguous, altering, subconscious or traditional contextual reflexes of spoken language, the centres of meaning which Ogden and Richards termed 'emotive' and which Empson treats under the rubric of 'value' and 'feel', fall outside the tight but exiguous mesh of logic. They belong to the pragmatic.

But it is its great untidiness that makes human speech innovative and expressive of personal intent. It is the anomaly, as it feeds back into the general history of usage, the ambiguity, as it enriches and complicates the general standard of definition, which give coherence to the system. A coherence, if such a description is allowed, 'in constant motion'. The vital constancy of that motion accounts for both the epistemological and psychological failure of the project of a 'universal character'.

Roughly stated, the epistemological obstacle is this: there could only be a 'real' and 'universal character' if the relation between words and the world was one of complete inclusion and unambiguous correspondence. To construct a formal universal syntax we would need an agreed 'world-catalogue' or inventory of all fundamental particulars, and we would have had to establish the essential, uniquely defining connection between the symbol and the thing symbolized. In other words, a *characteristica universalis* demands not only a correct classification of 'all primary units in the world' but requires proof that all such 'simples' have indeed been identified and listed. Once more, the image is that of Adam naming all that comes before him in a closed garden of perfect synonymy. As both Leibniz and Wittgenstein (after the *Tractatus*) found, the thing cannot be done; for if we had such a catalogue and classification to begin with, the 'universal character' would already exist and there would be no need to construct a new and logically conclusive idiom.

The most obvious difficulties, however, arise from the psychology of meaning. A logical grammar such as the universalists aim for has to ignore all differences between the way in which diverse languages, cultures, and individuals use words. In fact, 'meaning' is scarcely ever neutral or reducible to a static, unambiguous setting. Within any given language or period of history the rules of grammar are nothing more than very approximative, unstable summaries of regularities or 'majority' habits derived from actual speech. This truth is not invalidated by the possibility that the boundaries within which such regularities can change may be determined by deep-seated and perhaps universal constraints.

Natural language is local, mobile, and pluralistic in relation to even the simplest acts of reference. Without this 'multivalence' there would be no history of feeling, no individuation of perception and response. It is because the correspondence between words and 'things' is, in the logician's sense of the term, 'weak' that language is strong. Reverse these concepts, as artificial universal languages do, and the absence of any natural, complex strength in the ensuing modes of communication is obvious. What Esperanto or Novial does is to translate 'from the top'. Only the more generalized, inert aspects of significance survive. The effect is that of a photographic 'still' taken by a tourist on his first visit to a country whose actual forms of life, whose 'context of situation' (Firth's term for the 'dynamic and creative patterned processes of situation in which language behaviour is dominant') he does not grasp. There are conditions of 'translation' in which an Esperanto is of undisputed efficiency: but these are minimalist conditions. They abstract those imprecise and redundant energies which make possible the communication—always approximate—of what we as individuals, as participants in a particular milieu and family of remembrance are trying to say.

This is not to diminish the importance of the public elements of language, of the drive towards clarity and consensus. These also are deeply-rooted constants in the evolution of speech and, as I want to indicate in a moment, their role has, if anything, become greater in the course of history. The entire business of translation, the current search for universals in generative grammars, express a fundamental reaction against the privacies of individual usage and the disorder of Babel. If a substantial part of all utterances were not public or, more precisely, could not be treated as if they were, chaos and autism would follow.

Again we are dealing with an indispensable duality, with a dialectical relation between 'congruent opposites'. The tensions between private and public meaning are an essential feature of all discourse. The hermetic poem lies at one extreme, the S.O.S. or the road-sign at the other. Between them occur the mixed, often contradictory and in some degree indeterminate usages of normal speech. Vital acts of speech are those which seek to make a fresh and 'private' content more publicly available without weakening the uniqueness, the felt edge of individual intent. That endeavour is inherently dualistic and paradoxical. But if we listen closely, there will not be a poem, not a live statement from which this 'contradictory coherence' is absent.

4

Lastly, I want to consider a fourth duality or 'contrastive set', that of truth and falsity. The relations of natural language to the possible statement of truth and/or falsity seem to be fundamental to the evolution of human speech as we know it, and they alone, I believe, can direct us towards an understanding of the multiplicity of tongues. To speak of 'language and truth' or of 'language and falsity' is, obviously enough, to speak of the relations between language and the world. It is to inquire into the conditions of meaning and of reference and into the conditions which make reference meaningful to the individual and the interlocutor. Again translation—the transfer from one designative coherence to another—is the representative, because particularly visible, case. In another sense, questions about language and truth imply the whole of epistemology and, perhaps, of philosophy. In numerous philosophic systems, such as Platonism, Cartesianism or the critiques of Hume and of Kant, the topic of the status and representation of truth is the central issue. It would be instructive, though also reductive, to divide philosophies into those for which truth and falsity are elemental substances or properties, and those for which falsity is, as G. E. Moore held, only untruth, a privation or negation of truth.

Yet though the problem of the nature of truth and many of the metaphysical and logical moves made when the topic is discussed are as ancient as systematic philosophy itself, it can be said that the theme enters a new phase at the close of the nineteenth century. And it is a phase intimately related to the study of language. The modern style of inquiry stems from several sources. It is partly a reaction, ethical in its severity, against the seemingly solipsistic, unworriedly eloquent metaphysics which had dominated European philosophic argument from Schelling to Hegel and Nietzsche. The new direction also derives from a re-examination of the foundations of mathematics. To put it in a crassly schematic way: the turn of century witnessed a change from an 'outward', hypostatized concept of truth—as an absolute accessible to intuition, to will, to the teleological spirit of history—to a view of truth as a property of logical form and of language. This change embodied the hope that a strict formalization of mathematical and logical procedures would reveal itself as a transcription, idealized no doubt but none the less reproductive, of the mechanics of the mind. This is why a somewhat naïve mentalism continues to turn up in some of the most neutral, anti-metaphysical, or anti-psychological of modern logistics and analytics.

The history of 'the linguistic turn' is itself a broad subject. Even if we consider only the argument on 'truth', we can make out at least four main stages. There is the early work of Moore and Russell, then of Russell and Whitehead, with its explicit background in the logistics of Boole, Peano, and Frege. There are the attempts to establish semantic definitions of 'truth' made by Tarski, by Carnap, and by the Logical Positivists during the 1930s, attempts carried forward, in a highly personal vein, by Wittgenstein. A third focus is provided by 'Oxford philosophy' and, most notably, by the 1950 debate on 'truth' between Austin and P. F. Strawson and the extensive literature to which this exchange gave rise. There is a current phase strongly coloured by structural linguistics and of which Jerrold J. Katz's 'The Philosophical Relevance of Linguistic Theory' (1965) is a representative statement.[1] But even these very general partitions blur the facts. The example of Frege, of Russell, and of Wittgenstein cuts across different postulates and methodologies. Quine does not fit readily into any chronological rubric but his work on reference and on imputations of existence is among the most influential in the whole modern movement. Key figures—Wittgenstein is the salient instance—changed their positions in the course of work. Biographically and in point of substance, moreover, individuals and schools (more accurately, 'collaborative styles') overlap. There is something like an 'Austin mannerism' in much of recent analytic and linguistic philosophy even where Austin's conclusions may be challenged or not directly apposite.

It is also legitimate to think of the development of modern views on truth in terms of the difference between a formal model of language and a focus on natural language. This, in substance, is the distinction I have been emphasizing in this study. In his useful historical survey, Richard Rorty sees the essential differentiation as one between Ideal Language and Ordinary Language philosophers.[2] Very roughly put, the Ideal Language philosopher holds that genuine philosophical problems are muddles caused by the fact that 'historico-grammatical syntax' (the ways in which we actually speak) does not dovetail with 'logical syntax'. Such a syntax 'underlies' natural language; it can be reconstructed and made visible in a formal paradigm. This is the view of the early Russell, of Wittgenstein's *Tractatus*, of Carnap and of Ayer. It is the philosopher's job to look at philosophical problems in the framework of a rigorously constructed metalanguage in which all philosophic propositions will turn out to be statements about

[1] The key articles are reprinted in a number of anthologies. The following are of particular use: Max Black (ed.), *Philosophical Analysis* (New Jersey, 1950); A. J. Ayer *et al.*, *The Revolution in Philosophy* (London, 1956); R. R. Ammerman (ed.), *Classics of Analytic Philosophy* (New York, 1965). In the following discussion I have relied mainly on the two series of *Logic and Language* ed. by A. N. Flew (Oxford, 1951 and 1953), and on Richard Rorty's collection, *The Linguistic Turn* (University of Chicago Press, 1967).

[2] Cf. Rorty's Preface, op. cit.

syntax and interpretation. Problems that do *not* turn out to be syntactic and relational in this unambiguous sense are pseudo-dilemmas or archaic bugbears. They spring from the regrettable fact that normal speech and traditional ontology have the habit of muddling words and using what Ryle calls 'systematically mis-leading expressions' ('God exists' can be shown to be only a 'so-called existential statement' in which 'existent' is only a bogus predicate and that of which, in grammar, it is asserted is only a bogus subject).

The Ordinary Language approach is formulated in Strawson's critique of Carnap and his followers. Agreed that philosophical dilemmas have their source 'in the elusive, deceptive modes of functioning of unformalized linguistic expressions'. But how can we construct an ideal language without first describing accurately and exhaustively the procedures and confusions of ordinary discourse? If such description is possible, it may by itself resolve the perplexities and opaqueness thrown up by natural speech. A meta-linguistic model *may* be of some help—it externalizes, it 'profiles' the area of confusion—but it cannot do the job of normative elucidation. Similarly Austin held that there was not much point in reforming and tightening common usage until we know far more exactly what that usage is. Ordinary language may not be 'the last word', but it offers an immense terrain for us to get on with.

These contrasting approaches and the numerous 'mixed', intermediary strategies deployed by linguistic philosophers lead to different images of the shape and future of philosophy. It may be that all serious philosophy will be, in Wittgenstein's phrase, a kind of 'speech therapy', attending to, mending the infirmities of ordinary language and the spurious but vehement conflicts they provoke. Linguistic philosophy might, however, lead to a Copernican revolution of its own, substituting for the Kantian model of the *a priori* of cognition a new understanding of the internalized constraints, of the abstract orderings which make language itself possible. It would thus fulfil the long dream of a universal philosophic grammar. Conceivably empirical linguistics will develop to the point at which it can provide non-banal

formulations of the nature of truth and of meaning (this is clearly
implied in the aims of Chomsky and of 'deep structuralists').
Finally, as Rorty puts it, linguistic analysis may do so thorough a
job of exorcism that we might 'come to see philosophy as a
cultural disease which has been cured'.

Two points emerge. Linguistic philosophy comprises a substantial part of twentieth-century philosophy, particularly in
England and the United States. It has put the investigation of
formal or empirical grammars at the centre of logic, of epistemology, and of philosophic psychology. But it has viewed language
in a special way (Rorty suggests the covering term 'methodological
nominalism'). In so doing it has not only edged several branches of
traditional philosophy away from professional respectability, i.e.
aesthetics, theology, much of political philosophy. It has also distinguished itself sharply from other ways of conceiving and feeling
language. This distinction, with its scarcely concealed inference
of vacuity in the other camp, applies to Husserl, to Heidegger, to
Sartre, to Ernst Bloch. Consequently, there is historical and
psychological justification for setting 'linguistic philosophy' apart
from 'philosophy of language' (*Sprachphilosophie*). This separation
is damaging. It is doubtful whether Austin's well-known prognostication can be realized so long as the gap remains: 'Is it not
possible that the next century may see the birth, through the
joint labours of philosophers, grammarians and numerous other
students of language, of a true and comprehensive *science of
language*?'

'Truth' makes up a ubiquitous but also distinct topic in modern
linguistic analysis.[1] Several schemes have been put forward. What
we find in Moore, in Russell's early teachings on logical atomism
and propositions, and in the *Tractatus*, is a correspondence theory.
Language is in some way a one-to-one picture of the world,

---

[1] I have based my discussion on George Pitcher (ed.), *Truth* (New Jersey, 1964),
and Alan R. White, *Truth* (London, 1970). I have made use also of the following:
P. F. Strawson, 'On Referring' (*Mind*, LIX, 1950); Paul Ziff, *Semantic Analysis*
(Cornell University Press, 1960); A. J. Ayer, *Foundations of Empirical Knowledge*
(London, 1963); Rita Nolan, 'Truth and Sentences' (*Mind*, LXXVIII, 1969);
Ronald Jager, 'Truth and Assertion' (*Mind*, LXXIX, 1970); R. J. and Susan
Haack, 'Token-Sentences, Translation and Truth-Value' (*Mind*, LXXIX, 1970).

propositions 'are like' the things they are about. F. H. Bradley's *Essays on Truth and Reality* of 1914, together with the analyses of propositions made by Logical Positivists such as Schlick and C. G. Hempel, lead to what has been called a 'coherence theory' of truth. The crux here is internal consistency and a systematically coded relation between perception and object. (Logicians tell us that all coherence theories are vulnerable to Gödel's famous proof that no system of a certain order of complexity can demonstrate its own consistency without 'importing' new, external inferences, without recourse to additional principles whose own consistency is open to question.)

As its name connotes, the 'semantic theory' of truth addresses itself most immediately to the nature of the relations between grammar and reality. This approach originates mainly in Tarski's 'Der Wahrheitsbegriff in den formalisierten Sprachen', first published in Polish in 1933, and in Carnap's *Logische Syntax der Sprache* issued in Vienna in 1934 and translated into English three years later. Carnap's *Introduction to Semantics* (1942) gave wide currency to the semantic view.[1] Semantic definitions of truth are formulated in connection with ideal, artificial languages which are, in fact, generalized deductive systems of varying degrees of formal complexity. 'True' is a predicate which may legitimately occur in certain special kinds of sentences (called 'object-sentences' or 'token-sentences'). These are generated according to rigorous rules and formal constraints in the metalanguage. Usually, this metalanguage is transcribed in one or another convention of symbolic logic, and here there are often explicit links with Russell and Whitehead's *Principia Mathematica* and, ultimately, with Leibniz. Tarski seems to define 'truth' as the precise acceptability or admissibility of a certain statement within a definite formal language in which a two-valued (true/false) and not a many-

---

[1] A thorough introduction to the work of Tarski and Carnap may be found in W. Stegmüller, *Das Wahrheitsproblem und die Idee der Semantik: Eine Einführung in die Theorien von A. Tarski und R. Carnap* (Vienna, 1957). The following critiques are of particular use: Max Black, 'The Semantic Definition of Truth' (*Analysis*, VIII, 1948), and A. Pap, 'Propositions, Sentences, and the Semantic Definition of Truth' (*Theorie*, XX, 1954).

valued logic obtains. This notion and its treatment are technically abstruse but not I think, irrelevant to an understanding of questions of polysemy and ambiguity as they occur in translation. Carnap's strategy is less clear but also more suggestive as there runs through it a constant inference of possible extension from constructed languages to natural language and to the classification of the actual sciences.

Severe critiques have been made of each of these theories. In turn, these critiques lead to new approaches. Drawing on F. P. Ramsey's device of 'logical superfluity' ('true that p' is only another, redundant way of saying 'it is a fact that p'), Strawson has rejected the idea that propositions are 'like' the world. His approach deals with many sentences that are meaningful and intelligible without saying anything either true or false. There are, Strawson insists, numerous grammatical predicates which are satisfactory in themselves but have no application now or here. The relation being explored is that between 'all John's children are asleep' and the possibility, of which the speaker may be ignorant, that John has no children.

Other views on 'truth' have continued in the field. There is a pragmatic tradition associated with the doctrines of Peirce, William James, and F. C. S. Schiller. Its common-sense flavour is illustrated by the title of Schiller's best-known paper: 'Must Philosophers Disagree?' published in the *Proceedings of the Aristotelian Society* for 1933. Elements of this approach and a genius for disconcerting instances characterize the logic of Quine. There is the linguistic empiricism or materialism of the Marxists with its stress on 'what is out there'.[1] But no less than in other branches of recent philosophic investigation, it is the analytic positions which have been the most influential and actively pursued. The matter of truth has been one of the relations between 'words and words' more often than between 'words and things'.

This mode of discussion has been going forward for over half

---

[1] Cf. I. S. Narski, 'On the Conception of Truth' (*Mind*, LXXIV, 1965) with its references to Lenin and sanguine conclusion that '*truth is a progress*'.

a century. The layman, so far as he is able to follow even the general outlines of an exceedingly cloistered, frequently meta-mathematical debate, will be struck by several aspects. The literature contains a wealth of closely observed grammar. Whatever the future status of Anglo-American linguistic philosophy *qua* philosophy, the techniques of scrupulous 'listening to language' on which it is based and the models of speech-behaviour which it has elaborated, will stand. The examples of unclear meaning, of logical and substantive opaqueness which Moore, Wittgenstein, Austin select or contrive from natural language make for a wry poetry. Wittgenstein belongs to the history of hermetic and aphoristic practices in German literature as do Hölderlin and Lichtenberg. The finesse of Austin's acoustical sense for speech, his ability to spot the almost surrealistic turns of unguarded oddity in common diction were such that he would have been, had he so purposed, an acute philologist or literary critic. His antennae for the mask of words were like Empson's. Austin on 'pretending to be a hyena' in the make-believe, party-forfeit way, 'a very recent usage, perhaps no older than Lewis Carroll', is, as the reference plainly indicates, a bit of practical poetics. Time and again, the analytic study of 'truth' has provided ancillary insights into language *in extremis*, into the conditions of expressive means when these are at the limits of syntax. As a result of this whole philosophic movement, our discriminations between 'sentences', 'statements', 'propositions', 'references', 'postulates', 'predications', 'assents', 'affirmations', and many other crucial counters in the description of speech-acts ought to be more exact and substantial than before.

Simultaneously, however, the argument about 'truth' shows some of the radical constrictions in the entire 'linguistic analytic' mode. It has proceeded in disregard of experimental psychology and of what may be termed, in the general sense, information theory. Though it is, explicitly, a study of the conventions or necessities of relation between language and 'what is', linguistic analysis has taken little account of the progress made in our understanding of perception and cognition. We find no awareness that the problem of 'truth' and predication is to a large degree bound up with the procedures of the human perceptual systems. These are themselves intricate combinations of neurophysiological, ecological, and cultural–social factors.[1] The lack of awareness is the more telling as there are many points of mutual interest. Wittgenstein's dissatisfactions with the status of 'pain' and other internalized sensations correlate closely with questions about pain and other somatic data raised by psychologists and physiologists. A theory of language and of truth which does not keep in view the distinction between the relation of a perceptual stimulus to its causal source and the relation of a symbol to its referent—the latter depends on a linguistic community and social code—is in danger of being one-sided and artificial. Just as in the case of the models of deep structure proposed by generative grammars, there is in the analytic diagnosis of 'truth' a danger of confusion, of overlap between a purely idealized schema and reality. Max Black's objection to Tarski's semantic theory has wider bearing:

The 'open' character of a natural language, as shown in the fluctuating composition of its vocabulary, defeats the attempt to apply a definition of truth based upon enumeration of simple instances. The attempt is as hopeless as would be that of setting out the notion of 'name' by listing all the names that have ever been used.[2]

This criticism can be extended. Unquestionably the analytic rejection of any naïve theory of correspondence between word and object has been of philosophic use. Nevertheless, there is some psychological spuriousness about the idea that any better working model can be offered or, more cogently, that any philosophically more satisfactory model can be acted on. Michael Dummett puts the matter frankly:

[1] Cf. Jerome S. Bruner, *Toward a Theory of Instruction* (Harvard University Press, 1966), and James J. Gibson's pioneer work *The Senses Considered as Perceptual Systems* (New York, 1966), especially pp. 91–6. The possibility that sensory perceptions are 'culture-bound' and require 'translation' is examined in W. Hudson, 'The Study of the Problem of Pictorial Perception among Unacculturated Groups' (*International Journal of Psychology*, II, 1967), and Jan B. Deregowski, 'Responses Mediating Pictorial Recognition' (*Journal of Social Psychology*, LXXXIV, 1971).

[2] Max Black, 'The Semantic Definition of Truth', p. 58.

Although we no longer accept the correspondence theory, we remain realists *au fond*; we retain in our thinking a fundamentally realist conception of truth. Realism consists in the belief that for any statement there must be something in virtue of which either it or its negation is true: it is only on the basis of this belief that we can justify the idea that truth and falsity play any essential role in the notion of the meaning of a statement, that the general form of an explanation of meaning is a statement of the truth-conditions.[1]

There is no escape from this 'duplicity' so long as analyses of assertions, statements, propositions or belief in regard to 'truth' are divorced from any interest in the psychology and sociology of cognition. Only such interest will support Strawson's legitimate demand that the question to be asked is: 'How do we use the word "true"?'

But the restrictiveness of the analytic linguistic approach may lie even deeper. 'Any satisfactory theory of truth', declared Austin, using a term of which he was in other contexts chary (what *is* a 'theory of truth'?), 'must be able to cope equally with falsity.'[2] None of the accounts of truth given by modern linguistic philosophy seems to me to fulfil this requirement. Yet I believe that the question of the nature and history of falsity is of crucial importance to an understanding of language and of culture. Falsity is not, except in the most formal or internally systematic sense, a mere miscorrespondence with a fact. It is itself an active, creative agent. The human capacity to utter falsehood, to lie, to negate what is the case, stands at the heart of speech and of the reciprocities between words and world. It may be that 'truth' is the more limited, the more special of the two conditions. We are a mammal who can bear false witness. How has this potentiality arisen, what adaptive needs does it serve?

The set of intentional and linguistic procedures which lies between the theoretic absolutes of 'truth' and 'falsity' is so multiple and finely shaded that no logic, no psychology, and no semantics have given even a provisional account of it. There have

---

[1] Michael Dummett, 'Truth', reprinted in G. Pitcher (ed.), op. cit., pp. 106-7.
[2] J. L. Austin, 'Truth', reprinted in Pitcher, pp. 27-8.

been many analytic and behavioural probes into nodal points, into such formally and culturally salient areas as induction, argument by hypothesis, philosophic doubt. There have been grammatical investigations of optatives and subjunctives. The development of modal and many-valued logics has extended the treatment of propositions beyond categories of exclusive truth or falsity. There is a considerable technical literature on conditionality.[1] The logical status of hypotheticals has been often debated.[2] Some logicians see no particular problem in counter-factual assertions—'Napoleon did not die on St Helena'—and insist that they must not be confused with subjunctive conditionals. The real crux lies in the verification of all and any conditional statements.[3] Others incline to the view that subjunctive conditional sentences—'if Napoleon had won at Waterloo he would have continued as Emperor'—do pose a special and non-trivial question.[4] How may we best handle a category of statements which are assuredly intelligible but which cannot be said in principle to be either verifiable or falsifiable?

Yet, on the whole, there is hardly another branch of logical, philosophical inquiry at once so prolix and sterile. It may be that the logician is out of sorts from the start. Hume's admonition in the first Book of the *Treatise* inhibits him: all *hypothetical* arguments or 'reasonings upon a supposition' are radically infirmed by the absence of any 'belief of real existence'. Thus they are 'chimerical and without foundation'. The entire terrain is a muddle. 'Both *if* and *can*', writes Austin in his well-known paper on 'Ifs and Cans' (1956), are 'protean words, perplexing both grammatically and philosophically.' They 'engender confusion'.

---

[1] I have found the following of particular use: Stuart Hampshire, 'Subjunctive Conditionals' (*Analysis*, IX, 1948); M. R. Ayers, 'Counterfactuals and Subjunctive Conditionals' (*Mind*, LXXIV, 1965); K. Lehrer, 'Cans Without Ifs' (*Analysis*, XXIX, 1969); Bernard Mayo, 'A New Approach to Conditionals' (*Analysis*, XXX, 1970).

[2] Cf. D. Pears, 'Hypotheticals' (*Analysis*, X, 1950); Charles Hartshorne, 'The Meaning of "Is Going to Be"' (*Mind*, LXXIV, 1965); A. N. Prior, 'The Possibly-True and the Possible' (*Mind*, LXXVIII, 1969).

[3] This is the view taken by M. R. Ayers in 'Counterfactuals and Subjunctive Conditionals'.

[4] This is the position adopted by Stuart Hampshire in his 1948 article.

But looked at from a different view, it may be felt that they 'engender life', that fundamental energies of adjustment between language and human need lie precisely in the logically recalcitrant zone. Hypotheticals, 'imaginaries', conditionals, the syntax of counter-factuality and contingency may well be the generative centres of human speech. They carry the stress of the 'organic' in the notion of 'organization'. Unavoidably, the relation between these two terms is conceptually obscure: how do we cope with a 'protean stability', with a systematic open-endedness? Once again there is need of astonishment, of a susceptibility—the poet has it, the logician ought to have it—to the thought that things might have been otherwise, that a perfect clarity would have narrowed the field. It *is* remarkable, to put it soberly, that we are able to conceptualize and embody in language the limitless category of 'the impossible', that neither flying azure pigs nor furious green dreams pose any irreducible conceptual or semantic barriers. 'Impossibility' does modulate towards a blur: we are able to say, but not responsibly to conceive of, the proposition that '*a* is not *a*'. But one wants to know so much more, just at this apparently straightforward zero point where the laws of the system are violated, about what the distance of irresponsibility, of factitiousness is between the absent or insignificant concept and the perfectly coherent verbal form. No safety-wire in the publicly available grammar stops us from talking nonsense correctly. Why should this be? What defect or, on the contrary, what licence for reshaping, for expansion at the crowding edges is instrumental in this lack of constraints?

Counter-factual conditionals—'if Napoleon was now in the field, the business in Vietnam would take a different turn'—do more than occasion philosophical and grammatical perplexity. No less than future tenses to which they are, one feels, related, and with which they ought probably to be classed in the larger set of 'suppositionals' or 'alternates', these 'if' propositions are fundamental to the dynamics of human feeling. They are the elbow room of the mind, its literal *Lebensraum*. The difference between an artificial language such as FORTRAN, programmed by information and computer theorists, and natural language is one of vital ambiguities, of chimeric potentiality and undecidability. Given a vocabulary and a set of procedural rules (both subject to change), given the limitations of comprehensibility and certain performance boundaries (no endless sentences), we can *say anything*. This latent totality is awesome and should be felt as such. It well-nigh precludes applied logic—the parameters are too numerous, the possibilities of acceptable order too unstable and local. ('Es ist menschenunmöglich', not 'humanly possible', says Wittgenstein in the *Tractatus*, 4.002, to derive a language-logic, 'Sprachlogik', from natural language.) But this instability is perhaps the most telling of the evolutionary adaptations, of the reachings outward, that determine our humanity.

Ernst Bloch is the foremost metaphysician and historian of this determination. He conceives the essence of man to be his 'forward dreaming', his compulsive ability to construe 'that which is now' as being 'that which is not yet'. Human consciousness recognizes in the existent a constant margin of incompletion, of arrested potentiality which challenges fulfilment. Man's awareness of 'becoming', his capacity to envisage a history of the future, distinguishes him from all other living species. This Utopian instinct is the mainspring of his politics. Great art contains the lineaments of unrealized actuality. It is, in Malraux's formula, an 'anti-destiny'. We hypothesize and project thought and imagination into the 'if-ness', into the free conditionalities of the unknown. Such projection is no logical muddle, no abuse of induction. It is far more than a probabilistic convention. It is the master nerve of human action. Counter-factuals and conditionals, argues Bloch, make up a grammar of constant renewal. They force us to proceed afresh in the morning, to leave failed history behind. Otherwise our posture would be static and we would choke on disappointed dreams. Bloch is a messianic Marxist; he finds the best rudiments of futurity in dialectical materialism and the Hegelian–Marxist vision of social progress. But his semantics of rational apocalypse have general philosophic and linguistic application. More than any other philosopher, Bloch has insisted that 'reasonings upon a supposition' are not, as Hume in his exercise of systematic doubt ruled, 'chimerical and without foundation'. They are, on the contrary, the means for our survival and the distinctive mechanism of personal and social evolution. Natural selection, as it were, favoured the subjunctive.

In a genuine philosophic grammar and science of language, Bloch's *Geist der Utopie* and *Prinzip Hoffnung* would relate to Austin's 'Ifs and Cans'. The ontological and the linguistic–analytical approaches would coexist in mutual respect and be seen as ultimately collaborative. But we are still a long way from this consolidation of insight.

My conviction is that we shall not get much further in understanding the evolution of language and the relations between speech and human performance so long as we see 'falsity' as primarily negative, so long as we consider counter-factuality, contradiction, and the many nuances of conditionality as specialized, often logically bastard modes. *Language is the main instrument of man's refusal to accept the world as it is*. Without that refusal, without the unceasing generation by the mind of 'counter-worlds'—a generation which cannot be divorced from the grammar of counter-factual and optative forms—we would turn forever on the treadmill of the present. Reality would be (to use Wittgenstein's phrase in an illicit sense), 'all that is the case' and nothing more. Ours is the ability, the need, to gainsay or 'un-say' the world, to image and speak it otherwise. In that capacity in its biological and social evolution, may lie some of the clues to the question of the origins of human speech and the multiplicity of tongues. It is not, perhaps, 'a theory of information' that will serve us best in trying to clarify the nature of language, but a 'theory of misinformation'.

We must be very careful here. The cardinal terms are not only elusive; they are so obviously tainted with a twofold indictment, moral and pragmatic, Augustinian and Cartesian. 'Mendacium est enuntiatio cum voluntate falsum enuntiandi' ('A lie is the wilful utterance of an articulate falsehood'), says Saint Augustine in his *De mendacio*. Note the stress on 'enunciation', on the point at which falsity is enacted through speech. It is very nearly impossible to make neutral use of 'mis-statement', 'deception', 'falsehood', 'mis-prision', or 'unclarity', the latter being the special object of Cartesian criticism. The unclear, the ambiguously or obscurely stated is an offence both to conscience and reason. Swift's account of the Houyhnhnms compacts an ethical with a pragmatic and a philosophical condemnation:

And I remember in frequent Discourses with my Master concerning the Nature of Manhood, in other parts of the World; having occasion to talk of *Lying*, and *false Representation*, it was with much Difficulty that he comprehended what I meant; although he had otherwise a most acute Judgment. For he argued thus; That the Use of Speech was to make us understand one another, and to receive Information of Facts; now if anyone *said the Thing which was not*, these Ends were defeated; because I cannot properly be said to understand him; and I am so far from receiving Information, that he leaves me worse than in Ignorance; for I am led to believe a Thing *Black* when it is *White*, and *Short* when it is *Long*. And these were all the Notions he had concerning that Faculty of *Lying*, so perfectly well understood, and so universally practised among human Creatures.

Again we observe the close juncture of speech with verity, the view of truth as being a linguistic responsibility. Falsity, mis-correspondence with the actual state of affairs results from the enunciation of 'the Thing which was not'. The 'impropriety'— Swift's terminology is at once psychologically flat and adroitly comprehensive—is simultaneously moral and semantic. A lie 'cannot be properly said to be understood'. Of course there can be 'error', a colour-blindness, a smudge on the spectacles. Discriminations must be allowed according to a scale of intent, of sustaining or inhibiting circumstance. Nevertheless, though mistake and deliberate falsehood are differentiated, both are seen from the outset as privations, as ontological negatives. The entire gamut from black lie to innocent error is to be found on the left and shadow-side of language.

Yet how vast that side is and, *pace* Swift's irony, how imperfectly understood. The outrightness of moral and epistemological rebuke in Saint Augustine, in Swift—whose argument is cognate to that of Hume on 'chimeras'—is itself historical. The Greek view was far more qualified than the Patristic. One need only recall the enchanted exchanges between Athene and Odysseus in the *Odyssey* (XIII) to realize that mutual deception, the swift saying of 'things which are not' need be neither evil nor a bare tactical constraint. Gods and chosen mortals can be virtuosos of mendacity, contrivers of elaborate untruths for the sake of the verbal craft (a key, slippery term) and intellectual energy involved. The classical world was only too ready to document the fact that the Greeks took an aesthetic or sporting view of lying. A very ancient conception of the vitality of 'mis-statement' and 'mis-understanding', of the primordial affinities between language and dubious meaning, seems implicit in the notorious style of Greek oracles. In the *Hippias minor* Socrates enforces an opinion which is exactly antithetical to that of Augustine. 'The false are powerful and prudent and knowing and wise in those things about which they are false.' The dialogue fits only awkwardly in the canon and its purpose may have been purely 'demonstrative' or ironically *a contrario*. None the less, Socrates' case stands: the man who utters falsehood intentionally is to be preferred to the one who lies inadvertently or involuntarily. In the *Hippias minor*, the topic is referred to what was probably an allegoric commonplace, to a comparison between Achilles and Odysseus. The effect is, at best, ambivalent. 'For I hate him like the gates of death who thinks one thing and says another,' declares Achilles in Book IX of the *Iliad*. Opposed to him stands Odysseus, 'master deceiver among mortals'. In the balance of the myth it is Odysseus who prevails; neither intellect nor creation attenuate Achilles' raucous simplicity.

In short, a seminal, profound intuition of the creativity of falsehood, an awareness of the organic intimacy between the genius of speech and that of fiction, of 'saying the thing which is not', can be traced in various aspects of Greek mythology, ethics, and poetics. Gulliver's equation of the function of language with the reception of 'Information of Facts' is, by Socratic standards, arbitrary and naïve. This 'polysemic' awareness survives in Byzantine rhetoric and in the frequent allusions of Byzantine theology to the duplicities, to the inherently 'misguiding' texture of human speech when it would seek the 'true light'. But from Stoicism and early Christianity onward, 'feigning', whose etymology is so deeply grounded in 'shaping' (*fingere*), has been in very bad odour.

This may account for the overwhelming one-sidedness of the logic and linguistics of sentences. To put it in a crude, obviously figurative way, the great mass of common speech-events, of words spoken and heard, does not fall under the rubric of 'factuality' and truth. The very concept of integral truth—'the whole truth and nothing but the truth'—is a fictive ideal of the court-room or the seminar in logic. Statistically, the incidence of 'true statements'— definitional, demonstrative, tautological—in any given mass of discourse is probably small. The current of language is intentional, it is instinct with purpose in regard to audience and situation. It aims at attitude and assent. It will, except on specialized occasions of logically formal, prescriptive, or solemnized utterance, not convey 'truth' or 'information of facts' at all. We communicate motivated images, local frameworks of feeling. All descriptions are partial. We speak less than the truth, we fragment in order to reconstruct desired alternatives, we select and elide. It is not 'the things which are' that we say, but those which might be, which we would bring about, which the eye and remembrance compose. The directly informative content of natural speech is small. Information does not come naked except in the schemata of computer languages or the lexicon. It comes attenuated, flexed, coloured, alloyed by intent and the milieu in which the utterance occurs (and 'milieu' is here the total biological, cultural, historical, semantic ambience as it conditions the moment of individual articulation). No doubt there is a large spectrum of degree, of moral accent, between the imprecise shorthand of our daily idiom, the agreed falsity of social conventions, the innumerable white lies of mundane co-existence at one end and certain absolutes of philosophic, political non-truth at the other. The shallow cascade of mendacity which attends my refusal of a boring dinner en-gagement is not the same thing as the un-saying of history and lives in a Stalinist encyclopedia. Gnostic finalities of falsehood are not in common play. But between them these two polarities delimit what is, by all evidence, the larger part of private and social speech.

Linguists and psychologists (Nietzsche excepted) have done little to explore the ubiquitous, many-branched genus of lies.[1] We have only a few preliminary surveys of the vocabulary of falsehood in different languages and cultures.[2] Constrained as they are by moral disapproval or psychological malaise, these inquiries have remained thin. We will see deeper only when we break free of a purely negative classification of 'un-truth', only when we recognize the compulsion to say 'the thing which is not' as being central to language and mind. We must come to grasp what Nietzsche meant when he proclaimed that 'the Lie—and *not* the Truth—is divine!' Swift was nearer the heart of anthropology than he may have intended when he related 'lying' to the 'Nature of Manhood' and saw in 'false Representation' the critical difference between man and horse.

We need a word which will designate the power, the compulsion of language to posit 'otherness'. That power, as Oscar Wilde was one of the few to recognize, is inherent in every act of form, in art, in music, in the contrarieties which our body sets against gravity and repose. But it is pre-eminent in language. French allows *altérité*, a term derived from the Scholastic discrimination between essence and alien, between the tautological integrity of God and the shivered fragments of perceived reality. Perhaps 'alternity' will do: to define the 'other than the case', the counter-factual pro-

---

[1] Otto Lipmann and Paul Blaut, *Die Lüge in psychologischer, philosophischer, sprach- und literaturwissenschaftlicher und entwicklungsgeschichtlicher Betrachtung* (Leipzig, 1927) remains a pioneering work. There are points of considerable psychological and philosophic interest in René Le Senne, *Le Mensonge et le caractère* (Paris, 1930), and in Vladimir Jankélévitch, 'Le Mensonge' (*Revue de Métaphysique et de Morale*, XLVII, 1940), and *Du Mensonge* (Lyons, 1943). Jankélévitch returned to the theme, from a more epistemological point of view, in an article on 'La Méconnaissance' (*Revue de Métaphysique et de Morale*, new series, IV, 1963). Harald Weinrich's *Linguistik der Lüge* (Heidelberg, 1966) is a lucid but restricted introduction to an as-yet unmapped field. The most recent treatment is that of Guy Durandin, *Les Fondements du mensonge* (Paris, 1972).

[2] Cf. Samuel Kroesch, *Germanic Words for Deceiving* (Göttingen–Baltimore, 1923); B. Brotheryon, *The Vocabulary of Intrigue in Roman Comedy* (Chicago, 1926); W. Luther, *Wahrheit und Lüge im ältesten Griechentum* (Leipzig, 1935), an important, neglected beginning; Hjalmar Frisk, *Wahrheit und Lüge in den indogermanischen Sprachen* (Götenborg, 1936); J. D. Schleyer, *Der Wortschatz von List und Betrug im Altfranzösischen und Altprovenzalischen* (Bonn, 1961).

positions, images, shapes of will and evasion with which we charge our mental being and by means of which we build the changing, largely fictive milieu of our somatic and our social existence. 'We invent for ourselves the major part of experience,' says Nietzsche in *Beyond Good and Evil* ('wir erdichten . . .' signifying 'to create fictionally', 'to render dense and coherent through *poiesis*'). Or as he puts it in *Morgenröte*, man's genius is one of lies.

We can conceive of a signal system of considerable efficacy and scanning range which lacks the means to 'alternity'. A number of animal species possess the expressive and receptive equipment needed to communicate or exchange elaborate and specific information. Whether acoustically or by coded motion (the dancing bees) they can initiate and interpret cognitive, informative messages. They can also use camouflage, ruse, and beautifully exact manoeuvres of misdirection. Miming injury, the mother bird will try to lead the predator away from her brood. The line between such tactics of counter-factuality and lies or 'alternity' looks fluid. But the difference is, I think, radical. The un-truths of animals are instinctual, they are evasive or sacrificial reflexes. Those of man are voluntary and can be wholly gratuitous, non-utilitarian, and creative. To the question 'where is the water-hole?', 'where is the source of nectar?', an animal can give an answer in sound or motion. The answer will be a true one; it is a strictly constrained response to an 'information-stimulus'. Though making use of words, the Houyhnhnms will do likewise: they can only emit or interpret 'information of facts'. Swift's emblem remains one of elemental centaurs, of an instinctual ethic across the borders from man. It may be that the rubric of camouflage extends to silence, to a withholding of response. At a higher level of evolution, in the primate stage perhaps, the animal will refuse an answer (there is something less than human in Cordelia's loving reticence). But even here only a complex reflex is involved. Full humanity only begins with a reply stating 'the thing which is not': i.e. 'the water-hole is a hundred yards to my left' when it is actually fifty yards to my right, 'there is no water-hole around here', 'the water-hole is dry', 'there is a scorpion in it'. The series of possible false answers, of imagined and/or stated 'alternities' is limitless. It has neither a formal nor a contingent end, and that unboundedness of falsehood is crucial both to human liberty and to the genius of language.

When did falsity begin, when did man grasp the power of speech to 'alternate' on reality, to 'say otherwise'? There is, of course, no evidence, no palaeontological trace of the moment or locale of transition—it may have been the most important in the history of the species—from the stimulus-and-response confines of truth to the freedom of fiction. There is experimental evidence, derived from the measurement of fossil skulls, that Neanderthal man, like the newborn child, did not have a vocal apparatus capable of emitting complex speech sounds.[1] Thus it may be that the evolution of conceptual and vocalized 'alternity' came fairly late. It may have induced and at the same time resulted from a dynamic interaction between the new functions of un-fettered, fictive language and the development of speech areas in the frontal and temporal lobes. There may be correlations between the 'excessive' volume and innervation of the human cortex and man's ability to conceive and state realities 'which are not'. We literally carry inside us, in the organized spaces and involutions of the brain, worlds other than the world, and their fabric is pre-ponderantly, though by no means exclusively or uniformly, verbal. The decisive step from ostensive nomination and tautology—if I say that the water-hole is where it is I am, in a sense, stating a tautology—to invention and 'alternity' may also relate to the dis-covery of tools and to the formation of social modes which that discovery entails. But whatever their bio-sociological origin, the uses of language for 'alternity', for mis-construction, for illusion and play, are the greatest of man's tools by far. With this stick he has reached out of the cage of instinct to touch the boundaries of the universe and of time.[2]

---

[1] Cf. Philip H. Lieberman and Edmund S. Crelin, 'On the Speech of Neanderthal Man' (*Linguistic Inquiry*, II.2, 1971).

[2] While reading the original proofs of this chapter, I came across the following passage, also in galley, by Sir Karl Popper ('Karl Popper, Replies to my Critics' in *The Philosophy of Karl Popper*, ed. Paul Arthur Schilpp, La Salle, Illinois, 1974, pp. 1112–13):

At first the instrument probably had a banal survival value. It still carried with it the impulse of instinctual mantling. Fiction was disguise: from those seeking out the same water-hole, the same sparse quarry, or meagre sexual chance. To misinform, to utter less than the truth was to gain a vital edge of space or subsistence. Natural selection would favour the contriver. Folk tales and mythology retain a blurred memory of the evolutionary advantage of mask and misdirection. Loki, Odysseus are very late, literary concentrates of the widely diffused motif of the liar, of the dissembler elusive as flame and water, who survives. But one suspects that the adaptive uses of 'alternity' reached deeper, that the instrumentalities of fiction, of counter-factual assertion were bound up with the slowly evolving, hazardous definition of self. There is a myth of hand-to-hand encounter—a duel, a wrestling bout, a trial by conundrum whose stake is the loser's life—which we come across in almost every known language and body of legend. Two men meet at a narrow place, often a ford or thin bridge, at sundown, and each in turn tries to force or bar a crossing. They fight till morning but neither prevails. The outcome is an act of naming. Either the one combatant names the other ('thou art Israel' says the Angel to Jacob), or each of the two discloses his

'The development of human language plays a complex role within this process of adaptation. It seems to have developed from signalling among social animals; but I propose the thesis that what is most characteristic of the human language is the possibility of story telling. It may be that this ability too has some predecessor in the animal world. But I suggest that the moment when language became human was very closely related to the moment when a man invented a story, a myth in order to excuse a mistake he had made—perhaps in giving a danger signal when there was no occasion for it; and I suggest that the evolution of specifically human language, with its characteristic means of expressing negation—of saying that something signalled is not true—stems very largely from the discovery of systematic means to *negate* a false report, for example a false alarm, and from the closely related discovery of false stories—lies—used either as excuses or playfully.

If we look from this point of view at the relation of language to subjective experience, we can hardly deny that every genuine report contains an element of decision, at least of the decision to speak the truth. Experiences with lie detectors give a strong indication that, biologically, speaking what is subjectively believed to be the truth differs deeply from lying. I take this as an indication that lying is a comparatively late and fairly specifically human invention; indeed that it has made the human language what it is: an instrument which can be used for mis-reporting almost as well as for reporting.'

name to the other—'I am Roland,' 'I am Oliver brother of the fair Aude,' 'I am Robin of Sherwood forest,' 'I am Little John'. Several primordial themes and initiatory rites are implicit. But one is the crux of identity, the perilous gift a man makes when he gives his true name into the keeping of another. To falsify or withhold one's real name—the riddle set for Turandot and for countless other personages in fairy-tales and sagas—is to guard one's life, one's *karma* or essence of being, from pillage or alien procurement. To pretend to be another, to oneself or at large, is to employ the 'alternative' powers of language in the most thorough, ontologically liberating way. The Houyhnhnms and the Deity inhabit a tautology of coherent self: they are only what they are. As e. e. cummings put it:

> one is the song which fiends and angels sing:
> all murdering lies by mortals told make two.

Through the 'make-up' of language, man is able, in part at least, to exit from his own skin and, where the compulsion to 'otherness' becomes pathological, to splinter his own identity into unrelated or contrastive voices. The speech of schizophrenia is that of extreme 'alternity'.

All these masking functions are familiar to rhetoric and to the conventions of social discourse. Talleyrand's maxim 'La parole a été donnée à l'homme pour déguiser sa pensée' is a pointed commonplace. As is the philosophic belief, concisely argued in Ortega y Gasset's essay on translation, that there is some fundamental gap or slippage between thought and words. Lies, says Vladimir Jankélévitch in his study of 'Le Mensonge', reflect 'the impotence of speech before the supreme wealth of thought'. A crude dualism is at work here, an unanalysed notion of 'thought' as previous to or distinct from verbal expression. The identical point—language seen as a garment cloaking the true forms of 'thought'—is put forward in Wittgenstein's *Tractatus* (4.002): 'Die Sprache verkleidet den Gedanken. Und zwar so, dass man nach der äusseren Form des Kleides nicht auf die Form des bekleideten Gedankens schliessen kann; weil die äussere Form des Kleides nach ganz anderen Zwecken gebildet ist als danach, die Form des Körpers erkennen zu lassen.' The simile is not only epistemologically and linguistically misleading; it betrays a characteristic moral negative. Language commits larceny by concealing 'thought'; the ideal is one of total equivalence and empirical verifiability (cf. the Houyhnhnms). 'What is said is always too much or too little,' observes Nietzsche in the *Will to Power*, 'the demand that one should denude oneself with every word one says is a piece of *naïveté*.' Even here the pejorative image of disguise, of the false garb over the true skin is operative. Undoubtedly the linguistic resources of concealment *are* vital. It is difficult to imagine either the 'humanization' of the species or the preservation of social life without them. But these are, in the final analysis, defensive adaptations, body-paint, the capacity of the leaf-moth to take on the coloration of its background.

The dialectic of 'alternity', the genius of language for planned counter-factuality, are overwhelmingly positive and creative. They too are rooted in defence. But 'defence' here has a quite different meaning and gravity. At the central level the enemy is not the other drinker at the water-hole, the torturer seeking your name, the negotiator across the table, or the social bore. Language is centrally fictive because the enemy is 'reality', because unlike the Houyhnhnm man is not prepared to abide with 'the Thing which is'.

Can we particularize T. S. Eliot's finding that mankind will only endure small doses of reality? Anthropology, myth, psychoanalysis preserve dim vestiges of the ancient shock man suffered at his discovery of the universality and routine of death. Uniquely, one conjectures, among animal species, we cultivate inside us, we conceptualize and prefigure the enigmatic terror of our own personal extinction. It is only imperfectly, by dint of strenuous inattention, that we bear the knowledge of that finale. I have suggested that the grammars of the future tense, of conditionality, of imaginary open-endedness are essential to the sanity of consciousness and to the intuitions of forward motion which animate history. One can go further. It is unlikely that man, as we know him, would have survived without the fictive, counter-factual, anti-determinist means of language, without the semantic capacity, generated and stored in the 'superfluous' zones of the cortex, to conceive of, to articulate possibilities beyond the treadmill of organic decay and death. It is in this respect that human tongues, with their conspicuous consumption of subjunctive, future, and optative forms are a decisive evolutionary advantage. Through them we proceed in a substantive illusion of freedom. Man's sensibility endures and transcends the brevity, the haphazard ravages, the physiological programming of individual life because the semantically coded responses of the mind are constantly broader, freer, more inventive than the demands and stimulus of the material fact. 'There is only *one* world,' proclaims Nietzsche in the *Will to Power*, 'and that world is false, cruel, contradictory, misleading, senseless. . . . We need lies to vanquish this reality, this "truth", we need lies *in order to live*. . . . That lying is a necessity of life is itself a part of the terrifying and problematic character of existence.' Through un-truth, through counter-factuality, man 'violates' (*vergewaltigt*) an absurd, confining reality; and his ability to do so is at every point artistic, creative (*ein Künstler-Vermögen*). We secrete from within ourselves the grammar, the mythologies of hope, of fantasy, of self-deception without which we would have been arrested at some rung of primate behaviour or would, long since, have destroyed ourselves. It is our syntax, not the physiology of the body or the thermodynamics of the planetary system, which is full of tomorrows. Indeed, this may be the only area of 'free will', of assertion outside direct neurochemical causation or programming. We speak, we dream ourselves free of the organic trap. Ibsen's phrase pulls together the whole evolutionary argument: man lives, he progresses by virtue of 'the Life-Lie'.

The linguistic correlates are these: language is not only innovative in the sense defined by transformational generative grammar, it is literally creative. Every act of speech has a potential of invention, a capacity to initiate, sketch, or construct 'anti-matter' (the terminology of particle physics and cosmology, with its inference of 'other worlds' is exactly suggestive of the entire notion of 'alternity'). In fact, this *poiesis* or dialectic of counter-statement is even more complex, because the 'reality' which we oppose or set aside is itself very largely a linguistic product. It is made up of the metonymies, metaphors, classifications which man originally spun around the inchoate jumble of perceptions and phenomena. But the cardinal issue is this: the 'messiness' of language, its fundamental difference from the ordered, closed systematization of mathematics or formal logic, the polysemy of individual words, are neither a defect nor a surface feature which can be cleared up by the analysis of deep structures. The fundamental 'looseness' of natural language is crucial to the creative functions of internalized and outward speech. A 'closed' syntax, a formally exhaustible semantics, would be a closed world. 'Metaphysics, religion, ethics, knowledge—all derive from man's will to art, to lies, from his flight before truth, from his negation of truth,' said Nietzsche. This evasion of the 'given fact', this gainsaying is inherent in the combinatorial structure of grammar, in the imprecision of words, in the persistently altering nature of usage and correctness. New worlds are born between the lines.

Of course there is an element of defeat in our reliance on language and the imaginary. There are truths of existence, particularities of material substance which escape us, which our words erode and for which the mental concept is only a surrogate. The linguistic pulse of perception and counter-creation, of apprehension and 'alternity' is itself ambivalent. No one has come nearer to identifying the reciprocal motion of loss and creation in all utterance, in all verbalized consciousness, than Mallarmé in a compressed sentence in his preface to René Ghil's *Traité du Verbe* (1886): 'Je dis: une fleur! et, hors de l'oubli où ma voix relègue aucun contour, en tant que quelque chose d'autre que les calices sus, musicalement se lève, idée même et suave, l'absente de tous bouquets.' But as Mallarmé himself notes, in a preceding sentence, it is this absence which allows the human spirit its vital space, which enables the mind to construe essence and generality—*la notion pure*—beyond the narrows and shut horizons of our material condition.

In the creative function of language non-truth or less-than-truth is, we have seen, a primary device. The relevant framework is not one of morality but of survival. At every level, from brute camouflage to poetic vision, the linguistic capacity to conceal, misinform, leave ambiguous, hypothesize, invent is indispensable to the equilibrium of human consciousness and to the development

of man in society. Only a small portion of human discourse is
nakedly veracious or informative in any monovalent, unqualified
sense. The scheme of unambiguous propositions, of utterances
as direct pointers or homologous responders to a preceding
utterance, which is set out in formal grammars and in the extension of information theory to language study, is an abstraction.
It has only the most occasional, specialized counterpart in natural
language. In actual speech all but a small class of definitional or
'unreflective-response' sentences are surrounded, mutely ramified,
blurred by an immeasurably dense, individualized field of intention and withholding. Scarcely anything in human speech is
what it sounds. Thus it is inaccurate and theoretically spurious to
schematize language as 'information' or to identify language, be it
unspoken or vocalized, with 'communication'. The latter term will
serve only if it includes, if it places emphasis on, what is *not* said in
the saying, what is said only partially, allusively or with intent to
screen. Human speech conceals far more than it confides; it blurs
much more than it defines; it distances more than it connects. The
terrain between speaker and hearer—even when the current of
discourse is internalized, when 'I' speak to 'myself', this duality
being itself a fiction of 'alternity'—is unstable, full of mirage and
pitfalls. 'The only true thoughts,' said Adorno in his *Minima
Moralia*, 'are those which do not grasp their own meaning.'

Possibly we have got hold of the wrong end of the stick altogether when ascribing to the development of speech a primarily
informational, a straightforwardly communicative motive. This may
have been the generative impulse during a preliminary phase,
during a very gradual elaboration and vocalization of the truth-
conditioned signal systems of higher animals. One imagines a
transitional 'protolinguistic' stage of purely ostensive, stimulus-
determined 'speech' of the kind which recent investigators have
taught a chimpanzee.[1] Then, it may be towards the end of the last

[1] Cf. Philip H. Lieberman, 'Primate Vocalizations and Human Linguistic
Ability' (*Journal of the Acoustical Society of America*, XLIV, 1968); J. B. Lancaster,
'Primate Communication Systems and the Emergence of Human Language', in
P. C. Jay (ed.), *Primates* (New York, 1968); Allen R. and Beatrice T. Gardner,
'Teaching Sign Language to a Chimpanzee' (*Science*, CLXV, 1969). All the

Ice Age, occurred the explosive discovery that language is making and re-making, that statements can be free of fact and utility. In his *Einführung in die Metaphysik* (1953), Heidegger identifies this event with the true inception of human existence: 'Die Sprache kann nur aus dem Ueberwältigenden angefangen haben, im Aufbruch des Menschen in das Sein. In diesem Aufbruch war die Sprache als Wortwerden des Seins: Dichtung. Die Sprache ist die Urdichtung, in der ein Volk das Sein dichtet.' There is, to be sure, no evidence that this discovery, with which language as we know it truly begins, was explosive. But interrelated advances in cranial capacity, in the making of tools, and, so far as we can judge, in the lineaments of social organization do suggest a quantum jump. The symbolic affinities between words and fire, between the live twist of flame and the darting tongue, are immemorially archaic and firmly entrenched in the subconscious. Thus it may be that there is a language-factor in the Prometheus myth, an association between man's mastery over fire and his new conception of speech. Prometheus is the first to hold Nemesis at bay by silence, by refusing to disclose to his otherwise omnipotent tormentor the words which pulse and blaze in his own visionary intellect. In Shelley's *Prometheus Unbound* Earth celebrates this paradoxical victory, the articulation through silence of the powers of word and image:

> Through the cold mass
> Of marble and colour his dreams pass;
> Bright threads whence mothers weave the robes their children
> wear;
> Language is a perpetual Orphic song,
> Which rules with Daedal harmony a throng
> Of thoughts and forms, which else
> senseless and shapeless were.
>
> (412–17)

evidence, together with a powerful argument on the evolution of language out of the use of tools, is summarized in Gordon W. Hewes, 'An Explicit Formulation of the Relationship Between Tool-Usings, Tool-Making, and the Emergence of Language' (*Visible Language*, VII, 1973).

If we postulate, as I think we must, that human speech matured principally through its hermetic and creative functions, that the evolution of the full genius of language is inseparable from the impulse to concealment and fiction, then we may at last have an approach to the Babel problem. All developed language has a private core. According to Velimir Khlebnikov, the Russian futurist who thought more deeply than any other great poet about the frontiers of language, 'Words are the living eyes of secrecy.' They encode, preserve, and transmit the knowledge, the shared memories, the metaphorical and pragmatic conjectures on life of a small group—a family, a clan, a tribe. Mature speech begins in shared secrecy, in centripetal storage or inventory, in the mutual cognizance of a very few. In the beginning the word was largely a pass-word, granting admission to a nucleus of like speakers. 'Linguistic exogamy' comes later, under compulsion of hostile or collaborative contact with other small groups. We speak first to ourselves, then to those nearest us in kinship and locale. We turn only gradually to the outsider, and we do so with every safeguard of obliqueness, of reservation, of conventional flatness or outright misguidance. At its intimate centre, in the zone of familial or totemic immediacy, our language is most economic of explanation, most dense with intentionality and compacted implication. Streaming outward it thins, losing energy and pressure as it reaches an alien speaker.

In the process of external contact a pidgin must have arisen, an interlingua minimally resistant to current, predictable needs of economic exchange, of territorial adjustment or joint enterprise. Under certain circumstances of combinatorial advantage and social fusion, this 'amalgam at the border' will have developed into a major tongue. But at many other times and places contact will have atrophied and the linguistic separation between communities, even neighbouring, will have deepened. Otherwise it becomes exceedingly difficult to account for the proliferation of mutually incomprehensible tongues over very short geographical distances. In brief: I am suggesting that the outwardly communicative, extrovert thrust of language is secondary and that it may in substantial measure have been a late socio-historical acquirement. The primary drive is inward and domestic.

Each tongue hoards the resources of consciousness, the world-pictures of the clan. Using a simile still deeply entrenched in the language-awareness of Chinese, a language builds a wall around the 'middle kingdom' of the group's identity. It is secret towards the outsider and inventive of its own world. Each language selects, combines and 'contradicts' certain elements from the total potential of perceptual data. This selection, in turn, perpetuates the differences in world images explored by Whorf. Language is 'a perpetual Orphic song' precisely because the hermetic and the creative aspects in it are dominant. There have been so many thousands of human tongues, there still are, because there have been, particularly in the archaic stages of social history, so many distinct groups intent on keeping from one another the inherited, singular springs of their identity, and engaged in creating their own semantic worlds, their 'alternities'. Nietzsche came very close to unravelling the problem in a somewhat cryptic remark which occurs in his early, little-known paper 'Über Wahrheit und Lüge im aussermoralischen Sinne': 'A comparison between different languages shows that the point about words is never their truth or adequacy: for otherwise there would not be so many languages.' Or to put it simply: there is a direct, crucial correlation between the 'un-truthful' and fictive genius of human speech on the one hand and the great multiplicity of languages on the other.

Most probably there is a common molecular biology and neurophysiology to all human utterance. It seems very likely that all languages are subject to constraints and similarities determined by the design of the brain, by the vocal equipment of the species and, it might be, by certain highly generalized, wholly abstract efficacies of logic, of optimal form, and relation. But the ripened humanity of language, its indispensable conservative and creative force lie in the extraordinary diversity of actual tongues, in the bewildering profusion and eccentricity (though there is no centre) of their modes. The psychic need for particularity, for 'in-clusion' and invention is so intense that it has, during the whole of man's history until very lately, outweighed the spectacular, obvious material advantages of mutual comprehension and linguistic unity. In that sense, the Babel myth is once again a case of symbolic inversion: mankind was not destroyed but on the contrary kept vital and creative by being scattered among tongues. But in this sense also there is in every act of translation—and specially where it succeeds—a touch of treason. Hoarded dreams, patents of life are being taken across the frontier.

It follows, once again, that the poem, taking the word in its fullest sense, is neither a contingent nor a marginal phenomenon of language. A poem concentrates, it deploys with least regard to routine or conventional transparency, those energies of covertness and of invention which are the crux of human speech. A poem is maximal speech. 'Au contraire d'une fonction de numéraire facile et représentatif, comme le traite d'abord la foule,' writes Mallarmé in the preface to René Ghil, 'le Dire, avant tout rêve et chant, retrouve chez le poëte, par nécessité constitutive d'un art consacré aux fictions, sa virtualité.' There can be no more concise formula for the dynamics of language: 'a Saying'—un Dire—which is, above all, dream and song, remembrance and creation. It is with this conception that a philosophic linguistic must come to terms.

In considering the principal dualities which characterize natural language—the physical and mental, the time-bound and creator of time, the private and the public, truth and falsity—I have tried to suggest that a genuine linguistic will be neither exhaustive nor formally rigorous. It may be, on the analogy of a hologram, that the uses of recall, recognition, selection through contrastive scanning involved in even the simplest act of verbal articulation are a 'function' of the total state of the brain at any given moment. If this is so, the degree of relevant intricacy, the number of 'connections' and interactive 'fields' which would need to be mapped and statistically evaluated could be so large that we will never get very far beyond metaphoric, though perhaps predictive and even therapeutic approximations. In short: we do not have until now any general theory equipped to formalize let alone quantify a dynamic, open-ended system of an order of complexity even comparable to human speech (and I hope to indicate in the next chapter that the very notion of such a general theory is most likely illusory).

The 'depths' plotted by transformational generative grammars are themselves largely a disguised simile or a convention of notation. The procedures of diagnosis involved are severely reductive. This is so of the types of evidence they bring forward: the sentences which 'deep-structure' grammarians

use as specimens in their expositions are usually such as are little likely to be misinterpreted. And where they do touch upon ambiguity there is commonly an eccentricity and artificiality in the examples which may be symptomatic. The real hazards of language are conspicuously *not* represented. Samples taken from political, moral, religious, methodological *and linguistic* discussion would give a very different impression. Studies of language which avoid dealing with those features of language which have been most frustrating to our efforts to inquire into our deepest needs may justly be described as superficial.[1]

Such studies are superficial and reductive in another sense also. 'Chomsky's epigones', says Roman Jakobson, 'often know only one language—English—and they draw all their examples from it. They say, for instance, that "beautiful girl" is a transformation of "girl who is beautiful", and yet in some languages there is no such thing as a subordinate clause or "who is".'[2] Jakobson's example is, as it happens, a distortion of the transformational procedure, but the underlying charge is substantial. A profound bias towards 'monolingualism' pervades generative theories and their inference of universality. Whatever the sophistication of actual techniques (it can be over-estimated), the whole approach is at once 'rudimentary' and *a prioristic*. The disorders which it excludes, the 'non-acceptabilities' on which it legislates are among those springs of 'counter-communication' and 'alternity' which give language its primary role in our personal lives and in the evolution of the species.

This is my main point. Man has 'spoken himself free' of total

---

[1] I. A. Richards, *So Much Nearer* (New York, 1968), p. 95.
[2] Quoted in the *New Yorker*, 8 May 1971, pp. 79–80.

organic constraint. Language is a constant creation of alternative worlds. There are no limits to the shaping powers of words, proclaims the poet. 'Look,' says Khlebnikov, that virtuoso of extreme statement in his 'Decrees to the Planet', 'the sun obeys my syntax'. Uncertainty of meaning is incipient poetry. In every fixed definition there is obsolescence or failed insight. The teeming plurality of languages enacts the fundamentally creative, 'counter-factual' genius and psychic functions of language itself. It embodies a move away from unison and acceptance—the Gregorian homophonic—to the polyphonic, ultimately divergent fascination of manifold specificity. Each different tongue offers its own denial of determinism. 'The world', it says, 'can be other.' Ambiguity, polysemy, opaqueness, the violation of grammatical and logical sequences, reciprocal incomprehensions, the capacity to lie—these are not pathologies of language but the roots of its genius. Without them the individual and the species would have withered.

In translation the dialectic of unison and of plurality is dramatically at work. In one sense, each act of translation is an endeavour to abolish multiplicity and to bring different world-pictures back into perfect congruence. In another sense, it is an attempt to reinvent the shape of meaning, to find and justify an alternate statement. The craft of the translator is, as we shall see, deeply ambivalent: it is exercised in a radical tension between impulses to facsimile and impulses to appropriate recreation. In a very specific way, the translator 're-experiences' the evolution of language itself, the ambivalence of the relations between language and world, between 'languages' and 'worlds'. In every translation the creative, possibly fictive nature of these relations is tested. Thus translation is no specialized, secondary activity at the 'interface' between languages. It is the constant, necessary exemplification of the dialectical, at once welding and divisive nature of speech.

In turning now to interlingual transfers as such, to the actual business of the passage from one tongue to another, I am not moving away from the centre of language. I am only approaching this centre via a particularly graphic, documented direction. Even here, to be sure, the problems are too complex and various to allow any but a partial, intuitive treatment. Our age, our personal sensibilities, writes Octavio Paz, 'are immersed in the world of translation or, more precisely, in a world which is itself a translation of other worlds, of other systems'.[1] How does this world of translation work, what have men shouted or whispered to each other across the bewildering freedom of the rubble at Babel?

---

[1] Octavio Paz, Jacques Roubaud, Edoardo Sanguineti, Charles Tomlinson, *Renga* (Paris, 1971), p. 20.

Chapter Four

THE CLAIMS OF THEORY

I

THE literature on the theory, practice, and history of translation is large.[1] It can be divided into four periods, though the lines of division are in no sense absolute. The first period would extend from Cicero's famous precept not to translate *verbum pro verbo*, in his *Libellus de optimo genere oratorum* of 46 B.C. and Horace's reiteration of this formula in the *Ars poetica* some twenty years later, to Hölderlin's enigmatic commentary on his own translations from Sophocles (1804). This is the long period in which seminal analyses and pronouncements stem directly from the enterprise of the translator. It includes the observations and polemics of Saint Jerome, Luther's magisterial *Sendbrief vom Dolmetschen* of 1530, the arguments of du Bellay, Montaigne, and Chapman, Jacques Amyot to the readers of his Plutarch translation, Ben Jonson on imitation, Dryden's elaborations on Horace, Quintilian, and Jonson, Pope on Homer, Rochefort on the *Iliad*. Florio's theory of translation arises directly from his efforts to render Montaigne; Cowley's general views are closely derived from the nearly intractable job of finding an English transposition for the Odes of Pindar. There are major theoretic texts in this first phase: Leonardo Bruni's *De interpretatione recta* of *c.*1420, for example, and Pierre Daniel Huet's *De optimo genere interpretandi*, published in Paris in 1680 (after an earlier, less developed version of 1661). Huet's treatise is, in fact, one of the fullest, most sensible accounts ever given of the nature and problems of translation. Nevertheless, the main characteristic of this first period is that of immediate empirical focus.

[1] See the Selected Bibliography.

This epoch of primary statement and technical notation may be said to end with Alexander Fraser Tytler's (Lord Woodhouselee) *Essay on the Principles of Translation* issued in London in 1792, and with Friedrich Schleiermacher's decisive essay *Ueber die verschiedenen Methoden des Uebersetzens* of 1813. This second stage is one of theory and hermeneutic inquiry. The question of the nature of translation is posed within the more general framework of theories of language and mind. The topic acquires a vocabulary, a methodological status of its own, away from the demands and singularities of a given text. The hermeneutic approach—i.e. the investigation of what it means to 'understand' a piece of oral or written speech, and the attempt to diagnose this process in terms of a general model of meaning—was initiated by Schleiermacher and taken up by A. W. Schlegel and Humboldt. It gives the subject of translation a frankly philosophic aspect. The interchange between theory and practical need continued, of course. We owe to it many of the most telling reports on the activity of the translator and on relations between languages. These include texts by Goethe, Schopenhauer, Matthew Arnold, Paul Valéry, Ezra Pound, I. A. Richards, Benedetto Croce, Walter Benjamin, and Ortega y Gasset. This age of philosophic-poetic theory and definition—there is now a historiography of translation—extends to Valery Larbaud's inspired but unsystematic *Sous l'invocation de Saint Jérome* of 1946.

After that we are fully in the modern current. The first papers on machine translation circulate at the close of the 1940s. Russian and Czech scholars and critics, heirs to the Formalist movement, apply linguistic theory and statistics to translation. Attempts are made, notably in Quine's *Word and Object* (1960), to map the relations between formal logic and models of linguistic transfer. Structural linguistics and information theory are introduced into the discussion of interlingual exchange. Professional translators constitute international bodies and journals concerned mainly or frequently with matters of translation proliferate. It is a period of intense, often collaborative exploration of which Andrej Fedorov's *Introduction to the Theory of Translation* (*Vvedenie v teoriju perevoda*, Moscow, 1953) is representative. The new directions were set

out in two influential symposia: *On Translation* edited by Reuben
A. Brower and published at Harvard in 1959, and *The Craft
and Context of Translation: A Critical Symposium* which William
Arrowsmith and Roger Shattuck edited for the University of
Texas Press in 1961.

In many ways we are still in this third phase. The approaches
illustrated in these two books—logical, contrastive, literary,
semantic, comparative—are still being developed. Yet certain
differences in emphasis have occurred since the early 1960s.
The 'discovery' of Walter Benjamin's paper 'Die Aufgabe des
Übersetzers', originally published in 1923, together with the
influence of Heidegger and Hans-Georg Gadamer, has caused a
reversion to hermeneutic, almost metaphysical inquiries into
translation and interpretation. Much of the confidence in the
scope of mechanical translation, which marked the 1950s and
early sixties, has ebbed. The developments of transformational
generative grammars has brought the argument between 'universalist' and 'relativist' positions back into the forefront of
linguistic thought. As we have seen, translation offers a critical
ground on which to test the issues. Even more than in the 1950s,
the study of the theory and practice of translation has become a
point of contact between established and newly evolving disciplines. It provides a synapse for work in psychology, anthropology, sociology, and such intermediary fields as ethno- and
socio-linguistics. A publication such as *Anthropological Linguistics*
or a collection of articles on the *Psycho-Biology of Language* are
cases in point. The adage, familiar to Novalis and Humboldt, that
all communication is translation, has taken on a more technical,
philosophically grounded force. The papers read in the section on
the theory of translation at the Congress of the British Association
for Applied Linguistics in 1969, or those published two years later
in *Interlinguistica*, the *Festschrift* for Professor Mario Wandruszka,
himself perhaps the most influential of contrastive linguists, are
a fair example of the range and technical demands implicit in
current approaches to translation. Classical philology and comparative literature, lexical statistics and ethnography, the sociology

of class-speech, formal rhetoric, poetics, and the study of grammar are combined in an attempt to clarify the act of translation and the process of 'life between languages'.

However, despite this rich history, and despite the calibre of those who have written about the art and theory of translation, the number of original, significant ideas in the subject remains very meagre. Ronald Knox reduces the entire topic to two questions: which should come first, the literary version or the literal; and is the translator free to express the sense of the original in any style and idiom he chooses?[1] To limit the theory of translation to these two issues, which are in fact one, is to oversimplify. But Knox's point is apt. Over some two thousand years of argument and precept, the beliefs and disagreements voiced about the nature of translation have been almost the same. Identical theses, familiar moves and refutations in debate recur, nearly without exception, from Cicero and Quintilian to the present-day.

The perennial question whether translation is, in fact, possible is rooted in ancient religious and psychological doubts on whether there ought to be any passage from one tongue to another. So far as speech is divine and numinous, so far as it encloses revelation, active transmission either into the vulgate or across the barrier of languages is dubious or frankly evil. Inhibitions about decipherment, about the devaluation which must occur in all interpretative transcription—substantively each and every act of translation leads 'downward', to one further remove from the immediate moment of the *logos*—can be felt in Saint Paul. I Corinthians 14, that remarkable excursus on *pneuma* and the multiplicity of tongues, is ambivalent. If there is no interpreter present, let the alien speaker be silent. But not because he has nothing to say. His discourse is with himself and with God: 'sibi autem loquatur et Deo'. Moreover, where such speech is authentic, there must be no translation. He who has been in Christ and has heard unspeakable words—'arcana verba'—shall not utter them in a mortal idiom. Translation would be blasphemy (II Corinthians 12: 4). An even

[1] R. A. Knox, *On English Translation* (Oxford, 1957), p. 4.

more definite taboo can be found in Judaism. The *Megillath Ta'anith (Roll of Fasting)*, which is assigned to the first century A.D., records the belief that three days of utter darkness fell on the world when the Law was translated into Greek.

In most cases, and certainly after the end of the fifteenth century, the postulate of untranslatability has a purely secular basis. It is founded on the conviction, formal and pragmatic, that there can be no true symmetry, no adequate mirroring, between two different semantic systems. But this view shares with the religious, mystical tradition a sense of wastage. The vital energies, the luminosity and pressure of the original text have not only been diminished by translation; they have been made tawdry. Somehow, the process of entropy is one of active corruption. Traduced into French, said Heine, his German poems were 'moonlight stuffed with straw'. Or as Nabokov puts it in his poem 'On Translating "Eugene Onegin" ':

> What is translation? On a platter
> A poet's pale and glaring head,
> A parrot's screech, a monkey's chatter,
> And profanation of the dead.

Because all human speech consists of arbitrarily selected but intensely conventionalized signals, meaning can never be wholly separated from expressive form. Even the most purely ostensive, apparently neutral terms are embedded in linguistic particularity, in an intricate mould of cultural–historical habit. There are no surfaces of absolute transparency. *Soixante-dix* is not arrived at semantically by the same road as *seventy*; English can reproduce the Hungarian discrimination between the elder and the younger brother, *bátya* and *öcs*, but it cannot find an equivalent for the reflexes of associative logic and for the ingrained valuations which have generated and been reinforced by the two Hungarian words. 'Thus not even "basic notions", central points in a human sphere of experience, stand outside the area of arbitrary segmentation and arrangement and subsequent conventionalization; and the extent to which semantic boundaries as determined by linguistic form

THE CLAIMS OF THEORY 253

and linguistic usage coincide with absolute boundaries in the world around us is negligible.'[1]

This is the modern way of stating the argument from semantic dissonance. But the brief itself was hoary by the time Du Bellay argued it in his *Défence et illustration de la langue française* of 1549. It had been put already in Saint Jerome's epistles and prefaces. It had been reiterated, beautifully, in Dante's *Convivio*: 'nulla cosa per legame musaico armonizzata si può de la sua loquela in altra transmutare, senza rompere tutta sua dolcezza e armonia'. Nothing fully expressive, nothing which the Muses have touched can be carried over into another tongue without losing its savour and harmony. The strength, the *ingegno* of a language cannot be transferred. What Du Bellay did was to find an image of peculiar finality: 'Toutes lesquelles choses se peuvent autant exprimer en traduisant comme un peintre peut représenter l'âme avec le corps de celui qu'il entreprend tirer après le naturel.' The point is always the same: ash is no translation of fire.

Traditionally, the weight of the argument bears on poetry. Here the welding of matter and form is so close that no dissociation is admissible. Diderot's conclusion in the *Lettre sur les sourds et muets* (1751) was by no means novel. It is his phrasing, with its anticipation of modern 'semiology', which is striking: nothing will translate 'l'emblème délié, l'hiéroglyphe subtile qui règne dans une description entière, et qui dépend de la distribution des longues et des brèves.... Sur cette analyse, j'ai cru pouvoir assurer qu'il était impossible de rendre un poëte dans une autre langue; et qu'il était plus commun de bien entendre un géomètre qu'un poëte.' Again, when Rilke writes to Countess Sizzo in March 1922, there is nothing new in his contention that each word in a poem is semantically unique, that it establishes its own completeness of contextual range and tonality. What is interesting is his insistence that this applies to the most banal, grammatically flattened parts of speech, and that it divides a poem from all

---

[1] Werner Winter, 'Impossibilities of Translation', in William Arrowsmith and Roger Shattuck (eds.), *The Craft and Context of Translation* (Anchor Books, New York, 1964), p. 97.

current usage inside its own vernacular: '*Kein* Wort im Gedicht (ich meine hier jedes "und" oder "der", "die", "das") ist *identisch* mit dem gleichlautenden Gebrauchs- und Konversationswort; die reinere Gesetzmässigkeit, das grosse Verhältnis, die Konstellation, die es im Vers oder in künstlerischer Prosa einnimmt, verändert es bis in den Kern seiner Natur, macht es nutzlos, unbrauchbar für den blossen Umgang, unberührbar und bleibend. . . .' So drastic an apartness within a language will apply *a fortiori* to translation. The argument is implicit in Dr Johnson's Preface to the 1755 *Dictionary*; it is put once again by Nabokov, precisely two centuries later when he declares, with reference to English versions of Pushkin, that in the translation of verse anything but the 'clumsiest literalism' is a fraud. The modern Rumanian poet, Marin Sorescu, wittily sums up the whole catalogue of denial in a poem entitled 'Translation':

> I was sitting an exam
> In a dead language
> And I had to translate myself
> From man into ape.
>
> I played it cool,
> First translating a text
> From a forest.
>
> But the translation got harder
> As I drew nearer to myself.
> With some effort
> I found, however, satisfactory equivalents
> For nails and the hair on the feet.
>
> Around the knees
> I started to stammer.
> Towards the heart my hand began to shake
> And blotted the paper with light.
>
> Still, I tried to patch it up
> With the hair of the chest,
> But utterly failed
> At the soul.
>
> (T. Cribb's translation)

Which is Du Bellay's image exactly.[1]

Attacks on the translation of poetry are simply the barbed edge of the general assertion that no language can be translated without fundamental loss. Formally and substantively the same points can be urged in regard to prose. They take on a special intensity where philosophy is concerned. To read Plato or Kant, to grasp Descartes or Schopenhauer, is to undertake an elaborate, finally 'undecidable' task of semantic reconstruction. It is the unencumbered purity of philosophic thought 'that has made philosophy a model of Babylonian confusion. Many of its abstract concepts defy illustration. Some defy definition. Others are definable but not conceivable: "being" and "nothingness", the ὑπερούσιον of Plotinus, the Kantian *Transcendenz*, the *deitas* (as opposed to *deus*) of medieval mystics, all are "concepts" in name only. . . . The philosophical vocabulary has taken different turns even in the most closely related languages, with the result that many distinctions made in Greek or Latin or German are all but impossible to make in English.'[2] In the case of poetry such barriers are, at once, a contingent disadvantage and a symptom of integrity. But so far as philosophy goes, problems of untranslatability strike at the heart of the whole philosophic enterprise. As early as the *Cratylus* and the *Parmenides*, we are made to feel the tension between aspirations to universality, to a critical fulcrum independent of temporal, geographic conditions, and the relativistic particularities of a given idiom. How is the particular to contain and express the universal? The Cartesian mathematical paradigm and Kant's internalization of the categories of perception—the *a priori* of 'mind' before 'language'—are attempts to break out of the circle of linguistic confinement. But neither

---

[1] Or Leopardi's, when he writes in his vast commonplace book, the *Zibaldone*, for 27 July 1822: 'Ideas are enclosed and almost bound in words like precious stones in a ring. Truly they become incorporated in them like the soul in the body, so as to constitute one whole. Ideas are therefore inseparable from words, and if divided from them they are no longer the same. They evade our intellect and our powers of understanding; they become unrecognizable, which is what would happen to our soul if it were parted from our body.'

[2] E. B. Ashton, 'Translating Philosophie' (*Delos*, VI, 1971), pp. 16–17.

can be demonstrated from outside. Like all verbal discourse, philosophy is tied to its own executive means. To use Hegel's enigmatic but suggestive phrase, there is an 'instinct of logic' in each particular language. But this gives no guarantee that statements on universals will translate. No less than that of poetry, the understanding of philosophy is a hermeneutic trial, a demand and provision of trust on unstable linguistic ground.[1]

Between the most hermetic poem or metaphysics and the most banal prose, the question of translatability is only one of degree. Language, says Croce, is intuitive; each speech-act is, in any rigorous, exhaustive sense, unprecedented; it is instantaneously creative in that it has acted on, expanded, altered the potential of thought and sensibility. Strictly considered, no statement is completely repeatable (time has passed). To translate is to compound unrepeatability at second and third hand.[2] *L'intraducibilità* is the life of speech.

The case *for* translation has its religious, mystical antecedents as well as that against. Even if the exact motivations of the disaster at Babel remain obscure, it would be sacrilege to give to this act of God an irreparable finality, to mistake the deep pulse of ebb and flow which marks the relations of God to men even in, perhaps most especially in, the moment of punishment. As the Fall may be understood to contain the coming of the Redeemer, so the scattering of tongues at Babel has in it, in a condition of urgent moral and practical potentiality, the return to linguistic unity, the movement towards and beyond Pentecost. Seen thus, translation is a teleological imperative, a stubborn searching out of all the

[1] The problem of the translatability of philosophic texts has been of concern to I. A. Richards throughout his work, notably in *Mencius on The Mind*. There are invaluable discussions of particular problems in the *Journal and Letters of Stephen MacKenna*, ed. E. R. Dodds (London, 1936). Cf. also Johannes Lohmann, *Philosophie und Sprachwissenschaft* (Berlin, 1965), and Hans-Georg Gadamer, *Hegels Dialektik* (Tübingen, 1971). For a critical discussion of the entire hermeneutic approach, cf. Karl-Otto Apel, Claus von Bormann, *et al.*, *Hermeneutik und Ideologiekritik* (Frankfurt am Main, 1971). Though it does not deal directly with philosophy, Peter Szondi's essay 'Ueber philologische Erkenntnis' (*Die Neue Rundschau*, LXXIII, 1962) is an outstanding introduction to the problem of a 'science of understanding'.

[2] This thesis was developed by Croce in his *Estetica* (Bari, 1926).

apertures, translucencies, sluice-gates through which the divided streams of human speech pursue their destined return to a single sea. We have seen the strength, the theoretic and practical consequences of this approach in the long tradition of linguistic Kabbalism and illumination. It underlies the subtle exaltation in Walter Benjamin's view of the translator as one who elicits, who conjures up by virtue of unplanned echo a language nearer to the primal unity of speech than is either the original text or the tongue into which he is translating. This is 'the more final realm of language', the active adumbration of that lost, more integral discourse which, as it were, waits between and behind the lines of the text. Only translation has access to it. Until the undoing of Babel such access can only be partial. This is why, says Benjamin, 'the question of the translatability of certain works would remain open even if they were untranslatable for man'. Yet the attempt must be made and pressed forward. 'Every translation', urged Franz Rosenzweig when announcing his projected German version of the Old Testament, 'is a messianic act, which brings redemption nearer.'

The religious argument also had its intensely practical aspect. Much of the Western theory and practice of translation stems immediately from the need to disseminate the Gospels, to speak holy writ in other tongues—'variis linguis, prout Spiritus sanctus dabat eloqui illis' (Acts 2: 4). The *translatio* of Christ's message and ministry into the vulgate is a constant theme in Patristic literature and the life of the early Church. From Saint Jerome to Luther it becomes a commonplace, ceaselessly proclaimed and acted upon. No man must be kept from salvation by mere barriers of language. Each voyage of discovery brought with it the troubling presence of peoples whom distance and language had left ignorant of Christ's promise to man (Huet's work on translation directly reflects the theological puzzles posed by this, apparently, contingent banishment of primitive nations from the reach of the truth).[1] To translate Scripture into these literally darkened

---

[1] Cf. A. Dupront, *Pierre-Daniel Huet et l'exégèse comparatiste au XVIIe siècle* (Paris, 1930).

tongues is urgent charity. Each impulse towards reformation from inside the Church brings with it a call for more authentic, more readily intelligible versions of the holy word. There is a very real sense in which reformation can be defined as a summons to a fuller, more concrete translation of Christ's teachings both into daily speech and daily life. The ecstatic obviousness of the argument is manifest at a point where two master translators joined forces, in Tyndale's rendition of Erasmus's *Exhortations to the Diligent Study of Scripture* of 1529:

I would desire that all women should reade the Gospell and Paule's epistles, and I wold to god they were translated in to the tonges of all men. So that they might not only be read and knowne of the scotes and yryshmen, But also of the Turkes and saracenes. Truly it is one degre to good livinge, yee the first (I had almost sayde the cheffe) to have a little sight in the scripture, though it be but a grosse knowledge. . . . I wold to god the plowman wold singe a texte of the scripture at his plowbeme, and that the wever at his lowme with this wold drive away the tediousness of tyme.

By simple analogy the view that translation is essential to man's spiritual progress passed from the religious to the secular domain. Both had their common source in the learning and patronage of the Church. Though the quarrel over whether or not pagan texts should be read and translated at all is nearly as old as Christianity itself and flared up at frequent intervals, it was of course the Western Church which proved to be the great disseminator of the classics. The history of transfer and dissemination goes back at least to late twelfth- and thirteenth-century Toledo. In that meeting-point of Islamic, Christian, and Judaic intellect and sensibility, a number of scholar-translators, of erudite exegetists whose commentaries were a part of the translation into Latin of Hebrew and Arab texts (the latter often deriving from Greek originals), created a veritable interlingual centre. Throughout this seminal enterprise, Jews, either openly of the faith or converts, played a key role. It is in Toledo and southern France during brief moments of religious tolerance that the distinctive, perhaps decisive involvement of Jewish consciousness and polyglot inclinations with the conveyance and dissemination of ideas through-out Europe may be said to have begun.[1]

Though brief and building on an existing legacy, the papacy of Nicolas V (1447–55) marks a further significant step in the development of knowledge and argument via translation. Lorenzo Valla translated Thucydides, Guarino translated Strabo, Niccolò Perotti was payed 500 scudi for his Polybius, Valla and Pierro Candido Decembrio endeavoured to render the *Iliad* into Latin prose. More or less complete, more or less accurate versions of Xenophon and Ptolemy followed. The Aristotelian corpus was revised and completed. As Symonds puts it in his *Renaissance in Italy*, the whole of Rome had become 'a factory of translations from Greek into Latin'. The justification was proudly self-evident. Only translation could ensure that modern man would not be deprived of the wisdom and profit of the past. The *dignitas* of the human person, the transcendent reality of man's intellect, were affirmed by the fact that the new world could recognize itself in the excellence of the ancient. Though his interpretations were largely erroneous, Ficino found in Plato an enhancing mirror, a more splendid but fully recognizable image of his own and his contemporaries' features. A common humanity made translation possible.

In the two centuries between the reign of Pope Nicolas and Urquhart's Rabelais (1653), the history of translation coincides with and informs that of Western thought and feeling. No 'original' composition was more creative of new intellectual, social possibilities than were Erasmus's version of the New Testament (1516) or the Luther Bible (1522–34). We cannot dissociate the development of English sensibility in the Tudor, Elizabethan, and Jacobean periods from the new perspectives opened by Arthur Golding's translation of Caesar's *Gallic War* in 1565, by North's Plutarch in 1579, by Philemon Holland's Livy in 1600, and by the Authorized Version. The criteria, the hermen-

[1] See Danielle Jacquart, 'L'École des traducteurs', in *Tolède, XIIe–XIIIe: Musulmans, chrétiens et juifs: Le Savoir et la tolérance*, dirigé par Louis Cardilliac (Éditions Autrement: Série Mémoires No. 5; Paris, 1991), pp. 177–92.

eutic distances aimed for or unconsciously brought about by sixteenth- and seventeenth-century translators were various and sometimes contradictory. Antiquity was 'invented' more than it was discovered—it had, after all, been present, though sometimes surreptitiously, in the awareness of the Middle Ages—and this invention in turn led to new sight-lines on the present and the future. Translation provided the energies of Renaissance and Baroque Europe with an indispensable if largely fictive re-insurance. The exuberance of Rabelais, Montaigne, and, to a lesser extent, Shakespeare found in the classic precedent a ballast, a supple but steadying recourse to scale and order. But 'ballast' is too static an image. The Platonic, the Ovidian, the Senecan presence in European intellectual and emotional life of the late fifteenth and the sixteenth centuries was at once a guarantor that argument, fantasy, metaphor can be sustained at full pitch without muddle, that the human intellect can return from far places with the evidence of reasoned form, and an incitement to build against, to go beyond the classical achievement. (Galilean science, as Koyré has shown, depends on the same dialectical relation to its Aristotelian background: it works from and against the classical canon.)

Thus it was the Renaissance and Reformation translators, the line that stretches from Ficino's *Republic*, through Claude de Seyssel's Thucydides to Louis Le Roy, who principally made up the chronology, the landscape of reference in which Western literacy developed and whose obvious authority has only very recently been undermined. The confidence, the need for ideal echo were so great—'one conquered when one translated', said Nietzsche—that appropriation succeeded even where it was indirect. North's Plutarch is a recreative version not of the original Greek but of Jacques Amyot's French, published twenty years before. Latin and French models, themselves the outcome of a complicated iconographic and allegoric tradition which goes back to the late Middle Ages, play an important role in Chapman's uneven understanding of Homer (the first seven books of the *Iliad* appear in 1598). At a time of explosive innovation, and amid a real threat of surfeit and disorder, translation absorbed, shaped, oriented the necessary raw material. It was, in a full sense of the term, the *matière première* of the imagination. Moreover, it established a logic of relation between past and present, and between different tongues and traditions which were splitting apart under stress of nationalism and religious conflict. With its English, Latin, and Italian verse, with its at-homeness in Hebrew and Greek, Milton's book of poems of 1645 illustrates, supremely, the created contemporaneity of ancient and modern and the unified diversity—coherent as are the facets of a crystal—of the European community as they derive from two hundred years of translation.

In so extraordinary a period of actual performance, apologias for translation tend to have a triumphant or perfunctory air. It hardly seemed necessary to expand on Giordano Bruno's assertion, reported by Florio, that 'from translation all Science had its offspring'. When it was published in 1603, Florio's recasting of Montaigne included a prefatory poem by Samuel Daniel. Daniel's encomium is typical of innumerable pieces in praise of translation. But it is worth quoting from because it knits together the entire humanist case:

> It being the portion of a happie Pen,
> Not to b'invassal'd to one Monarchie,
> But dwell with all the better world of men
> Whose spirits are all of one communitie.
> Whom neither Ocean, Desarts, Rockes nor Sands,
> Can keepe from th' intertraffique of the minde,
> But that it vents her treasure in all lands,
> And doth a most secure commercement finde.
> Wrap Excellencie up never so much,
> In Hierogliphicques, Ciphers, Caracters,
> And let her speake never so strange a speach,
> Her Genius yet finds apt decipherers. . . .

Each time that a language-community and literature seeks to enrich itself from outside and seeks to identify its own strength contrastively, the poet will celebrate the translator's part in the 'intertraffique of the minde'. As Goethe, so much of whose work went towards the import into German of classical, modern

European and Oriental resources, wrote to Carlyle in July 1827: 'Say what one will of the inadequacy of translation, it remains one of the most important and valuable concerns in the whole of world affairs.' And speaking out of the isolation of the Russian condition, Pushkin defined the translator as the courier of the human spirit.

Nevertheless, if it is one thing to affirm the moral and cultural excellence of translation, it is quite another to refute the charge of theoretic and practical impossibility. Here again the essential moves are few and long established.

Not *everything* can be translated. Theology and gnosis posit an upper limit. There are mysteries which can only be transcribed, which it would be sacrilegious and radically inaccurate to transpose or paraphrase. In such cases it is best to preserve the incomprehensible. 'Alioquin et multa alia quae ineffabilia sunt, et humanus animus capere non potest, hac licentia delebuntur,' says Saint Jerome when translating Ezekiel. Not everything can be translated *now*. Contexts can be lost, bodies of reference which in the past made it possible to interpret a piece of writing which now eludes us. We no longer have an adequate *Rückeinfühlung*, as Nicolai Hartmann called the gift of retrospective empathy. In a sense which is more difficult to define, there are texts which we cannot *yet* translate but which may, through linguistic changes, through a refinement of interpretative means, through shifts in receptive sensibility, become translatable in the future. The source language and the language of the translator are in dual motion, relative to themselves and to each other. There is no unwobbling pivot in time from which understanding could be viewed as stable and definitive. As Dilthey was probably the first to emphasize, every act of understanding is itself involved in history, in a relativity of perspective. This is the reason for the commonplace observation that each age translates anew, that interpretation, except in the first momentary instance, is always reinterpretation, both of the original and of the intervening body of commentary. Walter Benjamin deflects the notion of a future translatability towards mysticism: one might speak of a life as 'unforgettable' even if all men had forgotten it and it subsisted only in 'the memory of God'; similarly there are works not yet translatable by man, but potentially so, in a realm of perfect understanding and at the lost juncture of languages. In fact, we are dealing with a perfectly ordinary phenomenon. The 'untranslatability' of Aristophanes in the latter half of the nineteenth century was far more than a matter of prudery. The plays seemed 'unreadable' at many levels of linguistic purpose and scenic event. Less than a hundred years later, the elements of taste, humour, social tone, and formal expectation which make up the reflecting surface, had moved into focus. Ask a contemporary English poet, or indeed a German poet, to translate—to read with anything like the required degree of response—Klopstock's *Messias*, once a major European epic. The angle of incidence has grown too wide. The argument against translatability is, therefore, often no more than an argument based on local, temporary myopia.

Logically, moreover, the attack on translation is only a weak form of an attack on language itself. Tradition ascribes the following 'proof' to Gorgias of Leontini, teacher of rhetoric: speech is not the same thing as that which exists, the perceptibles; thus words communicate only themselves and are void of substance.[1] Beside such radical, probably ironic, nominalism there is another main line of negation. No two speakers mean exactly the same thing when they use the same terms; or if they do, there is no conceivable way of demonstrating perfect homology. No complete, verifiable act of communication is, therefore, possible. All discourse is fundamentally monadic or idiolectic. This was a shopworn paradox long before Schleiermacher investigated the meaning of meaning in his *Hermeneutik*.

Neither of these two 'proofs' has ever been formally refuted. But their status is trivial. The very logicians who put the argument forward have shown this to be the case. They could not phrase their point if speech did not have a relationship of content to the real world (however oblique the relationship may be). And if communication at some level of expressive transfer was not possible, why would they seek to puzzle or persuade us with their

---

[1] Cf. K. Freeman, *Ancilla to the Pre-Socratic Philosophers* (Harvard University Press, 1957).

paradoxes? Like other bits of logical literalism, the nominalist and monadic refutations of the possibility of speech remain to one side of actual human practice. We *do* speak of the world and to one another. We *do* translate intra- and interlingually and have done so since the beginning of human history. The defence of translation has the immense advantage of abundant, vulgar fact. How could we be about our business if the thing was not inherently feasible, ask Saint Jerome and Luther with the impatience of craftsmen irritated by the buzz of theory. Translation *is* 'impossible' concedes Ortega y Gasset in his *Miseria y esplendor de la traducción*. But so is all absolute concordance between thought and speech. Somehow the 'impossible' is overcome at every moment in human affairs. Its logic subsists, in its own rigorous limbo, but it has no empirical consequences: 'no es una objeción contra el posible esplendor de la faena traductora.' Deny translation, says Gentile in his polemic against Croce, and you must be consistent and deny all speech. Translation is, and always will be, the mode of thought and understanding: 'Giacchè tradurre, in verità, è la condizione d'ogni pensare e d'ogni apprendere.'[1] Those who negate translation are themselves interpreters.

The argument from perfection which, essentially, is that of Du Bellay, Dr. Johnson, Nabokov, and so many others, is facile. No human product can be perfect. No duplication, even of materials which are conventionally labelled as identical, will turn out a total facsimile. Minute differences and asymmetries persist. To dismiss the validity of translation because it is not always possible and never perfect is absurd. What does need clarification, say the translators, is the *degree* of fidelity to be pursued in each case, the tolerance allowed as between different jobs of work.

A rough and ready division runs through the history and practice of translation. There is hardly a treatise on the subject which does not distinguish between the translation of common matter—private, commercial, clerical, ephemeral—and the re-creative transfer from one literary, philosophic, or religious text to

[1] G. Gentile, 'Il diritto e il torto delle traduzioni' (*Rivista di Cultura*, I, 1920), p. 10.

another. The distinction is assumed in Quintilian's *Institutiones oratoriae* and is formalized by Schleiermacher when he separates *Dolmetschen* from *Uebersetzen* or *Uebertragen* (Luther had used *Dolmetschen* to cover every aspect of the translator's craft). German has preserved and institutionalized this differentiation. The *Dolmetscher* is the 'interpreter', using the English word in its lower range of reference. He is the intermediary who translates commercial documents, the traveller's questions, the exchanges of diplomats and hoteliers. He is trained in *Dolmetscher-schulen* whose linguistic demands may be rigorous, but which are not concerned with 'high' translation. French uses three designations: *interprète*, *traducteur*, and *truchement*. The proposed discriminations are fairly clear, but the same terms cross over into different ranges. The *interprète* is the *Dolmetscher* or 'interpreter' in the common garden variety sense. But in a different context the name will refer precisely to the man who 'interprets', who elucidates and recreates the poem or metaphysical passage. The same ambiguity affects English 'interpreter' and Italian *interprete*: he is the helpful personage in the bank, business office, or travel bureau, but he is also the exegetist and recreative performer. *Truchement* is a complicated word with tonalities inclusive of different ranges and problems of translation. It derives from Arabic *tarjumān* (Catalan *torsimany*) and originally designates those who translated between Moor and Spaniard. Its use in Pascal's *Provinciales*, XV, suggests a negative feeling: the *truchement* is a go-between, whose rendering may not be disinterestedly accurate. But the term also signifies a more general action of replacement, almost of metaphor: the eyes can be the *truchement*, translating, substituting for the silent meanings of the heart. The *traducteur*, on the other hand, like the 'translator' or the *traduttore*, is fairly obviously Amyot rendering Plutarch or Christopher Logue meta-phrasing the *Iliad*.

Inevitably the two spheres overlap. Strictly viewed, the most banal act of interlingual conveyance by a *Dolmetscher* involves the entire nature and theory of translation. The mystery of meaningful transfer is, in essence, the same when we translate the next bill of lading or the *Paradiso*. None the less, the working distinction is obvious and useful. It is the upper range of semantic events which make problems of translation theory and practice most visible, most incident to general questions of language and mind. It is the literary speech forms, in the wide sense, which ask and promise most. I have tried to show that this is no contingent feature, no aesthetic preference. The poem, the philosophic discourse, embody those hermetic and creative aspects which are at the core of language. Where it addresses itself to a significant text, translation will engage this core.

In brief: translation is desirable and possible. Its methods and criteria need to be investigated in relation to substantive, mainly 'difficult' texts. These are the preliminaries. Theories of translation either assume them or get them out of the way briskly, with greater or lesser awareness of logical pitfalls. But what, exactly, are the appropriate techniques, what ideals ought to be aimed for?

When it is analysing complex structures, thought seems to favour triads. This is true of myths of golden, silver and iron ages, of Hegelian logic, of Comte's patterns of history, of the physics of quarks. The theory of translation, certainly since the seventeenth century, almost invariably divides the topic into three classes. The first comprises strict literalism, the word-by-word matching of the interlingual dictionary, of the foreign-language primer, of the interlinear crib. The second is the great central area of 'translation' by means of faithful but autonomous restatement. The translator closely reproduces the original but composes a text which is natural to his own tongue, which can stand on its own. The third class is that of imitation, recreation, variation, interpretative parallel. It covers a large, diffuse area, extending from transpositions of the original into a more accessible idiom all the way to the freest, perhaps only allusive or parodistic echoes. According to the modern view, the category of *imitatio* can legitimately include Pound's relations to Propertius and even those of Joyce to Homer. The dividing lines between the three types are necessarily blurred. Literalism will shade into scrupulous but already self-contained reproduction; the latter, at its upper range of self-sufficiency, tends to become freer imitation. Yet approximate though it is, this triple scheme has been found widely useful and it seems to fit broad realities of theory and technique.

All the terms in Dryden's exposition were current long before he used them. They were familiar to rhetoric and go back at least to Quintilian's differentiation between 'translation' and 'paraphrase'. But Dryden's analysis remains memorable. It did more than refute blind literalism or, as Dr Johnson puts it in his 'Life' of Dryden, 'break the shackles of verbal interpretation'. It laid down ideals and lines of discussion which are ours still.[1]

The 1680 Preface to *Ovid's Epistles, Translated by Several Hands* shows Dryden's genius at its best, which is compromise. The whole of Dryden's literary thought aims for the middle ground of common sense: as between Aristotelian dramaturgy and Shakespeare, as between the recent French models and the native tradition. In regard to translation he sought to trace a *via media* between the word-for-word approach demanded by purists among divines and grammarians, and the wild idiosyncracies displayed in Cowley's *Pindarique Odes* of 1656. Dryden's sensibility, both as theoretician and translator, was persuaded that neither could lead to the right solution. No less than the classic poet, the modern translator must stand at the clear, urbane centre.

He defined as *metaphrase* the process of converting an author word for word, line by line, from one tongue into another. The adverse example was Ben Jonson's translation of Horace's *Art of Poetry* published in 1640. Indeed, Ben Jonson and the role of Jonson as interpreter of Horace play a particular part throughout Dryden's critique. Both Jonson's results and common sense demonstrated that literalism was self-defeating. No one can translate both verbally and well. Dryden's simile retains its charm: ''Tis much like dancing on ropes with fettered legs: a man may shun a fall by using caution; but the gracefulness of motion is not to be expected: and when we have said the best of it, 'tis but a foolish task; for no sober man would put himself into a danger for the applause of escaping without breaking his neck.'

At the opposite extreme we find *imitation* 'where the translator (if now he has not lost that name) assumes the liberty not only to

[1] For a full discussion cf. W. Frost, *Dryden and the Art of Translation* (Yale University Press, 1955).

vary from the words and sense, but to forsake them both as he sees occasion'. Here the cautionary example is Cowley's extravagant transformation of Pindar and of Horace. Cowley in introducing his Pindar, had justified his practice on the ground that a man would be thought mad if he translated Pindar literally, and that the enormous distance between Greek and English would defeat any attempt at faithful yet elegant representation. Hence he had 'taken, left out, and added what I please'. No doubt pedants would carp, but 'it does not at all trouble me that the Grammarians perhaps will not suffer this libertine way of rendring foreign Authors, to be called Translation; for I am not so much enamour'd of the Name Translator, as not to wish rather to be Something Better, tho' it want yet a Name'. Cowley's hope is prophetic of twentieth-century ambitions, but Dryden will have none of it. The 'imitator' is no better, and often worse, than the composer who appropriates his theme from another and produces his own variations. This may well turn up scintillating stuff and it will show the translator to virtuoso advantage, but it is 'the greatest wrong which can be done to the memory and reputation of the dead'.

Dryden's use of *imitation*, which Pound and Lowell will adopt but with a positive inflection, is striking. The word has a long, intricate, often chequered history.[1] Its negative connotations go back to the Platonic theory of *mimesis* which, in the case of the figurative arts, occurs at two removes from the reality and truth of Ideas. The word takes on a positive value in Aristotle—with his reference to the universality and didactic importance of imitative instincts—and in Latin poetics. There it helps to express the dependent but also reinventive relations of Roman literature to the Greek precedent. Dryden's use seems to aim at Jonson and at what he found to be Jonson's particular readings of Horace.

---

[1] Cf. W. J. Verdenius, *Mimesis; Plato's Doctrine of Artistic Imitation and its Meanings to Us* (Leiden, 1949); Arno Reiff, *Interpretatio, imitatio, aemulatio* (Bonn, 1959); Göran Sörbom, *Mimesis and Art* (Uppsala, 1966). A discussion of the Horatian uses of *imitatio* may be found at the close of Vol. II of C. O. Brink's edition of *Horace on Poetry; the Ars Poetica* (Cambridge University Press, 1971). Ben Jonson's relations with classical aesthetics are discussed in Felix E. Schelling, *Ben Jonson and the Classical School* (Baltimore, 1898), and Hugo Reinsch, *Ben Jonsons Poetik und seine Beziehung zu Horaz* (Erlangen, Leipzig, 1899).

Jonson discusses *imitatio* in *Timbers*, a miscellany of critical observations published in 1641. 'Imitation' is one of the four requisites in a true poet. It is the capacity 'to convert the substance or riches of another poet to his own use.... Not to imitate servilely, as Horace saith, and catch at vices for virtue; but to draw forth out of the best and choicest flowers, with the bee, and turn all into honey; work it into one relish and savour; make our imitation sweet.' For Jonson creative ingestion is the very path of letters, from Homer to Virgil and Statius, from Archilochus to Horace and himself. It is Dryden, who is so deeply and successfully implicated in the same descent through appropriation, who gives the word a negative twist.

The true road for the translator lies neither through *metaphrase* nor *imitation*. It is that of *paraphrase* 'or translation with latitude, where the author is kept in view by the translator, so as never to be lost, but his words are not so strictly followed as his sense, and that too is admitted to be amplified, but not altered'. This, Dryden tells us, is the method adopted by Edmund Waller and Sidney Godolphin in their 1658 translation of Book IV of the *Aeneid*. What counts more, it is the approach which Dryden himself followed in his numerous translations, from Virgil, Horace, Ovid, Juvenal, Chaucer, and which he expounded in his criticism (notably in the Preface to *Sylvae* of 1685). Through *paraphrase* 'the spirit of an author may be transfused, and yet not lost'. Right translation is 'a kind of drawing after the life'. Ideally it will not pre-empt the authority of the original but show us what the original would have been like had it been conceived in our own speech. In the Preface to his translations from Virgil, issued in 1697, Dryden summarizes a lifetime of thought and practice:

On the whole matter, I thought fit to steer betwixt the two extremes of paraphrase and literal translation; to keep as near my author as I could. Without losing all his graces, the most eminent of which are in the beauty of his words; and those words, I must add, are always figurative. Such of these as would retain their elegance in our tongue, I have endeavoured to graft on it; but most of them are of necessity to be lost, because they will not shine in any but their own. Virgil has sometimes two of them in a line; but the scantiness of our heroic verse is not capable of receiving more than one; and that too must expiate for many others which have none. Such is the difference of the languages, or such my want of skill in choosing words. Yet I may presume to say ... that, taking all the materials of this divine author, I have endeavoured to make Virgil speak such English as he would himself have spoken, if he had been born in England, and in this present age.

Dryden has dropped the awkward, ambivalent term *imitation*. But the design remains the same. 'In England, and in this present age': these are the confines and ideals of the translator's craft. He can observe and achieve them only by holding the middle ground.

Goethe's involvement in translation was lifelong. His translations of Cellini's autobiography, of Calderón, of Diderot's *Neveu de Rameau* are among the most influential in the course of European literature.[1] He translated from Latin and Greek, from Spanish, Italian, English, French and Middle High German, from Persian and the south Slavic languages. Remarks on the philosophy and technique of translation abound throughout his work, and a number of Goethe's poems are themselves a commentary on or metaphoric treatment of the theme of translation. Deeply persuaded, as he was, of the continuity of life-forms, of the harmonious, though often hidden interweaving and cross-reference in all morphological reality, Goethe saw in the transfer of meaning and music between languages a characteristic aspect of universality. His best-known theoretical statement occurs in the section on translation in the lengthy prose addenda to the *West-Östlicher Divan* (1819). It has been endlessly quoted, but seems to me a more difficult, idiosyncratic treatment of the problem than is generally supposed.

Goethe's scheme, like Dryden's, is tripartite. But this time the

---

[1] Goethe's individual translations and relations to different languages are the object of a considerable monographic literature. It occupies entries 10081 to 10110 in Section XIII of Fascicule 8 of the *Goethe-Bibliographie*, ed. Hans Pyritz *et al.* (Heidelberg, 1963), pp. 781–3. Fritz Strich's well-known *Goethe und die Weltliteratur* (Berne, 1946) deals with the general theme of Goethe's relations to other literatures. But, so far as I am aware, we have not had until now a full-scale study of Goethe's translations and of their influence on his own writings and philosophy of form.

divisions are chronological as well as formal. Goethe postulates that every literature must pass through three phases of translation. But as these phases are recurrent, all may be found taking place simultaneously in the same literature though with respect to different foreign languages or genres.

The first order of translation acquaints us with foreign cultures and does so by a transference 'in our own sense'. It is best performed in plain, modest prose. Rendered in this way, the foreign matter will, as it were, enter our daily and domestic native sensibility (*nationelle Häuslichkeit*) imperceptibly. We will scarcely be conscious of the new and elevating currents of feeling which play about us. The second mode is that of appropriation through surrogate. The translator absorbs the sense of the foreign work but does so in order to substitute for it a construct drawn from his own tongue and cultural milieu. A native garb is imposed on the alien form. But the impulse to metamorphosis and entelechy which governs all living shapes, leads inevitably to a third category of translation. The highest and last mode will seek to achieve perfect identity between the original text and that of the translation. This identity signifies that the new text does not exist 'instead of the other but in its place' ('so dass eins nicht anstatt des andern, sondern an der Stelle des andern gelten solle'). This third mode requires that the translator abandon the specific genius of his own nation, and it produces a novel *tertium datum*. As a result, this type of translation will meet with great resistance from the general public. But it is the noblest. Its penetration of the foreign work, moreover, tends towards a kind of complete fidelity or 'interlinearity'. In this regard the third and loftiest mode rejoins the first, most rudimentary. The circle in which 'the foreign and the native, the known and the unknown move' is harmoniously closed.

Though very brief, or perhaps because of its concision, Goethe's model is intricate and not altogether clear. On the face of it, the first type of translation looks like straightforward mediation. It is almost the aim of the ordinary *Dolmetscher* and its purpose is essentially informative. Yet the example Goethe cites is that of Luther's Bible. Can he really have meant to say that Luther's immensely conscious, often magisterially violent reading is an instance of humble style, imperceptibly insinuating a foreign spirit and body of knowledge into German? The second manner, says Goethe, is in the root sense of the term *parodistic*. The French are past masters of this confiscatory technique, *vide* the innumerable 'translations' of the Abbé Delille. Goethe's slant here is obviously pejorative, and Delille's imitations are, on the whole, very poor. Yet the process which Goethe describes—the transformation of the original into the translator's current idiom and frame of reference—is surely one of the primary modes and indeed ideals of the interpreter's art. In addition to Delille, Goethe instances Wieland. Now as we know from other passages in Goethe's writings and conversations, such as *Zum brüderlichem Andenken Wielands*, Goethe prized the achievements of the author of *Oberon*. He knew that Wieland's imitations of Cervantes and Richardson, and his translations of Cicero, Horace, and Shakespeare had been instrumental in the coming of age of German literature. Goethe's critique is probably both moral and aesthetic. Undoubtedly the 'parodist' enriches his own culture and is invaluable to the spirit of the age. But he only appropriates what is concordant with his own sensibility and the prevailing climate. He does not enforce new, perhaps recalcitrant sources of experience on our consciousness. And he does not preserve the autonomous genius of the original, its powers of 'strangeness'.

Only the third class of translators can accomplish so much. Goethe's example here is Johann Heinrich Voss, whose versions of the *Odyssey* (1781) and *Iliad* (1793) Goethe rightly considered to be one of the glories of European translation and a principal instrument in the creation of German Hellenism. It is this third way which has brought Shakespeare, Tasso, Calderón, Ariosto, into reach of German consciousness, making of these 'Germanized strangers' (*eingedeutschte Fremde*) a crucial factor in Germany's linguistic, literary awakening. This third or 'metamorphic' approach is that which Goethe himself pursues in the *West-Östlicher Divan*. And the examples which he quotes or infers—Voss, Schlegel, Tieck, himself—are eloquent. Nevertheless it is very difficult to make out precisely what he is describing. The pivot is the distinction between 'instead of' and 'in the place of'. In the first alternative, which is presumably the 'parodistic' one, the original is diminished and the translation pre-empts a factitious authority. In the second case a symbiosis occurs, a fusion which somehow preserves the apartness, the uniqueness of the original while evolving a new and richer structure. Goethe and the Persian singer Hafiz conjoin their respective forces in a transformational encounter. This meeting and melting takes place 'outside' German and Persian—or, at least, 'outside' German as it has existed *until* the moment of translation. But both tongues are enriched through the creation of a new hybrid or, more precisely, entity.

Such paraphrase is unsatisfactory and leaves a good deal open to conjecture. There are aspects of Goethe's commentary which belong with his gnomic writings. The best one can say is that this account of the threefold motion of translation and of the ultimate circularity of the process (Benjamin's sense of 'interlinear' clearly derives from Goethe's) is deeply enmeshed in Goethe's central philosophic beliefs. Translation is an exemplary case of metamorphosis. It exhibits that process of an organic unfolding towards the harmonic integrity of the sphere or closed circle which Goethe celebrates throughout the realms both of spirit and of nature. In perfect translation as in the genetics of evolution there is a paradox of fusion and new form without the abolition of component parts. As Benjamin did after him, Goethe saw that the life of the original is inseparable from the risks of translation; entity dies if it is not subject to transformation. The final stanza of *Eins und Alles*, written in 1820, is one of the central statements we have of the need for translation:

> Es soll sich regen, schaffend handeln,
> Erst sich gestalten, dann verwandeln;
> Nur scheinbar stehts Momente still.
> Das Ewige regt sich fort in allen:
> Denn alles muss in Nichts zerfallen,
> Wenn es im Sein beharren will.

Among the many other triadic systems that of Roman Jakobson is worth noting.[1] It is far more comprehensive in scope than either Dryden's or Goethe's scheme. But something of the old framework is visible under the new 'semiotic' universality.

Adopting Pierce's theory of signs and meaning, Jakobson postulates that for us 'both as linguists and as ordinary word-users, the meaning of any linguistic sign is its translation into some further, alternative sign, especially a sign "in which it is more fully developed"' (the phrase derives from Peirce). Translation, therefore, is the perpetual, inescapable condition of signification. The translation of verbal signs falls into three classes. We *reword* when we translate a word-sign by means of other verbal signs within the same language. All definition, all explanation is, as Pierce's model shows, translation. *Translation proper*, or interlingual translation, is an interpretation of verbal signs by means of signs in some other language. Thirdly, says Jakobson, there is *transmutation*: in this 'intersemiotic' process verbal signs are interpreted by means of non-verbal sign systems (pictorial, gestural, mathematical, musical). The first two categories are, at crucial points, similar. Inside a language synonymy is only very rarely complete equivalence. 'Rewording' unavoidably produces 'something more or less'; definition through rephrasing is approximate and reflexive. In consequence the mere act of paraphrase is evaluative. 'Likewise, on the level of interlingual translation, there is ordinarily no full equivalence between code-units.' The difference is that whereas 'rewording' seeks to substitute one code-unit for another, 'translation proper' substitutes larger units which Jakobson calls *messages*. Translation is 'a reported speech; the translator recodes and transmits a message received from another source. Thus translation involves two equivalent messages in two different codes'. By using the neutral term 'involves' Jakobson side-steps the fundamental hermeneutic dilemma, which is whether it makes sense to speak of messages being *equivalent* when codes are *different*. The category of *transmutation*, on the other

[1] Roman Jakobson, 'On Linguistic Aspects of Translation', in Reuben A. Brower (ed.), *On Translation*.

hand, specifies a point which I made at the outset. Because it is interpretation, translation extends far beyond the verbal medium. Being in effect a model of understanding and of the entire potential of statement, an analysis of translation will include such intersemiotic forms as the plotting of a graph, the 'making' or 'arguing out' of propositions through dance, the musical setting of a text, or even the articulation of mood and meaning in music *per se*. I will be looking at some examples of such 'inter-mediary' transfer in my last chapter.

Jakobson concludes by saying that poetry, governed as it is by paranomasia—by the relationship between the phonemic and the semantic unit as in a pun—is 'by definition' untranslatable. Only 'creative transposition' is possible: from one poetic form into another in the same language, from one tongue into another, or between quite different media and expressive codes. But although poetry is, as always, the critical instance, every translation of a linguistic sign is, at some level, a 'creative transposition'. The two primary realities of language as I tried to define them are operative in this phrase: the creative and the masking. To 'transpose creatively' is to alternate the look and relation of things.

It can be argued that all theories of translation—formal, pragmatic, chronological—are only variants of a single, inescapable question. In what ways can or ought fidelity to be achieved? What is the optimal correlation between the A text in the source-language and the B text in the receptor-language? The issue has been debated for over two thousand years. But is there anything of substance to add to Saint Jerome's statement of the alternatives: *verbum e verbo*, word by word in the case of the mysteries, but meaning by meaning, *sed sensum exprimere de sensu*, everywhere else?

Whatever treatise on the art of translation we look at, the same dichotomy is stated: as between 'letter' and 'spirit', 'word' and 'sense'. Though the rendition of sacred texts poses a problem which is at once special and central to the whole theory of translation, there have in fact been very few absolute literalists. Translating from Latin in the mid-fifteenth century, Nicholas von Wyle demanded a total concordance, a matching of word to word:

'ain yedes wort gegen ain andern wort'. Even errors must be transcribed and translated as they are an integral part of the original.[1] Few, on the other hand, have carried the theory of complete mimetic freedom as far as Pound when he defines the poems in *Personae* as 'a long series of translations, which were but more elaborate masks'.[2]

Almost invariably we are presented with an argument from and for compromise. The ideal, the tactics of mediation between letter and spirit are worked out in the sixteenth and seventeenth centuries, from Étienne Dolet's *Manière de bien traduire d'une langue en aultre* of 1540 to Pierre-Daniel Huet's *De interpretatione* in its second, expanded version of 1680. French pre-eminence in the theory of translation during this period was no accident: it reflected the political and linguistic centrality of French culture during and after the break-up of European Latinity (a phenomenon which, of course, inspired the search for an agreed discipline of translation). Dolet's five rules for the translator may themselves go back to Italian grammarians and rhetoricians of the early sixteenth century and perhaps to Leonardo Bruni. They have the virtue of obviousness. The would-be interpreter must have a perfect grasp of 'the sense and spirit' of his author. He must possess knowledge in depth of the language of the original as well as of his own tongue. He ought, as Horace bids him, be faithful to the meaning of the sentence, not to the word order. It is mere superstition, says Dolet, '(diray ie besterie ou ignorance?) de commencer sa traduction au commencement de la clausule'. Fourthly our translator will aim for a version in plain speech. He will avoid the importation of neologisms, rare terms, and esoteric flourishes of syntax so beloved of sixteenth-century scholars and Latinists. The final rule applies to all good writing: the translator must achieve harmonious cadences (*nombres oratoires*), he must

---

[1] I owe this reference to Rolf Kloepfer, *Die Theorie der literarischen Uebersetzung. Romanisch-deutscher Sprachbereich* (Munich, 1967). Kloepfer, in turn, refers to a dissertation by Bruno Strauss on 'Der Uebersetzer, Nicholas von Wyle' (Berlin, 1911).

[2] Ezra Pound, *Gaudier-Brzeska: A Memoir* (London, 1916), p. 98.

compose in a sweet and even style so as to ravish the reader's ear and intellect.[1]

Dolet perished before he could expound these truisms in a more detailed, applied manner. A much less known, but interesting work, printed in Basle in 1559, gives us a complete picture of the standard, median approach which the humanists advocated in regard to translation. It is the *Interpretatio linguarum: seu de ratione convertendi & explicandi autores tam sacros quam prophanos* of Lawrence Humphrey (or Humfrey), a Puritan divine of considerable irascibility and learning who became Master of Magdalen College, Oxford. The *Interpretatio* runs to more than 600 pages and is one of the summarizing statements in the history of translation. Much of it is routine. But it also contains touches of originality and is notably tough-minded in its resort to practical examples. Like everyone before him, Humphrey divides translation into three modes: literalism, which he condemns as *puerilis & superstitiosa*, free or licentious adaptation, and the just *via media*. Humphrey's definition of the middle way is worth quoting because it elevates the banalities of compromise to the status of method: 'via media dicamus ... quae utriusque particeps est, simplicitatis sed eruditae, elegantiae sed fidelis: quae nec ita exaggerata est ut modum transeat, nec ita depressa ut sit sordida, sed frugalis, aequabilis, temperata, nec sordes amans, nec luxuriam, sed mundum apparatum.' Such poise between simplicity and learning, between elegance and fidelity, such exact observance of urbane elevation, neither emphatic on the one hand nor gross on the other, is contracted by Humphrey into the notion of 'aptitude'. The true translator will seek to attain 'plenitude, purity and propriety', but above all he seeks aptitude. He does so in choosing a text matching his own sensibility. The ideal of aptitude will govern his choice of an appropriate style. It will, most significantly, suggest which languages can or cannot be brought into fruitful contact. This is one of Humphrey's original points. He distinguishes between 'major' and 'trivial' tongues according to the

---

[1] Cf. Marc Chassaigne, *Étienne Dolet* (Paris, 1930), pp. 230–3, 272.

history, philosophy, and letters which they record and express. It is
solely between 'major' languages that the process of translation
is truly meaningful. Hence Humphrey's choice for analysis of
parallel texts in Hebrew, Greek, and Latin. But there can be
failures of aptitude even between major languages: thus, argues
Humphrey, Cicero is often uncertain and obtuse in his rendition
of Greek philosophic terms. Where he does his work well,
however, the translator is a man of utmost worth, a recognizer in
the full hermeneutic sense: 'si linguarum utilis sit cognitio, inter-
pretari utilissimum' (if the knowledge of languages is useful,
translation is most useful).

Huet knew the *Interpretatio linguarum*. He cites Humphrey,
together with More, Linacre, and Cheke as one of the few
Englishmen to have made a serious contribution to the matter of
translation. Huet's principle of stylistic accord is very close to
Humphrey's ideal of aptitude: 'Traduisez Aristote en périodes
cicéroniennes, vous faites une caricature; si vous imitez l'oiseau
intrus qui ne se bornant pas à déposer ses oeufs dans le nid
d'autrui, renverse à terre la couvée légitime, vous ne traduisez
plus, vous interpolez.' Like Humphrey, Huet approaches the
theory of translation from a point of practical need: the rendition
from Greek into Latin of an unpublished Commentary on
Matthew by Origen which he had come upon in the Royal Library
in Stockholm during a protracted, adventurous journey. Huet's
doctrine of the middle path between literalism and licence adds
nothing fundamentally new to that of his predecessors. The just
translator 'nativum postremo Auctoris characterum, quoad eius
fieri potest, adumbrat; idque unum studet, ut nulla eum detrac-
tione imminutum, nullo additamento auctum, sed integrum,
suique omne ex parte simillimum perquam fideliter exhibeat'
('copies the innate essence of his author to the extent to which this
is possible. His one study is faithfully to display his author whole,
taking nothing away and adding nothing'). But Huet's treatise, cast
in the guise of an imaginary conversation with three eminent
humanists, among them Isaac Casaubon, translator of Polybius
and a master-scholar of his age, is far more sophisticated than
Humphrey's. He was, as A. E. Housman puts it in his preface to

Manilius, 'a critic of uncommon exactness, sobriety, and male-volence'. Huet has a keen eye for the misuse of translation as self-enhancement; he speaks scathingly of translators who indulge their own *ingenium* at the expense of the text. He shows insight, albeit rudimentary, into the philosophic problem underlying all trans-lation: his *De Interpretatione* takes the term in its full cognitive sense. And though Huet's claim to have adequate command of Hebrew, Greek, Latin, Coptic, Armenian, Syriac, and all main European tongues may have been overstated, there can be little doubt that he was a polyglot and that his response to the quality of different languages was vivid. In at least one respect, moreover, the future Bishop of Avranches broke fresh ground. He devotes a part of his study to scientific translation. He sees in this one of the foremost tasks of civilization, and one that has been absurdly neglected. Among the rare exceptions, allows Huet, is the work of Jean Pena, himself a distinguished mathematician and the trans-lator of Euclid and of Theodosius Tripolita on spheres. Scientific texts confront the translator with particular demands. 'Ces choses s'enseignent et ne s'ornent point.' The translator may come up against technical locutions which defy any single, assured inter-pretation. In this event, advises Huet, it is best to leave the original expression as it is and to provide various possible readings and elucidations in the margin. At several points Huet's discussion concurs with the guide-lines laid down by Joseph Needham, three centuries later, in regard to the translation of Chinese scientific and mathematical terminology.[1]

The vocabulary, the methodological framework in which Herder, Schleiermacher, and Humboldt discuss the theory of translation is obviously new. The debate on translatability is now frankly and thoroughly a part of epistemology. The philological resources available to the comparative linguist are far more pro-fessional than any known to the seventeenth century. Now the dominant current is German. As has been often said by German poets and scholars, translation was the 'inmost destiny' (*innerstes*

---

[1] Though amateurish and long-winded, Léon Tolmer, *Pierre-Daniel Huet (1630–1721): Humaniste-Physicien* (Bayeux, 1949), is still the only full-scale treatment we have. Cf. in particular Chapter V.

Schicksal) of the German language itself.[1] The evolution of modern German is inseparable from the Luther Bible, from Voss's Homer, from the successive versions of Shakespeare by Wieland, Schlegel, and Tieck. Thus the theory of translation takes on an unprecedented authority and philosophic texture.

But beneath the new idiom and psychological finesse, the classical polarities remain unchanged. All that happens is that the dichotomy between 'letter' and 'spirit' is transposed into the image of the appropriate distance a translation should achieve between its own tongue and the original. Should a good translation edge its own language towards that of the original, thus creating a deliberate aura of strangeness, of peripheral opaqueness? Or should it naturalize the character of the linguistic import so as to make it at home in the speech of the translator and his readers? Herder marks these two alternatives by an adroit play on 'trans-lation': translations tend either to '*Ueber*setzung', aiming for as intimate a fusion with the original as is possible, or to 'Ueber*setzung*' where the emphasis falls on recreation (*setzen*) in the home tongue. Schleiermacher differentiates along these same lines when he divides *Dolmetschen* from true *Uebersetzen*. His originality lies in the lengths to which he, like Hölderlin, was prepared to go in seeking to recapture the structural, tonal elements of the foreign text. According to Schleiermacher, translation in depth demanded the modulation of one's own speech into the lexical and syntactic world of the original. Hence the 'Greek–German' of Hölderlin's Sophocles and of Schleiermacher's own versions of Plato. In practice, though not in theory, such symbiotic translations tend towards a special interlingua for translators, a transfer-idiom or hybrid such as J. J. Hottinger had called for, in 1782, in his curious tract *Einiges über die neuen Uebersetzerfabriken*.

Nevertheless, the old and obvious dualism remains. The very similes used by Florio, Dolet, Humphrey, and Huet remain active to this day. The relation of translator to author should be that of the portrait-painter to his sitter. A good translation is a new

[1] For extensive discussion of this theme, cf. the Proceedings of the Colloquium on Translation of the Bavarian Academy of Fine Arts, held in the summer of 1962 and published as *Die Kunst der Uebersetzung* (Munich, 1963).

garment which makes the inherent form familiar to us yet in no way hinders its integral expressive motion. Thus alone, says Florio in his preface to Montaigne, 'may sense keep form'. This retention of inner structure within exterior change is, in truth, analogous to '*Pythagoras* his *Metempsychosis*'. The identical formula, more drily phrased, is present in Schopenhauer. After observing querulously in chapter 35 of the *Parerga und Paralipomena*, that no amount of labour or genius would convert *être debout* into *stehen*, Schopenhauer concluded that no less was needed than a 'transference of soul'. 'The garment must be new, the inner form must be retained,' wrote Wilamowitz in his prefatory essay to Euripides' *Hippolytus* (1891): 'Jede rechte Uebersetzung ist Travestie. Noch schärfer gesprochen, es bleibt die Seele, aber sie wechselt den Leib: die wahre Uebersetzung ist Metempsychose.' The letter changes; the spirit is intact yet made new. Precisely as Saint Jerome had urged in his famous image of the sense taken captive, 'sed quasi captivos sensus in suam linguam victoris jure transposuit' ('he has carried meaning over into his own language, just like prisoners, by right of conquest') in the Preface to his version of the Book of Esther.

The question is: *how*? How may this ideal of mediation be achieved and, if possible, methodized? By what practical craft is the translator to produce that delicate moment of binary poise in which, to use Wolfgang Schadewaldt's formula, 'his mode of expression is already unmistakably Greek, yet still authentically German'?

There are, as we shall see, many demonstrations of the thing done; but very few diagnoses.

No translator has recorded with more scruple his inner life between languages or has brought a more intelligent intensity to the problem of 'letter' versus 'spirit' than did Stephen MacKenna. MacKenna gave his uncertain physical and mental health to the translation of Plotinus' *Enneades*. The five tall volumes appeared between 1917 and 1930. This solitary, prodigious, grimly unremunerative labour constitutes one of the masterpieces of modern English prose and formal sensibility. It is also a feat of 'learned poetics', of precise but recreative interpretation in which almost every facet of the business of translation is put to the test. The journal and correspondence of MacKenna, beautifully edited by E. R. Dodds, allow us to follow something of the penetrative process.

Like others who have thought the problem through, MacKenna favours a parallel text, but a free parallel. 'My total testimony,' he writes in 1919, 'would be that nothing could serve the classics more than superbly free translations—backed of course by the thoroughest knowledge—accompanied by the strict text. The original supplies the corrective or the guarantee; the reader, *I* find, understands the depths of his Greek or Latin much better for the free rendering—again, I think of a chaste freedom, a freedom based rigidly on a preservitude.'[1] MacKenna found himself unable to understand translations 'which would appear to satisfy the accepted ideas of "literalness": give me a free translation by a man of first-rate knowledge, and I'm often quite amused to find that out of the freedom I can reconstruct the Greek original almost verbatim.' He goes on to say in the same letter that literalism is itself a suspect hybrid of '(1) Liddell & Scott English or (2) a bastard English, a horrible mixture of Elizabethan, Jacobean, fairytale-ese, Biblicism and modern slang (not slang of word but, what is worse, of phrase or construction).'[2] In a monumental letter of 15 October 1926 MacKenna comes as close as he can to defining the proper modernity of a good translation from the classics. All style must be modern: 'Plato was modern to Plato.' If the translator looks at an old author when he sets to work, it is simply to suggest to him '*methods of phrasing which by analogy ought to be in* the language of today . . . even here of course one must be careful: it's as bad to be too ancient in phrase-mould as in actual word; or, not only too ancient, but too persistently terse and laboured'. To state the ideal, MacKenna borrows a phrase from Herbert Spencer: 'the great rule is I suppose this: "with a dignity adequate to the subject and its mood to *avoid* (or minimise) *friction*".'

[1] E. R. Dodds (ed.), *Journal and Letters of Stephen MacKenna* (London, 1936), pp. 154–5.
[2] Ibid., pp. 155–6.

But although he wrestled with the nature of translation as lucidly and with as firm a responsibility to the actual text as anyone ever has, MacKenna knew that there is in the art a large margin of obscurity, of 'miracle'. The metaphor of metempsychosis is implicit in an entry in his journal for 5 December 1907: 'Whenever I look again into Plotinus I feel always the old trembling fevered longing: it seems to me that I must be born for him, and that somehow someday I must have nobly translated him: my heart, untravelled, still to Plotinus turns and drags at each remove a lengthening chain.' Towards the latter stages of his work MacKenna could rightly say: 'what I have done with Plotinus is a miracle, the miracle of persistent resteadying of a mind that dips and tosses and disappears like cork on the waves of your Bay of Islands.'[1]

But the 'miracle' is never complete. Each translation falls short. At best, wrote Huet, translation can, through cumulative self-correction, come ever nearer to the demands of the original, every tangent more closely drawn. But there can never be a total circumscription. From the perception of unending inadequacy stems a particular sadness. It haunts the history and theory of translation. 'Wer uebersetzt,' proclaimed the German poet and pietist Matthias Claudius, 'der untersetzt.' His play on words, though elementary, is untranslatable. But the image is perennial. There is a special *miseria* of translation, a melancholy after Babel. Ortega y Gasset gives the best account of it. The theme itself, however, is as ancient as the art.

List Seneca, Saint Jerome, Luther, Dryden, Hölderlin, Novalis, Schleiermacher, Nietzsche, Ezra Pound, Valéry, MacKenna, Franz Rosenzweig, Walter Benjamin, Quine—and you have very nearly the sum total of those who have said anything fundamental or new about translation. The range of theoretic ideas, as distinct from the wealth of pragmatic notation, remains very small. Why should this be the case?

[1] Ibid., p. 187.

2

In the history and theory of literature translation has not been a subject of the first importance. It has figured marginally, if at all. The exception is the study of the transmission and interpretation of the Biblical canon. But this is manifestly a special domain, within which the matter of translation is simply a part of the larger framework of exegesis. There is no treatise on translation comparable in definition or influence to Aristotle's *Poetics* or Longinus on the sublime. It is only very recently (with the foundation of the International Federation of Translators in Paris in 1953) that translators have fully asserted their professional identity, that they have claimed a world-wide corporate dignity. Until then Valery Larbaud's description of the translator as the beggar at the church door was largely accurate: 'Le traducteur est méconnu; il est assis à la dernière place; il ne vit pour ainsi dire que d'aumônes.' Even today the financial rewards of translation are often ridiculously meagre when compared to the difficulty and importance of the work.[1] Though the *Index translationum* issued annually by UNESCO shows a dramatic increase in the number and quality of books translated, though translation is probably the single most telling instrument in the battle for knowledge and woken consciousness in the underdeveloped world, the translator himself is often a ghostly presence. He makes his unnoticed entrance on the reverse of the title-page. Who picks out his name or looks with informed gratitude at his labour?

On the whole it has always been so. It is doubtful whether Florio or North would have their modest place in English literature, at least so far as scholars and poets go, were it not for the uses Shakespeare made of Montaigne and of Plutarch. Chapman's version of Homer lives, under rather false colours as it happens, in Keats's sonnet. Who can identify the principal translators of Bacon, Descartes, Locke, Kant, Rousseau, or Marx? Who made

---

[1] For a witty account of the situation as it was in the late 1950s and early 1960s cf. Richard Howard, 'A Professional Translator's Trade Alphabet' in *The Craft and Context of Translation*. There is much material also in Walter Widmer, *Fug und Unfug des Uebersetzens* (Cologne–Berlin, 1959).

Machiavelli or Nietzsche accessible to those who had no Italian or German? In each of these cases the moment of translation is that of decisive meaning, the leap from a local to a general force. We speak of the 'immense influence' of *Werther*, of the ways in which the European awareness of the past was reshaped by the Waverley novels. What do we remember of those who translated Goethe and Scott, who were in fact the responsible agents of influence? Histories of the novel and of society tell us of the impact on Europe of Fenimore Cooper and Dickens. They do not mention Auguste-Jean-Baptiste Defaucompret through whose translations that impact is made. It remains a piece of pedantic lore that Byronism, certainly in France, Russia, and the Mediterranean is mainly the consequence of the translations of Amédée Pichot. It is the translations into French, English, and German by Motteux, Smollett, and Tieck respectively of Cervantes which constitute the life at large, the intensity in the literate imagination, of Don Quixote. Yet it is only lately that the translator—such as Constance Garnett, C. K. Scott Moncrieff, Arthur Waley—has begun emerging from a background of indistinct servitude. And even here his visibility is often that of a target: his role in making Dostoevsky or Proust available to us is underlined because it is felt that the work needs redoing.

It is obvious, when one stops to think of it, that intellectual history, the history of genres, the realities of a literary or philosophic tradition, are inseparable from the business of translation. But it is only in the last decades that we find close attention being paid to the history and epistemology of the transmission of meaning (what one would, technically, call a 'diachronic hermeneutic'). In what ways does the development of crucial philosophic, scientific, or psychological terms depend on successive translations of their initial or normative statement? To what degree is the evolution of western Platonism, of the image of 'the social contract', of the Hegelian dialectic in the communist movements, a result of selective, variant, or thoroughly mistaken translations? Koyré's investigations of the history of the translations of Copernicus, Galileo, and Pascal, Gadamer's inquiries into the theoretic and practical translatability of key terms in Kant

and Hegel, J. G. A. Pocock's study of the inheritance of the
vocabulary of politics from the Florentine Renaissance to Locke
and Burke, are pioneering efforts. There is until now only a
rudimentary understanding of the language-aspects of intellectual
history and of the study of comparative institutions. Yet they are
absolutely central. Without a grasp of the nature of translation
there can be no account of the current in the circuit. 'It is part of
the plural character of political society that its communication
networks can never be entirely closed, that language appropriate to
one level of abstraction can always be heard and responded to
upon another, that paradigms migrate from contexts in which they
had been specialized to discharge certain functions to others in
which they are expected to perform differently.'[1] This 'plural
character' determines the history of thought. The openness of the
networks, the migration of the paradigms are a direct function
of translation, first intralingual, then into other languages. It is
strange that this function should appear so largely anonymous or
accidental.

Granted, then, that translation is a focal but neglected topic.
Granted also, as William Arrowsmith and Roger Shattuck put it in
their preface to the papers of the University of Texas symposium,
that 'intelligent comments on translation . . . tend to be unavailable
or scattered, tucked away in odd corners, and their arguments
diffused. The crucial, comprehensive volume of pioneering
scholarship has yet to be written.'

But is 'translation' in fact a subject? Is the material of a kind and
internal order which theoretic analysis, as distinct from historical
scholarship and descriptive review, can deal with? It may be that
there is no such thing as 'translation' in the abstract. There is a
body of *praxis* so large and differentiated as to resist inclusion in
any unitary scheme. One can group and examine examples of
literary translation from Livius Andronicus' *Odyssey* to the present.
One can investigate the chequered history of the translation
of scientific and philosophic terms. It would be possible, and
fascinating, to assemble what records there are of the development

[1] J. G. A. Pocock, *Politics, Language and Time* (New York, 1971), p. 21.

of commercial, legal, diplomatic translation, to study the interpreter and his functions in economic and social history. Schools
for translators, such as are believed to have flourished in Alexandria
in the second century A.D. or in Baghdad, under the leadership of
Hunain ibn Ishaq, during the ninth century, would be worth
analysing and comparing. There is urgent justification for the
'stemmatic' review of major philosophic and literary texts, i.e. for
the recension of successive and interrelated translations of a given
original in order to provide the history of its diffusion, influence
and (mis)interpretation with a sound material basis. But each of
these areas—and almost everything remains to be done in them—
constitutes only an *ad hoc* and contingent definition: it circum-
scribes a local, empirical phenomenon or aggregate of phenomena.
There are no axiomatic categories.

We have seen that the theoretic equipment of the translator
tends to be thin and rule-of-thumb. What the historian or student
of translation brings is a more or less informed, a more or less
perceptive commentary on the particular instance. We collate and
judge this or that Arabic version of Aristotle or Galen. We contrast
Roy Campbell's reading into English of a Baudelaire sonnet with
the readings proposed by Robert Lowell and Richard Wilbur. We
set Stefan George's Shakespeare next to Karl Kraus's. We follow
the transformation of Racine's alexandrines into the hexameters
of Schiller's *Phädra*. We wonder at the recasting of Lenin on
empirio-criticism into Urdu and Samoyed. 'What is therefore
desperately needed,' say Arrowsmith and Shattuck, 'is patient,
persuasive elaboration of the principles appropriate to different
"genres" as each one has found historical expression, as well as an
awareness of their differing functions and their respective virtues
and limitations.' This is, unquestionably, a vital aim, and one that
demands great learning and linguistic tact. But such elaboration
cannot constitute a formal, theoretic study of the 'subject of
translation'. It does not lead to a systematic model of the general
structure and epistemological validity of the transfer of meaning
between languages.

It may be that no such model is possible. The limits of study
may be those determined by patient accumulation of descriptive

classes, by the gathering of practical hints ordered according to period, locale, and specific genre. To use a very rough analogy, the discipline of translation may be subject only to a Linnaean, not to a Mendelian type of formalization.

But even if we take the modest view, even if we regard the study of translation as descriptive–taxonomic rather than properly theoretic ('theoretic' meaning susceptible of inductive generalization, prediction, and falsifiability by counter-example), a severe difficulty arises. In the overwhelming majority of cases, the material for study is a finished product. We have in front of us an original text and one or more putative translations. Our analysis and judgement work from outside, they come after the fact. We know next to nothing of the genetic process which has gone into the translator's practice, of the prescriptive or purely empirical principles, devices, routines which have controlled his choice of this equivalent rather than that, of one stylistic level in preference to another, of word 'x' before 'y'. We cannot dissect, or only rarely. If only because it was deemed to be hack-work, the great mass of translation has left no records. There are no 'foul papers' for Urquhart's Rabelais. We have no drafts from Amyot's Plutarch.[1] We have only one brief set of notes from among the voluminous sketches, preliminary trials, and corrections which went into the preparation of the King James's Bible.[2] Pope's Homer is among the first great acts of translation available to us in manuscript.[3] But even after the eighteenth century documentation remains scarce. How many false starts, what arcs of association, what doodles of the brain and of the hand underlie Chesterton's uncannily evocative version of Du Bellay's 'Heureux qui comme Ulysse' or Goethe's rendition, which is a masterpiece, of Manzoni's 'Il Cinque maggio'?

It is only very recently, and this *is* a revolution in the subject,

---

[1] Cf. René Sturel, *Jacques Amyot* (Paris, 1908), pp. 357–424, 440–594.

[2] Cf. Ward Allen (ed.), *Translating for King James* (Vanderbilt University Press, 1969).

[3] Pope's Homer MSS are in the British Museum (Brit. Mus. Add. MS. 4807). Some short extracts from them are reproduced in Appendix C, Vol. X of the Twickenham Edition (London and Yale University Press, 1967).

that the 'anatomy' and raw materials of translation are becoming accessible to methodical scrutiny. We have Pound's letters to W. H. D. Rouse on translating Homer; Robert Fitzgerald's postscript to his *Odyssey*, trying to record specific motions of choice and discard; Nabokov's memoir, ironic and full of traps for the unwary yet deeply instructive, of how he rendered *Onegin* into English; Pierre Leyris's brief but acute remarks on his translations from Hopkins; Christopher Middleton's 'On Translating a Text by Franz Mon' published in the first number of *Delos* in 1968; John Frederick Nims's account of *métier* and ideals in his collection of *Poems in Translation*; Octavio Paz's work-notes for his Spanish version of Mallarmé's *Sonnet en 'ix'* in *Delos* 4. The Valery Larbaud archive in Vichy contains a wealth of material, as yet unexploited, on the work in progress which led to the remarkable French translations of *Moby Dick* and *Ulysses*. There is extant, though incomplete, some of the preliminary material for the French version of 'Anna Livia Plurabelle' undertaken by Samuel Beckett and his students, among them Sartre and Paul Nizan. Beginning in the 1920s, and in a more conscious, methodical way after the Second World War, translators have started preserving their drafts, rough papers, and successive *maquettes*. It is doubtful whether Michel Butor will destroy the work-sheets of his current attempt to find a French mirroring for *Finnegans Wake* or whether Anthony Burgess's efforts to do the same in Italian will not survive—notes, drafts, uncorrected proofs, final galleys and all— in the strongroom of some American university. The unformed fascinates us.

But although the new documentation will allow a much closer, more technically and psychologically substantiated look at the activities of the translator, at the actual executive modes of his art, analysis will remain at the descriptive and discrete level. The field is made neither formally rigorous nor continuous by an increase in the number and transparency of individual samples. It stays 'subject to taste and temperament rather than to knowledge'.[1] The inference, unmistakable in Arrowsmith's and Shattuck's pro-

---

[1] E. S. Bates, *Intertraffic, Studies in Translation* (London, 1943), p. 15.

gramme, of a progressive systematization, of an advance from local inventory and insight to generality and theoretic stability, is almost certainly erroneous. 'Translating from one language into another,' says Wittgenstein, 'is a mathematical task, and the translation of a lyrical poem, for example, into a foreign language is quite analogous to a mathematical *problem*. For one may well frame the problem "How is this joke (e.g.) to be translated (i.e. replaced) by a joke in the other language?" and this problem can be solved; but there was no systematic method of solving it.'[1] It is of extreme importance to grasp the distinction which Wittgenstein puts forward, to understand how 'solution' can coexist with the absence of any systematic method of solution (the full delicacy and complication of the idea is brought out by Wittgenstein's analogy with mathematics, a mathematics in which there are solutions but no systematic methods of solution). This distinction is, I believe, true not only of translation itself, but of the descriptions and judgements we can make of it. The rest of this book is an attempt to show this as clearly as possible, and to suggest the reasons why.

Obviously but also fundamentally, they are philosophical.[2] We have seen how much of the theory of translation—if there is one as distinct from idealized recipes—pivots monotonously around undefined alternatives: 'letter' or 'spirit', 'word' or 'sense'. The dichotomy is assumed to have analysable meaning. This is a central epistemological weakness and sleight of hand. Even during those periods in the history of thought when epistemology was acutely critical and self-critical, when the nature of the relations between 'word' and 'sense' came under stringent review, arguments on translation have proceeded as if the issue were trivial or resolved or of another jurisdiction. In whatever form it is put, *non verbum e verbo, sed sensum exprimere de sensu* assumes precisely that which requires demonstration. It predicates a literal meaning attached to verbal units, normally envisaged as single words in

[1] Ludwig Wittgenstein, *Zettel*, 698 (Oxford, 1967), p. 121.
[2] Previously, one would have said 'theological'. The change is one of terminological 'respectability'. But it is their rejection of this conventional change, and their refusal to allow the implicit differentiation, which give to the work on translation of Rosenzweig and Walter Benjamin its special depth and importance.

a purely lexical setting, which differs from, and whose straight-forward transfer will falsify, the 'true sense' of the message. Depending on the degree of logical sophistication available to him, the writer on translation will treat 'meaning' as more or less inherently transcendental. The underlying image is crude and, more often than not, left vague. 'Meaning' resides 'inside the words' of the source text, but to the native reader it is evidently 'far more than' the sum of dictionary definitions. The translator must actualize the implicit 'sense', the denotative, connotative, illative, intentional, associative range of significations which are implicit in the original, but which it leaves undeclared or only partly declared simply because the native auditor or reader has an immediate understanding of them. The native speaker's at-homeness, largely subconscious because inherited and cultural-specific, in his native tongue, his long-conditioned immersion in the appropriate context of the spoken or written utterance, make possible the economy, the essential implicitness of customary speech and writing. In the 'transference' process of translation, the inherence of meanings, the compression through context of plural, even contradictory significations 'into' the original words, get lost to a greater or lesser degree. Thus the mechanics of translation are primarily explicative, they explicate (or, strictly speaking, 'explicitate') and make graphic as much as they can of the sem-antic inherence of the original. The translator seeks to exhibit 'what is already there'. Because explication is additive, because it does not merely restate the original unit but must create for it an illustrative context, a field of actualized and perceptible ramifica-tion, translations are inflationary. There can be no reasonable presumption of co-extension between the source text and the translation. In its natural form, the translation exceeds the original or, as Quine puts it: 'From the point of view of a theory of translational meaning the most notable thing about the analytical hypotheses is that they exceed anything implicit in any native's dispositions to speech behavior.'[1]

This is unavoidable given the fact that the epistemological and

[1] W. van Orman Quine, *Word and Object*, p. 70.

formal grounds for the treatment of 'meaning' as dissociable from and augmentative to 'word' are shaky at best. The underpinning argument is not analytic but circular or, in the precise sense, circumlocutionary. It assumes an analysable understanding of the procedures by which 'meanings' are derived from, are internal to, or transcend 'words'. But it is just this understanding which translation claims to validate and enact (the circularity involved in the case makes the assertions of Whorf so central and vulnerable). To put it another way: from Cicero and Saint Jerome until the present, the debate over the extent and quality of reproductive fidelity to be achieved by the translator has been philosophically naïve or fictive. It has postulated a semantic polarity of 'word' and 'sense' and then argued over the optimal use of the 'space between'. This crude scheme undoubtedly reflects the ways in which we go about natural speech. It corresponds to that twofold motion of reference ('looking up') and expansive restatement which impels much of natural discourse. 'The intuitions,' allows Quine, 'are blameless in their way.' The theory of translation, so largely literary and *ad hoc*, ought not to be held to account for having failed to solve problems of meaning, of the relations between words and the composition of the world to which logic and metaphysics continue to give provisional, frequently contradictory answers. The fault, so far as the theory goes, consists of having manoeuvred *as if* these problems of relation were solved or as if solutions to them were inferentially obvious in the act of translation itself. *Praxis* goes ahead, must go ahead *as if*; theory has no licence to do so.

It is worth noting that the development of modern phenomenology has accentuated the areas of overlap between translation theory and the general investigation of sense and meaning. The conceptual claims, the idiom of Husserl, Merleau-Ponty and Emmanuel Levinas force on anyone concerned with the nature of translation a fuller awareness of, a more responsible discomfort at, notions of identity and otherness, of intentionality and signification. When Levinas writes that 'le langage est le dépassement incessant de la Sinngebung par la signification' (significance constantly transcends designation), he comes near to equating all speech-acts with translation in the way indicated at the outset of this study.[1] Phenomenological ontologies look very much like meditations on the 'transportability' of meanings.

But does this increasing reciprocity between epistemology and logic on the one hand, and the theory of translation–interpretation on the other, give any promise of systematic understanding? In fact, what do we mean here by 'understanding'?

Suppose we put the question in its strongest form: 'what, then, is translation?'; 'how does the human mind move from one language to another?' What sort of answers are being called for? What must be established for such answers to be plausible or, indeed, possible? The theory and analysis of translation have, until now, proceeded as if we knew, or as if the knowledge needed to make the question nontrivial were foreseeable given a reasonable time span and the current rate of progress in psychology, linguistics, or some other authenticated 'sciences'. I believe, on the contrary, that we do not know with any great precision or confidence what it is that we are asking and, concomitantly, what meaningful answers would really be like. A radical indeterminacy characterizes the question, conceivable answers, and our sense of the relation between them. To show this is to summarize all I have said so far.

3

A 'theory' of translation, a 'theory' of semantic transfer, must mean one of two things. It is either an intentionally sharpened, hermeneutically oriented way of designating a working model of *all* meaningful exchanges, of the totality of semantic communication (including Jakobson's intersemiotic translation or 'transmutation'). Or it is a subsection of such a model with specific reference to interlingual exchanges, to the emission and reception of significant messages between different languages. The preceding chapters have made my own preference clear. The 'totalizing' designation is the more instructive because it argues the fact that

[1] Emmanuel Levinas, *Totalité et infini* (The Hague, 1961), p. 273. Cf. also pp. 35–53, 179–83, 270–4.

all procedures of expressive articulation and interpretative reception are translational, whether intra- or interlingually. The second usage—'translation involves two or more languages'—has the advantage of obviousness and common currency; but it is, I believe, damagingly restrictive. This, however, is not the point. Both or either concepts of 'theory', the totalizing or the traditionally specific, can be used with systematic adequacy only if they relate to a 'theory of language'. This relation can be of two types. It is either one of complete overlap and isometry, i.e. 'a theory of translation is in fact a theory of language'. Or it can be one of strict formal dependence, i.e. 'the theory of language is the whole of which the theory of translation is a part'. The totality of Geometries comprehends, is perfectly homologous with, the study of the properties and relations of all magnitudes in all conceivable spaces. This is the first sort of relation. A particular geometry, projective geometry for example, derives rigorously from, is a part of, the larger science. This is the second sort. But it is possible neither to have a 'theory of projective geometry' nor a 'theory of geometrical meaning' without a 'theory of Geometry or Geometries' to begin with.

This platitude needs underlining. Even Quine lacks caution in his resort to the enhancing rubric of what is a genuine 'theory'. The bare notion of a mature theory of how translation is possible and how it takes place, of a responsible model of the mental attributes and functions which are involved, *presumes* a systematic theory of language with which it overlaps completely or from which it derives as a special case according to demonstrable rules of deduction and application. I can see no evasion from this truism. But the fact remains that we have no such theory of language (here again there has been no sufficiently stringent investigation of just what this phrase entails). The evidence available on key matters which such a theory would have to axiomatize and define is far from being in any stable, statistical comprehensive, or experimentally controllable state. In the main it consists of fragmentary data, rival hypotheses, intuitive conjectures, and bundles of images. On the crucial issues—crucial, that is, in regard to a systematic understanding of the nature of translation—linguistics is still in a roughly hypothetical stage. We have some measurements, some scintillating tricks of the trade and far-ranging guesses. But no Euclidean *Elements*.

Every understanding is actively interpretative. Even the most literal statement (what, actually, is a 'literal' statement?) has a hermeneutic dimension. It needs decoding. It means more or less or something other than it says. Only tautologies are coextensive with their own restatement. Pure tautologies are, one suspects, extremely rare in natural language. Occurring at successive moments in time, even repetition guarantees no logically neutral equivalence. Thus language generates—grammar permitting, one would want to say 'language is'—a surplus of meaning (meaning is the surplus-value of the labour performed by language). A fundamental asymmetry is operative in the process and means of linguistic signification. There may be a deep if elusive clue here to the question of origins about which, as we have seen, almost nothing sensible can be said. Asymmetry between means and yield may be a logical but also an evolutionary feature of language.

In an estimated 97 per cent of human adults language is controlled by the left hemisphere of the brain. The difference shows up in the anatomy of the upper surface of the temporal lobe (in 65 per cent of cases studied, the *planum temporale* on the left side of the brain was one-third longer than on the right).[1] This asymmetry, which seems to be genetically determined, is dramatized by the fact that the great majority of human beings are right-handed. Evidence for this goes back to the earliest known stone tools. No such cerebral unbalance has been found in primates or any other animal species. E. H. Lenneberg has suggested, in his *Biological Foundations of Language*, that there may be intricate bio-genetic and topological connections between asymmetry and the origins of speech. Perhaps the point can be put more generally.

It has been conjectured that hominids descended from the trees in the late Miocene or early Pliocene Ages. This move into level territory would entail an extraordinary enrichment and complica-

[1] Cf. Norman Geschwind and Walter Levitsky, 'Human Brain: Left–Right Asymmetries in Temporal Speech Regions' (*Science*, CLXI, 1968), and Norman Geschwind, 'Language and the Brain' (*Scientific American*, CCXXVI, 1972).

tion of social encounters. The archaic system of calls is no longer adequate and language comes to replace it. (Again a curious asymmetry or 'slippage' turns up: the human ear is most sensitive to sounds whose pitch corresponds to a frequency of about 3,000 cycles per second, whereas the ordinary speaking voice of men, women and children is at least two octaves lower in the scale. This may mean that call-systems and language coexisted, at least for a long time, on neighbouring frequencies.) Some anthropologists argue that the emergence of 'true language' was more sudden, that it coincided with the abrupt forward leap in the elaboration and diversity of tool-making towards the end of the last Ice Age. Neither hypothesis can be verified. But it might be that neither sees the full import of asymmetry. Pavlov's often-reiterated belief is worth recalling: the processes of learning and of language in men are different from those in animals. The upgrading in complexity is such as to make for a quantum jump. We are able to say so fantastically much more than we would need to for purposes of physical survival. We mean endlessly more than we say. The sources of superfluity, with their anatomical analogue in the asymmetries of the cortex, generate new surpluses. Asymmetry, in the central sense of which the configurations of the brain are the enacting form, was the trigger. It set in motion the dissonance, the dialectic of human consciousness. Unlike animal species we are out of balance with and in the world. Speech is the consequence and maintainer of this disequilibrium. Interpretation (translation) keeps the pressures of inventive excess from overwhelming and randomizing the medium. It limits the play of private intention, of plurality in meaning, at least at a rough and ready level of functional consensus. In an ambiguity which is at one level ontological and at another ironic, idiomatic level, political or social, we speak left and act right. Translation mediates; it constrains the constant drive to dispersion. But this too, of course, is conjecture.

Virtually everything we know of the organization of the functions of language in the human brain derives from pathology. It has been recorded under abnormal conditions, during brain surgery, through electrical stimulation of exposed parts of the brain, by observing the more or less controlled effects of drugs on cerebral functions. Almost the entirety of our picture of how language 'is located in' and produced by the brain is an extrapolation from the evidence of speech disorders followed by the study of dead tissue. This evidence, which dates back to Paul Broca's famous papers of the 1860s, is voluminous. We know a good deal about specific cerebral dominance, i.e. the unilateral control of certain speech functions by particular areas of the cortex. Damage to Broca's area (the third frontal gyrus on the left side) produces a characteristic aphasia. Articulation becomes slurred and elliptic; connectives and word endings drop away. Damage to the Wernicke area, also in the left hemisphere but outside and to the rear of Broca's area, causes a totally different aphasia. Speech can remain very quick and grammatical, but it lacks content. The patient substitutes meaningless words and phrases for those he would normally articulate. Incorrect sounds slip into otherwise correct words. The fascinating corollary to the aphasia described by Carl Wernicke, some ten years after Broca, is its suggestive proximity to the generation of neologisms and metaphor. In many known cases the results of verbal or phonemic paraphasia (ungoverned substitution) are almost inspired. There is a sense in which a great poet or punster is a human being able to induce and select from a Wernicke aphasia. The 'Sinbad the sailor' sequence from Joyce's *Ulysses* gives a fair illustration. But with a crucial difference: though aural reception of non-verbal sounds and of music may remain perfectly normal, a lesion in the Wernicke area will cut down severely on understanding. When both areas are intact but disconnected, the result is conduction aphasia. Fluent but abnormal speech continues, together with a large measure of comprehension. The patient is, however, incapable of repeating spoken language.

The study of these aphasias and of many other aspects of the neurophysiology of the brain does allow the construction of a possible model for the organization of speech. A division of functions takes place between Broca's area and Wernicke's depending on whether language is heard or read. When a word is read, for example, the angular gyrus located towards the rear of the left hemisphere receives a stimulus from the primary visual areas of the cortex. Having, as it were, passed through the 'transformer', this stimulus in turn arouses the corresponding auditory form of the word in the Wernicke area. If the word is to be spoken, the 'current' moves in the reverse direction, from Wernicke to Broca.[1]

Even to know so much or to have enough evidence to sustain such a model is a momentous achievement. Its therapeutic and cognitive implications are obvious. But it is by no means clear that a neurophysiological scheme and the deepening analysis and treatment of pathological states will lead to an understanding of the production of human speech. To know how a process is organized, to have a flowchart of sequential operations, is not, necessarily, to know the nature of the energies involved. A phenomenon can be mapped, but the map can be of the surface. To say, as do the textbooks, that the third frontal gyrus 'transforms' an auditory input into a visual–verbal output or feedback, is to substitute one vocabulary of images for another. Unlike the 'animal spirits' of Cartesian physiology, the new electro-chemical vocabulary allows and rationalizes medical treatment. This is an immense step forward. But it is an empirical and not, necessarily, analytic step. We do not know *what* it is we are talking about, though our discourse may induce profitable, experimentally verifiable techniques of treatment.

What are the dynamics of conceptualization? In what ways are sensory stimuli translated into, matched with appropriate verbal units? To what extent are visual, auditory, olfactory, and tactile perceptions themselves triggered and constrained by the (pre-set, self-correcting?) verbal matrix? How are words or units of information 'banked'? What is the electro-chemistry of scanning and of memory which ensures the right sequence of input, classification, recall, and emission? Does speech become organized,

---

[1] Cf. O. L. Zangwill, *Cerebral Dominance and Its Relation to Psychological Function* (London, 1960); T. Alajouanine, *L'aphasie el le langage pathologique* (Paris, 1968); A. R. Luria, *Traumatic Aphasia: Its Syndromes, Psychology and Treatment* (The Hague, 1970). For the intriguing suggestion that the limited capacities for speech of the right hemisphere could represent language at an exceedingly primitive level, cf. the report on the work of M. S. Gazzaniga in *New Scientist*, LIII, 1972, p. 365. The findings were first reported in *Neuropsychologia*, IX, 1972.

rule-governed at the interface between older and newer areas of the cortex? Is it, in some sense which we cannot even phrase adequately, an adaptive imitation of those much earlier, 'deeper' processes of encoding, replication, and punctuation which could parallel the genetic structure and transmission of organic forms? In what ways are the language-centres of the cortex subject to further evolution? (Can we even 'imagine' a more evolved mode of speech?)

An impressive amount of thought and experimental research is going into these problems at the present time. The mathematics of multi-dimensional interactive spaces and lattices, the projection of 'computer behaviour' on to possible models of cerebral functions, the theoretical and mechanical investigation of artificial intelligence, are producing a stream of sophisticated, often suggestive ideas. But it is, I believe, fair to say that nothing put forward until now in either theoretic design or mechanical mimicry comes even remotely in reach of the most rudimentary linguistic realities. The gap is not only one of utterly different orders of complexity. It seems rather as if the concept of a neurochemical 'explanation' of human speech and consciousness—the two are very nearly inseparable—were itself deceptive. The accumulation of physiological data and therapeutic practice could be leading towards a different, not necessarily relevant, sort of knowledge. There is nothing occult about this divergence. I have stressed throughout that the questions we ask of language and the answers we receive in (from) language are unalterably linguistic. We can neither formulate questions nor state replies outside the structures of language which are themselves the object of inquiry. It is not evident that the sciences, however advanced, will offer a reasonable procedure for arriving at an external view. We know no exit from the skin of our skin. This also, to be sure, is conjecture. What is certain is the fact that no model available at present or foreseeable in the fairly near future justifies any confident invocation of a 'theory of the generation of speech or of the transformation of cognitive material into semantic units'.

Zoologists report that the call-systems of gibbons have differentiated into what might be termed local 'dialects'. The signals emitted by whales and dolphins seem to show a certain degree of specificity and variation as between particular herds or schools. But there is no way of determining whether such phonetic variations, with their obvious utility for mutual recognition and territorial assertion, are in any way analogous to or a rudimentary stage of the differentiation in human speech forms. The diversity and mutual incomprehensibility of human tongues are, so far as we have any evidence, unique to man and inseparable from the existence of language as we know it. Nothing is known of their beginnings or fundamental aetiology.

I have sketched my own conviction. In significant measure, different languages are different, inherently creative counter-proposals to the constraints, to the limiting universals of biological and ecological conditions. They are the instruments of storage and of transmission of legacies of experience and imaginative construction particular to a given community. We do not yet know if the 'deep structures' postulated by transformational-generative grammars are in fact substantive universals. *But if they are, the immense diversities of languages as men have spoken and speak them can be interpreted as a direct rebellion against the undifferentiated constraints of biological universality. In their formidable variety 'surface structures' would be an escape from rather than a contingent vocalization of 'deep structures'.* Languages communicate inward to the native speaker with a density and pressure of shared intimation which are only partly, grudgingly yielded to the outsider. A major portion of language is enclosure and willed opaqueness. The intent is so ancient, its execution so remote from our public states of mind that we are not consciously aware of it. But it lives on in the layered fabric, in the tenacious quiddity of language, and becomes obvious when languages meet.

These points cannot be proved. I strongly feel that the hypothesis of 'alternity' and meta- or non-information is the one which describes most coherently the actual facts of linguistic diversity. It seems to me to take in more of semantic, historical, and psychological reality than other conjectures do. We will see how it forces itself upon one during the study of actual problems of translation, when one is concretely involved with the polysemic, hermetic nature of utterance. It is conceivable that we have misread the Babel myth. The tower did not mark the end of a blessed monism, of a universal-language situation. The bewildering prodigality of tongues had long existed, and had materially complicated the enterprise of men. In trying to build the tower, the nations stumbled on the great secret: that true understanding is possible only when there is silence. They built silently, and there lay the danger to God.

Whatever its causes, the multilingual condition invites or compels a certain percentage of mankind to speak more than one language. It also means that the exchanges of information, of verbalized messages on which history and the life of society depend, are in very large part interlingual. They demand translation. The polyglot situation and the requirements which follow from it depend totally on the fact that the human mind has the capacity to learn and to house more than one tongue. There is nothing obvious, nothing organically necessitated about this capacity. It is a startling and complex attribute. We know nothing of its historical origins, though these are presumably coincident with the beginnings of the division of labour and of trade between communities. We do not know whether it has limits. There are reliable records of polyglots with some measure of fluency in anywhere up to twenty-five languages. Is there any boundary other than the time span of individual lives? The study of the learning and development of speech in infants and young children is a large field.[1] Though Chomskyan theories greatly undervalue the role of environmental as against innate factors—surely it is clear that *both* are involved and interactive—generative grammars have given a powerful impetus to the investigation of how speech is acquired. There have also been inquiries into the linguistic growth of bilingual individuals.[2] But until now results have been either of

---

[1] For a lucid survey cf. M. M. Lewis, *Language, Thought and Personality in Infancy and Childhood* (London, 1963). Cf. also D. O. Hebb, W. E. Lambert, E. R. Tucker, 'Language, Thought and Experience' (*The Modern Language Journal*, LV, 1971).

[2] The most detailed study remains that of W. Leopold, *Speech Development of a Bilingual Child: a Linguist's Record* (Northwestern University Press, 1939–47).

the most general, intuitive sort, i.e. the ability to learn a second or third language with ease diminishes with age, or they have been fairly trivial statistics on the rates of acquisition of vowels, consonants, and phonemes during early years of life.[1] Neither the Chomskyan model of competence/performance, nor sociolinguistic surveys of multilingual children or communities tell us what is meant by 'learning a language' or by 'learning two or more languages', at the crucial level of the central nervous system.

Claims made towards a biochemical understanding of learning and of memory have recently been dramatic. From the point of view of the human brain the process of learning constitutes the most immediate environmental change. The research of Holger Hydén, of Steven Rose, and of other neurophysiologists and biochemists has shown that learning, which can be defined as repeated exposure to the stimulus of information, is accompanied by changed patterns of protein synthesis in the relevant areas of the cortex. There is evidence that a particular environmental change will activate a specific group or population of neurones. If the change is focused and sustained, as occurs during the reception and internalization of 'experience–information', corresponding alterations take place in the properties of these neurones. There are experimental grounds for believing that their configurations and patterns of assembly change. This 're-configuration' would provide the physical basis and organization of memory. When the stimulus weakens, becomes merely occasional, or is altogether absent, i.e. when the brain is no longer, or only rarely called upon to register and redeploy the given body of

[1] Cf. Roman Jakobson, 'Les lois phoniques du langage enfantin et leur place dans la phonologie générale', in N. S. Troubetzkoy, *Principes de phonologie* (Paris, 1949), and Helen Couteras and Sol Saporta, 'Phonological Development in the Speech of a Bilingual Child' in *Language Behavior*, compiled by J. Akin, A. Goldberg, G. Myers, J. Stewart (The Hague, 1970). Three special aspects of bilingual learning are examined respectively in W. E. Lambert, 'Measurement of the linguistic dominance of bilinguals' (*Journal of Abnormal Social Psychology*, L, 1955); M. S. Preston and W. E. Lambert, 'Interlingual Interference in a Bilingual Version of the Stroop Color-Word Task' (*Journal of Verbal Learning and Verbal Behavior*, VIII, 1969); and J. C. Yuille, A. Paivio, W. E. Lambert, 'Noun and Adjective Imagery and Order in Paired-Associate Learning by French and English Subjects' (*Canadian Journal of Psychology*, XXIII, 1969).

information, the neuronal changes dissipate and the neurones revert to their original, possibly undifferentiated or randomized grouping. Even as information is energy, so forgetting is entropy. There is also beginning to be some evidence as to couplings between the electrical activities of the cortex under stimulus and the subsequent biochemical events which seem to regulate the reception, the storage, and the retrievability of knowledge in and by the human brain.

Over the next years there *may* be a spectacular progress of insight into the biochemistry of the central nervous system. Though it is conceptually and practically extremely difficult to isolate a single type of stimulus from the fact of stimulation as such (environment interconnects at every point), refinements in microbiology may lead to correlations between specific classes of information and specific changes in protein synthesis and neuronal assembly. At the biochemical level, the idea that we are 'shaped' by what we learn could take on a material corollary. On present evidence, however, it is impossible to go beyond rudimentary idealizations. The neurochemistry of language-acquisition, the understanding of the changes in RNA which may accompany the 'storage' of a language in the memory centres and synaptic terminals of the cortex, necessitate models of a complexity, of a multi-dimensionality beyond anything we can now conceive of. Information can be conceived of as environment. The learning process and the ordered 'stacking' of memory must themselves constitute a dynamic, multi-directional phenomenon. The brain is never a passive tympanum. The act of internalization, however subconscious or reflexive, presumably triggers an immensely ramified field of associative recognitions, relocations, and serial impulses. Reasoning by analogy most probably has its counterpart in neuronal mechanisms through which a new unit of input is tagged and 'inserted' in its proper location. One must think of the cortex as an active space in which stimulus and response, continuity and change, inheritance and environment are totally reciprocal, totally definitional of each other.

By 'environment', moreover, much more is intended than the neurochemistry of stimulus acting on innate bio-genetic structures. Learning and memory are conditioned, at every level, by social and historical agencies. Information is neither in substance nor conceptually value-free. Ideology, economic and class circumstance, the historical moment do much to define the content, the relative hierarchies, the sheer visibility of knowledge as knowledge, of information or experience as worth recording. These categories are not permanent. Different societies, different epochs expose the central nervous system to different fields of stimulation. This is decisively the case in regard to language. A theory of the generation of language based on a conjectural postulate of innate competence and on the performance of an 'ideal speaker–listener relation' is no more than naked abstraction. The interface between the neurochemistry of language-learning and language-recall on the one hand, and the socio-historical framework in which an actual human being uses natural language on the other, is no remote, external boundary. The cortex and the 'world outside' in which language can be seen as a form of work, of social production, of economic and ideological exchange, cannot be meaningfully separated. Together they make up the generative environment of consciousness, the fabric of consciousness which is also environment.[1] But the number of parameters and variants is so great, and the modes of interaction are, by all evidence, so complex, that we cannot systematically represent or analyse them with the resources now available or, it may be, foreseeable.

Introspectively, one draws pictures. Thus one describes oneself as 'looking for' a word. Whenever it is baffled or momentarily vacuous, the search, the act of scanning, suggests circuitry. The relevant sensation or, more cogently, the vulgarized images we

---

[1] It is on this point that Marxist critiques of Chomskyan linguistics as an 'empty mentalism' no less naïvely-deterministic than the theories of Skinner have been most telling. Cf. F. Rossi-Landi, *Il linguaggio come lavoro e come mercato* (Milan, 1968); J. Kristeva: Σημειωτική. *Recherches pour une sémanalyse* (Paris, 1969), particularly pp. 280–5; Denis Slakta, 'Esquisse d'une théorie lexico-sémantique: pour une analyse d'un texte politique' in *Langages*, XXIII, 1971; Augusto Ponzio, 'Grammatica transformazionale e ideologia politica' in *Idéologie*, XVI–XVII, 1972. For a summary statement and full bibliography, cf. F. Rossi-Landi, *Ideologies of Linguistic Relativity* (The Hague, 1973).

make up of what are subliminal processes, leave one with a compelling notion of nervous probes 'trying this or that connection', recoiling where the wire is blocked or broken and seeking alternative channels until the right contact is made. The sensation of a 'near-miss' can be tactile. The sought word or phrase is a 'micromillimeter away from' the scanner; it is poised obstinately at the edge of retrieval. One's focus becomes excited and insistent. It seems to press against a material impediment. The 'muscles' of attention ache. Then comes the breach in the dam, the looked-for word or phrase flashing into consciousness. We know nothing of the relevant kinetics, but the implication of a correct location, of a 'slotting into place' is forceful, if only because of the muted but unmistakable impression of release, of a calming click which accompanies the instant of recall. When the right word is found, compression gives, and a deep-breathing currency—in the dual sense of 'flow' and 'integrated routine'—resumes. In contrast, under the spur of stimulants or histrionic occasion, or in the strange weightless tension of tiredness of mind, resistance seems to diminish in the verbal circuits and synapses multiply. Every bell chimes. Homonyms, paronomasia, acoustic and semantic cognates, synecdochic sets, analogies, associative strings proliferate, undulating at extreme speed, sometimes with incongruous but pointed logic, across the surfaces of consciousness. The acrostic or cross-word yields faster than our pencil can follow. We seem to know even more than we had forgotten, as if central sediments of memory or reserves normally unrecorded, because lightly imprinted or laid down without deliberate marking, had been galvanized. At yet another level of banal experience there are short-circuits and wires fuse. The identical morpheme, tonal combination, or atrophied phrase forces itself on the inner ear, insistently, like a bulb going on and off pointlessly. Some part of the memory current is trapped. Dreams, one suspects, may be attempts at associative context, pictorializations seeking to provide an *ad hoc* rationality, around crossed wires of blocked subconscious speech.

Penumbral as they are, and awkwardly dependent on the patronage of a contingent body of metaphor—that of electric circuits and storage batteries, or, at a mildly more dignified remove, of holograms and data-banks—all these sub-articulate sensations of tensed search, of decompression after the find, of lowered resistance under certain conditions, of wires crossed or fused, do point towards a spatial matrix, towards orderings in dimensionality. Language would seem to have or inhabit volume.

For the polyglot this impression is reinforced. He 'switches' from one language to another with a motion that can have a lateral and/or a vertical feel. As he moves from his native tongue to one acquired later, the impression of a steepening slope, of more constrained apertures, can be visceral. With constant recourse, the gradient levels. This is a common observation. As is the truism that neglect, the lying fallow, even of one's first language, though in this case to a lesser degree, will cause a certain dimming, a recession of vocabulary and of grammatical nuance from immediate recall. A mixed, contingent usage of two languages, on the other hand, can create interference effects, the phrase being sought in one idiom being 'crowded out' or momentarily screened by a phrase in the other. Impressionistic and banal as they are, these experiences, with their frequent aura of a deep-seated muscular or, at least, neurophysiological embodiment, again point to localization. The different languages known and used by the polyglot would somehow be 'spatialized' in his cortex. Very recent work with bilingual schizophrenics ('schizophrenia' being itself an unsatisfactory, catch-all term) may provide a similar clue. Patients who hear 'voices' or report hallucinations will locate these phenomena in only one of their two languages. Questioned in the other or 'safe' tongue, their answers and introspective testimony reveal no pathological interference. The implications are that functional brain damage in certain types of schizophrenia is limited to one area of verbal expression while leaving others intact, and that different areas can therefore be regarded as containing or mapping different languages.[1]

What is certain is that the immediacy, the retrievability of

[1] The experimental work has been done by R. E. Hemphill of the Groote Schuur Hospital in Capetown. It was reported in *The Times* of London for 10 January 1972, p. 3.

different tongues in the speech-acts of the polyglot is, in crucial part, a function of the environment. Different moods, different social settings, different locations strongly modify the sense of linguistic priority. When I have spent a few days in a country in which one of my 'first' languages is native, I not only find myself re-entering that language with a strong sensation of recollected fluency and central logic, but soon have my dreams in it. In a short time-interval the language which I have been speaking in another country takes on a tangible shell of strangeness. It has shifted both horizontally and in regard to centrality (there is a depth of burial and a very different depth of focal, natural recourse). This susceptibility of linguistic 'placing' to the influence of the surrounding social, psychological, and acoustical milieu is, by itself, sufficient to refute the more extreme theories of transformational-generative innateness. The external world 'reaches in' at every instant to touch and regroup the layers of our speech.

'Layers' is, of course, a piece of crass shorthand. It may mean nothing. The spatial organization, contiguities, insulations, synaptic branchings between, which account for the arrangement of different languages in the brain of the polyglot, and especially of the native bilingual, must be of an order of topological intricacy beyond any we can picture. I harbour the feeling that the reticulations of interlingual contact and transfer in my own mind, as in that of any polyglot, belong to at least two principal hierarchies. The one seems to draw on the objective analogies ('cross-echoes') and mnemonically salient contrasts between phonetic units in the several languages. The other would appear to be based on a prodigiously tangled and private network of associations between morphemes or semantic units on the one hand and the circumstances of my own life on the other. This second topology operates irrespective of formal linguistic barriers. In other words, at least one of the modes of spatialization of phonetic, grammatical and semantic material in my consciousness interleaves the languages I know according to criteria of proximity or antithesis, of cognateness or exclusion, which are wholly personal and interlingual. Thus one of the 'languages' inside me, probably the richest, is an eclectic cross-weave whose patterns are unique to myself though the fabric is quite palpably drawn from the public means and rule-governed realities of English, French, German, and Italian. Moving 'between' languages, moreover, in what I obscurely apprehend as a complex, highly energized zone of modulation and indeterminacy, I register contiguities, correspondences, short-cuts which are based not only on speech-sounds, on patterns of meaning, on associations particular to my own life, but on word-shapes and tactile values. The implicit phenomenon is general but little understood. Words have their 'edge', their angularities, their concavities and force of tectonic suggestion. These features operate at a level deeper, less definable than that of either sound or semantics. They can, in a multilingual matrix, extend across and between languages. When we learn a new language, it may be that these modes of evocative congruence are the most helpful. Often, as we shall see, great translation moves by touch, finding the matching shape, the corresponding rugosity even before it looks for counterpart of meaning. It was probably the mellifluous convexity of *quamve* (cf. German *Qualm*) followed by the literal sharpness—acoustic as well, of course—of *bibistis*, and reinforced by *aquam*, itself a less 'liquid' word than *quamve*, which set off Pound's traverse in the *Homage to Sextus Propertius*: 'what water has mellowed your whistles?' Poets can even smell words.

Yet all these are only naïve pictures, made up of impressions, half-realized metaphors, and analogies with counters as obvious as electronics. It is very likely that the internalization of language and of languages in the human mind involves phenomena of ordered and ordering space, that temporal and spatially-distributive hier-archies are involved. But no topologies of n-dimensional spaces, no mathematical theories of knots, rings, lattices, or closed and open curvatures, no algebra of matrices can until now authorize even the most preliminary model of the 'language-spaces' in the central nervous system. These allow the autonomous existence of single languages while, at the same time, making possible the acquisition of other languages and the most intense degree of mutual penetration. They permit languages to recede from either the 'surface' or the 'centre' of immediate fluency, and then allow their return. The membranes of differentiation and of contact, the dynamics of interlingual osmosis, the constraints which preserve equilibrium between the blandness of mere lexical, public usage and the potentially chaotic prodigality of private invention and association, the speed and delicacy of retrieval and of discard involved in even the barest act of paraphrase or translation—all these are of a class of intricacy and evolutionary uniqueness of which we can, at present, offer no adequate image let alone systematic analysis.[1]

To summarize: we have no working model of the fundamental neurochemistry and historical aetiology of human speech. We have no anthropological evidence as to the causes or chronology of its thousandfold diversification. Our models of the learning process and of memory are ingenious but also of the most preliminary, conjectural kind. We know next to nothing of the organization and storage of different languages when they coexist in the same mind. How then can there be, in any rigorous sense of the term, a 'theory of translation'?

In view of the claims put forward by linguistics since the late 1950s I have, in the foregoing chapters, tried to show that the study of language is not now a science. In closing the abstract portion of this work, I am tempted to go further. Very likely, it never will be a science. Language is, at vital points of usage and understanding, idiolectic. When an individual speaks, he is effecting a partial description of the world. Communication depends on a more or less complete, more or less conscious translation of this partiality, on a matching, more or less perfunctory, with other 'partialities'. A 'complete translation', i.e. a definitive insight into and generalization of the way in which any human being relates word to object would require a complete access to him on the part of his interlocutor. The latter would have to experience a 'total mental change'. This is both logically and substantively a meaningless notion. It could never be shown to have taken place. All discourse, all interpretation of discourse

[1] The most sophisticated attempt made so far to provide such an analysis is that by René Thom. See his *Stabilité structurelle et morphogénèse* (Reading, Mass., 1972), pp. 124–5, 309–16.

works at a word-for-word and sentence-for-sentence level. There
is no privileged access to underlying totality.

What then are we dealing with as we now turn to material,
sociological, cultural aspects of translation? In Wittgenstein's
terms we will look at 'solutions', often inspired and crucially
helpful to our understanding of languages and of the history of
feeling; but we shall not be looking at a universal, an axiomatic
or externally verifiable 'method of solution'. Every interlingual
transfer, says Quine, is ruled by a principle of indeterminacy.
'There can be no doubt that rival systems of analytical hypotheses
can fit the totality of speech behavior to perfection, and can fit the
totality of dispositions to speech behavior as well, and still specify
mutually incompatible translations of countless sentences insus-
ceptible of independent control.'[1] We have seen that the reasons

[1] W. van Orman Quine, *Word and Object*, p. 72. Though formulated in an
altogether different philosophical idiom, Wittgenstein's pronouncements on
translation in the *Investigations* (23, 206, 243, 528) are closely parallel to Quine's
view of indeterminacy. Quine's thesis on the formally indeterminate plurality of
equally valid translations of given sentences has generated much controversy. Cf.
the exchange between R. Kirk, 'Translation and Indeterminacy' (*Mind*, LXXVIII,
1969), and A. Hyslop, 'Kirk on Quine on Bilingualism' (*Mind*, LXXXI, 1972). The
most searching critique so far is that made by John M. Dolan in 'A Note on
Quine's Theory of Radical Translation' (*Mechanical Translation and Computer
Linguistics*, X, 1967). Dolan sets out to show by a rigorous analysis of Quine's
premisses 'that the theory is, at best, an incomplete account and, thus, does not
follow from the analysis intended to support it'. Dolan's critique and his suggestion
that his argument undermines some part of Quine's well-known misgivings over
the distinction between analytic and synthetic, are impressive. But they seem to me
to strengthen the 'empirical–descriptive' or 'empirical–intuitive' elements of
Quine's model. The latter still seems to account more satisfactorily than any other
put forward by a logician for the indeterminacy in the translation of 'non-observa-
tional occasion sentences' and for the actual conformities observed in the tacit
analytical hypotheses of bilinguals. In short, Dolan's refutation makes more graphic
precisely the anthropological–linguistic situation which Quine posits. Professor
Dummett's critical treatment of Quine's account of indeterminacy in M. Dummett,
Frege: Philosophy of Language (London, 1973), pp. 612–23, appeared too late for me
to profit from. I would draw attention only to Dummett's crucial remark (p. 617)
that there is in Quine's model of the multiplicity of different possible translations
nothing which would prevent us from ascribing this 'apparent incompatibility to
equivocation'. This, exactly, is the point I have tried to make. But what strikes
Professor Dummett and Quine's other professional critics, fairly no doubt, as a
systematic flaw, seems to me to be part of the realism and psychological acumen of
Quine's exposition.

lie in the very nature of language and linguistic diversity, that they are inseparable from the functions of non-information, privacy, and poetics which are the creative attributes of human speech.

An error, a misreading initiates the modern history of our subject. Romance languages derive their terms for 'translation' from *traducere* because Leonardo Bruni misinterpreted a sentence in the *Noctes Atticae* of Aulus Gellius in which the Latin actually signifies 'to derive from, to lead into'. The point is trivial but symbolic. Often, in the records of translation, a fortunate misreading is the source of new life. The precisions to be aimed at are of an intense but unsystematic kind. Like mutations in the improvement of the species, major acts of translation seem to have a chance necessity. The logic comes after the fact. What we are dealing with is not a science, but an exact art. Some examples follow.

Chapter Five

THE HERMENEUTIC MOTION

I

THE hermeneutic motion, the act of elicitation and appropriative transfer of meaning, is fourfold. There is initiative trust, an investment of belief, underwritten by previous experience but epistemologically exposed and psychologically hazardous, in the meaningfulness, in the 'seriousness' of the facing or, strictly speaking, adverse text. We venture a leap: we grant *ab initio* that there is 'something there' to be understood, that the transfer will not be void. All understanding, and the demonstrative statement of understanding which is translation, starts with an act of trust. This confiding will, ordinarily, be instantaneous and unexamined, but it has a complex base. It is an operative convention which derives from a sequence of phenomenological assumptions about the coherence of the world, about the presence of meaning in very different, perhaps formally antithetical semantic systems, about the validity of analogy and parallel. The radical generosity of the translator ('I grant beforehand that there must be something there'), his trust in the 'other', as yet untried, unmapped alternity of statement, concentrates to a philosophically dramatic degree the human bias towards seeing the world as symbolic, as constituted of relations in which 'this' can stand for 'that', and must in fact be able to do so if there are to be meanings and structures.

But the trust can never be final. It is betrayed, trivially, by nonsense, by the discovery that 'there is nothing there' to elicit and translate. Nonsense rhymes, *poésie concrète*, glossolalia are untranslatable because they are lexically non-communicative or deliberately insignificant. The commitment of trust will, however, be tested, more or less severely, also in the common run and process of language acquisition and translation (the two being intimately connected). 'This means nothing' asserts the exasperated child in front of his Latin reader or the beginner at Berlitz. The sensation comes very close to being tactile, as of a blank, sloping surface which gives no purchase. Social incentive, the officious evidence of precedent—'others have managed to translate this bit before you'—keeps one at the task. But the donation of trust remains ontologically spontaneous and anticipates proof, often by a long, arduous gap (there are texts, says Walter Benjamin, which will be translated only 'after us'). As he sets out, the translator must gamble on the coherence, on the symbolic plenitude of the world. Concomitantly he leaves himself vulnerable, though only in extremity and at the theoretical edge, to two dialectically related, mutually determined metaphysical risks. He may find that 'anything' or 'almost anything' can mean 'everything'. This is the vertigo of self-sustaining metaphoric or analogic enchainment experienced by medieval exegetists. Or he may find that there is 'nothing there' which can be divorced from its formal autonomy, that every meaning worth expressing is monadic and will not enter into any alternative mould. There is Kabbalistic speculation, to which I will return, about a day on which words will shake off 'the burden of having to mean' and will be only themselves, blank and replete as stone.

After trust comes aggression. The second move of the translator is incursive and extractive. The relevant analysis is that of Heidegger when he focuses our attention on understanding as an act, on the access, inherently appropriative and therefore violent, of *Erkenntnis* to *Dasein*. *Da-sein*, the 'thing there', 'the thing that is because it is there', only comes into authentic being when it is comprehended, i.e. translated.[1] The postulate that all cognition is aggressive, that every proposition is an inroad on the world, is, of course, Hegelian. It is Heidegger's contribution to have shown that understanding, recognition, interpretation are a compacted, unavoidable mode of attack. We can modulate Heidegger's insistence that understanding is not a matter of method but of

---

[1] Cf. Paul Ricœur, 'Existence et herméneutique' in *Le Conflit des interprétations* (Paris, 1969).

primary being, that 'being consists in the understanding of other being' into the more naïve, limited axiom that each act of comprehension must appropriate another entity (we translate *into*). Comprehension, as its etymology shows, 'comprehends' not only cognitively but by encirclement and ingestion. In the event of interlingual translation this manoeuvre of comprehension is explicitly invasive and exhaustive. Saint Jerome uses his famous image of meaning brought home captive by the translator. We 'break' a code: decipherment is dissective, leaving the shell smashed and the vital layers stripped. Every schoolchild, but also the eminent translator, will note the shift in substantive presence which follows on a protracted or difficult exercise in translation: the text in the other language has become almost materially thinner, the light seems to pass unhindered through its loosened fibres. For a spell the density of hostile or seductive 'otherness' is dissipated. Ortega y Gasset speaks of the sadness of the translator after failure. There is also a sadness after success, the Augustinian *tristitia* which follows on the cognate acts of erotic and of intellectual possession.

The translator invades, extracts, and brings home. The simile is that of the open-cast mine left an empty scar in the landscape. As we shall see, this despoliation is illusory or is a mark of false translation. But again, as in the case of the translator's trust, there are genuine borderline cases. Certain texts or genres have been exhausted by translation. Far more interestingly, others have been negated by transfiguration, by an act of appropriative penetration and transfer in excess of the original, more ordered, more aesthetically pleasing. There are originals we no longer turn to because the translation is of a higher magnitude (the sonnets of Louise Labé after Rilke's *Umdichtung*). I will come back to this paradox of betrayal by augment.

The third movement is incorporative, in the strong sense of the word. The import, of meaning and of form, the embodiment, is not made in or into a vacuum. The native semantic field is already extant and crowded. There are innumerable shadings of assimilation and placement of the newly-acquired, ranging from a complete domestication, an at-homeness at the core of the kind which cultural history ascribes to, say, Luther's Bible or North's Plutarch, all the way to the permanent strangeness and marginality of an artifact such as Nabokov's 'English-language' *Onegin*. But whatever the degree of 'naturalization', the act of importation can potentially dislocate or relocate the whole of the native structure. The Heideggerian 'we are what we understand to be' entails that our own being is modified by each occurrence of comprehensive appropriation. No language, no traditional symbolic set or cultural ensemble imports without risk of being transformed. Here two families of metaphor, probably related, offer themselves, that of sacramental intake or incarnation and that of infection. The incremental values of communion pivot on the moral, spiritual state of the recipient. Though all decipherment is aggressive and, at one level, destructive, there are differences in the motive of appropriation and in the context of 'the bringing back'. Where the native matrix is disoriented or immature, the importation will not enrich, it will not find a proper locale. It will generate not an integral response but a wash of mimicry (French neo-classicism in its north-European, German, and Russian versions). There can be contagions of facility triggered by the antique or foreign import. After a time, the native organism will react, endeavouring to neutralize or expel the foreign body. Much of European romanticism can be seen as a riposte to this sort of infection, as an attempt to put an embargo on a plethora of foreign, mainly French eighteenth-century goods. In every pidgin we see an attempt to preserve a zone of native speech and a failure of that attempt in the face of politically and economically enforced linguistic invasion. The dialectic of embodiment entails the possibility that we may be consumed.

This dialectic can be seen at the level of individual sensibility. Acts of translation add to our means; we come to incarnate alternative energies and resources of feeling. But we may be mastered and made lame by what we have imported. There are translators in whom the vein of personal, original creation goes dry. MacKenna speaks of Plotinus literally submerging his own being. Writers have ceased from translation, sometimes too late, because the inhaled voice of the foreign text had come to choke their own. Societies with ancient but eroded epistemologies of ritual and symbol can be knocked off balance and made to lose belief in their own identity under the voracious impact of premature or indigestible assimilation. The cargo-cults of New Guinea, in which the natives worship what airplanes bring in, provide an uncannily exact, ramified image of the risks of translation.

This is only another way of saying that the hermeneutic motion is dangerously incomplete, that it is dangerous because it is incomplete, if it lacks its fourth stage, the piston-stroke, as it were, which completes the cycle. The a-prioristic movement of trust puts us off balance. We 'lean towards' the confronting text (every translator has experienced this palpable bending towards and launching at his target). We encircle and invade cognitively. We come home laden, thus again off-balance, having caused disequilibrium throughout the system by taking away from 'the other' and by adding, though possibly with ambiguous consequence, to our own. The system is now off-tilt. The hermeneutic act must compensate. If it is to be authentic, it must mediate into exchange and restored parity.

The enactment of reciprocity in order to restore balance is the crux of the *métier* and morals of translation. But it is very difficult to put abstractly. The appropriative 'rapture' of the translator—the word has in it, of course, the root and meaning of violent transport—leaves the original with a dialectically enigmatic residue. Unquestionably there is a dimension of loss, of breakage—hence, as we have seen, the fear of translation, the taboos on revelatory export which hedge sacred texts, ritual nominations, and formulas in many cultures. But the residue is also, and decisively, positive. The work translated is enhanced. This is so at a number of fairly obvious levels. Being methodical, penetrative, analytic, enumerative, the process of translation, like all modes of focused understanding, will detail, illumine, and generally body forth its object. The over-determination of the interpretative act is inherently inflationary: it proclaims that 'there is more here than meets the eye', that 'the accord between content and executive form is closer, more delicate than had been observed hitherto'. To class a source-text as worth translating is to dignify it immediately and to involve it in a dynamic of magnification (subject, naturally, to later review and even, perhaps, dismissal). The motion of transfer and paraphrase enlarges the stature of the original. Historically, in terms of cultural context, of the public it can reach, the latter is left more prestigious. But this increase has a more important, existential perspective. The relations of a text to its translations, imitations, thematic variants, even parodies, are too diverse to allow of any single theoretic, definitional scheme. They categorize the entire question of the meaning of meaning in time, of the existence and effects of the linguistic fact outside its specific, initial form. But there can be no doubt that echo enriches, that it is more than shadow and inert simulacrum. We are back at the problem of the mirror which not only reflects but also generates light. The original text gains from the orders of diverse relationship and distance established between itself and the translations. The reciprocity is dialectic: new 'formats' of significance are initiated by distance and by contiguity. Some translations edge us away from the canvas, others bring us up close.

This is so even where, perhaps especially where, the translation is only partly adequate. The failings of the translator (I will give common examples) localize, they project as on to a screen, the resistant vitalities, the opaque centres of specific genius in the original. Hegel and Heidegger posit that being must engage other being in order to achieve self-definition. This is true only in part of language which, at the phonetic and grammatical levels, can function inside its own limits of diacritical differentiation. But it is pragmatically true of all but the most rudimentary acts of form and expression. Existence in history, the claim to recognizable identity (style), are based on relations to other articulate constructs. Of such relations, translation is the most graphic.

Nevertheless, there is unbalance. The translator has taken too much—he has padded, embroidered, 'read into'—or too little—he has skimped, elided, cut out awkward corners. There has been an outflow of energy from the source and an inflow into the receptor altering both and altering the harmonics of the whole system. Péguy puts the matter of inevitable damage definitively in his critique of Leconte de Lisle's translations of Sophocles: 'ce que la réalité nous enseigne impitoyablement et sans aucune exception, c'est que toute opération de cet ordre, toute opération de déplacement, sans aucune exception, entraine impitoyablement et irrévocablement une déperdition, une altération, et que cette déperdition, cette altération est toujours considérable.'[1] Genuine translation will, therefore, seek to equalize, though the mediating steps may be lengthy and oblique. Where it falls short of the original, the authentic translation makes the autonomous virtues of the original more precisely visible (Voss is weak at characteristic focal points in his Homer, but the lucid honesty of his momentary lack brings out the appropriate strengths of the Greek). Where it surpasses the original, the real translation infers that the source-text possesses potentialities, elemental reserves as yet unrealized by itself. This is Schleiermacher's notion of a hermeneutic which 'knows better than the author did' (Paul Celan translating Apollinaire's *Salomé*). The ideal, never accomplished, is one of total counterpart or re-petition—an asking again—which is not, however, a tautology. No such perfect 'double' exists. But the ideal makes explicit the demand for equity in the hermeneutic process.

Only in this way, I think, can we assign substantive meaning to the key notion of 'fidelity'. Fidelity is not literalism or any technical device for rendering 'spirit'. The whole formulation, as we have found it over and over again in discussions of translation, is hopelessly vague. The translator, the exegetist, the reader is *faithful to* his text, makes his response responsible, only when he endeavours to restore the balance of forces, of integral presence, which his appropriative comprehension has disrupted. Fidelity is ethical, but also, in the full sense, economic. By virtue of tact, and tact intensified is moral vision, the translator–interpreter creates a condition of significant exchange. The arrows of meaning, of cultural, psychological benefaction, move both ways. There is,

---

[1] Charles Péguy, 'Les Suppliants parallèles' in *Oeuvres en prose 1898–1908* (Paris, 1959), I, p. 890. This analysis of the art of poetic translation first appeared in December 1905. Cf. Simone Fraisse, *Péguy et le monde antique* (Paris, 1973), pp. 146–59.

ideally, exchange without loss. In this respect, translation can be pictured as a negation of entropy; order is preserved at both ends of the cycle, source and receptor. The general model here is that of Lévi-Strauss's *Anthropologie structurale* which regards social structures as attempts at dynamic equilibrium achieved through an exchange of words, women, and material goods. All capture calls for subsequent compensation; utterance solicits response, exogamy and endogamy are mechanisms of equalizing transfer. Within the class of semantic exchanges, translation is again the most graphic, the most radically equitable. A translator is accountable to the diachronic and synchronic mobility and conservation of the energies of meaning. A translation is, more than figuratively, an act of double-entry; both formally and morally the books must balance.

This view of translation as a hermeneutic of trust (*élancement*), of penetration, of embodiment, and of restitution, will allow us to overcome the sterile triadic model which has dominated the history and theory of the subject. The perennial distinction between literalism, paraphrase and free imitation, turns out to be wholly contingent. It has no precision or philosophic basis. It overlooks the key fact that a fourfold *hermeneia*, Aristotle's term for discourse which signifies because it interprets, is conceptually and practically inherent in even the rudiments of translation.

Though they deny it, phrase-books and primers are full of immediate deeps. Literally: *J'aime la natation* (from *Collins French Phrase Book*, 1962). Word-for-word: 'I love natation', which is mildly lunatic though, predictably, Sir Thomas Browne used the word in 1646. 'I like to go swimming' (omitting the nasty problem of differential strengths in *aimer* and 'like'). 'Swimming' turns up in *Beowulf*; the root is Indo-European *swem*, meaning to be in general motion, in a sense still functional in Welsh and Lithuanian. *Nager* is very different: through Old French and Provençal there is a clear link to *navigare*, to what is 'nautical' in the governance and progress of a ship. The phrase-book offers: *je veux aller à la piscine*. 'Swimming-pool' is not wholly *piscine*. The latter is a Roman fish-pond; like *nager* it encodes the disciplined artifice, the inter-position before spontaneous motion, of the classical order. 'I want to go . . .' / *je veux aller*. . . . 'Want' is ultimately Old Norse for 'lack', 'need', the felt register of deprivations. The sense 'to desire' comes only fifth among the rubrics which follow on the word in the *OED*. *Vouloir* is of that great family of words, of which the Latin *vel-* and *vol-* (cognate with the root of Sanskrit *var*) are best known, signifying volition, focused intent, the advance of 'will' (its cognate). The phrase-book is uneasily aware of the profound difference. '*I want* should not be translated by *je veux*. In French this is a very strong form, and when used to express a wish creates the unfortunate impression of giving a blunt and peremptory order rather than of making a polite request.' But the matter is not basically one of differing forces of demand. 'Want' as Shakespeare almost invariably adumbrates, speaks out of concavity, out of absence and need. In French this zone of meaning would be circumscribed by *besoin, manque*, and *carence*. But *j'ai besoin d'aller nager* is instantaneously off-pitch or obscurely therapeutic.

'It looks like rain' / *le temps est à la pluie*. No attempt here at bare literalism or point-to-point carry. 'Rain' has no established cognates outside the Teutonic. The grammar of the phrase is elisive and infers futurity. 'It' stands for an aggregate of sensory contexts, ranging from the indefinably atmospheric to the broadest markers of cloud, scent, or abrupt silence in the foliage. 'It' is also purely syntactical, an ambiguous but indispensable member of the verb-phrase. Though 'looks like' is in this case only casually visual, an ensemble of phenomena generates the expectation that there will be rain. The tag involves an entire machinery of lazy prophecy, of probabilistic habit. The French counterpart— phrase-books tend to be primly archaic—is of matching semantic density. Leaving aside a cosmogony—it is no less—in which 'time' is homologous with 'weather', there is the grammar of *être à la pluie*. Here also there is contraction: the idiom elides intervening steps of conjecture: 'the weather is such that it leads to the inference that. . . .' A highly-compacted argument about contiguity inheres in *est à*, almost as if we were saying 'the hands of the clock are at. . . .' But the odd turn of 'possession', of time/weather being assigned to, being owned by the rain (i.e. *ceci est à moi*) is there, vestigially at least. It is abetted by the fact that *pluie* is not only or principally 'rain' but *pluvia*. The Latin has a figurative weight which accords with possession. The entire complex is more threatening. *Faire la pluie et le beau temps* is, as Saint-Simon or the Cardinal de Retz knew, to determine fortune in the affairs of state. 'Rain' soaks us 'to the skin' whereas *la pluie* penetrates *jusqu'aux os*. The Roman personification, cavern-mouthed as on a baroque fountain, is latent in the word. The literal mythologists who contrived the Jacobin calendar knew it when they named 20 January to 19 February *pluviôse*. I do not know just how, but these differences in presentness relate to the curious differences in tense. To know whether it will rain, we listen to the weather 'forecast'; the Frenchman listens to the *bulletin météorologique*. *Bulletins* are in essence retrospective; there may be apologia and falsehood in them—the Napoleonic usage—but no augury. Thus they connote degrees of certitude quite alien to 'forecast'. *Le temps est à la pluie* has a resigned yet also subtly acquiescent assurance entirely lacking from the ephemeral clairvoyance of 'it looks like rain'. The gravities differ, which allows Verlaine to play with and against banality when he sets Rimbaud's *Il pleut doucement sur la ville* in epigraph to his own enigmatically desolate

> Il pleure dans mon cœur
> Comme il pleut sur la ville.

('Rain on the city', 'rain in the city', 'rain down on': each is false. But why?)

Das Kind ist unter die Räder gekommen. Though it signifies violent, presumably sudden mishap and aims at instant communication, the German phrase encodes a fairly elaborate gesture of fatality. 'The child has been run over', which is the equivalent offered by the 'teach yourself' manual, hardly reflects the cautionary dispassion of the original. In the German phrasing the wheels have a palpable right of way; somehow the child has interrupted their licit progress. The grammatical effect is undeniably apologetic and even accusing: the syntactic neutrality of *das Rad* together with the near-passivity of the verb form edges the onus of guilt towards the child. The wheels have not culpably 'gone over it'; it is the child which has 'come to be under them'. 'Undergo' would be

inadmissible as translation, but it in fact conveys the accusatory
hint. *L'enfant s'est fait écraser* is even stronger in implicit blame.
Any attempt at giving a naïve equivalence in English would generate a sense of volition: 'the child has had itself run over'. The
French idiom intends nothing so crass. But the nuance of indictment is there and more, perhaps, than a nuance. It results from
the fact that *se faire* plus an infinitive can function as a kind of
passive without losing altogether the substratum of purposeful
action. For what may be obscurely historical or legislative reasons,
both the German and the French expression suggest the stance
of the coachman or driver. The English phrase is scrupulously
equitable. Thus no exact transfer is available.

Notoriously, the absence of the article in Russian can lead to
pluralities and ambiguities which English misses or renders by
expansive paraphrase. But the problem may arise as dramatically
with regard to French. Genesis 1 : 3 is a well-known instance.
Fiat lux. Et facta est lux has a memorable sequentiality. The
phonetic and grammatical exterior proclaim a phenomenon at
once stunning and perfectly self-evident (Haydn's setting of
the words in the *Creation* precisely communicates the effect of
supremely astounding platitude). Italian *Sia luce. E fu luce* uses five
words as against six and is, in that sense, even more lapidary. But
the initial sibilant, the soft *c* and the stress on gender in *luce*
(where Latin *lux* was, at least for part of its history, masculine),
musicalizes the imperiousness of the Vulgate. *Es werde Licht. Und
es ward Licht* is perfectly concordant with the Latin except in one
detail. The semantically elusive *Es* has to be there. *Werde Licht*
would misrepresent the whole tenor and significance of the
Creator's illocution. The *Es* preserves the mystery of creation
without previous substance. 'Let there be light: and there was
light' in the Authorized Version, or ' "Let there be light", and
there was light' in the New English Bible, expand on the Latin.
There are now eight words in the place of six. And the punctuation is lightened. The purpose, presumably, is to give a sense of
instant consequence. But the omission of the full-stop together
with lower-case 'and' sacrifice the Latin pedal point. In the
original the note of cosmic command is fully held while the

division into two short sentences makes for a dynamic surge. This is exactly what is called for: an instant of pent breath above a groundswell of complete certitude.

The French version is also eight words long and opts for a punctuation precisely medial between the two English variants. *Que la lumière soit; et la lumière fut.* But much has altered. Latin, Italian, German, and English preserve the characteristically Hebraic repetition of the cardinal word 'light' at the climax of the sentence(s). In each of the four cases the word-order is powerfully imitative of the action expressed. 'Light' has its pride of place in God's order and realization. In the French text the drama of accomplishment, of shattering obviousness is that of the verb: it turns on the movement from the imperative subjunctive of *soit* to the perfectedness of *fut* (purely acoustically this is counter-productive, in so far as *soit* is more sonorous, more evocative of accomplished harmony than is *fut* with its clipped vowel-sound). But the major difference comes with the use of the definite article. 'Let there be light, and the light was.' The diminution of impact is obvious. *Es werde das Licht. Und es ward das Licht* is possible in a way the English is not. It is weaker, more oddly specific and inferential of some Plotinian discrimination between effulgences, but just possible. Indeed, in the German Bible the article comes with the third designation: *Und Gott sah, dass das Licht gut war.* The Authorized Version also introduces the article at this same point: 'And God saw the light. . . .' But neither Italian, German, nor English admit of the article when rendering God's fiat and its primal fulfilment. The difference from the French version is profound. The syntax of the Deity and of accomplishment make for an effect of balance, of equation rather than of tautological majesty. The definite article posits conceptual essence before phenomenality. *Que la lumière soit* has an 'intellectuality'[1] altogether

[1] Mario Wandruzska, 'Drückt sich darin eine besondere Sehweise aus, eine besondere geistige Auffassung der Dinge, die gewissermassen den Begriff des Lichts schon vor dem ersten Schöpfungstag voraussetzt, eine besondere französische Intellektualität, die von Anfang an jede Erscheinung schon auf ihren Begriff zurückbezieht?' in *Sprachen: Vergleichbar und Unvergleichbar* (Munich, 1969), p. 187. Cf. also Henri Meschonnic, *Pour la poétique II* (Paris, 1973), pp. 436–53.

lacking from either the blank imperative of *Fiat lux* or the unforced immediacy of 'Let there be light' (*Que lumière soit*, on the other hand, could only be a wicked parody of Claudel). All these are crude approximations to a theory of central, complex difference. 'There was light there' differs from 'there was a light there' in uncommitted generality and scale: dawn, say, as against a lamp. French demands the one form: *Il y avait de la lumière*. In French, phenomenal appearance, epiphany are categorized and conceptually prepared-for as they are not necessarily in English. This is not a question of poorer means, but of metaphysical insistence. Again, a word-for-word transference would damage essential evidence.

These are the commonplaces of contrastive linguistics, of language instruction and of the humourists who produce *Fractured French* or *La Plume de ma tante*. The point at issue is this: far from being the most obvious, rudimentary mode of translation, 'literalism' or as Dryden called it, *metaphrase*, is in fact the least attainable. The true interlinear is the final, unrealizable goal of the hermeneutic act. Historically, practically, the interlinear and *mot-à-mot* may indeed be a crude device. But rigorously conceived, it embodies that totality of understanding and reproduction, that utter transparency between languages which is empirically unattainable and whose attainment would signal a return to the Adamic unison of human speech. Only Walter Benjamin saw this when he wrote that ideally 'literalness and freedom must without strain unite in the translation in the form of the interlinear version. . . . The interlinear version of the Scriptures is the archetype or ideal of all translation'. *Verbum e verbo* would be the Utopian moment in which all speech is immediate to meaning (logical in that it contains and makes explicit the *logos*).

In actual practice, of course, something else is meant. The language-primer, the interlinear school-text of Cicero or Xenophon is not a translation but a contingent lexicon. It sets a dictionary equivalent from the target-language above each word in the source-language. Strictly defined, a word-for-word interlinear is nothing else but a total glossary, set out horizontally in discrete units and omitting the criteria of normal syntax and word-order in the language of the user. In fact it is ordinarily a compromise between mere lexicality and some transposition or elaboration so as to achieve an acceptable sentence:

| Être, | ou | ne pas | être, | c'est | la question |
| To be, | or | not | to be, | that is | the question |

would be the strict interlinear. The French school-text adds *là* (*c'est là la question*) thus modifying the exact sequence in order to attain correctness. In this case, as it happens, even the word-for-word scheme succeeds in conveying something of the motion of the original and almost the entire sense. With an increase in the number of verbal units, with grammatical complication and with the appearance of ambiguity and pluralism of possible meanings, such congruence between literalness and understanding becomes statistically less and less probable. The next lines of the soliloquy promptly defeat any attempt at word-for-word transfer.

These are the crucial parameters throughout the early history of automatic translation. The translation machine attempts to maximize the coincidence between a word-for-word interlinear and the reconstitution of actual meaning. It hopes, as it were, to locate 'rows of words' of which the mere superscription with a lexical equivalent will make adequate sense. The machine is no more than a dictionary 'which consults itself' at very high speed. In its primitive versions, the automatic translator offers one lexical counterpart for every word or idiom in the original. More sophisticated mechanisms can suggest a number of possible definitions from which the human reader of the print-out will select the most apposite. This procedure is not in any complete hermeneutic sense an act of translation. The machine's evaluation of context is wholly statistical: how many times has the given word appeared before in this particular text or body of similar texts, and do the words which immediately precede or follow it match a prepared unit in the programme? But it would be wrong to underestimate either the interest or potential utility of machine-literalism. Statistical bracketings and memory-bound recognitions of the kind employed by the machine are very obviously a part of the inter-pretative performance in the human brain, certainly at the level of routine understanding. A large mass of scientific literature, moreover, is susceptible to more or less automatic lexical transfer. 'A monolingual reader, expert in the subject-matter of the text being translated, should find it possible, in most instances, to extract the essential content of the original from this crude translation, often more accurately than a bilingual layman.'[1] Because mathematical and logical symbols are, wherever possible, 'monosemic'—they have a single agreed meaning independent of local context—because a mass of scientific, taxonomic, technological nomenclature is rigorously standardized, automatic translation can go a long way by purely lexical means. 'H_2O consists of two units of hydrogen for one of oxygen' is the kind of sentence that is at once tautological and informative. It can be translated word-for-word into a host of tongues even if the automatic glossary is crude (i.e. if 'consists of' is part of a general 'box' which includes 'is made of', 'is built of', 'is an aggregate of', etc.). The nearer the tautological ideal is to the passage—the more stringently and linearly it follows on a set of definitions and unequivocally sequential derivations—the better the chances for accurate automatic translation. But although such linearity is absolute only in mathematics or symbolic logic, much of scientific, technical, and, perhaps, even commercial documentation approaches the model. In all these forms of coding, there are strong biases towards definitional constraint and a conventional limitation of semantic possibility (in a chemical paper, *valence* will hardly ever mean the kind of damask used for the frame of a canopy or bedstead). The theory and practice of automatic translation has, of course, attempted to go far beyond the lexical, word-for-word design. But that design has its powerful utilitarian function and illustrates a contemporary adaptation of the ancient, despised trot.

But this is not what translators of poetry, philosophy, or

[1] A. G. Oettinger, 'Automatic (Transference, Translation, Remittance, Shunting)', in R. Brower (ed.), *On Translation*, pp. 257–8. For an up-to-date view of the limitations of the automatic lexicon, cf. Paul L. Garvin, *On Machine Translation* (The Hague, 1972), pp. 118–23. By comparing Garvin's treatment with Y. Bar-Hillel's 'Can Translation be Mechanized?' (*Journal of Symbolic Logic*, XX, 1955), one obtains a general view of the changing climate in the field.

Scripture have meant when they claimed to be literalists. On the contrary. They have adhered, or claimed to adhere, to a word-for-word technique in the name of ideal penetration, of a submission to the original so manifest and humble that it will elicit the entirety of meaning intact. In self-denial, the translator submerges his own sensibility and the genius of his own language in that of the original. Where this fusion occurs—Roy Campbell speaks of it in regard to his translations from Saint John of the Cross—the initial hermeneutic move of trust, of *élancement*, comes to dominate the whole enterprise. The translator does not aim to appropriate and bring home. He seeks to remain 'inside' the source. He deems himself no more than a transcriber. But what happens in practice?

We recall that Dryden applied the term *metaphrase* to what he took to be the gross literalism of Ben Jonson's treatment of the *Ars poetica*. Published posthumously in 1640, Jonson's Horace probably dates back to the first decade of the century. Though *Timber* and the conversations with Drummond of Hawthornden show that Jonson was steeped in Horace's poetics, not much is known about the actual composition and purpose of this translation. Consider a famous passage in the original (350–60):

> [. . .] nec semper feriet quodcunque minabitur arcus.
>   verum ubi plura nitent in carmine non ego paucis
>   offendar maculis, quas aut incuria fudit
>   aut humana parum cavit natura. quid ergo est?
>   ut scriptor si peccat idem librarius usque,
>   quamvis est monitus, venia caret; ut citharoedus
>   ridetur chorda qui semper oberrat eadem:
>   sic mihi qui multum cessat fit Choerilus ille,
>   quem bis terve bonum cum risu miror, et idem[1]
>   indignor quandoque bonus dormitat Homerus; . . .

Ben Jonson translates:

> Not alwayes doth the loosed bow hit that
> Which it doth threaten: Therefore, where I see

---

[1] The transmitted reading here is *et*, but Jonson must have read *at* with a full point ending the sentence.

> Much in a Poëm shine, I will not be
> Offended with a few spots, which negligence
> Hath shed, or humane frailty not kept thence.
> How then? why, as a Scrivener, if h' offend
> Still in the same, and warned, will not mend,
> Deserves no pardon; or who'd play and sing
> Is laught at, that still jarreth in one string:
> So he that flaggeth much, becomes to me
> A *Choerilus*, in whom if I but see
> Twice, or thrice good, I wonder: but am more
> Angry, if once I heare good *Homer* snore.

Pope's variant in the *Essay on Criticism* becomes:

> Whoever thinks a faultless piece to see,
> Thinks what ne'er was, nor is, nor e'er shall be.
> In every work regard the writer's end,
> Since none can compass more than they intend;
> And, if the means be just, the conduct true,
> Applause, in spite of trivial faults is due.

In *Hints from Horace* Byron writes:

> Where frequent beauties strike the reader's view,
> We must not quarrel for a blot or two,
> But pardon equally to books or men,
> The slips of human nature, and the pen.

Ben Jonson's is, obviously, a translation in a sense in which Pope's and Byron's imitative commentaries are not. It is, moreover, unmistakably plain and attentive to the original. Presumably, it is the awkward, heavily Latin fabric of the seventh and eighth lines or the endeavour to preserve the original word-order through clumsy *enjambement* which Dryden found unacceptable. Nevertheless, Jonson's Horace is by no means a word-for-word interlinear. For one thing the *Ars poetica* runs to only 476 lines whereas Jonson's recasting requires 679. For another, it is, Nabokov would say, 'begrimed or beslimed by rhyme', and the structure of the Latin sentence is often sacrificed to the needs of English. Thus *quodcunque minabitur arcus* concisely closes the first line of the passage whereas Jonson not only adds the epithet 'loosed' but carries the entire motion into the following verse. The famous contrast between Choerilus' occasional virtues and Homer's rare nods is considerably altered in Jonson's version. Horace ends on a rhetorical question: 'Am I, then, to be angry whenever good Homer drops off?' Jonson's affirmative is either an arbitrary change or a misconstruction.

The case for literalism is far more drastic in Browning's *Agamemnon*. Browning had incorporated a translation of Euripides' *Heracles* in *Aristophanes' Apology*. It is a middling specimen of the Victorian mode of lyric sublimity but deserves to be remembered for its inspired reading of line 1142: ἤ γὰρ συνήραξ᾽ οἶκον ἤ βάκχευσ᾽ ἐμόν; as 'Did I break up my house or dance it down?' Four years later, in 1877, Browning published his version of Aeschylus. He called it 'a transcription' and set out to be 'literal at every cost save that of absolute violence to our language'. Browning purposed 'the very turn of each phrase' to be 'in as Greek a fashion as English will bear'. The notorious textual difficulty of the original and the elevation of Aeschylus' tone were to make this attempt the more arduous but illuminating. The result has generally been judged to be unreadable and Browning himself termed it a 'somewhat tiresome, perhaps fruitless adventure'.[1] Take the pronouncement of Kassandra (Browning insisted on the *K*) in lines 1178–97:

Well then, the oracle from veils no longer
Shall be outlooking, like a bride new-married:
But bright it seems, against the sun's uprisings
Breathing, to penetrate thee: so as, wave-like,

---

[1] One of the few balanced views of Browning's experiment is that put forward by Reuben Brower in his article on 'Seven Agamemnons' in *On Translation*. For an exhaustive analysis of the philological and stylistic aspects of Browning's Aeschylus, cf. Robert Spindler, *Robert Browning und die Antike* (Leipzig, 1930), II, pp. 278–94. Spindler is useful in that he shows in minute detail to what extent, and within what limits of grammatical displacement, Browning adhered to his contract of complete fidelity.

To wash against the rays a woe much greater
Than this. I will no longer teach by riddles.
And witness, running with me, that of evils
Done long ago, I nosing track the footstep!
For—the same roof here—never quits a Choros
One-voiced, not well-tuned since no 'well' it utters:
And truly having drunk, to get more courage,
Man's blood—the Komos keeps within the household
—Hard to be sent outside—of sister Furies:
They hymn their hymn—within the house close sitting—
The first beginning curse: in turn spit forth at
The Brother's bed, to him who spurned it hostile.
Have I missed aught, or hit I like a bowman?
False prophet am I,—knock at doors, a babbler?
Henceforth witness, swearing now, I know not
By other's word the old sins of this household!

The first thing to be said is that the Greek text is uncertain:
emendations have been proposed at several important points in the
original (i.e. 1181, 1182, 1187, 1196). The prophetess, moreover,
is speaking in mantic riddles (ἐξ αἰνιγμάτων), certainly
throughout the first six and a half lines. Herbert Weir Smyth, in
his Loeb Library version of 1926 offers:

Lo now, no more shall my prophecy peer forth from behind a veil like a
new-wedded bride; but 'tis like a rush upon me, clear as a fresh wind
blowing against the sun's uprising so as to dash against its rays, like a
wave, a woe mightier far than mine.

Lattimore, in 1953, reads:

No longer shall my prophecies like some young girl
new-married glance from under veils, but bright and strong
as winds blow into morning and the sun's uprise
shall wax along the swell like some great wave, to burst
at last upon the shining of this agony.

Comparison is not altogether to Browning's disadvantage. Neither
Smyth nor Lattimore achieves convincing sense or anything like a

normal English phrase-structure. Lattimore's 'to burst at last upon the shining of this agony' is not only meaningless, but throws away the vital point. As Mazon shows, in his helpful gloss, Kassandra is overwhelmed by the sense of a second catastrophe—the death of Agamemnon—even more terrible than the first—her own impending doom. Hence the simile of successive waves, for which Mazon instances parallels in the *Prometheus*, 1015, and in Plato's *Republic*, 472a. Browning, at this stage, is also impenetrable. But as the passage unfolds into relative lucidity, Browning's curious eleven-syllable metre and clotted phraseology occasionally communicate an *aural* density vital to Greek drama and to much of Victorian poetry but entirely absent in the later versions. 'They hymn their hymn—within the house close sitting—' exactly conveys ὑμνοῦσι δ'ὕμνον δώμασιν προσήμεναι as Lattimore's euphonious 'Hanging above the hall they chant their song of hate' does not. And the 'babbler' knocking at doors is right (Mazon gives *une radoteuse*) where 'some swindling seer who hawks his lies' is at once too literal (ψευδόμαντις) and too 'poetic'. At one or two points, in fact, Browning's violent literalism and commitment to Aeschylaean obscurity produce results more persuasive than any other versions. Both Smyth—'Oh, but he struggled to win me, breathing ardent love for me'—and Lattimore—'Yes, then he wrestled with me, and he breathed delight'—seek to reproduce the 'physicality', the panting violence of line 1206:

᾽Αλλ᾽ ἦν παλαιστὴς κάρτ᾽ ἐμοὶ πνέων χάριν

In both we get something of the image of the inflamed, triumphant wrestler. But the second half of Browning's transcription is finer, more Aeschylaean in motion and mystery:

But he was athlete to me—huge, grace breathing.

Like Nabokov's actual translation of *Eugene Onegin*—'In fact, to my ideal of literalism I have sacrificed everything (elegance, euphony, clarity, good taste, modern usage and even grammar) that the dainty mimic prizes higher than truth'—Browning's experiment remains a curio.[1] But literalism of this lucid, almost desperate kind, has within it a creative pathology of language. Intent on submerging himself totally in the original, prepared not to incorporate his appropriations fully into his own speech and culture, the translator hangs back at the frontier. More or less deliberately, he produces an 'interlingua', a centaur-idiom in which the grammar, the customary cadence, the phrasing, even the word-structure of his own tongue are subjected to the vocabulary, syntax, phonetic patterns of the text which he is translating or, more exactly, seeking to inhabit and only transcribe. He works 'between the lines' and a rigorous interlinear is exactly that: a no-man's-land in psychological and linguistic space. To translate word-for-word, to attempt a 'Greek English'—Browning's term— is to carry the process of intermediation to an extreme of theor- etical and technical violence in the hope of fusion (particles collid- ing and fusing with each other when they have been thrust from their respective orbits). The psychological and formal risks are considerable. At work between his own language and that of the source-text, the literalist exposes himself to vertigo. He may, in Benjamin's haunting image, find language so wrenched from its hinges, so forced and traversed, that its gates will slam shut behind him enclosing him in utter strangeness or silence.

At the trivial level, this strangeness will produce the mass of 'translationese', the slipshod farrago of *franglais* or of teutonisms

[1] I stress 'actual translation'. Taken together with the Commentary, Nabokov's production is a masterpiece of baroque wit and learning. According to the hermeneutic model I have put forward, Nabokov's 'Pushkin' represents a case of 'over-compensation', of 'restitution in excess'. It is a 'Midrashic' reanimation and exploration of the original text so massive and ingenious as to become, consciously or not, its rival. Such 'rival servitude' is probably central to Nabokov's attitude to the Russian language which he, in part, deserted, and to his own eminent but also ambivalent location in the Russian literary tradition. But all this, though it may be fascinating in itself and instructive for the student of translation, does not refute Alexander Gerschenkron's judgement: 'Nabokov's translation can and indeed should be studied, but despite all the cleverness and occasional brilliance it cannot be read' ('A magnificent Monument?', *Modern Philology*, LXIII, 1966, p. 340). 'Nabokovians' tend never to refer to this decisive article in which Gerschenkron, himself a virtuoso of Russian, meets the master on his own ground of literal exactitude.

which make up the general run of commercial and pulp translation. Texts concocted of unexamined lexical transfers, of grammatical hybrids which belong neither to the source nor to the
target language are the inter-zone or rather limbo in which
the rushed, underpaid hack translator works.[1] At a slightly more
elevated plane, we find the codified strangeness of most translations from the Persian, the Chinese, or the Japanese *haiku*. This
constitutes the 'moon in pond like blossom weary' school of
instant exotica. It can prove contagious even in great craftsmen
such as Waley. Creative dislocation towards an interlingual,
inherently unstable 'mid-speech' is a rarer, more demanding
occurrence.

Chateaubriand's prefatory *Remarques* to his translation of
Paradise Lost (1836) are of the most vivid formal and pragmatic
interest. Pushkin studied them closely when examining the
possibilities of a modern epic. 'What I have undertaken is a literal
translation in the strongest sense of the term, a translation which a
child and a poet will be able to follow line by line, word for word,
as if they had an open dictionary in front of them.' Chateaubriand
has made a tracing of the original ('J'ai calqué le poëme de Milton
à la vitre'). In order to do so this great master of the exigent
musicalities of French grammar has had to retain nominative
absolutes ('Thou looking on...'); he has been compelled to
use ablative absolutes without the auxiliary verb they require in
French; he has resorted to archaicisms and formed new words,
particularly negatives such as *inadoré* or *inabstinence*. Coming to
'many a row of starry lamps.../ Yielded light /As from a sky,'
Chateaubriand has written 'Plusieurs rangs de lampes étoilées...
émanent la lumière comme un firmament.'

Or je sais qu'*émaner* en français n'est pas un verb actif; un firmament
n'émane pas de la lumière, la lumière *émane d'un firmament*: mais traduisez
ainsi, que devient l'image? Du moins le lecteur pénètre ici dans le génie
de la langue anglaise; il apprend la difference qui existe entre les régimes
des verbes dans cette langue et dans la nôtre.

---

[1] For a representative *sottisier* of examples as between French and German, cf.
Walter Widmer, *Fug und Unfug des Uebersetzens*, pp. 57–70.

Dupré de Saint-Maur's version of *Paradise Lost* preserves the integrity of French grammar but is insipid and inaccurate. Luneau de Boisjermain's reading is a violently ungrammatical interlinear but, paradoxically, 'en suivant le mot à mot, elle fourmille de contresens'. Chateaubriand's translation, in a highly cadenced prose, is based on a coherent strategy. Its motion is one of diachronic reversal: it seeks to work upstream to the philological and cultural sources common to Milton's epic and to classic French. As did Milton, so Chateaubriand bases his choice of words and phrases on the precedent of Virgil, Seneca, Lucretius, the Vulgate and the Italian poets of the Renaissance and the Baroque. He meets the English text half-way in time as well as in linguistic space. Take the famous depiction of Satan after the end of Beelzebub's speech in Book I: 'He scarce had ceased when the superior fiend...'

Beelzebuth avait à peine cessé de parler, et déjà le grand Ennemi s'avançait vers le rivage: son pesant bouclier, de trempe éthérée, massif, large et rond, était rejeté derrière lui; la large circonférence pendait à ses épaules, comme la lune dont l'orbe, à travers un verre optique, est observé le soir par l'Astronome toscan, du sommet de Fièsole ou dans le Valdarno, pour découvrir des nouvelles terres, des rivières et des montagnes sur son globe tacheté. La lance de SATAN (près de laquelle le plus haut pin scié sur les collines de Norwège pour être le mât de quelque grand vaisseau amiral, ne serait qu'un roseau) lui sert à soutenir ses pas mal assurés sur la marne brûlante. . . .

Chateaubriand not only matches Milton's Latinity in *circonférence*, in *orbe*, in *verre optique* but goes, as it were, 'behind' Milton to a point of common origin in *marne*—a modernization of Old French or Breton-Celtic *marle* from which Milton's 'burning marle' directly derives. In *trempe éthérée* the dislocation is subtle: the phrase is, in French, difficult to conceptualize and nearly an oxymoron; surprisingly, moreover, *trempe* is of Walloon origin (Littré gives *treinp*); nevertheless, the words achieve not only a literal proximity to Milton but a deceptive aural, visual Latinism. In translations, as in word-play, false etymologies can take on a momentary truth. The sentence then unfolds into one of Milton's euphonious, sinuous leviathans of dependent clauses, both relative and adverbial: 'Nathless he so endur'd, till on the beach / Of that inflamed sea, he stood and call'd. . . .' Milton's sentence deploys the complex chain of images which leads from the thick strewing of the leaves at Vallombrosa, to the sedge scattered and afloat on the Red Sea, and which comes to rest, via the destruction of Pharaoh's host, on the triumphant 'syllogism' of

> so thick bestrown
> Abject and lost lay these, covering the flood,
> Under amazement of their hideous change.

The serpentine intricacy and menace of the original is crucially grammatical: it is enacted via the sequence when/whose/while/who/so. At one point Chateaubriand's literalism takes him too far: Milton's Etrurian shades which 'High over-arched imbower . . .' do not, one feels, carry the specific, at this juncture mildly discordant suggestion of a 'cradle' (*Les ombrages étruriens décrivent l'arche élevée d'un berceau* . . .). But Chateaubriand sacrifices the normal articulation of a French sentence-structure and exactly parallels Milton: *ainsi/quand/dont/tandis qu'ils/qui/ainsi*: in so doing he achieves that subordination of all syntactic considerations to pulse which marks the original. In *Paradis perdu* Chateaubriand's idiom is a French under immediate pressure of Latin—as, of course, is so much of ordinary French and of Chateaubriand's own style. But it is also a French which suggests that it has behind it an equivalent to an Authorized Version. As is often pointed out, no such equivalent exists. But its imaginary felt presence is unmistakable when French masters translate those works of English poetry and prose in which the Bible is a shaping precedent. Chateaubriand's Milton seems to lead, in turn, to Proust's translation of Ruskin's *Bible of Amiens* (1904), notably in the section *Interprétations*,[1] and to the version of Conrad's *Typhoon*

---

[1] Though Jean Autret, *L'Influence de Ruskin sur la vie, les idées et l'œuvre de Marcel Proust* (Geneva, 1955) contains much valuable information, two questions remain to be investigated: the points at which affinities between Proust and Chateaubriand are based on a common interest in English language and literature, and the extent to which stylistic displacements in Proust's several translations from Ruskin anticipate his own idiom as a novelist.

which Gide prepared in 1916–18. Each shares a measure of deliberate strangeness.

A peculiarly illuminating, intentional strangeness can result when a writer, particularly a lyric writer, translates his own work into a foreign language or is instrumental in such translation. The hermeneutic model is, in this event, one of essential donation but also of Narcissistic trial or authentication. The writer makes a gift of his own work to another language yet seeks in the copy the primary lineaments of his own inspiration and, possibly, an enhancement or clarification of these lineaments through reproduction. Again, the mirror acts as independent witness. One approach to Broch is to read the sum of his fiction and philosophy as an extended metaphor of translation: between present tense and death, between classic values and modern chaos, between verbal expression on the one hand and music and mathematics on the other. Broch's 'Some Remarks on the Philosophy and Technique of Translation' were probably set down in the late 1940s or very early 1950s. They make for a characteristically dense paper in which the defining terms are 'Logos' and 'Archetype'. Every language, he says, contains both, but whereas the 'Logos' is the universal principle of relational meaning (logic), 'Archetypes' are the specific, language-contextual embodiments of the universal process of symbolism and symbolization. 'Archetypes' are never fully translatable; but the 'logical' underlies all human tongues. It constitutes a 'meta-syntax' which makes translation possible (Broch's 'meta-syntax' anticipates the 'deep structures' of trans-formational generative grammars). Thus all translation operates in a mediating zone between the final autonomy of context-bound 'archetypes' and the universals of logic. Ultimately, the validity of a translation will depend on an undemonstrable assumption of universality or harmonic similitude in the human spirit. Broch calls this third term which, as it were, authenticates the acts of exchange between two languages, *tertium comparationis*.

He was himself exceptionally fortunate in his translators. In *The Sleepwalkers* Edwin and Willa Muir excelled even their own standards. Broch's collaboration with Jean Starr Untermeyer during five years of work on the English recasting of *The Death of*

Virgil was symbiotic. It produced a text which is, in many respects, indispensable to the original. Together, the German and the English versions achieve a contrapuntal coherence which at once elucidates and confirms *Der Tod des Vergil*. Being a lyric dramatization of the theme of the limits of human speech, Broch's fable is itself a 'translation at risk', an attempt to locate and test the edges of inarticulacy. Transfer into another language multiplies the risks but also verifies the possibility of the scheme. The Broch–Untermeyer version moves very far towards the German form with its endless spiralling sentences, mass of composite words and emphatic substantives through which Broch tries to express a simultaneity of physical and metaphysical meanings. But the uses of German in the book had themselves moved away from the normal architecture of the language into areas of experimental disjunction (*Lockerung*) and musicality. So English and German meet in a 'meta-syntax' as do those waves 'steel-blue and light, ruffled by a soft, scarcely perceptible cross-wind' in the well-known opening sentence or chord. Near the close of the 'Fire' section, Virgil's febrile but ordered reverie turns to the mystery of sense and symbol. Only in the voice of death shall these be integrally united. The passage I have in mind begins: 'Denn sie, Stimme der Stimmen, ausserhalb jeglicher Sprache, gewaltiger als jede, gewaltiger sogar als die Musik . . .' (pp. 236–7 in Vol. II of the *Gesammelte Werke*):

For this voice of all voices was beyond any speech whatsoever, more compelling than any, even more compelling than music, than any poem; this was the heart's beat, and must be in its single beat, since only thus was it able to embrace the perceived unity of existence in the instant of the heart's beat, the eye's glance; this, the very voice of the incomprehensible which expresses the incomprehensible, was in itself incomprehensible, unattainable through human speech, unattainable through earthly symbols, the arch-image of all voices and all symbols, thanks to a most incredible immediacy, and it was only able to fulfil its inconceivably sublime mission, only empowered to do so, when it passed beyond all things earthly, yet this would become impossible for it, aye, inconceivable, did it not resemble the earthly voice; and even should it cease to have anything in common with the earthly voice, the earthly word, the earthly language, having almost ceased to symbolize them, it could serve to disclose the arch-image to whose unearthly immediacy it pointed, only when it reflected it in an earthly immediacy: image strung to image, every chain of images led into the terrestrial, to an earthly immediacy, to an early happening, yet despite this—in obedience to a supreme human compulsion—must be led further and further, must find a higher expression of earthly immediacy in the beyond, must lift the earthly happening over and beyond its this-sidedness to a still higher symbol; and even though the symbolic chain threatened to be severed at the boundary, to fall apart on the border of the celestial, evaporating on the resistance offered by the unattainable, forever discontinued, forever severed, the danger is warded off, warded off again and again. . . .

Few concessions are made to the natural breaks and lucidities of English (though a narrative past tense is substituted for Broch's immediate 'mystical' present). 'Arch-image', 'threatened to be severed at', 'evaporated on the resistance', and many other units abandon the norm of English word-usage or grammar. Taken 'straight', this bit of prose suggests Gertrude Stein seeking to transcribe and perhaps parody Kant. But it is hardly meant to stand alone. It forces us back to the original which it in turn illuminates; its own opaqueness induces the original to declare itself more fully. It poses echoing questions as does a critical exegesis. In this interlinear—between the lines of the German text, between the semantic lines of English and of German, between both languages and an unknown but clearly postulated tongue which can transcend the constraints of imprecise objective reference—we come close to the poets' dream of an absolute idiolect. Here is a *tertium datum* unique to its occasion and which refuses to serve either as example or canonic mould. There is from the bilingual weave of *The Death of Virgil* (1945) no necessary return to either English or any German text except Broch's own. The final sentence of the book seeks to take us to 'the word beyond speech'.

Reference to meaning or language 'beyond speech' can be a heuristic device as at the end of Wittgenstein's *Tractatus*. It can be a conceit, often irritating, in epistemology or mysticism. But it can also serve as a metaphor, almost technical, through which to convey a genuine experience. The writer feels that there is a formal or substantive gap between his intentions, between the pressures of incipient shape or apprehension which he undoubtedly registers, and the means of expression available to him in the language. More generally, and without regard to the dubious psychology and logical inconsistencies involved, he feels that there is an authentic range of consciousness, of perceptual immediacy, which lies beyond articulate expression but which is none the less, or perhaps pre-eminently, numinous. If we are to allow that this invocation of transcendence is more than a rhetorical turn and tactic of sublimity, the writer must give hostages. His accomplished work must be of a stature to justify the presumption that he has in fact mastered the available language and executive forms and that he has already extended both to the utmost of intelligibility. One must have covered the ground before asserting, credibly, that valid though inaccessible data lie beyond the bounds. The entirety of the *Commedia* underwrites the felt need, the scruple of Dante's successive statements, from Canto X to XXXIII of the *Paradiso*, that language is failing him, that the light of ultimate meaning lies past speech. Having arrived at what he feels to be the irremediable limits of the word, the poet in whom such feeling is now a genuine tragic imperative will fall silent. Or he may be impelled to a drastic overreaching, to a transcendence of coherent discourse which is not, as in many surrealists, histrionic and opportune, but which puts reason and life itself at risk. The silences, the insanities, the suicides of a number of great writers are rigorous affirmations of an experience of the boundaries of language. In Hölderlin there can be no doubt either as to the preceding mastery or the totality of the transcendent risk. And it is precisely via Hölderlin's translations that the case for 'the word beyond speech' is put most visibly.

In modern hermeneutics the poetry, letters, and translations of Hölderlin occupy a privileged place. Heidegger's ontology of language is partly based on them, and it is from Hölderlin that Walter Benjamin deduces much of his theory of 'the logos' and of

translation.[1] The philosophic and philological literature which
has grown up around Hölderlin's often fragmentary and private
versions of Homer, Pindar, Sophocles, Euripides, Virgil, Horace,
Ovid, and Lucan is extensive and itself of extreme difficulty.[2] This
is due in part to the intrinsic density of the material. Hölderlin
is among the most taxing poets in literature. His elevation and
opaqueness are even further concentrated in a number of his
translations. But it is due also to historical and psychological
complications, to the difficulty which German sensibility, since
Goethe and Schiller, has experienced in coping with Hölderlin's
idiosyncratic radicalism and collapse of reason. Hölderlin's
translations are unquestionably of the first importance. They
represent the most violent, deliberately extreme act of hermeneutic
penetration and appropriation of which we have knowledge.
Particularly in his readings of Pindar and of Sophocles, Hölderlin
compels us to experience, as in fact only a great poet can, the

[1] Heidegger's *Erläuterungen zu Hölderlins Dichtung* were gathered in 1951. Beda
Allemann's *Hölderlin und Heidegger* (Zürich and Freiburg, 1954) explores the rela-
tionship between the ontologist and the poet but tends to reconstrue Hölderlin in
Heideggerian terms. Walter Benjamin's 'Zwei Gedichte von Friedrich Hölderlin'
dates back to 1914–15 (but was first published in 1955). Benjamin's essay on
'The Task of the Translator' reaches its visionary apex with specific reference to
Hölderlin's versions of Pindar and of Sophocles.

[2] The pioneering work was Norbert von Hellingrath's *Pindaruebertragungen
von Hölderlin* (Jena, 1911), followed by Günther Zuntz's dissertation *Ueber
Hölderlins Pindar-Uebersetzung* (Marburg, 1928). Two basic works came next:
Lothar Kempter's *Hölderlin und die Mythologie* (Zürich and Leipzig, 1929) and
Friedrich Beissner's *Hölderlins Uebersetzungen aus dem Griechischen* (Stuttgart,
1933). Pierre Bertaux's *Hölderlin. Essai de biographie intérieure* (Paris, 1936)
brilliantly placed the translations in the context of the poet's work as a whole. Since
then detailed treatments have proliferated. I have drawn on the following: Meta
Corsen, 'Die Tragödie als Begegnung zwischen Gott und Mensch, Hölderlin's
Sophoklesdeutung' (*Hölderlin-Jahrbuch*, 1948–9); Hans Frey, 'Dichtung, Denken
und Sprache bei Hölderlin' (Dissertation, Zürich, 1951); Wolfgang Schadewaldt,
'Hölderlin's Uebersetzung des Sophokles' (*Hellas und Hesperien*, Zürich and
Stuttgart, 1960); Karl Reinhardt, 'Hölderlin und Sophokles' in J. C. B. Mohr (ed.),
Hölderlin, Beiträge zu seinem Verständnis in unserm Jahrhundert (Tübingen, 1961);
M. B. Benn, *Hölderlin and Pindar* (The Hague, 1962); Jean Beaufret's admirable
Preface to Hölderlin, *Remarques sur Oedipe/Remarques sur Antigone* (Paris, 1965);
Rolf Zuberbühler, *Hölderlins Erneuerung der Sprache aus ihren etymologischen
Ursprüngen* (Berlin, 1969). The translations themselves have been assembled in
Volume V of the *Grosse Stuttgarter Ausgabe* but textual problems remain. Little in
the literature, moreover, looks closely at Hölderlin's translations from the Latin.

limits of linguistic expression and the barriers between languages which impede human understanding. These pressed on him intolerably, and it is their unsparing 'concreteness', the physical resistance they generate, which make Hölderlin's translations so fascinating and bewildering. Here I will touch only on their paradoxical literalness, on Hölderlin's attempt to achieve a cultural, verbal interlinear, a mid-zone between antique and modern, Greek and German. Again we see that literalism is not, as in traditional models of translation, the naïve, facile mode but, on the contrary, the ultimate.

With a vehemence which carried him beyond the metaphoric, Hölderlin came to regard all writing as a translation or transcription of encased, hidden meanings. Already his early, comparatively open poetry represents an attempt to renew German through a return to its ancient sources of hidden force. Hölderlin uses the *figura etymologica* (the reinterpretation of the meaning of words according to their supposed etymology) as does Heidegger: he is seeking to 'break open' modern terms in order to elicit their root-significance. He draws on Luther's idiom and on the vocabulary of the Pietist movement. He enlists Swabian forms and reverts to the Old High German or Middle High German meanings and connotations of words. Hölderlin was not alone in so doing. His etymologizing is part of an anti-Enlightenment tactic of linguistic nationalism and numinous historicism. Herder and Klopstock were direct, influential forerunners. But Hölderlin pressed further. He was trying to move upstream not only to the historical springs of German but to the primal energies of human discourse. These he located in the elemental compactness of the individual term. Hölderlin's view was, in a sense, the reverse of the Aristotelian assertion that 'names are of a finite number whereas objects are infinite'. For Hölderlin, the name, if closely pressed, would reveal a corresponding, previously perhaps unperceived, substantive presence. Thus the more difficult, the more opaque the word, the deeper, the more energized its charge of potential revelation: 'das schwere Wort wird zum magischen Träger des Tiefsinns'.[1]

---

[1] Rolf Zuberbühler, op. cit., p. 22.

This charge, moreover, might be intensified or made manifest by linguistic fusion, by a direct transfer of verbal units between languages. In Hölderlin *res vera* becomes *wahrer Sache*, *unstädtisch* is made of the haunting ἄπολις, and the enclitic γάρ is rendered by the rather enigmatic *nemlich* throughout the late hymns. Different tongues were erratic blocs wrenched from the unity of the *logos*. To weld their elements, even imperfectly, even at the risk of momentary incoherence, meant a partial return to the lost unity of meaning.

The compulsion of meaning out of mystery by means of a lyric violence of expression seemed most obviously realized in Pindar. Klopstock's translation of Horace's *Ode*, II. iv, and the imitation of Horace's 'quem tu, Melpomene' (IV. iii) which Klopstock had published in 1747

> Wen des Genius Blick, als er geboren ward,
>   Mit einweihendem Lächeln sah,
> Wen, als Knaben, ihr einst Smintheus Anakreons
>   Fabelhafte Gespielinnen,
> Dichtrische Tauben umflogt . . .

not only prefigured Hölderlin's own techniques of translation,[1] but confirmed his paradigm of the absolute poet. Hölderlin rendered six Olympian and ten Pythian *Odes* either whole or in part. Most likely, these 2,000 lines of translation, probably set down in early 1800, were a private experiment. As if in express defiance of

---

[1] Klopstock's example seems obvious in the prosodic structure and phonetic imitations of Hölderlin's treatment of Horace's *Ode* II. vi. Cf.

> unde si Parcae prohibent iniquae
> dulce pellitis ovibus Galaesi
> flumen et regnata petam Laconi
>   rura Phalantho.

with Hölderlin's

> Lassen mich dahin nicht die neidischen Parzen
> So will ich suchen den Galesusstrom,
> Den lieblichen mit den wolligen Schafen,
> Und die Felder, vom Spartaner
>   Phalantus beherrscht.

in which the distribution of sibilants, liquids, and fricatives is strikingly imitative of the Latin.

Cowley's famous warning that 'if a man should undertake to translate Pindar word for word, it would be thought that one mad man had translated another', Hölderlin strove for utmost literalism. He used such devices as hyperbaton, the separation of object from predicate, the isolation of epithets either preceding or following on their substantive, the asymmetry of predicates and attributes, in order to produce a 'German–Greek' intelligible to German speakers but intensely representational of Pindar's 'rushing darkness'.[1] Though there are eloquent passages, the close, for instance, of the Third Pythian,

> Klein im Kleinen, Gross im Grossen
> Will ich sein; den umredenden aber immer mit Stimme
> Den Dämon will ich üben nach meinem
> Ehrend dem Geschick.
> Wenn aber mir Vielheit Gott edle darleiht,
> Hoffnung hab' ich Ruhm zu
> Finden hohen in Zukunft.
> Nestor und den Lykischen
> Sarpedon, der Menge Sage,
> Aus Worten rauschenden
> Baumeister wie weise
> Zusammengefügete, erkennen wir.
> Die Tugend aber durch rühmliche Gesänge
> Ewig wird.
> Mit wenigem aber zu handeln ist leicht.

the translation is, to a large extent (and even in this instance), forced and unconvincing. But the trial run proved fruitful. Hölderlin's late hymns are 'Pindaric' not only in certain rhetorical aspects—their openings seem to reflect the Sixth Nemean and their codas are often reminiscent of the Third Pythian—but in a much deeper vein of spiritual mimesis. Pindar's strict metrical regularity, which Hölderlin understood only vaguely, seemed to liberate Hölderlin's prosodic impetus. He took from Pindar a vision of lyric poetry as an act of almost oracular celebration and

---

[1] Cf. M. B. Benn, op. cit., pp. 143–4.

disclosure, and a technique of rare dense speed. Paradoxically
unimpeded by frequent misunderstandings of the original Greek,
these experiments in total penetration and similitude lead both
to Hölderlin's crowning poems and to his appropriations of
Sophocles. Hölderlin seemed to derive from his work on Pindar
the (reckless) confidence that he could pierce to the core of meaning in ancient Greek, that he could break through the barriers of
linguistic, psychological remoteness to a 'pre-logic' or universality
of inspiration. He made of the act of understanding and restatement an archaeology of intuition. He went deeper than any
philologist, grammarian, or rival translator in his obsessive search
for universal roots of the poetic and of language (again, as with the
speech-mystics of the seventeenth century and the Pietists, the
borrowed image of the 'root of words' is being used literally).

Hölderlin's *Umdichtung* of Sophocles (the German word allows
the exactly apposite double meaning of 'poetic transformation' and
of 'con-densation' or 'compaction around an object'), together
with the gnomic commentaries which accompany it, have been
thoroughly studied.[1] To Hölderlin's contemporaries, *Ödipus der
Tyrann* and *Antigone* seemed either wildly misconceived or farcical.
The small circle which took note of them at all inclined to see
in these versions symptoms of the mental disorder which soon
enveloped the poet in silence. Modern commentators, on the
contrary, have judged Hölderlin's text to be not only the ultimate
in reconstitutive understanding of Sophocles but an unequalled
penetration of the meaning of Greek tragedy as a whole.[2] In his
grasp of the nature of the Divine presence and event in tragic

[1] Cf. W. Schadewaldt, op. cit., pp. 766–824. But despite extensive investigation,
large uncertainties remain. The extent and quality of Hölderlin's knowledge of
Greek are still problematic, as are the probably crucial relations of his own
treatment of Sophocles to that of Hegel. The whole topic of the role of *Oedipus* and
Antigone, especially the latter, in the growth of German idealism, and in the works
of Hegel, Kierkegaard, and Schopenhauer, demands thorough analysis. It may
emerge that Hölderlin's appropriations were somewhat less eccentric than it would
seem. Hegel also was planning a translation of Sophocles and Kierkegaard's
'reconstruction' of Antigone in *Either/Or* is more extravagant than anything in
Hölderlin. cf. my *Antigones* (Oxford, 1984).
[2] This is true not only of Benjamin and Heidegger but of such classicists as
Reinhardt and Schadewaldt.

drama, Hölderlin has 'come closer to Sophocles than any other translator'.[1] These drastic differences of opinion reflect the enigmatic nature of Hölderlin's enterprise. The texts as we know them seem to incorporate different levels of intention. There are, notably in the treatment of *Oedipus*, elements of straightforward, almost pedantic translation, aspects of what was to be a public version of Sophocles' complete tragedies. There are in both plays bursts of private hermeneutic violence, attempts to wrench meaning out of its Greek carapace by force of word-for-word transposition. But there is also, principally in the *Antigone*, a programme of translation as enhancement, as corrective reconstitution derived both from an intimate reading of the original poet's spirit (a reading which Sophocles himself could not have achieved) and from the perspective of subsequent history. As Hölderlin writes in the often-cited letter to Wilmans of 28 September 1803, translation as he conceives it is emendation, externalization, a bodying forth of implicit meanings (*ein Herausheben*), but it is also correction: 'ihren Kunstfehler, wo er vorkommt, verbessern'. Such correction and improvement is possible, indeed mandatory, because the translator's vision of the original is diachronic; time and the evolution of feeling have given to his echo a power of fulfilment. The correction made by the translator is latent in the original; but only he can realize it. One cannot exclude entirely the thought that there is in this visionary pre-emption a touch of madness. But the strategy of interpretative excess and linguistic dislocation is integral to Hölderlin's finest, sanest poetry and critical exegesis.

Hölderlin's 're-presentational mutation' of line 20 in *Antigone* (Schadewaldt speaks aptly of *Neusprechen* and *Nachsprechen*) reads:

Was ist's, du scheinst ein rotes Wort zu färben?

At a straightforward level this is nonsense, and was felt to be so by Hölderlin's first readers. Confronted by Antigone's abrupt intimations of nearing calamity, Ismene asks: τί δ'ἔστι; δηλοῖς γάρ τι καλχαίνουσ' ἔπος. 'What is it? Clearly some news, some asser-

[1] W. Schadewaldt, op. cit., p. 822.

tion, torments you' (Mazon translates: *quelque propos*). Yet the intentions behind Hölderlin's version are unmistakable and, to a significant degree, justified. He believed that the antique sense of words, particularly in tragic drama, had a material aura and consequence lacking in modern epistemology. A prophecy, an oracular dictum, a formula of anathema in Greek tragedy carried with it a literal fatality. Speech did not stand for or describe the fact: it was the fact. Antigone is not only adumbrating mental anticipations of menace and of blood: she is darkening, making more sanguinary, words which are already deeds of revolt and of suicide. καλχαίνουσ' does mean 'making red'. Being uttered—dyed red—the *epos* of Antigone has become fatal, ineluctable gesture. An anthropology, a contrastive linguistics of the role of discourse in ancient and modern society underlies and necessitates Hölderlin's literalism, his paradoxical attempt to understand and even improve on the original while proceeding word-for-word. The tactic is violent and often absurd, but much recent thinking about speech habits in primitive cultures and the strength of physical mandate in, say, ancient Hebrew, does bear out Hölderlin's point.[1]

Because they embody speech-acts even more 'involuntary', even more primal than those of the protagonists, choral lyrics are to Hölderlin the essence of dramatic being. Schiller's mirth when he and Goethe listened to a reading of the choruses in Hölderlin's *Antigone*, his urbane assurance that his sometime disciple had been deranged when writing them, are well known. The impression of wilful chaos must have been scandalous, and that of obscure violence ought to be so still:

> Vater der Erde, deine Macht
> Von Männern, wer mag die mit Uebertreiben erreichen?
> Die nimmt der Schlaf, dem alles versinket, nicht
> Und die stürmischen, die Monde der Geister
> In alterloser Zeit, ein Reicher,
> Behältst der Olympos
> Marmornen Glanz du,

[1] Cf. Isaac Rabinowitz, ' "Word" and Literature in Ancient Israel' (*New Literary History*, IV, 1972).

Und das Nächste und Künftige
Und Vergangne besorgst du.
Doch wohl auch Wahnsinn kostet
Bei sterblichen im Leben
Solch ein gesetztes Denken.

It is precisely through *Uebertreiben* (exaggeration), through a clear response to the risks of *Wahnsinn* (insanity, false meaning) that the poet seeks to recapture the power and meaning of *Antigone* 604–14. But it is impossible to judge his performance without an understanding of the rigorous, though paradoxical, logic of transformation which it enacts. Schiller's reflex was not erroneous but, at a central level, irrelevant.

Hölderlin's theory of language is based on the search for the numinous, perhaps sacred *Grund des Wortes*. It is in the individual word that the elemental energies of immediate signification are literally embodied. The hermeneutic recapture of original intent at the sentence-level is illusory because all sentences are context-bound and their analysis involves us in a dilemma of infinite regression. Only the word can be circumscribed and broken open to reveal its organic singularity. As Hellingrath was the first to show, this 'verbal monism' or monadism governs not only Hölderlin's translations of Sophocles but the *harte Fügung* (marmoreal fabric) of his late, greatest hymns. The stylistic criterion is one expressed in section XXII of Dionysius of Halicarnassus' *De compositione verborum*: 'the words should be like columns firmly planted and placed in strong positions so that each word should be seen on every side, and that the parts should be at appreciable distances from one another' (W. Rhys Roberts's translation). Thus the elisions which characterize the syntax of Hölderlin's *Antigone* and of the *Anmerkungen* to the play, the mute spaces between words, invite us to view the individual word 'in the round', to go 'behind it'. Connectives, the inherent causal bias in idiomatic sentence-structures, create a deceptive surface and façade of logic. The essence of Sophoclean language, as of all authentic tragedy, resides 'in dem faktischen Worte, das mehr Zusammenhang, als ausgesprochen, schicksalsweise vom Anfang bis zu Ende gehet . . .'

Articulate such relations (*Zusammenhang*), provide them with an enforced smoothness and linearity, and you will have betrayed the literally daemonic potency of definition, of action, encased in the human word.

Towards the close of his creative career, Hölderlin developed what can only be called a mystical dialectic. He viewed the function of the poet, indeed of any human being seeking to 'essentialize' his condition, in terms of vehement encounter with an opposite principle. These dialectical collisions involve antithetical ideals, concepts, polarities which Hölderlin designated either by names of his own coinage or by titles to which he ascribed novel, often private meanings. Antique and modern, organic and *Aorgisch*, Oriental and Hesperidean, light and dark, the communicative and the inarticulate were to clash in a dialectic of conflict and mediation. Of such agonistic confrontations the most important was that between human and divine. Hölderlin's mature theory of poetry and of tragic drama turns on a terribly private yet philosophically ambitious model of interaction between man and God. Only by challenging the autonomy of the divine, by invading the 'space of the gods', can man accomplish his own transcendent potential and simultaneously force the gods to observe and fulfil their own ambiguous contiguities to the mortal order. The tragic agent—Hölderlin is thinking mainly of Oedipus and Antigone but also of the Sophoclean Ajax—enmeshes himself in deliberate polemic intimacy with the gods. He becomes, in Hölderlin's famous but obscure terminology, an *antitheos* whose challenge to the divine, whose perilous proximity to the gods is at once a blasphemy, a suicidal *hubris* and an ultimate assertion of the dialectical reciprocity of existence of men and gods ('wo einer, in Gottes Sinne, wie *gegen* Gott sich verhält'). Antigone's invocation of 'my Zeus' in Hölderlin's celebrated but debatable reading of line 450 is simultaneously an act of arbitrary appropriation, an incursion into the 'absent' realm of divine justice, and a desperate affirmation of the relevance of that realm to the survival of mankind and society.

Hölderlin's inmost sense of this sacrificial dialectic, in which collision and even mutual destruction are the necessary means to proper definition and distance, is impossible to paraphrase satisfactorily. The conception is dynamic and, therefore, in some measure intelligible and verifiable in the motion of his late poems, in the advance—at a certain level intentional or consciously gauged—of sense towards insanity, of statement towards silence. But as the commentaries on *Oedipus* and *Antigone* show, Hölderlin himself found it extremely difficult to phrase, let alone explain, his ontology and mythology of cosmic encounters. This, I suggest, is the point at which the concept and activity of translation became crucial.

Hölderlin's genius reaches its final realization in translation because the clash, mediation, and dialectic fusion of Greek and German were to him the readiest, most tangible enactment of the collisions of being. The poet brings his native tongue into the charged field of force of another language. He invades and seeks to break open the core of alien meaning. He annihilates his own ego in an attempt, both peremptory and utterly humble, to fuse with another presence. Having done so he cannot return intact to home ground. In each of these hermeneutic motions, the translator performs an action deeply analogous with that of Antigone when she trespasses on the sphere of the gods. The translator also is an *antitheos* who does violence to the natural, divinely sanctioned division between languages (what right have we to translate?) but who affirms, through this rebellious negation, the final, no less divine, unity of the *logos*. In the implosive shock and blaze of real translation, both tongues are destroyed and meaning enters, momentarily, into a 'living darkness' (the image of Antigone's burial). But a new synthesis emerges, a unison of fifth-century Attic with early-nineteenth-century German. It is a 'strange' idiom because it belongs integrally to neither language. Yet it is charged with currents of meaning more universal, nearer the sources of all human speech, than either Greek or German. Thus, for the late Hölderlin, the poet comes closest to his own true tongue when he translates. Beyond the fusion that comes of great translation—but in a sense which is now concrete and to which the poet has earned legitimate access—lies silence. Perfect coherence is speechless and unspoken.

We find ourselves here at the far limits of any rational theory or practice of linguistic exchange. Hölderlin's is the most exalted, enigmatic stance in the literature of translation. It merits constant attention and respect by virtue of the psychological risks implied and because it produced passages of an intensity of understanding and 're-saying' such as to make commentary impertinent. Take, for instance, the version of the Chorus in *Antigone* 944 ff.:

> Der Leib auch Danaes musste,
> Statt himmlischen Lichts, in Geduld
> Das eiserne Gitter haben.
> In Dunkel lag sie
> In der Totenkammer, in Fesseln;
> Obgleich an Geschlecht edel, o Kind!
> Sie zählete dem Vater der Zeit
> Die Stundenschläge, die goldnen.

At one level Hölderlin must have known that he was re-inventing, that Sophocles' Danae is 'guarding the fruit of Zeus' golden rain'. But at another level, he was meshing into a single mysterious image both the elements of gold and Olympian visitation and his own conception of how man marks time in tragic agony (*das Zählen der Zeit im Leiden*).[1] The result is at once less and more than translation.

Charged as it is with stylistic genius and interpretative audacity, Hölderlin's art of translation always derives from literalism, almost, in fact, from a literalism not only of the single word but of the letter. As he says in the first version of *Patmos*, God loves best those who tend, who are guardians of the 'firm letter' (*der feste Buchstab*). Paradoxically, therefore, the most exalted vision we know of the nature of translation derives precisely from that programme of literalism, of word-for-word metaphrase which traditional theory has regarded as most puerile.

---

[1] Cf. the discussion of this passage in K. Reinhardt, op. cit., pp. 94–8.

2

Ordinarily translation, even literary translation, moves on no such wilful, lofty plane. It aims to import and to naturalize the content of the source-text and to simulate, so far as it is able, the original executive form of that content. The conceptual short-cut is, canonically, of the kind stated by Dryden in his definition, itself broadly traditional, of *paraphrase*: 'to produce the text which the foreign poet would have written had he been composing in one's own tongue'. But even if we allow—as we are compelled to if discussion is to proceed—the identification for purposes of extraction and transfer of a 'content', that is to say of a potentially extensible body of meaning which can be separated out from the unique ensemble of the original phonetic–syntactic–semantic context, the proposed manoeuvre is more awkward, more inherently problematic, than might appear. What I have called the third move in the hermeneutic of appropriation, the portage home of the foreign 'sense' and its domestication in the new linguistic–cultural matrix, is almost never a linear, point-to-point carry. It instances, in a strong version and at different levels of strategic fiction, the issue of 'alternity', the diacritical externalization (*mise en relief*) of language differences which test or, more often, concatenate different possibilities and versions of being. The proposition 'the foreign poet would have produced such and such a text had he been writing in my language' is a projective fabrication. It underwrites the autonomy, more exactly, the 'meta-autonomy' of the translation. But it does much more: it introduces an alternate existence, a 'might have been' or 'is yet to come' into the substance and historical condition of one's own language, literature, and legacy of sensibility. This augmentative, challenging, or nostalgic function (the records of translation show an abundance of each of these modes) is clarified by the problem of chronology. Strictly speaking, every act of translation except simultaneous translation as between earphones, is a transfer from a past to a present. As we have seen at the outset, the hermeneutic of import occurs not only across a linguistic–spatial frontier but also requires a motion across time. What ordinary translation tries to do is 'to produce the text which the foreign poet would have written had he been working in one's own speech now, or more or less now'. The latitude of 'more or less', the elasticity of assumed contemporaneity is, as we shall see, one of the persistent, functional aspects of the whole construct of understanding and restatement.

There can be a refusal of latitude. The translator may claim that it is impossible to convey meaning adequately both across the barrier of language-differences and across time. He may insist on pure horizontality. This can be achieved either by translating only contemporary matter or by seeking to match the date of the receptor-language to that of the source. Though writing today, the translator aims to translate Spenser into sixteenth-century Castilian, he produces a version of Marivaux in eighteenth-century Russian, he renders Pepys's journals into seventeenth-century Japanese. This synchronicity has the charm of utter logic. It is (probably) absurd, but for reasons which are not trivial. Let us suppose that the translator is in fact able to produce a matching vocabulary and grammar: by dint of lexical and syntactical erudition he *can* translate *Werther* into a Dutch or a Bengali of the 1770s. No more recent idiom, no subsequent phrasing is used. But can this artifice of retrospection apply to his own sense of the text, either in the original or in his transcription? All context is diachronic and the field of meaning, of tonality, of associative range is in motion. The translator may choose the right word and grammatical turn, but he knows its later history; inevitably, the spectrum of connotations is that of his own age and locale. Even where he finds the precise chronological equivalent, the objects or facts of feeling referred to are embedded in his own modern perception of them. Consequently they will function either as antiques which, obviously, they were not at the time the original reference was made, or they will have altered. In short, the dilemma is that of Borges's fable: even facsimile is an illusion when time has passed. The phonetic sign, the word, may have remained stable, being arbitrary, but its meanings, the *signifié*, do not.

Certain trials of synchronicity are, nevertheless, among the most revealing episodes in the history and theory of translation. We find several in the 1820s, presumably at the suggestion of romantic historicism, of the attempt, graphic in historical writing from Herder to Michelet, to penetrate and recapture the authentic consciousness or 'inscape' of a circumstantial past. Leopardi intended to translate Herodotus into medieval Italian. Paul-Louis Courier's experiments at reproducing Herodotus and Longus in Renaissance French are a case of ambiguous but highly suggestive 'arbitrary contemporaneity': Courier seeks to rediscover the classical text as it was rediscovered and made European by the sixteenth-century humanists. Rossetti's *Early Italian Poets* and *Dante and his Circle* appear in 1861 and 1874 respectively. Here the intended synchronicity is again a hybrid. Rossetti is seeking to cast his own pictorial, poetic style in a mould of Italian medievalism; but he is also continuing a practice of archaicism, largely conventionalized and Spenserian, which derives from the ballads, from Augustan imitations of the *Faerie Queene*, and from Keats. The resulting idiom is at once reconstructive and normative in that it aspires to make of the old manner a modern ideal. Thus we obtain a rather blurred ancientness. Dante to Cavalcanti:

> Guido, I wish that Lapo, thou, and I,
>> Could be by spells conveyed, as it were now,
>> Upon a barque, with all the winds that blow
> Across all seas at our good will to hie.
> So no mischance nor temper of the sky
>> Should mar our course with spite or cruel slip;
>> But we, observing old companionship,
> To be companions still should long thereby.
> And Lady Joan, and Lady Beatrice,
>> And her the thirtieth on my roll, with us
>> Should our good wizard set, o'er seas to move
> And not to talk of anything but love:
> And they three ever to be well at ease,
> As we should be, I think, if this were thus.

In effect, neither Rossetti's own style nor his penetration of the original are sufficient to create the illusion of concordance.[1] Comparison of Rossetti's *Cavalcanti* with Pound's is to the latter's advantage.[2]

Émile Littré's approach to Dante was of a different order of rigour and intellectual power. To the great lexicographer and historian of the French language, problems of historical linguistics seemed obviously related to those of translation. Littré made this point in a remarkable essay published in the *Journal des débats* for January 1857.[3] His remarks were occasioned by two recent versions of the *Commedia*, that of A. Mesnard and that of Lamennais. Lamennais had completed his translation in 1853. He had set out to transfer the original into the French of Rabelais and Amyot. Though fear of incomprehension had made him give up this plan, Lamennais kept his translation literal and archaic. It endeavours to be 'précis, concis, primitif'. It remains almost unknown but has a fierce psychological interest: banished from the priesthood, Lamennais produced a Ghibelline recension more unforgiving than the original.[4] Wishing to re-educate the French ear to the nobility of its ancestral speech, Littré translated one book of the *Iliad* into thirteenth-century French. But he quickly saw the lack of logic in the proceeding. So he reverted to Dante. By reproducing the *Commedia* in the *langue d'oïl* which Dante

---

[1] On Rossetti's trivialization of Dante's vocabulary of love, see Nicolette Gray, *Rossetti, Dante and Ourselves* (London, 1947), pp. 34–8. Cf. also R. J. Morse, 'Rossetti and Dante' (*Englische Studien*, LXVIII, 1933), R. C. Simonini, 'Rossetti's Poems in Italian' (*Italica*, XXV, 1948), and G. Hough, *The Last Romantics* (London, 1961), pp. 71–82.

[2] Cf. A. Paolucci, 'a careful examination of the two translations will show, I think, that in spite of the faults we have noted, Pound's translation recaptures the nostalgic mood of the Italian more effectively than Rossetti's version.' ('Ezra Pound and Rossetti as Translators of Guido Cavalcanti', *Romanic Review*, LI, 1960, p. 263.)

[3] This article, which is among the neglected classics of the nineteenth-century theory of translation, was reprinted, with some alterations, as pp. 394–434 in Vol. I of Littré's *Histoire de la langue française* (Paris, 1863). For a general sketch of Littré's views on language, cf. Alain Rey, *Littré: L'Humaniste et les mots* (Paris, 1970).

[4] Cf. F. Duine, *La Mennais: Sa vie, ses idées, ses ouvrages* (Paris, 1922), pp. 300–6.

himself had known, Littré would not only induce the reader to study and savour 'notre vieil idiome' but perhaps bridge the gap of essential meaning between Dante's world and the modern. Littré hoped that a version in the *langue d'oïl* of the thirteenth and fourteenth centuries would give his interpretation that precise distance to Virgil and that unison with Latin Christendom which determined the spirit of Dante's epic. *L'Enfer mis en vieux langage François* appeared in 1879. It was almost still-born and has been remembered, if at all, as the diversion of an eccentric scholar.[1] Only the philologist and medievalist can judge of Littré's success in creating a synchronic replica. But the effects are often striking:

> Peu sont li jor que li destins vous file,
>   Li jor qu'avez encor de remanent;
> Ne les niez à suivre sans doutance
> Le haut soleil dans le monde sans gent.
>
>   Gardez queus vostre geste et semance;
> Fait vous ne fustes por vivre com la beste,
> Mais bien por suivre vertu et conoissance.
>
>   Mi compagnon, par ma corte requeste,
> Devinrent si ardent à ce chemin,
> Que parti fussent maugré mien com en feste.
>
>   Ore, tornant nostre arriere au matin,
> O rains hastames le vol plein de folie,
> Aiant le bort sempre à senestre enclin.
>
>   Jà à mes ieus monstroit la nuit serie
> Le pole austral; et li nostre ert tant bas,
> Que fors la mer il ne se levoit mie.

At several points this replication of Ulysses' narrative (XXVI, 114–29) becomes a direct tracing or *calque* (*gent/gente, semance/semenza, vol plein de folie/folle volo, fors la mer/fuor del marin*). But elsewhere, the distance between Littré's archaic diction and the *Commedia* is greater than it would be, at the verbal level at least, between Dante

[1] One of the few favourable reviews is that by Francesco d'Ovidio in *Nuovi studii danteschi* (Milan, 1907). For the background to Littré's experiment, cf. Lucien Auvray, 'Dante et Littré' in *Mélanges de philologie, d'histoire et de littérature offerts à Henri Hauvette* (Paris, 1934).

and modern French. *Doutance, corte requeste, arriere au matin, rains* have an antiquarian specificity which pertains entirely to the early history of French and which counteracts the directness of Dante's 'new style'. By a Borges effect, it is Dante who appears to be translating Littré whose *Enfer* is older than the *Inferno* and related to the *chanson de geste* rather than the Virgilian epic. Here is some lost 'harrowing of Hell' which came down to Dante via the masters of the Provence.

The fiction of a 'lost native source' obsessed Rudolf Borchardt. Why had Dante not written in medieval German? Or, more urgently, why had thirteenth-century German literature and civilization, poised as they were between the Teutonic north and the Mediterranean, in vital contact both with the pagan marches to the east and with Gallic Latinity, not produced a *Comedia divina* (the archaic spelling is Borchardt's)? This hypothetical question engaged Borchardt, a somewhat enigmatic scholar–poet inclined to a pan-European mystique, from 1904 to 1930. He arrived at the conviction that the *Commedia* is missing in German. Dante's absence from the history of the German language and of German sensibility in the period 1300–1500 destroyed deep logical and material affinities between German feudalism and the 'classical' Christendom of the Provence and of Tuscany. Far from being a sovereign renewal of German, the idiom of Luther was in many respects a defeat. Unlike medieval German, Luther's *Neuhochdeutsch* was often helpless before the concreteness and sensuous force of the Biblical original. After Luther, argues Borchardt, came Opitz and Gottscheid and with them a palsied neo-classicism and bureaucratic academicism alien to fundamental strains in the German genius. Borchardt advanced this view in a survey of German Dante-translations since Schlegel's pioneering efforts of 1794–9 ('Dante und deutscher Dante', 1908). He developed it further in two 'Epilegomena zu Dante' published respectively in 1923 and 1930. But Borchardt's obsession strove beyond theory. The past was not immutable. Even as the human mind can dream a future so it can reshape the past. Adopting Novalis's celebrated definition of the translator as 'the poet of poetry', Borchardt conceived of translation as having a unique authority against time and the banal contingency of historical fact. By virtue of 'creative retransformation' (*Rückverwandlung*), the translator could propose, indeed enact an alternative development for his own language and culture. True archaicism, explains Borchardt in a letter to Josef Hofmiller of February 1911, is not antiquarian pastiche, but an active, even violent intrusion on the seemingly unalterable fabric of the past. The 'archaicist' enforces his will on the past, discarding from history or adding to it in the perspective of hindsight. The passage is astonishing:

der genuine Archaismus greift in die Geschichte nachträglich ein, zwingt sie für die ganze Dauer des Kunstwerks nach seinem Willen um, wirft vom Vergangenen weg was ihm nicht passt, und surrogiert ihr schöpferisch aus seinem Gegenwartsgefühl was es braucht; wie sein Ausgang nicht die Sehnsucht nach der Vergangenheit, sondern das resolute Bewusstsein ihres unangefochtenen Besitzes ist, so wird sein Ziel nicht ihre Illusion, sondern im Goethischen Sinne des Wortes die Travestie.

This was Borchardt's method in 'travestying' Dante, in making *Dante Deutsch* as his title blankly proclaims. Borchardt's medium is a fiction of arrested and redirected time, a personal *Frühneuhochdeutsch* with elements ranging from the fourteenth century to Luther. It contains bits of High, Low, and Middle High German, Alemannic, Alpine dialects, *termini technici* from the vocabulary of mines and quarries (*teufe, stollen, zeche, guhr, sintern*) and wordforms and grammatical devices coined by Borchardt.[1] He had no illusion as to its fictive character:

Die Sprache in die ich übertrug, kannte ich weder als solche noch konnte es sie als solche gegeben haben; das Original warf erst ihren Schatten gegen meine innere Wand: sie entstand, wie eine Dichtersprache entsteht, *ipso actu* des Werkes. Die italienische Wendungen, genau befolgt, ergaben ein Deutsch, das zwischen 1250 und 1340 im ganzen Oberdeutschland sehr leidlich verstanden worden wäre.[2]

But to make of this linguistic fiction a possible 'might have been',

[1] Cf. the authoritative study of Borchardt's language in Hans-Georg Dewitz, '*Dante Deutsch': Studien zu Rudolf Borchardts Uebertragung der 'Divina Comedia'* (Göppingen, 1971), pp. 167–222.

[2] Rudolf Borchardt, *Gesammelte Werke* (Stuttgart, 1959), II, p. 522.

an alternity with potential consequences for the present and future of the German spirit, was the object of the exercise. That which had never been might still become (*Ungeschenes immer noch geschehen*).

Though it was noticed by Hesse, Curtius, Vossler, and Hofmannsthal, *Dante Deutsch* has remained largely ignored. Its texture is as difficult and in some ways as secretive as the vision of potential history which it embodies. It is, however, certainly so far as the *Inferno* and the *Purgatorio* are concerned, a work of peculiar genius. Borchardt 'relived' Dante with an almost pathological intensity; his reading of the poem as 'ein Hochgebirge Epos', a traverse of alpine chasms and escarpments, is at once singular and convincingly sustained. It is interesting to set Borchardt's version of Ulysses' narrative beside Littré's:

> 'Brüder, die mir durch hundert tausend wüste
> 　　fährden bis her in untergang gefronet:
> 　　dieser schon also winzigen, dieser rüste,
> Die unser sinnen annoch ist geschonet,
> 　　wollet nicht weigeren die auferschliessung
> 　　—der sonne nach—der welt da nichts mehr wohnet!
> Betrachtet in euch selber eure spriessung!
> 　　ihr kamt nich her zu leben gleich getier,
> 　　ja zu befolgen mannheit und entschliessung.'
> In den gefährten wetzete ich solchen gier
> 　　mit diesem kurzen spruch nach fahrt ins weite,
> 　　dass ich sie dann nicht mögen wenden schier.
> Und lassend hinter uns des ostens breite,
> 　　schufen uns ruder schwingen toll zu fliegen,
> 　　allstunds zubüssend bei der linken seite.
> Alls das gestirn des andern poles siegen
> 　　sah schon die nacht, und unsern abgesunken,
> 　　als thät er tief in meeres grunde liegen.

There are admirable nuances: *untergang* for *occidente* (with the premonitory touch of disaster), *auferschliessung* with its delicate suggestion of the image of outward motion latent in *esperienza*, *mannheit* for *virtute*—an equivalence which restores the force of etymology—*toll zu fliegen* in which Borchardt simulates both the phonetic and semantic relations of the original, *tief in meeres grunde liegen* which exactly mirrors the quiet menace of *del marin suolo*. Through these precisions, the translator renders the principal intent of Dante's text, the inference of catastrophe in the midst of the bracing thrust of Ulysses' summons. For all its abruptness (Borchardt valued *Schroffheit*), this version produces a more immediate fluency of rhyme and linked motion than perhaps any other. It sustains the stabbing beat; indeed, the eighth line could come directly out of one of Brecht's settings of Villon. It 'punches' in the same way. And notice how *gier*, although subterraneously as it were, gives an effect, both tactile and tonal, which exactly matches *acuti* at the corresponding point in Dante's verse.

But the detail counts for less than does the bizarre logic of the whole. Here the hermeneutic of appropriation is meant not only to enrich the translator's native inheritance but to change it radically. Translation is made metamorphosis of the national past. All tongues and literatures are treated as a common store of being from which we may draw at will in order to countermand the errors, the lacunae of reality. An English Flaubert, an Italian Rabelais, a French Edward Lear—these are fantastications. But Borchardt reminds us that such fantastications are given substance in the act of translation (the actual idiom, 'an English Flaubert' signifying a translated book or edition, confirms this mutation). The intransigence of *Dante Deutsch* shows that no language, no linguistically-informed sense of personal or social identity is unaltered by what it imports.

As a rule, to be sure, only the translator of a contemporary text synchronizes. What is the use of rendering Dante into modes of French or of German all but inaccessible to the readers who need a translation in the first place? But although a total reconstructive archaicism of the kind produced by Littré or Borchardt is rare, archaicism to some degree and a displacement of style towards the past are pervasive in the history and craft of translation. The translator of a foreign classic, of the 'classics' properly speaking, of scriptural and liturgical writings, of historians in other languages, of philosophic works, avoids the current idiom (or certainly did so until the modernist school). Explicitly or by unexamined habit, with stated intent or almost subconsciously, he will write in a vocabulary and grammar which predate those of his own day. The parameters of linguistic 'distancing', of historical stylization are endlessly variable. Translators may opt for forms of expression centuries older than current speech. They may choose an idiom prevalent only a generation back. Most frequently, the bias to the archaic produces a hybrid: the translator combines, more or less knowingly, turns taken from the past history of the language, from the repertoire of its own masters, from preceding translators or from antique conventions which modern parlance inherits and uses still for ceremony. The translation is given a patina.

English Homers have persistently been 'aged'. In Pope the process is a subtle one and the effect is often caused by imitations of Dryden's selections from the *Iliad*.[1] In the nineteenth century archaicism becomes vehement and often absurd. Persuaded that 'For the power of preserving the charms, while veiling the blemishes of rhyme, no metre existing in the English language can bear comparison with the Spenserian,' P. H. Worsley in 1861–2 produced an *Odyssey* on the model of the *Faerie Queene*. Thus from Book XXI and the great turn to vengeance:

> Meantime the king was handling the great bow,
> Turning it round, now this way and now that,
> To prove it, if the horn or timber show
> Print of the worm. They, marvelling much thereat,
> Spake one to other, leaning as they sat:
> 'Surely the rogue some pilfering expert is
> In bows and arrows, which by fraud he gat—
> Or would the varlet mould a bow like this?
> So featly doth he feel it with his hands, I wis.'

It is easy to deride this sort of misprision, to see in the kind of books it produces the all too literal 'print of the worm'. But the

---

[1] Cf. H. A. Mason, *To Homer Through Pope* (London, 1972), pp. 170–7.

archaic convention was dominant. The difference lay in the poetic distance chosen: William Morris's *Odyssey* (1887) is part Norse saga, part Tennyson, and part archaeology:

'Lo here, a lover of bows, one cunning in archery!
Or belike in his house at home e'en such-like gear doth lie;
Or e'en such an one is he minded to fashion, since handling it
   still,
He turneth it o'er, this gangrel, this crafty one of ill!'
And then would another one be saying of those younglings
   haughty and high:
'E'en so soon and so great a measure of gain may he come by
As he may now accomplish the bending of the bow.'
So the Wooers spake; but Odysseus, that many a rede did know,
When the great bow he had handled, and eyed it about and
   along,
Then straight, as a man well learned in the lyre and the song,
On a new pin lightly stretcheth the cord, and maketh fast
From side to side the sheep-gut well-twined and overcast:
So the mighty bow he bended with no whit of labouring. . . .

T. E. Lawrence termed the *Odyssey* a 'novel'; unlike bookish traducers he could bring to the task of translation direct experience of hand-to-hand combat, he had built and sailed rafts, he had travelled incognito in enemy country avoiding watch-fires. But what is more retrograde than 'T. E. Shaw's' 1932 version of Homer, what could be more 'literary' in the trivial sense? It is not physical immediacy that Lawrence achieves but a farrago of Victorian Orientalism in the manner of Doughty, of Biblical pastiche, and scout-master heroics:

The bronze-headed shaft threaded them clean, from the leading helve onward till it issued through the portal of the last ones.
Then he cried to Telemachus, 'Telemachus, the guest sitting in your hall does you no disgrace. My aim went true and my drawing the bow was no long struggle. See, my strength stands unimpaired to disprove the suitors' slandering. In this very hour, while daylight lasts, is the Achaeans' supper to be contrived: and after it we must make them a different play, with the dancing and music that garnish any feast.' He frowned to him in

warning: and Telemachus his loved son belted the sharp sword to him
and tightened grip upon his spear before he rose, gleaming-crested, to
stand by Odysseus, beside the throne.

This to translate a poet who, as Matthew Arnold had urged, is
neither 'quaint' nor 'garrulous' but always 'rapid', 'plain' and
'direct' in word and thought.

In the translation of a philosophic text every literary device is or
ought to be expressly analytic. Carried to extremes this intentionality produces Heidegger's notorious rendition of Parmenides'
'τὸ γὰρ αὐτὸ νοεῖν ἐστίν τε καὶ εἶναι' as 'Zusammengehörig sind
Vernehmung wechselweise und sein', when the simple, straightforward reading is: 'thinking is the same thing as being'. Philosophic
translation should seek to fix meaning uniquely and to render
logical sequence transparent. To produce a 'dated' version of a
philosophic original is gratuitous unless the time-distance chosen
specifically elucidates and makes unmistakable the sense, the
technical status of the text. Readings of the *Timaeus* as an analogue
to the Pentateuch, hermetically transmitted via a 'Mosaic–Orphic'
tradition, or as a prefiguration of Trinitarian and Christological
motifs, are at least as old as the Middle Ages.[1] Jowett's stated
purpose when he published his translation of the *Dialogues* in 1871
was to achieve the greatest possible clarity consonant with the
exact meaning of the Greek. He was fully aware that 'it is difficult
to explain a process of thought so strange and unaccustomed to us,
in which modern distinctions run into one another and are lost
sight of'. But he was confident that he had rendered the *Timaeus*
faithfully because here, more than at any other point in his doctrine, Plato had expounded 'the goodness of God'. In Jowett's
version this exposition takes on an intense colour of Victorian
Christianity. Stylistic details accumulate to produce an unmistakable and consistent effect. Plato's frequent references to 'the god'
or 'the demiurge' are rendered by 'God'; Jowett uses the formula
'thus he spake'; he puts 'Lucifer' for the 'Morning Star'. In 36e

---

[1] Cf. Henri de Lubac, *Exégèse médiévale: les quatre sens de l'Écriture* (Paris,
1959–64), IV, pp. 189, 215.

we find: 'Now when the Creator had formed the soul according to his will...' where F. M. Cornford's translation of 1937 reads: 'When the whole fabric of the soul had been finished to its maker's mind...' Jowett translates 'Such was the mind and thought of God in the creation of time' (38c) where F. M. Cornford, with a striking reversal of capitalization, reads 'In virtue, then, of this plan and intent of the god for the birth of Time...' In the later sections of the *Timaeus*, Plato uses 'the god' and 'the gods' almost at random, and even combines the two in the same sentence (in 71a). But Jowett retains 'God'. As Cornford points out,[1] the resulting distortions of the tone of the original, and of its logical emphasis, are far from trivial. Plato was no monotheist; he believed in the divinity of the whole of phenomenal nature and attributed a divine status to the heavenly bodies. Jowett's 'Christianization' of the dialogue, moreover, misses a central aspect of Plato's teaching on creation. The 'demiurgus' (Thomas Taylor's translation of 1804)[2] operates on materials which pre-exist. Plato's cosmic builder is resolutely conceived in the image of a human craftsman, not of an omnipotent Deity in the Judaic–Christian vein.

Jowett's commitment to a monotheistic programme of universal goodness and order is not in doubt. He had, as Swinburne's 'Recollections of Professor Jowett' remind us, long harboured the project of translating and editing a 'Child's Bible'. But the terminological bias in his *Timaeus* is not, one suspects, consciously doctrinal. It springs from a specific archaicism. Aiming at elevation and harmonious beat, Jowett follows the model of the Authorized Version. This is the more striking when one compares the 1871

[1] Cf. F. M. Cornford, *Plato's Cosmology: The Timaeus of Plato Translated with a Running Commentary* (London, 1937), pp. 34–9 and 280 for a critical discussion of the differences between Plato's demiurge and the Creator of Genesis or the God of the New Testament. In a footnote to Plato's use of ἔμενεν in 42e, Cornford notes how important it is to resist Biblical suggestions and to differentiate the 'confinement within his own nature' of the Platonic cosmic architect from the 'rest' of God in Genesis 2:2.

[2] Though openly Neoplatonic in its view of the *Timaeus*, Taylor's translation, with its attempt to adapt contemporary scientific–technological terms, is in certain respects closer than Jowett's to the flavour of the Greek. Taylor's *Timaeus and Critias or Atlantis* has been reissued with a preface by R. Catesby Tagliaferro (New York, 1944).

text of Plato with the versions of Saint Paul to the Thessalonians, Galatians, and Romans which Jowett had published in 1855 and 1859. In his treatment of Pauline Greek, Jowett, obviously conscious of the proximity of the King James Bible, strove to achieve his own more modern, scholarly turn. But when he came to Plato, and to the *Timaeus* in particular, he found the Biblical precedent irresistible. The resulting texture is not a straightforward echo of Jacobean English. It is a 'semi-archaicism' in which the language of 1611 is filtered through that of the later seventeenth century and that of the Victorian poets. Only extensive quotation could show this 'layering' in detail but extracts from 40a–d will illustrate the prevailing rhythm:

Thus far and until the birth of time the created universe was made in the likeness of the original, but inasmuch as all animals were not yet comprehended therein, it was still unlike. Therefore, the creator proceeded to fashion it after the nature of the pattern in this remaining point. . . . Of the heavenly and divine, he created the greater part out of fire, that they might be the brightest of all things and fairest to behold. . . . Vain would be the attempt to tell all the figures of them circling as in dance, and their juxtapositions, and the return of them as in revolutions upon themselves, and their approximations, and to say which of these deities in their conjunctions meet, and which of them are in opposition, and in what order they get behind and before one another, and when they are severally eclipsed to our sight and again reappear, sending terrors and intimations of the future to those who cannot calculate their movements— to attempt to tell of all this without a visible representation of the heavenly system would be labour in vain.

The tonality is not Plato's Greek so much as it is the 'Biblical-baroque' of the nineteenth century as produced from the time of Coleridge's gloss on *The Rime of the Ancient Mariner* to the prose of Hardy.

The archaic reflex extends far beyond the presumed solemnity and apartness of the classics. The bulk of literary, historical, philosophical translation, even where it concerns fiction, political writings, or plays intended for production, shows symptoms of retreat from current speech. When we score a translation as being lifeless, as being cast in 'translationese', what we are usually condemning is the patina. In terms of the hermeneutic model there are two principal reasons for archaicism. The first is implicit in the dynamics and techniques of understanding. In seeking to penetrate the sense and logic of form of the original, the translator proceeds archaeologically or aetiologically. He attempts to work back to the rudiments and first causes of invention in his author. Discussing his own version of Virgil's *Eclogues*, Valéry says: 'Le travail de traduire, mené avec le souci d'une certaine approxima- tion de la forme, nous fait en quelque manière chercher à mettre nos pas sur les vestiges de ceux de l'auteur; et non point façonner un texte à partir d'un autre; mais de celui-ci, remonter à l'époque virtuelle de sa formation.' Thus the period flavour in so much of translation, whether or not it is accurately mimed, may be a legitimate consequence of the reconstructive method. I shall come back to this point.

But there is also a second, tactical motive. The translator labours to secure a natural habitat for the alien presence which he has imported into his own tongue and cultural setting. By archaicizing his style he produces a *déjà-vu*. The foreign text is felt to be not so much an import from abroad (suspect by definition) as it is an element out of one's native past. It had been there 'all along' awaiting reprise. It is really a part of one's own tradition temporarily mislaid. Master translations domesticate the foreign original by exchanging an obtrusive geographical–linguistic dis- tance for a much subtler, internalized distance in time. The German reader of the Wieland–Schlegel–Tieck Shakespeare experiences the flattering impression of looking back on something entirely his own. The remoteness is that of his own historical past. Coming across *La Chanson du vieux marin* in its 1911 guise, the French ear could readily suppose that Valery Larbaud, that well- known hunter of literary oddities, had resuscitated a poem of the type popularized by Victor Hugo's *Odes et ballades*. The strange- ness of the text does not stem from the distance between French and English, but from differences of sensibility between modern French verse and the conventions of early Romanticism. Archaicism internalizes. It creates an illusion of remembrance which helps to embody the foreign work into the national repertoire.

In the history of the art very probably the most successful domestication is the King James Bible. Though many aspects of this collaborative enterprise are now understood, and though the general programme of the commissions of translation is well documented, the details of composition, of amendment, of theoretical debate, if any, remain obscure. Only one set of working papers has until now turned up, and although it is among the most fascinating primary sources in the entire history of translation, it is also brief.[1] There had been more than fifty English Bibles by 1611. The panels of scholars and divines who began work in 1604 had been expressly charged to base their text on the Bishops' Bible of 1568 and to consult Tyndale's Bible, Matthew's Bible, Coverdale's Bible, and Whitchurch's Bible. In the event, they went back much further, to the Middle English Gospels and Psalter and to the Wycliffite Bible. Whether the 'antique rightness of the phrasing' in the Authorized Version, as A. C. Partridge puts it, is due to deliberate stylistic policy, whether it is the achievement of Miles Smith, one of the two final editors, or whether it mainly reflects the influence of the genius of Tyndale, the greatest of English Bible translators, is not certain. But the pervasive patina, the sense of an idiom grounded in Tudor rather than in Jacobean usage and speech-rhythms are decisive. They ensured the remarkably rapid acceptance of the 1611 translation as not only canonic, but as somehow native to the spirit of the language and as a document uniquely inwoven with the past of English feeling. Though John Selden accused the translators of being antiquarian, they were in fact, as David Daiches has shown, heirs to Reuchlin and Erasmus in their standards of up-to-date scholarship.[2] What countless readers then and since have experienced in turning to their work is

[1] Cf. Ward Allen (ed.), *Translating for King James: Notes Made by a Translator of King James's Bible*. Prof. Allen's discovery in 1964 of the notes taken by John Bois during the final revision of *Romans* through *Revelation* at Stationers' Hall in London in 1610–11 is not only of extreme interest in itself, but holds out the possibility that further material may come to light.

[2] Cf. David Daiches, *The King James Version of the English Bible: An Account of the Development and Sources of the English Bible of 1611 with Special Reference to the Hebrew Tradition* (University of Chicago Press, 1941), particularly Chapter IV.

an unequalled feeling of 'at-homeness'; they have found a native presence in what is, in obvious truth, a remote, entirely alien world of expression and reference. By choosing or achieving almost fortuitously a dating some two to three generations earlier than their own, the translators of the Authorized Version made of a foreign, many-layered original a life-form so utterly appropriated, so vividly out of an English rather than out of a Hebraic, Hellenic, or Ciceronian past, that the Bible became a new pivot of English self-consciousness. The archaicism was 'not a phenomenon of vocabulary alone, but a complex of historical factors, impossible to isolate'.[1] They include archaic weak plurals, the inflexion of the second- and third-person singular of verbs, the use of past participles of verbs, the preservation of the idiomatic verb *wot*, weak preterites such as *shaked*, the common Middle English assimilation of the preterite and past participle inflexions to the stem-final *t* of weak verbs (the Ark 'was *lift* up above the earth' in Genesis 7 : 17), and numerous words which had dropped out of current speech or were rapidly becoming obsolescent at the turn of the century.[2] Far from being static or merely ornamental, this archaicism embodied the vitality, the logic of a cumulative tradition. This 'ingestion' and transmutation of Hebrew, Greek, and Latin sources into English sensibility, where it continues to play a part more immediate than that of Scripture in any other European community, more linguistically central and theologically diffuse, would not have occurred had the scholars and editors of 1604–11 laboured to be 'modern'. It was by looking back that they justified the proud definition contained in the Preface: 'Translation it is that openeth the window, to let in the light; that breaketh the shell, that we may eat the kernel.'

The translator can manipulate anachronisms for special effects. In his imitations from Villon, Basil Bunting interleaves legendary matter, already archaic to Villon himself, with nineteenth- and twentieth-century allusions:

---

[1] A. C. Partridge, *English Biblical Translation* (London, 1973), p. 138.

[2] I am following Prof. Partridge's detailed discussion of these points in op. cit., pp. 115–38.

> Abélard and Eloïse,
> Henry the Fowler, Charlemagne,
> Genée, Lopokova, all these
> Die, die in pain.
>
> And General Grant and General Lee,
> Patti and Florence Nightingale,
> Like Tyro and Antiope
> Drift among ghosts in Hell. . . .

The feeling is one of macabre universality, of violent up-dating, but also of dream-like unreality. The remembrance of death, death itself perhaps, is 'nothing, save a fume / Driving across a mind'.

In Marianne Moore's La Fontaine (1954), the control of temporal distances is complex and brilliant. Though Miss Moore professes herself to be entirely Pound's disciple—'the natural order of words, subject, predicate, object; the active voice where possible; a ban on dead words, rhymes synonymous with gusto'— her practice is in fact highly idiosyncratic. It comprises extreme verbal decorum, often precisely attuned to the speech-habits of women in nineteenth-century New England, an arch pleasure in the Latinate and technical term, tricks of compaction which owe something to the elided grammar of Emily Dickinson, and a pattern of enjambment and prim caesura which are the mark of her own verse. The cunning plainness of the *Fables*, La Fontaine's admixture of colloquial with neo-classical modes, exactly suited Marianne Moore's gifts. Take one of the best-known fables (III. xi):

> Certain renard gascon, d'autres disent normand,
> Mourant presque de faim, vit au haut d'une treille
>   Des raisins mûrs apparemment,
>   Et couverts d'une peau vermeille.
> Le galand en eût fait volontiers un repas;
>   Mais comme il n'y pouvait atteindre:
> 'Ils sont trop verts,' dit-il, 'et bons pour des goujats.'
>   Fit-il pas mieux que de se plaindre?

La Fontaine's archaicisms are slight and ironic: thus *galand* retains connotations of mirth (from Old French *galler*) and cunning. The omission of *ne* in direct interrogation had been condemned by grammarians even before La Fontaine; but here, as in other *Fables*, the poet uses it to get an effect of wry concision. *Goujat* is an ancient designation of men-at-arms, beautifully apposite here because its origin is probably Gascon, like that of the fox, and because it is exactly the kind of rough colloquialism which La Fontaine introduces into his polished setting. Now Miss Moore:

> A fox of Gascon, though some say of Norman descent,
> When starved till faint gazed up at a trellis to which grapes
> were tied—
> > Matured till they glowed with a purplish tint
> > As though there were gems inside.
> Now grapes were what our adventurer on strained haunches
> chanced to crave,
> > But because he could not reach the vine
> He said, 'These grapes are sour; I'll leave them for some
> knave.'
> > Better, I think, than an embittered whine.

Though retaining the identical number of lines and simulating La Fontaine's syntactical movement closely, the translator takes liberties. She invents the gems, which give to the fox's pursuit an additional, intrusive rapacity. On the other hand, she accurately renders through the use of *Matured* the emphatic, now obsolete meaning of La Fontaine's *apparemment* (i.e. 'clearly', 'with every evidence'). And by translating *verts* as *sour*, Miss Moore makes the statement proverbial in exactly La Fontaine's manner. Aesop's original underlies both the French and the English version with equal force. The effect of the whole is unquestionably modern, even American (the 'Now' structure in line five). But it is also ironically ceremonious in the neo-classical sense. 'When starved till faint', 'chanced to crave', 'knave' are faintly archaic as are certain touches in the French. Thus the old presses on the new with a delicate authority, meshing two levels of time and two styles.

Anachronism need not be retrospective. The translator may telescope time violently so as to produce a shock of contemporaneity. In his licentious but numbingly powerful variations on Book XIX of the *Iliad*, published in 1967, Christopher Logue depicts the speed and surging course of Achilles' magical horses by drawing on a most present image of uncanny thrust:

> The chariot's basket dips. The whip
> fires in between the horses' ears,
> and as in dreams or at Cape Kennedy they rise,
> slowly it seems, their chests like royals, yet,
> behind them in a double plume the sand curls up. . . .

The reference, moreover, invokes not only a blazing but stately motion; it also intones the perfectly appropriate note of impending, heroic death. The translator can modernize not only to induce a feeling of immediacy but in order to advance his own cause as a writer. He will import from abroad conventions, models of sensibility, expressive genres which his own language and culture have not yet reached. Reversing Borchardt's conceit of a 'lost past', he makes of translation an incitement for the future. It was the peculiar genius of Pound's Propertius and Cavalcanti to use antique material, to treat it with verbal archaicism, but to make syntax and motion programmatic and modernist. Pound's versions of Latin and of Provençal are meant to exemplify new possibilities in the stress-patterns, in the manners of address, in the segmentation of English and American verse. The translations of Khlebnikov's 'etymological' poems by Paul Celan, Hans Carl Artmann, and Magnus Enzensberger are, in the context of German poetry, a futurist manifesto. Ted Hughes's adaptation of Seneca's *Oedipus* in 1968 closely prefigures the idiom of *Crow* published two years later. Through translation of this order the past of other languages and literatures is made native to one's own and radical. When Celia and Louis Zukofsky render Catullus'

> Caeli, Lesbia nostra, Lesbia illa,
> illa Lesbia, quam Catullus unam

> plus quam se atque suos amavit omnes,
> nunc in quadriviis et angiportis
> glubit magnanimi Remi nepotes.

by

> Caelius, Lesbia new star, Lesbia a light,
> all light, Lesbia, whom Catullus (o name
> loss) whom his eyes caught so as avid of none,
> none else—slunk in the driveways, the dingy parts
> glut magnanimous Remus, his knee-high pots.

they are, at one level, spinning off puerile acrostics, but they are, at another, abusing their source with strategic intent. They are trying to instance possible procedures for American poetry now and tomorrow and hinting, confusedly, at a theory of immediate universal understanding.[1]

Such reversals, dislocations, arbitrary collages of historical chronology are negations or reorderings of actuality. They introduce an alternative past into the development of one's own language and code of perception, or they project possible futures. Like the multiplicity of languages, like the fact that different languages have not evolved synchronically, the treatment of time in translation as a strategical variable reflects that fundamental drive to free invention, to alternity which impels human speech. The translator imports new and alternative options of being.

3

The first move towards translation, which I have called 'initiative trust', is at once most hazardous and most pronounced where the translator aims to convey meaning between remote languages and cultures. Quine defines 'radical' translation as that of the language of a hitherto untouched people. The linguist will proceed, will commit himself to an expectation of understanding 'by intuitive judgement based on details of the native's behavior: his scanning

---

[1] *Catullus translated by Celia and Louis Zukofsky* (London, 1969).

movements, his sudden look of recognition, and the like'.[1] But even this 'radical' case is privileged. Often the interpretative act will have nothing but written, probably incomplete material to go on. There is no living informant left and no gestural or social context. The palaeographer or anthropological linguist decodes out of silence. On what then does he base his assumption that there is a sense to be extracted and retrieved, more or less substantively, into and via his own speech? (The two segments or moments of this assumption are closely related but not identical: it is logically conceivable that a translator, having gained great mastery over a source-language will conclude 'I understand this text but find no way of restating it in my own native tongue.')

The underlying postulate is both broadly pragmatic and idealistic. Its experiential base, roughly unexamined and conventional, is the knowledge—*how*, in fact, could a counter-example be demonstrated?—that no entirely undecipherable or entirely untranslatable body of speech has ever turned up; that all interlinguistic contacts, literary, anthropological, even archaeological have yielded or will, on strong statistical grounds, be sure to yield a set of communicable if not exhaustive or necessarily unambiguous meanings. The idealistic premiss is one of universal homology and rationality. It can take diverse forms: ecumenical, Cartesian, anthropological. But the conclusion is the same: the similarities between men are finally much greater than the differences. All members of the species share primal attributes of perception and response which are manifest in speech utterances and which can therefore be grasped and translated. Darwin found the differences between the Selk'nam and Yamana Indians of the Tierra del Fuego and civilized man 'greater than that between a wild and a domesticated animal', but they are not sufficient to preclude communication. On the contrary: that which is in fact linguistically and culturally most remote may, at moments, strike us as most poignant and cognate to our own consciousness. Though the thought processes in 'primitive' verbalization may differ widely from our own (itself a disputed point) we can never-

[1] W. V. O. Quine, *Word and Object*, p. 30.

theless 'easily understand them as records of human life; we can without much difficulty appreciate their imaginative and emotional strength; we can even feel something of their strictly poetical appeal'.[1] Wordsworth's anxious traveller made apprehensive of fatality as the moon drops suddenly behind the beloved's cottage is kin to the near-Stone Age hunter of the Andaman Islands when he sings:

> From the country of the Yerewas the moon rose;
> It came near; it was very cold,
> I sat down, Oh, I sat down,
> I sat down, Oh, I sat down.[2]

The referential apparatus of different languages and cultures is not the same, and overlap is never all-inconclusive. But wherever they are on earth, at whatever economic-social level, men read the cold of the moon in the same way or in ways sufficiently akin so that they can modulate to a mutual recognition. Where intricate, high cultures are involved, the premiss of congruent rationality gains in strength. The objectivity of the external world is invoked to validate a postulate of common understanding. 'I hope we have shown,' writes Joseph Needham with the authority conferred by an epochal achievement in cultural relation, 'that across the very great barrier of the ideographic and alphabetic languages, and across the time distance of ten or twenty centuries, minds trained in the observation and experimental study of Nature, and in the techniques which utilize her gifts, can still communicate.'[3] As we have seen, the axiom of deep structures and constraints put forward by generative transformational grammars has sought to give a demonstrable expression to the pragmatic and idealistic premiss of universal communication.

It has been my argument that neither the empirical nor the theoretical premiss is beyond cavil. A considerable portion of the

---

[1] C. M. Bowra, *Primitive Song* (London, 1963), p. 26.
[2] Bowra cites this text from C. B. Kloss, *In the Andamans and Nicobars* (London, 1903), p. 189.
[3] Joseph Needham, 'The Translation of Old Chinese Scientific and Technical Texts' in *Aspects of Translation*, p. 87.

anthropological evidence for verifiable communication between native informant and linguistic observer is thought to be suspect. We are growing wary of the hermeneutic circularity which may subvert the decipherment of a message from the past or from cultural–social contexts radically alien to our own. The assumption that speech habits and the conventions of concordance between word and object have not altered 'across the time distance of ten or twenty centuries' is one that causes increasing discomfort. If the impetus to signification is frequently and, in part at least, originally internal, if meaning is quite often kept from the outside questioner or communicated only in part, the whole issue of the status and extent of conveyed and translated sense remains open. Nothing in Quine's famous model of stimulation and stimulus meaning logically or materially excludes the notion of a tribe which would have agreed among its members to deceive the linguist-explorer. Schoolboy coteries, fraternal lodges, craft guilds proceed in just this manner. Quine's 'Gavagai' might not be the passing rabbit but the derisive *double-entendre* or nonsense locution chosen by the native speaker to conceal from the inquirer the actual, possibly numinous name of the animal. Quine's scheme requires an additional axiom of good faith, of initiative trust on both sides. The fact that such good faith may not be *wholly* forthcoming does not entail that the anthropologist's lexicon would be valueless. It would, at certain points, be a lexicon or grammar of the surface, containing, without being cognizant of the situation, marks of a special code of concealment or ironic play. All of us have come up against 'language blanks or blankness' in familial and social discourse inside our own culture. We think we have understood where we have, in reality, been proffered only conventional tokens or duplicities. How much likelier it is that the recorder–translator of remote speech forms will be similarly deceived or 'put off'.[1]

Are there, then, conclusive refutations of the intuitive confidence in the decidability and transmission of meaning which

---

[1] A very similar point seems to have been made by Wittgenstein. Cf. Allan Janik and Stephen Toulmin, *Wittgenstein's Vienna* (New York, 1973), p. 228.

initiates every act of translation? Are there 'untranslatabilities' caused by the remoteness from each other of phonetic structures and cultural contexts? Are there definitive answers to the programme which Pound set for himself in his article of 1913 on 'How I Began': 'I would know what was accounted poetry everywhere, what part of poetry was "indestructible", what part could *not be lost* by translation, and—scarcely less important—what effects were obtainable in *one* language only and were utterly incapable of being translated' (a question which is pertinent but logically naïve because if such effects are limited to *one* language no outside observer could fully ascertain or demonstrate their existence)?

A fair number among the most admired, influential Western translations relate to remote languages and to cultures radically alien to our own: FitzGerald's *Rubáiyát*, Goethe's versions of Hafiz, Waley's selections from Chinese, Japanese, and Mongolian, the Authorized Version itself. Some of the most persuasive translations in the history of the *métier* have been made by writers ignorant of the language from which they were translating (this would be so notably where rare, 'exotic' languages are involved). North's Plutarch derives not from the Greek but from the French of Amyot; Pound had no Chinese when he translated from Fenollosa's manuscript the poems in *Cathay*; Donald Davie's adaptation of Mickiewicz's *Pan Tadeusz* is based entirely on an English prose version by G. R. Noyes; Auden and Robert Lowell work at one or two removes from the Russian when translating Pasternak or Voznesensky. Yet in many of these cases it is not only the common reader, without any personal knowledge of the language being translated, who feels convinced; it is the rare Anglo-Saxon competent in Chinese or in Polish and, on striking occasions, the original poet or native speaker to whom the English text is shown for judgement. The relevant mechanics of penetration and transfer are obviously intricate and special; but they suggest a more general theory.

The difficulties of translating Chinese into a Western language are notorious. Chinese is composed mainly of monosyllabic units with a wide range of diverse meanings. The grammar lacks clear tense distinctions. The characters are logographic but many contain pictorial rudiments or suggestions. The relations between propositions are paratactic rather than syntactic and punctuation marks represent breathing pauses far more than they do logical or grammatical segmentations. In older Chinese literature it is almost impossible to demarcate prose from verse: 'If they have developed as more or less separate entities in the West, they coalesce and merge in Chinese; in fact, it would not be incorrect to say that the genius of Chinese prose is verse.'[1] No grammar or dictionary is of very much use to the translator: only context, in the fullest linguistic–cultural sense, certifies meaning. Yet despite these 'impossibilities', Chinese draws the Western translator to its litera- ture. In English the history of attempted transfer is extensive: it runs at least from Du Halde's *A Description of the Empire of China* published in 1738–41 to the present.[2] The oddity lies in the fact that so many of the best-known translators have no Chinese. Bishop Percy, whose translations appeared in 1761, worked from an earlier English manuscript and from the Portuguese. Stuart Merrill, Helen Waddell, Amy Lowell, Witter Bynner, Kenneth Rexroth have used prose trots, previous translations, French versions, the word-by-word aid of sinologists, to arrive at their results. Paradoxically, scandalously perhaps, these constitute an ensemble of peculiar coherence and they are, in one or two cases, superior in depth of recapture to translations based on actual knowledge of the original. The notorious challenge is, of course, that of *Cathay* (1915).

This collection is, one feels, not only the best inspired work in Pound's uneven canon, but the achievement which comes nearest

[1] Achilles Fang, 'Some Reflections on the Difficulty of Translation' in *On Translation*, pp. 120–1.

[2] The novice, i.e. almost everyone, will find invaluable pointers in Arthur Waley, 'Notes on Chinese Prosody' (*Journal of the Royal Asiatic Society*, April 1918); I. A. Richards, *Mencius on the Mind, Experiments in Multiple Definition* (London, 1932); Arthur Waley, *Introduction to Chinese Painting* (London, 1933); Arthur Waley, *The Way and its Power: A Study of the Tao Tê Ching and its Place in Chinese Thought* (London, 1934); Robert Payne, *The White Pony, An Anthology of Chinese Poetry from the Earliest Times to the Present Day, Newly Translated* (New York, 1947); Roy Earl Teele, *Through a Glass Darkly: A Study of English Translations of Chinese Poetry* (Ann Arbor, 1949); James J. Y. Liu, *The Art of Chinese Poetry* (Chicago, 1962).

to justifying the whole 'imagist' programme. The 'Song of the Bowmen of Shu', 'The Beautiful Toilet', 'The River Merchant's Wife: A Letter', 'The Jewel Stairs' Grievance', the 'Lament of the Frontier Guard', 'Taking Leave of a Friend' are masterpieces. They have altered the feel of the language and set the pattern of cadence for modern verse (Waley's translations into *vers libre* derive from the immediate precedent of Pound). But these are also, at many points, acute transmissions of the Chinese, reconstructions of extreme delicacy and rightness. Fenollosa misreads the first two characters in the second line of Li Po's 'Ku Feng (After the Style of Ancient Poems) No. 14'; he distorts the meaning of line twelve and mistakes the function of the war-drums; he blurs the end of the poem through erroneous, confusing glosses. Pound's 'Lament of the Frontier Guard' respects the literal surface but also penetrates beneath it to restore what Fenollosa has missed or obscured. Waley's version of 'The Song of Ch'ang-kan' is closely inspired by *Cathay* but aims to correct Pound's linguistic errors. In fact Pound's 'While my hair was still cut straight across my forehead' turns out to be more exact and pictorially informative than Waley's 'Soon after I wore my hair covering my forehead', and Pound's famous solecism 'At fourteen I married My Lord you' communicates precisely the nuance of ceremonious innocence, of special address from child to adult, which constitutes the charm of the original and which Waley misses. Thus on sinological grounds alone 'The River Merchant's Wife: A Letter' is closer to Li Po than is Waley's 'Ch'ang-kan'.[1] How were these 'translucencies'—Eliot's term—achieved by a translator ignorant of Chinese and working from an often defective transcription of and commentary on the source-text?

As Eliot and Ford Madox Ford saw, Pound's search for imagist intensity, his theory of emotional concentration through *collages* and the intersection of different planes of allusion, coincided

---

[1] These examples derive completely from Wai-lim Yip, *Ezra Pound's 'Cathay'* (Princeton University Press, 1969), pp. 84–94. Cf. also Earl Miner, 'Pound, Haiku, and the Image' (*Hudson Review*, IX, 1956); Achilles Fang, 'Fenollosa and Pound' (*Harvard Journal of Asian Studies*, XX, 1957); Hugh Kenner, 'Ezra Pound and Chinese' (*Agenda*, IV, 1965).

perfectly with what he took to be the principles of Chinese poetry and ideograms. To this one must add the incalculable stroke of what Pound himself called 'divine accident', the facility, always crucial to Pound's career, to enter into alien guise, to assume the mask and gait of other cultures. Pound's genius is largely one of mimicry and self-metamorphosis. 'Even when he is given only the barest details, he is able to get into the central consciousness of the original author by what we may perhaps call a kind of clairvoyance.'[1] This insinuation of self into otherness is the final secret of the translator's craft.

But the penetration of *Cathay* across remoteness and linguistic intermediacy is part of a more general phenomenon of hermeneutic trust. The China of Pound's poems, of Waley's, is one we have come fully to expect and believe in. It matches, it confirms powerful pictorial and tonal anticipations. *Chinoiserie* in European art, furniture and letters, in European philosophical–political allegory from Leibniz to Kafka and Brecht, is a product of cumulative impressions stylized and selected. Erroneously or not, by virtue of initial chance or of method, the Western eye has fixed on certain constants—or what are taken to be constants—of Chinese landscape, attitude, and emotional register. Each translation in turn appears to corroborate what is fundamentally a Western 'invention of China'.[2] Pound can imitate and persuade with utmost economy not because he or his reader knows so much but because both concur in knowing so little. Hence the familial, almost stemmatic resemblance between various European translations from the Chinese, a resemblance certainly greater than that between the Chinese texts and poetic schools themselves. Judith Gautier's 'Le Départ d'un ami' in *Le Livre de Jade* (1867) differs from Pound's 'Taking Leave of a Friend' in verbal detail, but the conventions of melancholy and cool space are precisely analogous:

> Par la verte montagne, aux rudes chemins, je vous reconduis
> jusqu'à l'enceinte du Nord.

---

[1] Wai-lim Yip, op. cit., p. 88.
[2] The phrase is that of Hugh Kenner. See his 'The Invention of China' (*Spectrum*, IX, 1967).

L'eau écumante roule autour des murs, et se perd vers l'orient.
C'est à cet endroit que nous nous séparons . . .

D'un long hénissement, mon cheval cherche à rappeler le
vôtre . . .
Mais c'est un chant d'oiseau qui lui répond! . . .

(This final addition is not only gratuitous—the Chinese simply has 'Neigh, neigh goes the horse at parting'—but mars the stylization by introducing a European motif of ironic dialectic.) Exactly the same focus is achieved again in Hans Bethge's version of Wange-Wei: 'Der Abschied des Freundes' (included in *Die chinesische Flöte* of 1929):

> Wohin ich geh? Ich wandre in die Berge,
> Ich suche Ruhe für mein einsam Herz,
> Ich werde nie mehr in die Ferne schweifen,—
> Müd ist mein Fuss, und müd ist meine Seele,—
> Die Erde ist die gleiche überall,
> Und ewig, ewig sind die weissen Wolken . . .

Mahler's setting of these lines in *Das Lied von der Erde* is, in terms of mode and instrumentation, yet another Western 'invention of China'. But all these translations are in fact related silhouettes of an intensely complex, varied original. The converse is true when Chinese artists sketch European or American cities and landscapes. These emerge delicately, characteristically uniform. New York shimmers on vague waters, like a vertical Venice. At best we can make out the criteria of suppression, formalization, and emblematic emphasis on which these images are based.

All English versions of the *Arabian Nights*, even Edward Powys Mathers's which is taken entirely from the French of J. C. Mardrus, display the same rose-water tint. French, German, Italian, English renditions of Japanese *haiku* are intimately related and come out in hushed monotone. In other words: the more remote the linguistic–cultural source, the easier it is to achieve a summary penetration and a transfer of stylized, codified markers. The Western translator from Arabic, Urdu, or Ainu is in a peculiar

sense circumventing, 'getting behind' the language of the original
with its local densities, idiomatic variables, and historical–stylistic
accidence. He is viewing his source, often via an intermediate
paraphrase, as a feature, almost non-linguistic, of landscape,
reported custom, and simplified history. In Pound's imitations of
China, in Logue's Homer, ignorance of the relevant language is a
paradoxical advantage. No semantic specificity, no particularity of
context interposes itself between the poet-translator and a general,
cultural-conventional sense of 'what the thing is or ought to be
like'. Whatever the archaeologists may tell us, we have come to
envision antique statuary as pure white marble; and time's erosion,
having worn away the original loud colours, affirms our misprision.

4

'Translucencies' are much more difficult to achieve at close
quarters. The innocence of great distance, the conventionally
negotiated immediacy of exoticism are unavailable. The translator
is now working with a source-text from a language and/or a
cultural milieu proximate to his own. This vicinity can be a fact
of historical, geographical contiguity; often it stems from the
common etymological origins and related development of his own
native speech and that of the original. In this situation, which is
statistically almost standard, the translator responds, feels himself
answerable to far more than the bare phonetic–syntactic object
before him. His hermeneutic incursion, i.e. the thrust of understanding into the neighbouring or kindred language and cultural
context, is complicated by a legacy of mutual contact. Understanding is attended by a body of assumption and nearly instinctual
prognostication. The Western Arabist or translator of primitive
song travels light. The European translator of a European 'foreign'
text, the Slavist translating from a branch of Slavic works towards
his source via concentric circles of linguistic–cultural self-
consciousness, presumptive information, and recognition. These
obviously illuminate and explicate the source-text; they generate
criteria of comparison and analogy whereby to assess the degree of
understanding and 'transferability'. But they also make the text to

be translated denser, more opaque (literally *verdichtet*). Therefore the relations of the translator to what is 'near' are inherently ambiguous and dialectical. The determining condition is simultaneously one of elective affinity and resistant difference.

This matter of 'difference' is crucial and takes us as close as anything will to a reasonable sense of the untranslatable. All differentiation is reciprocal, and operates in both directions. As Jacques Derrida puts it, a difference can only be thought dually: 'qu'*à partir* de la présence qu'il diffère et *en vue* de la présence différée qu'on vise à se réapproprier'.[1] The French translator experiences English as *different from* French. The experiencing of this 'difference from' is itself a personal, psychological manifold extending from an indistinct somatic basis (the phonetics, the sensory 'feel', the savour, the velocities, the pitch and stress system of the two tongues) the whole way to the most abstract, intellectualized awareness of semantic contrast. But the difference is also reactive on the individual and society; it defines conversely. English 'differs from' French as it does not from German or from Portuguese. The German- or Portuguese-speaker experiences this difference in regard to his own language and, with complexly variable modulations, in regard to languages of which he will have a less certain grasp. Each 'differing from' is diacritical in a generalized formal, historical sense but also inexhaustibly specific. The frontiers between languages are 'alive'; they are a dynamic constant which defines either side in relation to the other but no less to itself. This is the enormously complex topology which lies behind the old tag that knowledge of a second language will help clarify or deepen mastery of one's own. To experience difference, to feel the characteristic resistance and 'materiality' of that which differs, is to re-experience identity. One's own space is mapped by what lies outside; it derives coherence, tactile configuration, from the pressure of the external. 'Otherness', particularly when it has the wealth and penetration of language, compels 'presentness' to stand clear.

Working at the point of maximal exposure to embodied dif-

---

[1] Jacques Derrida, *Marges de la philosophie*, p. 9.

ference, the translator is forced to realize, to make visible, the perimeters, either spacious or confined, of his own tongue, of his own culture, of his own reserves of sensibility and intellect. The French translator of an English text is led to externalize, to enact beyond conscious control, a certain redefinition, indeed reacquisition of French. This redefinition generates a 'French', i.e. a construct of analogy, metaphrase, innovation, more or less concealed incapacity, hybrid locution, which is not the same 'French' as that produced by the French translator from, say, German. In this sense a 'translationese' can be a specifically biased, disoriented (*désaxé*) but by no means trivial version of a language. Each differentiation entails its own dynamic of internal regrouping, even as each frontier zone between nations has its own special character of exaggerated national assertion and, at the same time, of amalgam with elements over the border (hence the questions regarding the internal topology of the multilingual). The difference of English from French for the French-speaker, of French from English for the English-speaker—the terms can cross over on either side of the equation, being the reverse and obverse of the diacritical contact—is at every linguistic point so dense and plural as to deny formal description. Differences between languages as experienced by speakers on either and both sides of the cut are made up of elements of congruence, disjunction, partial overlap, imitation, refusal, graduated inter-mediacy, which are historical and symbolic, inherited and idio-syncratic, planned and unconscious. Chinese or Swahili are 'immensely' different from French. But this immensity is decep-tively categorical and thin. It is mainly inert 'in-difference' across an all but vacuous space. A 'close distance', on the other hand, as between French and English, is wholly energized by interactive differentiation. The more charged the proximity, the stronger the impulse to defensive self-definition, to the conservation of integral form. How then is the French translator *of* and *from* English (though both prepositions allow only the accusative one feels an elusive distinction between them) to make his version of the source-text translucent while resisting the impulse to assertive autonomy? Only by using a metaphoric 'calculus' which can simultaneously, coextensively, integrate and differentiate.

In his foreword to the 1959 Pléiade edition of Shakespeare, Gide makes classic allowances for demarcation. The 'Latin spirit' stumbles without the support of logic; Shakespearean imagery, on the other hand, vaults across pedestrian relation. To make his point Gide entrenches himself in erudite preciosity: 'Un appesantissement de tardigrade couvre en claudicant l'espace que le vers shakespearien a franchi d'un bond' (*tardigradus, claudicare*). Modern French lacks that *plaisante plasticité* still shown by the language of Ronsard and Montaigne who are Shakespeare's counterparts. French substantives and epithets allow of no inflection; the French word-order is, therefore, unyielding when compared to the suppleness of English. Elizabethan speech is surrounded by an aura of evocation—for which Gide proposes the term *harmoniques*—elusive not only in the particular case, but generally recalcitrant to the bias towards precision, towards definite nomination, in French. Often the sense of the original is indeterminate; native English speakers, scholars of Shakespeare and Elizabethan usage offer widely divergent interpretations. What is the French translator to make of *Antony and Cleopatra* v. i, 52 'A poor Egyptian yet...' in which *yet* could be rendered as *pourtant, encore, jusqu'à présent, désormais, de nouveau, en plus*, etc. or might, by a change in punctuation, be attached to the clause following: 'yet the queen my mistress...'? It is probably a part of Shakespeare's strategy, and of the strategy of spoken drama as such, to allow indecision, to let different possibilities of meaning 'hover' around the principal axis. But the translator must choose or inflate into explicative paraphrase; and the French translator is induced, by the grain of his own language and mental habits, to make his choice damagingly exact. Yet, at once, Gide qualifies. The incandescent imagery, the blaze of discordant metaphor which is produced by the Shakespearean text (like sparks, says Gide, struck from the hooves of a galloping horse) will teach us 'neither to reason well nor to write correctly'. French classical authors, on the contrary, enjoin 'extraordinary virtues'. The delight of the child in Shakespeare is natural ('l'enfant peut se passionner, se sentir le cœur tout gonflé d'émotions sublimes'). But the corollary is plain: there is in the French classic practice a contrasting adultness.

Gide's differentiations are at once individual and exemplary of a prolonged historical dialectic. They express his ascetic rationalism and the puritanic fastidiousness of his mature style. But they are also fully representative of a debate on linguistic–cultural values which dates back to the earliest printed French translations of Shakespeare, Pierre-Antoine de La Place's four-volume edition of 1745–6. Voltaire's advocacy of Shakespeare had begun in 1726, in the eighteenth of the *Lettres philosophiques*. He had polemicized vigorously on behalf of Shakespeare's 'strong and fertile genius'. But only forty years later, scandalized at the success and consequences of his own arguments, Voltaire wrote to the Comte d'Argental: 'as the height of calamity and horror, it was I who in the past first spoke of this Shakespeare; it was I who was the first to point out to Frenchmen the few pearls which were to be found in this enormous dunghill. It never entered my mind that by doing so I would one day help the effort to trample on the crowns of Racine and Corneille in order to wreathe the brow of this barbaric mountebank.' This effort was to culminate in the 'bardolatry' of Stendhal's *Racine et Shakespeare*, in Berlioz's proclamation that the 'lightning-flash of Shakespeare's genius revealed the whole heaven of art to me', in Victor Hugo's roster of the ultimate sublime in which the author of *Hamlet* stands beside Orpheus, Isaiah, Aeschylus, and Jesus. Each of these exaltations was programmatic. As Voltaire had seen, a French celebration of Shakespeare has to be dissociative from, subversive of Corneille, Racine, and Molière (though the Romantics sought to rescue the latter for the pantheon). The process of differentiation is implicitly one of polemic self-examination. And because of the magnitude of the Shakespearean presence—'Shakespeare c'est le drame,' wrote Victor Hugo flatly—the polemic, the impulse to self-scrutiny, went far beyond questions of literary genre. The French language itself can be experienced, it has been so experienced by writers and translators, as an 'absence of Shakespeare'.

The evolution of standard modern French contains an aesthetic, one could almost say a social–political ethic, of retrenchment. Possibilities of verbal prodigality, of grammatical exuberance, of metaphoric licence present in fifteenth- and sixteenth-century speech and writing were suppressed or relegated to the argotic and eccentric by the centralizing neo-classicism of seventeenth-century reform. Though regional speech forms continued an unbroken life, their challenge to the capital was never sufficient to alter the academic–bureaucratic norm. French can muster pomp and ceremony even in excess of English; but its altitudes are characteristically abstract and of a dry, generalized grandeur peculiarly grounded in elision. Examine Bossuet in vaulting progress and the underlying retraction (*l'idéal de la litote*) is unmistakable. The alternative register of concrete profusion, of a 'gestural' rather than Cartesian-grammatical logic, of deliberate conjunction between proper and low idiom, has always existed. It is manifest in Rabelais, in Céline, and, more obliquely, in Claudel. But it has been, since Montaigne, a rather isolated, often parodistic strain which draws its energies from the evident domination of the classic. This domination, articulate in the didactic, public authority of French syntax, constrains even the seeming free flights of modernism and surrealism. The criteria of thrift, lucidity, articulate sequence which organize the laconic encompassing of Racine also organize the executive means of poetry from Mallarmé to Char. The testing provocation comes from without, from the 'distant vicinity' of Shakespeare. Voltaire's change of front, the extremism of the Romantics, the to and fro of Gide point to a shared awareness of the 'Shakespearean gap' in French. French literature provides no figure as immediately universal (a fact aggravated by all but fitful Anglo-Saxon immunity to Racine). More disturbingly: the French ear apprehends in Shakespeare's uses of language those potentialities of 'totality' once vital but long since eroded in correct French speech. The bleak pontification of Gide's critique—Shakespeare can teach neither 'right reason' nor 'correct style'—is revealing of a profoundly unsettling provocation. Did French literature, sensibility, even social existence forgo, abrogate certain chances of largesse, experiment, emotional discovery realized in Shakespeare and the Shakespearean composites in English life and the English language?

Yet the 'Shakespearean absence' is not an unqualified loss. The modal completeness of French literature (major performances in every genre), the continuous strength but also originality of French literary movements and periods from the thirteenth century to

today suggest, diacritically, that a Shakespeare in the history
of one's language and letters can be an ambiguous providence.
A Shakespearean presence seems to consume certain energies
of form and perception through its own finality. It may fatally
debilitate, again by virtue of complete exploitation, the genre
in which it is realized (the subsequent course of English verse
drama). It may lead either to perpetual imitation—the problem
of freshness in the English iambic pentameter—or to laboured,
ultimately sterile exercises in repudiation (Pound's *Cantos* are at
one level an attempt to establish a repertoire of rhetorical tone
and imagery emancipated from Shakespeare). There is a sense
fictive, because obviously unverifiable, but also intuitively suggestive, in which Balzac's triumphant construction of a social *summa*,
Baudelaire's dramatization of the radical discord between art and
society, Rimbaud's notations of disorder—so distinct from, so
unencumbered by Shakespearean enactments of madness—were
made possible and necessary by Shakespeare's absence from
French or, more precisely, by the pressures of felt alternative
which his 'absent presence' brought to bear on French consciousness. Conversely, if there is no Proust in the English novel, I mean
no novelist who has made prose fiction inclusive of the uttermost
of philosophic intelligence and, at the same time, of unbounded
social, sexual, aesthetic exploration, Shakespeare's central inherence in the language, in the very notion of English literature
may, at some level, be a contributory cause. Certain reaches and
deeps have never again been worth simulating.

It is this dialectic of differentiation, multiplied, complicated by
personal and temporal circumstance, which locates the French
translator in regard to a Shakespearean text. He moves towards
that text through compact spaces of language, of culture and of
almost visceral defensiveness.[1]

---

[1] The history of French Shakespeare translations is catalogued in M. Horn-
Monval, *Les Traductions françaises de Shakespeare* (Paris, 1963). Cf. also Albert
Dubeux, *Les Traductions françaises de Shakespeare* (Paris, 1928); Pierre Leyris,
'Pourquoi retraduire Shakespeare?' in *Oeuvres complètes de Shakespeare* (Paris,
1954); C. Pons, 'Les Traductions de "Hamlet" par des écrivains français' (*Études
anglaises*, XIII, 1960); and the issue on 'Shakespeare in France' of the *Yale French
Studies*, XXXIII, 1964. See also P. Brunel, *Claudel et Shakespeare* (Paris, 1971).

Cleopatra's lament over Antony (IV. xv. 63 ff.) is quintessential of Shakespeare's late supremely-charged economy:

> The crown o'th' earth doth melt. My lord!
> O, withered is the garland of the war,
> The soldier's pole is fall'n: young boys and girls
> Are level now with men: the odds is gone,
> And there is nothing left remarkable
> Beneath the visiting moon.

These successive propositions display Cleopatra's bounding pace, her impatience with contingency. But a subtle closeness meshes each motion. If 'crown' sustains the imperial theme and relates obviously to 'the garland of the war', it also announces the spatial, cosmological image which connects 'earth' to 'pole' (the word may, as in *Hamlet* and *Othello*, stand for 'lode-star') and joins both to the visitations of the moon. More plainly, 'pole' conveys the picture both of Antony's spear or baton of command and of the wreathed maypole with its ancient connotations of centrality—the world's ritual axis—and of celebration. The festival theme is operative in 'crown' and 'garland' but also in the reference to 'young boys and girls'. Such, however, is the compaction of the passage, that this reference to the immature and to 'boys' in particular immediately evokes Antony and Cleopatra's scorn for the 'boy' Caesar. 'Odds' can signify both 'advantage' and 'peculiar distinction'. With Antony's eclipse the world literally declines into flat inertia and the cold of a lunar phase. Charmian's instant rejoinder—'O, quietness, lady!'—is concisely two-fold: it begs calm of the distraught queen but also proclaims the lifeless state of being.

Gide's choice of prose when translating the play has individual as well as formal-historical attributes. It relates, no doubt, to a personal sense of limitation. But it also engages the traditional dilemma of the disparities between available prosodies. The *alexandrin*, native to, all but inseparable from, the French conception of heroic, lyrically elevated theatre, is inapposite to English blank verse. The oppositions between pentameter and decasyllabics, on the other hand, seem to underline all the differences which separate quantitative from qualitative metrics. But a French prose translation of Shakespeare also embodies the whole mechanism of dialectical differentiation and self-definition. Until well into the twentieth century French high drama is written in verse. The challenging though flawed exception is Musset's *Lorenzaccio* (1833), a play modelled, precisely, on Shakespeare. The position, therefore, is one of psychological and technical inversion. The 'Shakespearean absence' in French tragic drama is, from one point of view, related to the absence of prose. The *alexandrin* would seem to exclude from the means of dramatic realization the 'roughage' of sensory location, of humour, of idiomatic violence, available to Elizabethan verse but also, presumably, to French prose. Molière's *Don Juan* gives a glimpse, but no more, of what might have been. To render Shakespearean poetry into the strongest possible French prose is to argue a vital alternative for French drama. In other words: the strategy is internal and aims at crucial inhibitions in French linguistic sensibility and habits of literary form. But the strategy is also one of 'critical export'. Prose, French prose signally, tests for systematic design. It searches out weak logic and makes vagueness self-betraying. In respect of both syntactic structures and cultural feeling, a French prose version of *Antony and Cleopatra* is a pointed scrutiny.

La couronne de l'univers se dénoue. Seigneur! La guirlande du combat se fane et l'étendard est abattu. A présent, les enfants et les hommes se valent. Tout s'égalise, et la lune en visitant la terre ne saura plus quoi regarder.

Though the difference in word-count is insignificant (forty as against forty-four), Gide's reading, especially through its taut cadence, is meant to exemplify criteria of extreme concision. It is stringently alert to the expansionist latitude prevalent in literary translation. It avoids explanatory paraphrase. Thus Gide selects only one of the several strands of linked image and inference in the original. It is that of martial grandeur. *La couronne de l'univers se dénoue* eliminates the topographical concreteness, the intimations at once material and emblematic, in 'the crown of the earth melting'. *Dénoue* points clearly to a laurel wreath. This figuration is systematically developed in *guirlande du combat* and *l'étendard est abattu*. Yet Gide, who is sacrificing for rigour, is himself evasive in *guirlande du combat*: the expression has no natural meaning in French, it only translates and it less than translates, *combat* being diminutive of 'war'. *Les enfants* drastically (needlessly?) curtails 'young boys and girls', suppressing the sarcastic swerve towards Caesar. At the close of the speech Gide distorts. He personifies the moon: it is 'she'—the feminine being, at this point so emphatic and symbolically laden in French—who will find nothing to look upon. Shakespeare's formulation exhibits, in word-order and muffled cadence, the inertness of a supine globe. Gide shifts the weight of activity to the moon. The whole distribution of feelings is altered. Charmian's 'Du calme, Madame!' not only trivializes; it omits the deadening fall towards extinction which is the cumulative sense and effect of Cleopatra's lament.

Yet even these liberties and abrogations are only outward difficulties. The sinew of Cleopatra's logic is physical. She constantly speaks her physical being. Bereft of Antony, the world is 'no better than a sty'. The 'melting' of the earth's crown, the garland 'withered', the pole 'fallen', the 'levelling' of manhood, the moon's 'visitation' have an undeniable concreteness. Their sensory implication gives body, in the literal meaning of the word, to the elusive abruptness and wide sweep of Cleopatra's images. Moreover, though they are finely tuned, the undertones of sexuality are nevertheless insistent. It would be unrealistic and a trivialization of the density of Shakespeare's method to neglect the cumulative erotic force of successive touches. The allusion to physical failure, the sense of a cadence from radiant virility to impotence, are graphic in 'melting' and 'withering'. There is almost a direct sexual rhetoric in 'The soldier's pole is fall'n'. The 'levelling' of boys and girls with men, which follows at once, enforces the motif of erotic pathos, of a world in which there is no longer to be found the critical difference between man and boy. One asks also, though only conjecturally, whether there is not a pertinent hint of feminine sexuality in the 'visiting moon'.

Again, Gide's imperceptions may have private facets. But they reside, more significantly, with constraints imposed by formal expectations and the language-matrix. The order of 'physicality', of poetic logic founded on the authority and knit of the human body, which organizes the style of Cleopatra, is alien to French high theatre. The dramaturgy of Racine may fairly be termed discourse without body. It accomplishes extreme intensities of transubstantiation and 'bodies forth' a last violence of thought and feeling. But it is at no stage somatic. This 'in-' or transubstantiation is centrally distinctive of French speech where the latter is elevated, public, and 'correct'. It would be a vulgar simplification to say that good French enacts, bears the imprint of, a Cartesian mind–body dualism. But in no other European tongue is this dualism so native. Hence, one imagines, the fluent coincidence between the Pierre Leyris–Elizabeth Holland translation and the original of Prospero's 'These our actors . . . were all spirits and / Are melted into air . . .' (a conceit which the French translators, indicatively, trace back to Pindar and the Attic tragedians):

Ces acteurs, je vous l'ai dit déjà, étaient tous des esprits; ils se sont fondus en air, en air impalpable. Pareillement à l'édifice sans base de cette vision, les tours coiffées de nuages, les palais fastueux, les temples solennels, le grand globe lui-même avec tous ceux qui en ont la jouissance se dissoudront, comme ce cortège insubstantiel s'est évanoui, sans laisser derrière eux la moindre vapeur. Nous sommes faits de la même étoffe que les songes et notre petite vie, un somme la parachève. . . .

Insubstantiality is the keynote common to source and translation. The historical, social development of civil French—'civil' in the political and academic sense also—is one of metaphrase, taboo, circumlocution, calculated to keep at an orderly remove the intrusion of bodily presence and functions. *L'univers se dénoue* (with its witty but imported suggestion of formal *dénouement*), *l'étendard est abattu*, the contraction of 'young boys and girls' into a neutral term, the mutation of the moon into a reflective observer, accomplish impulses of 'mentalism', of desexualization wholly inherent in French rhetoric, in the 'alternity' of world-views which the French language sets out. Gide's translation is, consequently, one of deficit. But 'the absent' also has its dialectical converse and positive. There is nothing in English drama to match the exhaustive purity of *Bérénice* (witness Otway's attempt at adaptation). The totality of shock, of spiritual crisis, which Racine generates via the introduction of a single material touch— a chair—into his fields of pure energy, are extrinsic to English sensibility, and the language will not cope. Robert Lowell makes Jacobean melodrama of *Phèdre*. The hermeneutic of the translator's (partial) return to his own native tongue is one of vulnerability.

This is the case, to be sure, on both sides of the 'trans-action'. In the fifth chapter of Part II of *Madame Bovary*, Flaubert describes Léon's idealizing adoration of Emma and the latter's serene languor following on the birth of her first child. Léon relinquishes even the vaguest of carnal hopes:

Mais, par ce renoncement, il la plaçait en des conditions extraordinaires. Elle se dégagea, pour lui, des qualités charnelles dont il n'avait rien à obtenir; et elle alla, dans son cœur, montant toujours et s'en détachant, à la manière magnifique d'une apothéose qui s'envole. C'était un de ces sentiments purs qui n'embarrassent pas l'exercice de la vie, que l'on cultive parce qu'ils sont rares, et dont la perte affligerait plus que la possession n'est réjouissante.

Emma maigrit, ses joues pâlirent, sa figure s'allongea. Avec ses bandeaux noirs, ses grands yeux, son nez droit, sa démarche d'oiseau et toujours silencieuse maintenant, ne semblait-elle pas traverser l'existence en y touchant à peine, et porter au front la vague empreinte de quelque prédestination sublime? Elle était si triste et si calme, si douce à la fois et si réservée, que l'on se sentait près d'elle pris par un charme glacial, comme l'on frissonne dans les églises sous le parfum des fleurs mêlé au froid des marbres. Les autres même n'échappaient point à cette séduction.

A complete reading of this passage, and it is not evident what 'completeness' signifies in this context or how it is to be shown, poses difficulties even for the native speaker. The grammatical articulations are numerous and delicate. They formalize a constant interplay between rhetorical amplitude and elision. Both paragraphs contain unstable, possibly illusory or falsely posited relations between Léon's internalized image of Emma, Emma as a physical presence, and the indefinite 'spectatorial' pronoun *on*. The transitions from one focus to another are of extreme subtlety.

The modulation from *que l'on se sentait* to *comme l'on frissonne* is on the margin of normal logic. We know from the manuscript that the printed version represents the end of a process of experiment and elision calculated to achieve a particular effect of chill fluidity. The alternance of grammatical number at the close of this same sentence is no less deliberate. *Parfum des fleurs* and *froid des marbres* are strictly parallel in regard to syntax and to the construction, singular followed by plural, but from the phonetic point of view they are chiastic: the sequence of voiced labials and fricatives being partly reversed (*par/mar, fleurs/froid*) with *mêlé* as the sharply vocalized fulcrum. By putting *marbres* in the plural, Flaubert obtains the twofold connotation of cold stone and of a sepulchre or effigy.

These points are straightforward. But what is to be said of the conjunction in *sa démarche d'oiseau et toujours silencieuse maintenant*? Obviously, the *et* acts as a copula between two members of the sentence. But as the latter as a whole is governed by the preposition *avec*, the simple connective comes to play a rather intricate, in some sense 'anti-grammatical' role. Analytically we would read [*avec*] *sa démarche d'oiseau*, in which case *et toujours silencieuse maintenant* is in descriptive, qualifying apposition to *démarche*. But the odd jolt which the ear experiences in hearing the sentence points to the possibility that the prepositional sequence has been left behind. Read thus, *et* initiates an elided predicative movement *et* [*étant*] *toujours silencieuse maintenant* with direct reference not to *démarche* but to Emma. The formal undecidability between the two readings is, of course, willed. Flaubert uses the economy of a certain syntactic duplicity to achieve a maximal richness of suggestion and correlation. Or consider the muted imbalance between the conditional and the indicative in the otherwise ornately rounded, almost neo-classical turn of *dont la perte affligerait plus que la possession n'est réjouissante*. The strict classicist would, one supposes, enclose the symmetry by writing *plus que la possession n'en est réjouissante*. Flaubert deflects the equilibrium to a purpose.

Though the text shifts from 'key' to 'key' with intense rapidity, the vocabulary binds it close. *Renoncement, qualités charnelles, mon-*

tant, magnifique, apothéose, purs, exercice, bandeaux noirs, prédestination sublime, belong to a cumulative 'liturgical' series. They prepare the evocation of the church with its funereal aroma of flowers and marble. Emma's *dégagement* initiates the trope of high ascent in *montant* and *s'envole* which, in turn, establishes the logic of *démarche d'oiseau*. The phonetic organization realizes the same impression of multiple but interactive strands. The vowel sequence and the *a, é, i, o* pattern at the climax of the second sentence act out the flight to apotheosis. The *i* sounds in *maigrit* and *pâlirent* (already undermined by the long *â*) play uncertainly against the opaque weight of *joues* and *s'allongea*. The contrast is figurative of Emma's decline, yet hints at the requisite histrionic note. The distribution of vowel sounds together with the sequence of voiceless stops, liquids, sibilants, and fricatives in the penultimate sentence is so closely plotted that only a full-scale phonological analysis would be adequate. Notice only the gradations of openness and contraction in the *a* sounds of the crucial series: *calme, charme, glacial, marbre* (the *a* element in the diphthong in *froid* being relevant also). Yet however minute, phonetic analysis would account for only a small part of Flaubert's executive means. The cadence of these two paragraphs is wholly intentional and fulfilled (*durchkomponiert*). Unfortunately, the metrics of prose and notations for stress patterns in prose remain rudimentary. One can point to Flaubert's primarily aural modes of interval and punctuation; to the frequent presence in his prose of 'spectral' alexandrines; to the evident sonority of the peroration, a sonority subverted by the sibilant-nasal pattern of *séduction*. But these are platitudes. The acoustic cunning of the two paragraphs embodies, is rigorously wedded to, a planned tonality, and we lack exact means of paraphrasing, let alone formalizing, the ways in which 'tone' is a function of sound, of grammar and of idiom, but also more.

Already a first audit shows that these paragraphs exhibit locally the counterpoint of pomp and deflation which governs the entire novel. Léon's imaginings of Emma are couched in a jargon of romantic sanctification. Emma herself exudes an aura of ethereal sublimity. Yet Léon's idealization and Madame Bovary's actual

deportment are at every point undermined. Léon caresses sentiments of purity, of disinterested adoration with the same vulgar
indulgence which will mark his later conduct. In a draft version
Flaubert made the point obvious by characterizing Léon's feelings
as *presque désintéressée*. Emma's disincarnation, on the other hand,
is a cliché of frustrated appetite. Set at the close of the paragraph
the allusion to *jouissance* strikes the full note of ironic deflation and
sexuality. The latter aspect is reinforced by *cette séduction*, a phrase
which places Emma's pallid silences in an ambiguous tactical light.
Homais's inane tribute, immediately following the passage—'C'est
une femme de grands moyens qui ne serait pas déplacée dans une
sous-préfecture'—not only completes the cumulative effect of
ironic correction but makes of *moyens* the precise marker of
ambivalence. Emma Bovary's movements, even when in genuine
pain, are 'means'.

Past these superficial features lies the interplay between the
abstract and the physical. In the actual depiction of Emma, the
terms modulate from physical notations, already 'disembodied'
by virtue of phonetics and cadence, to hollow spirituality. The
insubstantial and the sensual are, in turn, astutely melded in the
closing simile: the scent of flowers and the marble cold are at once
impalpable and strangely 'epithelial'—we feel them under our
skin.

At this level of enlistment, language seems to communicate
simultaneities of meaning and of inference which are obviously
initiated by the writer and, up to a point, deliberately crafted, but
which then become self-augmenting. Each time we return to a
significant passage in *Madame Bovary* or in any other major text,
we learn to hear more of its contained possibilities, more of the
pulse of relation which gives it 'internality'. Where language is
fully used *meaning is content beyond paraphrase*. This is to say that
where even the most thorough paraphrase stops, meaning begins
uniquely. This uniqueness is determined by the conjunction of
typographical, phonetic, grammatical facts with the semantic
whole. Because it is not the passage itself, all paraphrase—
analytic, hermeneutic, reproductive—is fragmentary (even where
it is wordier than the original). Paraphrase predicates a fiction: it

proceeds as if 'meaning' were divisible from even the barest detail and accident of oral or written form, as if any utterance could ever be a total stand-in for any other. This fiction is, of course, indispensable to human communication, to the conventions of approximate equivalence which underlie everyday speech. But a passage of serious poetry or prose reminds us that this fiction, however fundamental to man and society, has a limited status. Where language is charged to the full, paraphrase is less and less 'like the thing itself'. Meaning, on the contrary, is more and more 'what comes next'. The direction of comprehension, therefore, will not be lateral—a slide from *a* to *b*, from text to interpretation, from source to translation along horizontal lines—but ingressive. We learn to listen. To do so acutely we must discipline our own attention. We discard the static of ready explanation, of scattered association, of personal commentary, in order to listen totally. The need for self-effacement, for submissive scruple, is imaged in 'understand'. The more receptive our listening inward, the better the chance that we shall hear a force and logic of expression more central than 'meaning'. Indeed, unless we are very careful in our terminology, 'meaning' will carry a stubborn implication of transferability, of equivalence in another form. It is only when we apprehend the 'meaning of meaning', the expressive totality integral to a given set of verbal, syntactic, language-specific units, that we understand fully. It is then, in Heidegger's terms, that we hear 'language speak' (*die Sprache sprechen*), that we separate its own 'saying' from our accidence, as does the poet.

How is the translator of *Madame Bovary* to persuade us that he has listened?

Marx's daughter, Eleanor Marx Aveling, published her translation in 1886. It was for a long time the sole English version and was taken up in the Everyman's Library. George Moore had been instrumental in the project, but Eleanor Marx was principally inspired by what she took to be the radical posture of Flaubert's book. Here was a statement of the condition of women under the suffocating regime of bourgeois hypocrisy and mercantile ideals. Here, as in Ibsen's *Doll's House*, which the Avelings helped introduce to a circle of London readers, was a revolutionary exposure of the falsity of marriage and of family relations in a repressive capitalist system. The book had been prosecuted for obscenity in the courts of Napoleon III. Eleanor Marx saw in this prosecution a nakedly political attempt to silence an artist who, by sheer honesty of vision, had laid bare the cant and corruption of life in the Second Empire. Thus the translator brought to her task an explicit programmatic 'set'. She approached the text almost entirely via context, via what she felt to be a shared sphere of moral–political intention. Kindred circumstance was to overcome an inherently formal, therefore insubstantial linguistic differentiation.

Read now, what is frequently an imperceptive version is steadied by its period flavour. 'To bear on her brow the vague impress of some divine destiny' is not exact but does suggest the appropriate idiom. Emma's 'aquiline nose' on the other hand is of a ready piece with her 'soaring' and her 'bird-like walk'. If the singular 'marble' misses the calculated richness of Flaubert's connotation, 'seduction' is retained and rightly placed. What is lacking is the controlling dialectic of the passage, the ironic undercut of the pathos. The translator has identified herself with Emma (there was, of course, to be a tragic concurrence in real life). All semantic options are decided in the heroine's favour. It is she herself who is 'always silent now'. 'Black hair' not only leaves untranslated the correct meaning and histrionic implications of *bandeaux noirs* but, together with 'aquiline nose', underlines the impression of Emma's nobility. Léon's adoration sets Madame Bovary 'on an extraordinary pinnacle'—a translation more pictorial and unambiguous than Flaubert's *en des conditions extraordinaires*. And when, in a passage which follows, the novelist reveals that Emma's *lèvres si pudiques* conceal sexual frustration, Eleanor Marx transcribes *pudiques*, admittedly a word of extreme complexity with a covert edge of nastiness, by the straightforward 'chaste'.

Gerard Hopkins's translation of 1948 is, linguistically, better informed. It reflects a deliberate attitude towards problems of technique and verbal fabric. The translator lightens the pace so as to achieve both transparency of motion and the relevant note of disembodiment. Léon comes to think of Emma as 'disincarnate'.

She is 'untrammelled by the flesh' and 'ever winging upwards like a radiant goddess'. Where Hopkins resorts to archaism or a rhetorical key, he does so to simulate the original. Léon's is 'the sort of emotion a man cultivates for its very rarity, convinced that its loss would outweigh in misery what possession might give of joy'. *Et toujours silencieuse maintenant* is resolved by [with] 'her new moods of silence'. The indefinite *on* is suppressed. It is Léon's heart which feels 'an icy charm', it is he who is shivering as in church, it is 'others besides himself' who are affected by Emma's 'witchery'. These are considerable liberties, and again *bandeaux noirs* is missed. But at times innovation succeeds: emotion 'detached from mundane affairs' is in apt correspondence to *qui n'embarrassent pas l'exercice de la vie*; 'some predestined blessedness' is at once more accurate and suggestive than 'some divine destiny'. 'So sad she was, so calm, so sweet . . .' not only suggests a native classic but uses English prosody to mime the suspect pathos of the source. Yet the distance to Flaubert remains problematic. The 'presence' of *Madame Bovary* in Hopkins's version is that of a 'world classic' naturalized, in part at least, by previous translation and by the role which the work has played in English fiction after James. Hence a contradictory ambience difficult to define but characteristic of the hermeneutic structure of alienness and appropriation. Hopkins's perspective is both too near and too far. It almost postulates the reader's access to the original in order to ensure its own freedom. Clearly Hopkins has gone much beyond Eleanor Marx in discarding extraneous commitments to the political, social context. He listens closely. But a good deal of what he hears is layered resonance—in the history of the modern novel, in the changes of sensibility brought on, to a certain degree, by Flaubert himself. The result is, at times, a deceptive ease of transfer. We do not feel the resistant particularity of the 'other'. But great translation must carry with it the most precise sense possible of the resistant, of the barriers intact at the heart of understanding. Stefan George's poem 'Das Wort' communicates, more exactly than any other literary or linguistic text, the reality of the frontier (*born*, *landes saum*), and of the likelihood that words will break in transit:

Wunder von ferne oder traum
Bracht ich an meines landes saum
Und harrte bis die graue norn
Den namen fand in ihrem born—
Drauf konnt ichs greifen dicht und stark
Nun blüht und glänzt es durch die mark ...
Einst langt ich an nach guter fahrt
Mit einem kleinod reich und zart
Sie suchte lang und gab mir kund:
'So schläft hier nichts auf tiefem grund'
Worauf es meiner hand entrann
Und nie mein land den schatz gewann ...
So lernt ich traurig den verzicht:
Kein ding sei wo das wort gebricht.

We must not trust the translation whose words are entirely 'un-broken'. As with a sea-shell, the translator can listen strenuously but mistake the rumour of his own pulse for the beat of the alien sea.

Yet 'mis-taking', to grasp in place of, to transliterate, as it were, between seizure and surrogation, is indispensable. We have seen that serious understanding depends on a linguistic and cultural experiencing of resistant difference. But the transcendence of difference, the process of internalizing the probabilities of non-communication, of acute doubt as to whether the thing can be done at all, demands *Wahlverwandschaft* (elective affinity). At close linguistic–cultural quarters the translator often finds himself in a state of recognition. The hermeneutic and praxis of his de-cipherment and subsequent restatement are those of mirrors and *déjà-vu*. He has been here before he came. He has chosen his source-text not arbitrarily but because he is kindred to it. The magnetism can be one of genre, tone, biographical fantasy, conceptual framework. Whatever the bonding, his sense of the text is a sense of homecoming or, as the sentimental tag precisely puts it, of a home from home. Poor translation follows on negative 'mistaking': erroneous choice or mechanical, fortuitous circum-stance have directed the translator to an original in which he is not at home. The alienness is not one of differentiation undergone, circumscribed as a moment in the dialectic of transit, but a muddled, vacant disaccord which can, in fact, be independent of linguistic difference. Thus there are within our own tongue and culture numerous works with which we have no just relation, which leave us cold. Positive 'mistaking' on the contrary generates and is generated by the feeling of at-homeness in the other language, in the other community of consciousness. The point is a central one. Translation operates in a dual or dialectical or bipolar energy-field (one's preference between these terms being simply a question of meta-language). Resistant difference—the integral and historical impermeability, apartness of the two languages, civilizations, semantic composites—plays against elective affinity—the translator's pre- and recognition of the original, his intuition of legitimate entry, of an at-homeness momentarily dislocated, i.e. located across the frontier. At close quarters, say as between two European languages, the charge is maximal at both poles. The shock of difference is as strong as that of familiarity. The translator is held off as powerfully as he is drawn in. Translucency comes of the unresolved antinomy of the two currents, of the vital swerve into and away from the core of the original. Some such picture seems to obtain in the micron spaces between high-energy particles drawn together by gravity but kept apart by repulsion.

But notice how 'positive mistaking', the translator's recognition or Narcissism on which the business depends for half its logic, sets odd psychological traps. Once the translator has entered into the original, the frontier of language passed, once he has certified his sense of belonging, why go on with the translation? He is now, apparently, the man who needs it least. Not only can he hear and read the original for himself, but the more unforced his immersion the sharper will be his realization of a uniquely rooted meaning, of the organic autonomy of the saying and the said. So why a translation, why the circumvention which is the way home (the third movement in the hermeneutic)? Undoubtedly translation contains a paradox of altruism—a word on which there are stresses both of 'otherness' and of 'alteration'. The translator performs for others,

at the price of dispersal and relative devaluation, a task no longer
necessary or immediate to himself. But there is also a proprietary
impulse. It is only when he 'brings home' the simulacrum of the
original, when he recrosses the divide of language and community,
that he feels himself in authentic possession of his source. Safely
back he can, as an individual, discard his own translation. The
original is now peculiarly his. Appropriation through understanding and metamorphic re-saying shades, psychologically as
well as morally, into expropriation. This is the dilemma which I
have defined as the cause of the fourth, closing movement in the
hermeneutic of translation. After completing his work, the genuine
translator is *en fausse situation*. He is in part a stranger to his
own artifact which is now radically superfluous, and in part a
stranger to the original which his translation has, in varying
degrees, adulterated, diminished, exploited, or betrayed through
improvement. I will come back to the consequent need for compensation, for a restoration of parity. This need is obsessive in the
distances, at once resistant and magnetic, of Hobbes to Thucydides,
of Hölderlin to Sophocles, of MacKenna to Plotinus, of Celan to
Shakespeare, of Nabokov to Pushkin.

An elective affinity can be national. The best documented
example is the German identification with Shakespeare. From the
first mention of Shakespeare's name in a German text in 1682 to
the present, the process of incorporation has been continuous.[1] It
has modified German literature, the development of the German
stage, the habits of rhetorical and informal reference which shape
a national style and sensibility. *Die Shakespearomanie*, as Grabbe
termed it in 1827, could reach grotesque extremes: I have men-

---

[1] Only a careful look at the *Jahrbuch der Deutschen Shakespeare-Gesellschaft*, an
index for whose first ninety-nine volumes appeared in 1964, can give a fair
impression of the relevant literature. Albert Cohn's *Shakespeare in Germany in the
Sixteenth and Seventeenth Centuries* (London and Berlin, 1865), and Rudolf Genée's
Geschichte der Shakespearschen Dramen in Deutschland (Leipzig, 1871) remain useful.
Roy Pascal's *Shakespeare in Germany* (Cambridge University Press, 1937) is a good
introduction to the main trends for the period 1740–1815. Joseph Gregor,
Shakespeare, Der Aufbau eines Zeitalters (Vienna, 1935) is interesting because of its
untroubled assumption of a central authority, textual, theatrical, psychological in
the German–Austrian interpretation of Shakespeare.

tioned before the claims made, in the 1880s, that Shakespeare himself was of 'Flemish–Teutonic' descent. Enthusiasm often went with misreading. The nineteenth-century German public and pedagogues saw in Shakespeare a tragedian of middle-class morality, a more inspired version of Diderot and Lessing. Goethe, in his revealingly-entitled essay *Shakespeare und kein Ende*, came to the conclusion that Shakespeare is, above all, a poet to be read; staged, his plays are full of weakness and crudity. Goethe's productions of Shakespeare in Weimar—notoriously the *Romeo and Juliet* of 1811—drastically amended the infirm original. German philosophic readings of Shakespeare, German schools of dramaturgy, made of their idol a Platonist and a radical materialist, a universal humanist and a bellicose nationalist, a bourgeois moralist and an advocate of pandemic sensuality, a symbolist so arcane as to have defied all previous unriddling and a naturalist in the manner of Hauptmann or Wedekind.

But common to these antithetical projections was the conviction formulated by Gundolf in his *Shakespeare und der Deutsche Geist* (1927) that the Elizabethan playwright is 'wie kein anderer das menschgewordene Schöpfertum des Lebens selbst'.[1] The phrase explicitly parallels the metaphor of Christ's incarnation, the descent of the supreme life-giving agency into the guise of man. Extravagant as it is, Gundolf's idiom closely conveys the experience of the inherence of Shakespeare in the vital core and creative means of the German language. The point had already been made by Friedrich Schlegel in his *Geschichte der alten und neuen Literatur* (1812). As Schlegel said, German Shakespeare translations had transformed the native tongue and the range of national consciousness. From Wieland on, but particularly in the A. W. Schlegel–Dorothea Tieck–Baudissin versions as they appeared between 1797 and 1833, the German language, in attempting to penetrate and represent Shakespeare, had realized its own modern potential and limitations. Through A. W. Schlegel's genius for *Entsagung* (the renunciation of self in the enveloping authority of the original), writes Gundolf, the German tongue had literally

---

[1] Friedrich Gundolf, *Shakespeare und der Deutsche Geist* (Berlin, 1927), p. vi.

embodied Shakespeare's *Seelenstoff*, his *anima* or 'soul-substance': 'so ward die Möglichkeit einer deutschen Shakespeare-übertragung verwirklicht worin der deutsche Geist und die Seele Shakespeares durch ein gemeinsames Medium sich ausdrückten, worin Shakespeare wirklich deutsche Sprache geworden war.'[1] *Ueber-tragung*—carry-over, appropriative transport, followed by total symbiosis. The English text has not been translated into the German language, says Gundolf, it *has become that language*. Thus the translator transmutes the original into its own true self (Mallarmé's 'Tel qu'en Lui-même l'éternité le change' which, of course, is also based on the *topos* of translation). The notion is, at one level, absurd, at another of the greatest philosophic–linguistic interest. 'Shakespeare' was somehow hidden inside the accidental husk of English. The teleology of his full meaning, of the 'mean-ing of his meaning', the realization of his complete historical–spiritual presence, lay with German. The space between the German translator and the Shakespearean original is, as it were, just inside the mirror. How can there be translucency at this negated distance?

Sonnet 87 is closely plotted. It clearly illustrates Shakespeare's habit of exploiting a specialized area of language, in this instance legal and fiscal, while generating in depth a more intimate, con-crete statement—here a crucial gesture in the power relations between the speaker, the mistress, and the 'rival poet' of the preceding set. This motion inward from a technical façade, with the raw hurt and irony of the primary utterance held in check by an accentuated conventionality of idiom and grammatical turn, poses pitfalls for the reader and translator. The drama lies in the syntax, in the syntactic pressure of private need and private taunt which is trapped within yet also declared by the vocabulary. The effect of containment and of delayed shock is, in part, achieved by *rallentando*: being in some degree technical, the language of the sonnet impedes us from facile empathy. So does the sinewy, contracted word-order. This, also, is something the translator will watch for.

---

[1] Gundolf, *Shakespeare und der Deutsche Geist*, p. 351.

Farewell thou art too deare for my possessing,
And like enough thou knowst thy estimate,
The Charter of thy worth gives thee releasing:
My bonds in thee are all determinate.
For how do I hold thee but by thy granting,
And for that ritches where is my deserving?
The cause of this faire guift in me is wanting,
And so my pattent back again is swerving.
Thy selfe thou gav'st, thy owne worth then not knowing,
Or mee to whom thou gav'st it, else mistaking,
So thy great guift upon misprision growing,
Comes home againe, on better judgement making.
    Thus have I had thee as a dreame doth flatter,
    In sleepe a King, but waking no such matter.

There are obvious nodes and pluralities. 'Deare' signifies both 'expensive' and 'cherished'. 'Possessing' initiates the sustained duality of sexual and economic reference. 'Estimate' is ironic and works several rather intricate ways: 'assessment' is relevant as well as 'self-esteem'. 'Charter', used similarly in *Othello* (a play peculiarly apposite to Sonnet 87), has implications both of 'contract' and of 'privilege' or 'freedom granted'. 'Bonds', as often in the canon, bridges different areas of experience and discourse: here the legal–economic and the erotic–personal. Sometimes, though I am not certain whether in the present case, the echo of 'bounds', limitations of self and action, is pertinent. 'Determinate' pulls us back to the vocabulary of law and of conveyancing in particular. J. Dover Wilson in the *New Shakespeare*, citing Tucker Brooke, states that lines 5–8 are 'based on the legal principle that a contract is unenforceable if it lacks a valuable consideration'. The use of 'swerving' is odd and powerful: one recalls a whole cluster of Shakespearean imagery relating to 'bias' and to swift motion out of natural balance. 'Mistaking' comprises a perfectly evident but grave pun—'taken amiss', 'accepted in error'. 'Misprision' is again a piece of legal terminology but at the same time a word with drastic overtones, at once psychological and bodily. 'Judgement' unobtrusively concludes the forensic theme. 'No such matter' may be richer than it looks, inviting a modulation from 'no such substance' to 'a thing of no importance'. Past these obvious knots there are possibilities of depth characteristic of Shakespeare. Both times 'guift' has a peculiarly metallic, ambiguous sheen (delicately underlined by the repeated 'thou gav'st'). Here, as so often in Shakespeare—cf. the spectrum of *kind* in *Lear*—one asks whether a precise, though utterly 'natural', unforced, etymological awareness is not in play. Old Norse and Old English 'gift' signify the payment made for a bride; the German hononym means 'poison'. In 'wanting' as we have seen before, 'lack' and 'need' are simultaneously active. In short, at almost every moment in the sonnet, Shakespeare's language is exhaustive both of a range of semantic fields—anti-Petrarchan, erotic, monetary, judicial—and of its own stored history.

Stefan George's views of Shakespeare are not always easy to make out.[1] But clearly he saw in the Elizabethan master the incarnation of esoteric grandeur and essential Platonism which defined George's own image of philosophic art. Thus his *Umdichtung* of the Sonnets, first published in 1909, was a feat of intense self-projection. George declared his version to be 'anti-romantic'; now, and for the first time, the German reader was to gain access to the inner meaning of the text. He was to be initiated into the Platonic allegory latent in the original but somehow masked by the conventions of Elizabethan discourse and the misprision of subsequent interpreters. The translation or, rather, 'realization via restatement' must show in what ways Shakespeare's passionate love for and self-bestowal on his young male friend constitute the central truth of the entire sonnet sequence:

> Lebwohl! zu teuer ist dein besitz für mich
> Und du weisst wohl wie schwer du bist zu kaufen...
> Der freibrief deines werts entbindet dich...
> Mein recht auf dich ist völlig abgelaufen.
>
> Wie hab ich dich, wenn nicht durch dein gewähren?

[1] Cf. O. Marx, *Stefan George in seiner Uebertragung englischer Dichtung* (Amsterdam, 1967).

Verdien ich was von deinen schätzen allen?
Aus mir ist nicht dein schenken zu erklären ...
So ist mein gnadenlehn anheimgefallen.

Du gabst dich damals, deinen wert nicht sehend—
Vielleicht auch dem du gabst, mich, anders nehmend ...
Dein gross geschenk, aus irrtum nur entstehend,
Kehrt heimwärts bessrem urteil sich bequemend.

So hatt ich dich wie träume die beschleichen—
Im schlaf ein fürst, doch wachend nichts dergleichen.

The translation of the first quatrain aims at extreme fidelity. *Teuer* is cognate of 'deare' and carries the same twofold meaning. If 'du weisst wohl wie schwer du bist zu kaufen' departs from literal closeness, it nevertheless communicates Shakespeare's mournful irony and the decisive hint of mendacity or venality in the beloved. *Freibrief* is beautifully near, containing the relevant implications both of contract and of freedom. Via *carta*, whose Italian meanings George would be alert to, *brief* relates richly to Shakespeare's 'Charter'. The first blindness comes with 'Mein recht auf dich' which is almost a denial of the concentrated ambivalence—the investment and the servitude—in 'My bonds in thee'. Already George is licensing his controlling image of the Platonic master's 'rights in' the loved youth. Quatrain two is, in every sense, difficult. The bitter directness of the poet's query is at once veiled and underlined by the technical idiom. We are meant to slow down, to observe the strength of personal hurt and offence as it plays against the disciplining confines of the Petrarchan and legalistic armature. George follows the original word-order and *gewähren* retains the intimations of legalism and of condescension required by the text. *Verdienen*, however, has too diffuse a wealth of suggestion: though it mirrors the touch of servitude in 'de/serving' (*Ver/dienen*), it also signifies 'to earn', a motif absent from Shakespeare's phrasing and intent. Line seven is evaded. *Erklären* in George's language conveys particular values of 'elucidation', of the self-unfolding of the privileged individual in the process of amorous initiation. The entire force of 'wanting', its emotional coherence in and through contrary directions of meaning, gets lost. The closing, on the other hand, is ingenious. 'Swerving' goes by default but George's characteristic amalgam *gnadenlehn* sustains the contractual theme. By yoking together 'grace' and 'loan', 'mercy' and 'bestowal', this awkward contrivance articulates the ambiguous stance of the speaker. *Anheimgefallen* is no less complex and pertinent: it conveys the motion of disaster, of 'dis/grace', and also prepares the ironic reversion of 'Comes home againe' in line twelve.

At the start of the third quatrain, George's Platonic bias betrays the material intensity of the original, an intensity made palpable and disturbing by the compression of Shakespeare's syntax. *Sehend*, with its Neoplatonic and Petrarchan stress on sight, the most noble, spiritualized of senses, misses the crucial conjunction of intellectual with sexual 'knowledge'. *Anders nehmend* is again abstract: it elides the association of 'mis-judgement' with 'spurious erotic possession' operative in *mis/taking*. George is now entrapped both by his general strategy—Shakespeare the Platonist and hermeticist—and by the necessities of rhyme. *Sich bequemend*, with its domestic, faintly unctuous undertones, is not only damaging in itself, but misses the extreme subtlety of Shakespeare's suggestion, the plural fabric of possible readings in 'on better judgement making'. Albeit the beloved's valuation of the lover's worth was erroneous, of this error came selfish increase. If the 'great guift' has judged its recipient and found him 'wanting', Shakespeare's concept here being one of radical psychological clairvoyance, it too has been judged and the quality of its reversion is ambivalent. George's negation of the grosser, more opaque elements in the sonnet now inhibits him from apprehending, at their full measure, the more arcane. His growing awareness of this paradox may account for the discord in his treatment of the final couplet. The latter happens to be less perfunctory than in many of the Sonnets. Monarchs whom their dreams flatter or divert momentarily from their true estate recur at key points in Shakespeare. The motif of sexual entry during the beloved's slumber ('Thus have I had thee') is at once farcical and Neoplatonic or Gnostic. If it is present at all in Sonnet 87, its presence is faint. *Beschleichen* is as drastically wrong as would be Iachimo in *A Midsummer Night's Dream*. As if unsettled by his general allegorizing gloss, George makes a move towards the physical and trivializes Shakespeare's bitter constraint. The last line is feeble: *nichts dergleichen* comes of the search for rhyme.

Karl Kraus's *Nachdichtung* of the Sonnets—the distinction between this term and George's *Umdichtung* being one which declares a polemic difference as to hermeneutics and translation— was composed between October 1932 and mid-January 1933. Kraus's laconic afterword makes plain that his own version was conceived as a critique of George's. By 'doing violence' both to the English sense and to German verbal and grammatical usage, George had produced a 'unique abortion'. Kraus aims to embody in the German language and in German poetry ('language' is deliberately placed ahead of 'poetry') a 'hitherto inaccessible' but central part of Shakespeare's genius. Contrary to George, Kraus sees this genius as professional, under compulsion of occasion, and often uneven. There are in the Sonnets both magnificences and weaknesses, spiritual heights and journeywork. Like Kraus himself, Shakespeare had to use what came to hand:

> Leb wohl! Zu hoch stehst du im Preis für mich,
> und weisst, dass du vor allen auserkoren.
> Nach deines Wertes Rechte frei, zerbrich
> den Bund; mein Recht auf dich hab ich verloren.
>
> Wenn nicht geschenkt, wie wärst du meine Habe?
> War durch Verdienst solch Reichtum mir beschert?
> Da ich in nichts bin würdig deiner Gabe,
> gehört sich's, dass sie wieder dir gehört.
>
> Du gabst dich, weil du deinen Wert nicht kanntest,
> vielleicht auch weil den meinen du verkannt;
> drum wieder wird, da deinen Sinn du wandtest,
> was mein durch Irrtum war, dir zugewandt.
>
> So warst du mein durch eines Traumes Macht:
> ich schlief als Fürst, zum Nichts bin ich erwacht.

Kraus begins uncertainly. He reads 'estimate' at face value and therefore misses the note of venality and legalism in lines 3–4. George's *abgelaufen*, with its mercantile connotation, is much to be preferred to the characteristically romantic set *auserkoren/verloren*. *Bund*, on the other hand, rescues the dual sense of 'unison' and 'bondedness' in the English text. With the fifth line, Kraus achieves concentration and momentum. The discord between *geschenkt* and *Habe* dramatizes the prevailing imbalance of the plot. *Habe* touches the requisite chord of violent possession. *In nichts* exaggerates the lover's abasement but prepares the literal extinction ('no such matter') inferred at the close. Though at some remove from the original, the two uses of *gehört* in line eight, with their equivocation on 'propriety' and 'property', are genuinely Shakespearean. The very faint colloquialism, the nuance of Viennese in the movement of the line and in *gehört sich's*, are legitimate. The third quatrain confirms Kraus's grip. The strong play on *kennen* and *verkennen*, its strength augmented by the four close-linked assonant rhymes, shows the translator's complete awareness of what is going on in the sonnet. Kraus's tautness images a situation of legal rigour and formalized reciprocity (the bond of seeming love is revealed as merely contractual; error of heart is reduced to tort). Kraus also echoes the retarding zigzagging construction of Shakespeare's sentence. The sinuous tonality of *zugewandt*, prefigured in the preceding verse, renders exactly the lover's notion of ambiguous reversion. The handling of the coda is free. The significant touch of 'flattery' is omitted, and placed where it is, *erwacht* is too positive. *Zum Nichts*, however, is admirable and recalls to mind Heidegger's insistence that German *Nichts* has content, that it is not blank nothingness. It is at the very last that Kraus goes wrong. By using the first-person singular—*ich schlief... bin ich erwacht*—he suggests a scenario such as that of Christopher Sly's awakening in *The Taming of the Shrew*. The ironies in Sonnet 87 are of an entirely different order: the poet recognizes the indifference of the beloved but hints that the mistaking of true love has extinguished the being of the proud beloved as well as of himself. The grammatical 'suspension'

between alternate directions of reference is essential and profoundly dramatic.

Sonnet 87 is not among the *Einundzwanzig Sonette* of
Shakespeare translated by Paul Celan and published in 1967.
Celan's techniques and philosophy of translation, moreover, are
of such intricacy as to defy any but extensive treatment.[1] At one
level Celan strives to reconstitute Shakespeare's meaning or,
more precisely, the rhetorical, prosodic, topical 'means of his
meaning'—and often he does so with succinct conviction. But the
elective affinity which leads Celan to Shakespeare is at once more
compelling and more problematic. Celan seems to test his own
capacity for meaning, his own imperative need for and distrust of
finished poetic utterance against the Shakespearean precedent. It
is at this point that Celan's acutely paradoxical, unresolved, and
finally self-destructive coexistence with the German language is
relevant. By virtue of his translations from Russian, French, and
English, Celan could displace German into a position of salutary
strangeness. He could approach it with therapeutic dispassion as a
raw material fatally his own yet also contingent and potentially
hostile. All of Celan's own poetry is translated *into* German. In
the process the receptor-language becomes unhoused, broken,
idiosyncratic almost to the point of non-communication. It
becomes a 'meta-German' cleansed of historical–political dirt
and thus, alone, usable by a profoundly Jewish voice after the
holocaust. To consider Celan's Shakespeare translations separately from the rest of his work is, therefore, almost impossible. I
want to look at one example only in which, characteristically,
Celan makes of his recomposition of Shakespeare's meaning an
active image of the process of translation itself and, specifically, of
that dialectic of appropriation and indemnity which constitutes the
last, most difficult moment in the hermeneutic scheme.

By omitting the gesture of nomination and direct address in line
five—'I grant, sweet love, thy lovely argument'—Celan turns

---

[1] See the instructive, though exaggeratedly fine-spun essay by Peter Szondi,
'Poetry of Constancy—Poetik der Beständigkeit: Celans Uebertragung von
Shakespeares Sonett 105' in *Celan-Studien* (Frankfurt, 1972).

Sonnet 79 into a meditation on poetry and on the dependencies of the poet on the object or occasion which inspire him. The use of repetition where there is none in the original—

> But now my gracious numbers are decay'd,
> And my sick Muse doth give another place.

> Doch jetzt, da will mein Vers kein Vers mehr sein,
> die Muse, siech, ist fort-, ist fortgezogen.—

is thematic. Repetition is the purest concentrate of translation. To repeat identically is to translate along the axis of time (repetition comes after, however closely). To repeat 'freely', as does Celan, is to exemplify the entire dialectic of secondariness and potential invention which binds the translator to and divorces him from his source. Read thus, lines 7–14 become an exegesis on the exchange of meanings, on the enigmatic equivalence of poet and object, of poem and translation:

> Yet what of thee thy poet doth invent
> He robs thee of, and pays it thee again:

> He lends thee virtue, and he stole that word
> From thy behaviour; beauty doth he give,
> And found it in thy cheek: he can afford
> No praise to thee but what in thee doth live.

> Then thank him not for that which he doth say,
> Since what he owes thee thou thyself dost pay.

Celan focuses on the 'allegory of language'. The poet has extracted from his source its life-spirit, *der Geist*—a word wholly of a different world of meaning from Shakespeare's but of a world possibly inevitable *after* Shakespeare and in the language of Kant and Hegel. But the poet/translator appropriates in order to restore: *Der Dichter nahms, es wiederzuerstatten*, in which *erstatten* carries its full force of 'compensation' through and by means of 'restatement' (as in *einen Bericht erstatten*). Where Shakespeare speaks of 'virtue' stolen from 'behaviour', Celan is drastically ontological:

> Er leiht dir Tugend. Dieses Wort, er stahls
> dir, deinem Sein.

He dislocates line twelve so as to achieve the same effect of philosophic totality. Disregarding the Petrarchan trope and symmetry in the original, Celan hypostatizes 'but what in thee doth live' into life itself:

> Er leiht dir Tugend. Dieses Wort, er stahls
> dir, deinem Sein. Er kann dir Schönheit geben:
> sie stammt von dir—er raubte, abermals.
> Er rühmt und preist: er tauchte in dein Leben.

The commerce between meanings, between poets, which is translation, is preceded by violent and total incursion. *Er tauchte in dein Leben*: we plunge into the life, into the integral being of the source attempting (vainly?) to break through the Narcissus-image which meets us at the surface and, it may be, continues to meet us at considerable depth.

Celan presses home this 'meaning of Shakespeare's meaning' and his relations to that meaning in a closing couplet whose rhymes emphatically echo the crucial designation of *deinem Sein*:

> So dank ihm nicht für seiner Worte Reihn:
> was er dir schuldet, es ist dein und dein.

The final repetition arrests yet also opens, as into an endless mirror-sequence, a verse of mysterious perfection. In a manner which entirely negates paraphrase it expresses the hermeneutic of compensation, the ways in which a true translation restores to the original—after rapine: *er stahls, er raubte*—what was its own, but what is also, having only been latent, more than its own (the simple act of repetition, *dein und dein*, is strongly augmentative). There could be no denser statement of reciprocity at close quarters.

In each case, George's, Kraus's, Celan's, the consequence of translation is more and less than translucency. The translator proceeds via a prodigality of theoretic, cultural, and linguistic presupposition. The context in which his interpretation and rendition occur is so 'over-determined' as to blur perspective and the scruple of distance. This context is no less than the entire corpus of German Shakespeare translations (the translator translates after and against his predecessors almost as much as he translates his source). The context is also the interiority, which is psychologically authentic though it may be arbitrary and falsely acquisitive as well, of Shakespeare's works inside the German-speaker's sense of his own language and of its literary modes. It is, finally, the particular abrogations or extensions of self which carry the translator, notably when he is himself a writer of some stature, to the original. The resulting representation is over-informed and over-informing; it has, in Keats's phrase, a 'palpable design' on its object. It finds before it seeks.

Thus the translator at close quarters is at every point under contradictory stress. He is aware that he will always know too little about his source-text because there is a sense in which he 'knows what he does not know'. This is to say that his experience of the 'other' language and 'other' culture is so abundant, so collusive, as to suggest to him a strong sense of the total context. He recognizes the 'infinite regression', the formally undecidable compass of historical information, linguistic sensibility, local ambience which could bear on the meaning of the work which he is translating. On the other hand, he 'knows too much'. He brings to the performance of translation a deceptive bias to transparency. The apparatus of critical comparison, cultural familiarity, immersive identification with which he works proliferates and can do so unconsciously. He knows more or better than his author. Pound can make *Cathay* spare and translucent because he, and his Western readers, know next to nothing of the original. The English translator of Flaubert, the German translator of Shakespeare are drawn into a complex space of recognition. The organization of his own sensibility is in part a product of that which he is about to translate. Hence the paradox of restoration and homecoming which Celan elicits from Sonnet 79. Where translation takes place at close cultural–linguistic proximity, therefore, we can distinguish two main currents of intention and semantic focus. The delineation of 'resistant difficulty', the endeavour to situate precisely and convey intact the 'otherness' of the original, plays against 'elective affinity', against immediate grasp and domestication. In perfunctory translation these two currents diverge. There is no shaping tension between them, and paraphrase attempts to mask the gap. Good translation, on the contrary, can be defined as that in which the dialectic of impenetrability and ingress, of intractable alienness and felt 'at-homeness' remains unresolved, but expressive. Out of the tension of resistance and affinity, a tension directly proportional to the proximity of the two languages and historical communities, grows the elucidative strangeness of the great translation. The strangeness is elucidative because we come to recognize it, to 'know it again', as our own.

Theoretically, therefore, translation at great distance turns out to be the trivial case. What merits wonder is the fact that there can be serious alternity of meaning and expressive form inside the same language-family and cultural lattice. In the relevant instance, the exceptional translator is able both to affirm and to deny Wallace Stevens's curious assertion that 'French and English constitute the same tongue'. As in the 'strange' physics of very high energy, attraction and repulsion are simultaneously most intense at proximity. Ovid's *Metamorphoses* are themselves a fable of constant translation, of the tragic or ironic changes of identity into new form. Their influence on the Italian epic and on Italian lyric poetry has been extensive, certainly from Boccaccio to Tasso.[1] The Italian language, furthermore, is intimately Latin in its phonetics, derivations, syntactic structure and matrix of historical, cultural reference. In translating Ovid, Salvatore Quasimodo avails himself both of Ovid's ubiquity in Italian writing and art from the late Middle Ages to the Baroque, and of the intimate kinship between the two languages (I have italicized some of the obvious homologies):[2]

---

[1] Cf. A. F. Ugolini, *I cantari italiani d'argomento classico* (Geneva, 1933), and E. Parattore (ed.), *Atti del Convegno internazionale ovidiano, Sulmona, maggio 1958* (Rome, 1959).

[2] Salvatore Quasimodo, *Dalle Metamorfosi di Ovidio* (Milan, 1966).

et: 'Fer opem, Galatea, *precor*, mihi; ferte, parentes',
dixerat 'et *vestris* periturum admittite *regnis*'.
Insequitur Cyclops partemque e *monte* revulsam
mittit et *extremus* quamvis pervenit ad illum
angulus e saxo, totum tamen obruit Acin.
At nos, quod fieri *solum* per *fata* licebat,
fecimus ut vires assumeret Acis *avitas*.
Puniceus de mole cruor manabat et intra
temporis exiguum rubor *evanescere* coepit
fitque color primo turbati fluminis imbre
purgaturque mora. . . .

('Oh, Galatea, help me, I pray you. Help me, my
parents, and convey me, whom am doomed to perish,
to your kingdom.' The Cyclops pursued him and
hurled a chunk of rock ripped from the mountain-side.
Only the merest corner touched Acis, but still it was
enough to bury him quite. But I did the only thing
which fate allowed: I caused Acis to take on his
ancestral powers. Crimson blood came trickling from
beneath the stony heap. After a short time, its reddish
colour began to fade; it turned to the colour of a
stream swollen by early rains; and then, in a little
while, grew altogether clear.)

(*Metamorphoses*, XIII. 880–90)

'*Aiuto*, Galatea, ti *prego, aiuto*, o padre, o madre,
nel *vostro regno* accogliete il figlio prossimo alla morte.'
E il Ciclope *l'insegue*, e staccato un pezzo di *monte*
lo lancia sul fuggiasco. *Solo* un *estremo*
della rupe lo colse, ma fu per lui la morte.
E perché Aci riprendesse la forza dell'*avo*
feci quello che potevo ottenere del *fato*.
Dalla rupe scorreva sangue vivo, ma ecco, quel rosso
comincia a *svanire* come *colore* di *fiume*
che torbido di pioggia schiarisce a poco a poco.

Yet how divergent is the effect. Quasimodo's text is actually only half a line longer than Ovid's, but the impression throughout is one of loosening. This is at many points a question of phonetic values: *mittit* as against *lo lancia sul fuggiasco*; *perché Aci riprendesse la forza del'avo* in lieu of the lapidary *ut uires assumeret Acis auitas*; the onomatopoeic *che torbido di pioggia schiarisce a poco a poco* expanding on *imbre purgaturque mora*. But the divergence is due also to more deliberate causes. Often, indeed, Quasimodo finds an Italian word which represents an alternative to an obvious Latinism. *Pezzo di monte* avoids *sasso* (Ovid's *saxo*); *solo un estremo* leads away from *angolo* (Latin *angulus*); *sangue vivo* bypasses the suggestion of *rubro* which is nakedly Latin; *obruit* would evoke *rovinare* if Quasimodo had not put *ma fu per lui la morte* which looks antique and monumental but in fact is not, being vaguely operatic. And even where an exact correspondence is unavoidable—*evanescere* into *svanire*—the vowel change is sufficient to alter the flavour and very nearly to define that striving for distance, for autonomous space, which relates modern Italian to its Latin bone and nerve-structure.

In short: at every moment in this passage we find the dialectic of resistance within extreme affinity which makes the task of understanding and restatement across close linguistic–cultural divisions so challenging; as can be the task of understanding or communication between two human beings too nearly involved in each other's unspoken purpose.

5

The final stage or moment in the process of translation is that which I have called 'compensation' or 'restitution'. The translation restores the equilibrium between itself and the original, between source-language and receptor-language which had been disrupted by the translator's interpretative attack and appropriation. The paradigm of translation stays incomplete until reciprocity has been achieved, until the original has regained as much as it had lost. 'Pour comprendre l'autre,' wrote Massignon in his famous study of the 'internal syntax' of Semitic tongues, 'il ne faut pas se l'annexer, mais devenir son hôte.'[1] This dialectic of trust, of reciprocal enhancement is, in essence, both moral and linguistic. It makes of the language of translation a language which has its own status of vulnerability, of unhousedness, of elucidative strangeness because it is an instrument of relation between the foreign tongue and one's own. The inner mechanism of compensation, the offertory turn of the translator towards the original which he had penetrated, appropriated and left behind, is probably impossible to formalize. But it has numerous concrete, historical realizations.

Translation recompenses in that it can provide the original with a persistence and geographical–cultural range of survival which it would otherwise lack. Given the facts of modern literacy, the Greek and Latin classics owe to the translator their partial escape from silence. Translation into a world-language can make a general force of texts written in a local tongue. Kierkegaard, Ibsen, Strindberg, Kazantzakis have been given their impact by translation. Translation can illuminate, compelling the original, as it were, into reluctant clarity (witness Jean Hyppolite's translation of Hegel's *Phenomenologie*). It can, paradoxically, reveal the stature of a body of work which had been undervalued or ignored in its native guise: Faulkner returned to American awareness after he had been translated and critically acclaimed in France. In every such case there has been compensation and echo has turned to benefaction. But what I mean by 'radical equity', by the 'equalizing transfer' which completes the hermeneutic cycle, is at once more universal and specific. Though its roots are moral, and though its performance may involve the whole philosophy of understanding and of culture, 'fidelity'—which is the enactment and expression of reciprocity—is finally technical. It is a bond of adequacy as between text and text, taking 'adequacy' in its strongest sense.

A bad translation is one which is inadequate to its source-text for reasons which can be legion and obvious. The translator has misconstrued the original through ignorance, haste, or personal limitation. He lacks the mastery of his own language required for

---

[1] Quoted in Henri Meschonnic, *Pour la poétique II*, p. 411.

adequate representation. He has made a stylistic or psychological blunder in choosing his text: his own sensibility and that of the author whom he is translating are discordant. Where there is difficulty the bad translator elides or paraphrases. Where there is elevation he inflates. Where his author offends he smooths. Ninety per cent, no doubt, of all translation since Babel is inadequate and will continue to be so. Its inadequacy falls under one or more of the obvious rubrics which I have pointed to. But the entire range of inadequacies can be unified and made more precise. Translation fails where it does not compensate, where there is no restoration of radical equity. The translator has grasped and/or appropriated less than is there. He traduces through diminution. Or he has chosen to embody and restate fully only one or another aspect of the original, fragmenting, distorting its vital coherence according to his own needs or myopia. Or he has 'betrayed upward', transfiguring the source into something greater than itself. In each case the imbalance caused by the initial motions of trust, decipherment, and appropriative use remains unrighted. The translation outweighs the original or is outweighed by it; or there is a bypassing, a more or less perfunctory similitude instead of the taut meshing of resistance with affinity.

The common imbalance, of course, is that of diminution. The translation is 'irresponsible' towards the original in that it restores less than the original contains and, often, less than the translator has in fact understood. When Priam enters Achilles' tent in the black of night to beg for Hector's body (*Iliad* XXIV. 477 ff.), Homer combines and gives complete expression to a number of motifs which have done much to shape the history of Western feeling. A different yet intimately inwoven doom lies on both men. With Hector's death Troy stands condemned and Priam's own life is now destined to a cruel end. Achilles, however, is also fatally marked. The slaying of Hector is the climax of his own brief course. There is, therefore, a deep bond of imminent ruin between the suppliant and the conqueror. Looking upon each other, the mankiller and the aged king experience a sense or vision of chiasmic exchange: before Priam's wondering gaze, Achilles becomes the lost Hector and all the warrior-sons lost in battle; to

Achilles, on the other hand, Priam evokes Peleus, the old father left behind and soon to be robbed of *his* son and guard. The scene dramatizes unutterable sorrow and a tragic, universal authority of human waste. Yet in midst of desolation there is hunger and a need for sleep. The body mutinies against the rhetoric and sovereignty of despair. Achilles bids Priam join him in a finely prepared meal. The meat crackles on the spit and it is time to stop weeping. Only Rabelais has ever matched the scope, the implacable sanity of Homer's tragi-comic view of life. Even Niobe fell to her food after all her children had been done to death. If the translator misses or attenuates this mystery of common sense, he will have failed Homer.

Chapman's version of 1611 has its splendours: Priam appears 'So unexpected, so in night,' (a phrase whose concise genius resists grammatical analysis) 'and so incredible'. There is something of the Jacobean tragic mode in Achilles' assurance to Priam that Troy 'Shall finde thee weeping roomes enow'. Chapman is convincing when he structures the entire scene around Achilles' 'large man-slaughtring hand' which Priam kisses though it is marked with Hector's blood and which, later in the night, carves the 'silver-fleec't sheepe' and hands a choice 'browne joynt' to the regal guest. But Chapman's style is, notoriously, over-elaborate and unsteady. There are baroque ingenuities out of place ('He shall be tearful, thou being full'). Where Homer advances lightly Chapman convolutes. And remaining oratorical, he misses the desolate intimacy of the encounter, the parity of anguish which encloses both actors in a common darkness.

Hobbes's *Iliad* of 1676 is the pastime of a very old man embittered by what he took to be the inadequate reception of his philosophical–political life-work. What fascinates Hobbes, as it did when he was translating Thucydides, is the relentless poise of the classic Greek view of human conflict. Homer alone has realized the ideal of 'Justice and Impartiality' which ought to govern heroic poetry. And Hobbes, in his commentary on the poem, adds superbly: 'For neither a Poet nor an Historian ought to make himself an absolute Master of any man's good name.' Long before Matthew Arnold, moreover, and in explicit contrast to Chapman, Hobbes felt that the essence of Homeric verse was one of speed. Hence his choice of decasyllabic lines often bone-spare. But Hobbes was no poet and the result is almost ludicrously thin:

> Come then old man and lay your grief away,
> And for the present think upon your meat,
> And weep for Hector when you come to Troy,
> For true it is your loss of him is great.

Pope's treatment of Priam's supplication (published in 1720) is acutely informed. His express use of the term 'suppliant' under-lines his awareness of the ritual quality of the whole action. No less than Chapman, he fixes on the motif of Achilles' hands: 'That Circumstance of Priam's kissing the Hands of Achilles is inimitably fine; he kiss'd, says Homer, the hands of Achilles, those terrible, murderous hands that had robb'd him of so many Sons: By these two Words the Poet recalls to our Mind all the noble actions perform'd by Achilles in the whole Ilias; and at the same time strikes us with the utmost compassion for this unhappy King, who is reduced so low as to be oblig'd to kiss those Hands that had slain his Subjects, and ruin'd his Kingdom and Family.' For Pope 'inimitably' carries its full weight of specific inhibition. At his best, Homer is beyond the reach of even the most inspired translation. Characteristically, Pope aims at the high places by creating a 'secondary classicism', a lyric invocation of those traditional ornaments and allusions of which the Homeric epic is itself the ultimate source:

> War, and the Blood of Men, surround thy Walls!
> What must be, must be. Bear thy Lot, nor shed
> These unavailing sorrows o'er the Dead;
> Thou can'st not call him from the Stygian Shore,
> But thou alas! may'st live, to suffer more!

Pope mediates through Virgil and Milton. This organic classicism makes for the strength of his reading, but also for its decorative inflation:

> Where round the Bed whence Achelous springs
> The wat'ry Fairies dance in Mazy Rings,
> There high on Sipylus his shaggy Brow,
> She stands her own sad Monument of Woe;
> The Rock for ever lasts, the Tears for ever flow!

Given the pressure for elegance and the deliberate density of literary echo (in this case Miltonic and, distantly, Shakespearean), Pope is profoundly out of key when he comes to the pivotal motif of food and sleep after high sorrow:

> But now the peaceful Hours of sacred Night
> Demand Refection, and to Rest invite. . . .

There could hardly be a more drastic inadequacy of sensibility or style than the decorous Latinity of 'Demand Refection'. Homer's moral clarity, a clarity made unworriedly expressive of moral values by the physical directness of bodily need and presence, is utterly trivialized. Having flinched from the energy which transcends taste, Pope irreparably dissipates meaning.

What persuaded Cowper to devote his domestic genius to the *Iliad* (1791), rather than to some other ancient classic, is not obvious, though there is a manifest aim to be more rigorous, truer to the sinewy simplicities of the original than was Mr. Pope. Cowper's is a thoroughly Miltonic Homer. He announces in his Preface that 'no person familiar with both can read either without being reminded of the other'. The consequence is often ungainly pastiche. *Paradise Lost* mingles with *Samson Agonistes*:

> But since the powers of heaven brought on thy land
> This fatal war, battle and deeds of death
> Always surround the city where thou reign'st.
> Cease, therefore, from unprofitable tears,
> Which, ere they raise thy son to life again,
> Shall, doubtless, find fresh cause for which to flow.

Published in 1951, Richmond Lattimore's *Iliad* has been both widely praised and criticized; through schools and the general reader, its influence has been considerable. It seeks to suggest something of the formulaic techniques discovered in the original by Milman Parry. It embodies the sum of modern textual and historical scholarship. Its 'free six-beat line' is intended to reproduce the free flow and oral characteristics of Homer's narrative. It does not evade the strength of the obvious:

> Now you and I must remember our supper.
> For even Niobe, she of the lovely tresses, remembered
> to eat, whose twelve children were destroyed in her palace....
>
> But she remembered to eat when she was worn out with
> weeping ...
> Come then, we also, aged magnificent sir, must remember to
> eat....

This is, fairly straightforwardly, what the Greek says. Whence the incongruities, the persistent impression of flattening? Striving for a 'timeless', unobtrusively lucid idiom, Lattimore has, in fact, fallen into a peculiar cadence, part Longfellow, part Eisenhower. 'Tall Priam' is subtly but also decisively wrong for Πρίαμος μέγας; 'He had just got through with his dinner' is similarly accurate but off-key. 'Aged magnificent sir' is corrosively wrong; an insinuation of the ridiculous—American undergraduate first approaching his Oxbridge tutor—is inescapable. Though Lattimore is incisive at the close of the passage—Achilles *butchers* the sheep *fairly*, an ambiguity which rightly points both to justice and to handsome valour—the Lattimore version as a whole is already a period piece. Its intended timelessness has turned parochial. This, precisely, is what Homer is not.

But the quest is incessant still. Even since the first edition of this study in 1975, English and American translations of Homer have continued to be published. Nineteen-ninety proved to be something of an *annus mirabilis*. Christopher Logue issued the next segment of his wildly personal but often inspired readings of the *Iliad*. *Kings* recounts, in the full sense of this term, Homer's First and Second Book. Robert Fagles's complete *Iliad*, with invaluable introductory material by Bernard Knox, combines to a high degree a learned approach with a commitment to the essentially oral, 'voiced' pulse and design of the epic. In 1990 also, Allen Mandelbaum, a specialist in the translations of epics into nervous modernity, issued his *Odyssey*. Nor would any comprehensive survey want to exclude Derek Walcott's *Omeros*, a twentieth-century epic narrative out of the sea-worlds of the West Indies whose internal and explicit uses of Homer, of both *Iliad* and *Odyssey*, are 'transmutations' of the distant source, as Roman Jakobson defined that order of relation.

None of the translations I have quoted (and there are, at a very rough count, more than 200 complete or selected English renditions of the *Iliad* and *Odyssey* from 1581 to the present) is adequate to the original. None restores the balance of equity, though Pope's is unquestionably an epic in its own right. In his imitation of Book XIX, Christopher Logue pictures Achilles' helmet:

> though it is noon the helmet screams against the light,
> scratches the eye, so violent it can be seen
> across three thousand years.[1]

This trick of blinding vision across time is both a definition of the classic and of the task of the translator. To make visible in its own light. Not to dim to our own.

Magnification is the subtler form of treason. It can arise from a variety of motives. Through misjudgement or professional obligation, the translator may render an original which is slighter than his own natural powers (Baudelaire translating Thomas Hood's 'The Bridge of Sighs'). The source may have become numinous or canonic and later versions exalt it to an alien elevation. This is certainly the case at many points in the Authorized Version. In the Psalms, for example, the formulaic, literalist texture of the Hebrew idiom is frequently distorted to baroque magnificence. Or compare the King James's version of the Book of Job with that of M. H. Pope published in the Anchor Bible in 1965. The translator may be working in a context of decorum loftier than that of his author: Shakespearean translation between

---

[1] Christopher Logue, *Pax*, p. 19.

the 1770s and the late nineteenth century is put askew by constraints of refinement and heroic posture. Too often, the translator
feeds on the original for his own increase. Endowed with linguistic
and prosodic talents, but unable to produce an independent, free
life-form, the translator (Pound, Lowell, Logue, even Pasternak)
will heighten, overcrowd, or excessively dramatize the text which
he is translating to make it almost his trophy.

The most interesting examples of 'transfiguration' from both
a technical and cultural point of view, however, are those in
which a 'betrayal upward' takes place, as it were, unwittingly. The
translator produces a piece of work which surpasses the original in
stylistic quality or in emotional scope. Such instances may be
comparatively rare but they are also seminal. Implausible as the
notion will seem in a context of Anglo-Saxon values, it can, I am
persuaded, be reasonably maintained that Schlegel and Tieck have
improved on numerous stretches of foolery, bawdy, and verbal
farce in Shakespeare's comedies (see their versions of *The Two
Gentlemen of Verona, As You Like It,* and *The Merry Wives of
Windsor*). Christopher Marlowe transmutes Ovid's *Amores* II. 10
into poetry of genius. Santayana's translation of Théophile
Gautier's poem 'L'Art' is a greater thing than the original. Yet
however brilliant the yield, the process is one of 'overcompen-
sation', and the cardinal balance is broken. 'A translator is to be
like his author,' wrote Dr Johnson in reference to Dryden, 'it is
not his business to excel him.' Where he does so, the original is
subtly injured. And the reader is robbed of a just view.

Louise Labé was a poetess of naïve intensity. She adopted the
most shopworn of Petrarchan means but gave to them a frank
physical import. Coming from a woman this literalism gives her
language and rhetorical turns an almost childish force of demand:

> Baise m'encor, rebaise moy et baise.
> Donne m'en un de tes plus sauoureus,
> Donne m'en un de tes plus amoureus:
> Ie t'en rendray quatre plus chaus que braise.
>
> Las, te pleins tu? ça que ce mal i'apaise,
> En t'en donnant dix autres doucereus.

Ainsi meslans nos baisers tant heureus
Iouissons nous l'un de l'autre à notre aise....

Sixteenth-century *baiser* does not signify, as it does in current French, complete sexual intercourse; but the carnal vivacity, the 'heat' of the poem are unmistakable. There is an intense suggestion of 'oven-sweetness' (*plus chaus que braise, dix autres doucereus*); the imperatives are those of a child asking for a fresh-baked biscuit. This is verse to melt in the mouth. Rilke translates thus:

Küss mich noch einmal, küss mich wieder, küsse
mich ohne Ende. Diesen will ich schmecken,
in dem will ich an deiner Glut erschrecken,
und vier für einen will ich, Überflüsse

will ich dir wiedergeben. Warte, zehn
noch glühendere, bist du nun zufrieden?
O dass wir also, kaum mehr unterschieden,
glückströmend in einander übergehn....

Though the rhyme-scheme is freer than in the original, Rilke's version is, formally, ingenious. The shift from complaint (*te pleins tu?*) to satisfaction (*bist du nun zufrieden?*) remains faithful to the inference of an intimate, smothered exchange between the lovers. But almost immediately, Rilke aggrandizes the sonnet and sets it in a more solemn register. The implication of infinity in *ohne Ende* is exhilarating and baroque but ruins the ordinariness, the chambered warmth of Louise Labé's setting. *An deiner Glut erschrecken* is again a violent augment. There is nothing of menacing ardour in the French text; we may indeed burn our lips on something freshly-baked, but the experience is not one of terror. The second quatrain turns wholly on the drowsy sibilance of the rhymes: *apaise/doucereus/heureus/aise*. Sensuality drifts into repose. Rilke pitches the situation much higher: precisely as in Donne's *Exstasie*, the lovers relinquish their own singleness of being and melt into a Platonic unison. The small magic of the original is shattered. We are no longer *à notre aise*, the crucial note; the candid but domestic eroticism of *iouissons* is gone. The lift, the philosophic energy of Rilke's lines are quite beyond the resources of Louise Labé. They prepare the eloquent movement of dissipation, sensual and spiritual, with which Rilke closes: *wenn ich, aus mir ausbrechend, mich vergeude*, where the original has the merely playful *Si hors de moy ne fay quelque saillie*. But although it is a more important poem, or rather because it is, Rilke's translation diminishes its source.

Jules Supervielle's is a very distinctive but minor presence. His 'Chanson' is shapely but not free from banality and what are, after Verlaine, stock phrases:

> Jésus, tu sais chaque feuille
> Qui verdira la forêt,
> Les racines qui recueillent
> Et dévorent leur secret,
> La terreur de l'éphémère
> A l'approche de la nuit,
> Et le soupir de la Terre
> Dans le silence infini.
> Tu peux suivre les poissons
> Tourmentant les profondeurs,
> Quand ils tournent et retournent
> Et si s'arrête leur cœur....

Celan, in a way which is unique to his own genius, at once contracts and magnifies:

> Jesus, du kennst sie alle:
> das Blatt, das Waldgrün bringt,
> die Wurzel, die ihr Tiefstes
> aufsammelt und vertrinkt....

By singularizing both leaf and root, Celan gives the invocation a formidable immediacy. *Ihr Tiefstes* has that precise duality of abstraction and image which *secret* lacks; it is borne out by the accuracy of *vertrinken*, where *dévorer* strikes one as accident or purely sonorous. *L'éphémère* is indistinctly portentous and the lines following are banal. Not so with Celan:

> die Angst des Taggeschöpfes,
> wenn es sich nachthin neigt,
> das Seufzen dieser Erde
> im Raum, der sie umschweigt.

Taggeschöpfes, nachthin, umschweigen are densities which Celan has made peculiarly his own. Beyond Supervielle, the translation fulfils the intention of gravity, of a dragging dark (an intention weakened in the original by the banality of *infini*). Celan's term *wühlen abgrundwärts* is a grammatical–tonal motion more precise, more ominous than that realized in the French. And the translator even surpasses Supervielle's finest stroke: *Et si s'arrête leur cœur.* The German performs Jesus' gliding descent into the deeps and gives to the implicit contrast between divine eternity and the brief life of organic forms a mysterious location in time and action:

> Du kannst den Fisch begleiten,
> dich wühlen abgrundwärts
> und mit ihm schwimmen, unten,
> und länger als sein Herz. . . .[1]

After this it is almost impossible to go back to Supervielle; translation of this order being, in one sense, the cruellest of homages.

Consider, finally, the exaltations suffered by the Owl and the Pussy-Cat in Francis Steegmuller's version,[2] exaltations which derive throughout from the contrastive phonetics and semantics of French and English. *Miel roux* is distinguished as Edward Lear's 'some honey' is not; *une lettre de crédit* has a potential logic and elegance entirely denied to 'Wrapped up in a five-pound note'. But the gap truly widens with the next line: 'The Owl looked up to the stars above'. In the French text both the verb and object take wing: *Le hibou contemplait les astres du ciel*. The 'stars above' are

---

[1] *Jules Supervielle: Gedichte: Deutsch von Paul Celan*, was published in Frankfurt in 1968. A full edition of Celan's translations from French (including Simenon), English, and Russian is needed. Only when it is available will it be possible to investigate the interrelation of the 'original poet' with the 'restater' of genius.

[2] *Le Hibou et la Poussiquette: Edward Lear's 'The Owl and the Pussycat' freely translated into French by Francis Steegmuller* (London, 1961).

familial, *les astres du ciel* inevitably orbed and portentous. In Lear
the Owl sings to 'a small guitar'; Steegmuller omits the epithet.
Now his tone mounts:

> 'Ô Minou chérie, ô Minou ma belle,
> Ô Poussiquette, comme tu es rare,
>> Es rare,
>> Es rare!
> Ô Poussiquette, comme tu es rare!'

The phonetic facsimile is cunning: You are/*Es rare*. But the
elevation is obvious. Although it mimes the sound of the original,
the *quette* in *Poussiquette* has overtones of *coquetterie*, of diminutive
elegance much beyond the back-yard ecstasies in Lear. And *rare*
is, by definition, more choice than 'beautiful'. In the following
stanza, the translator transfigures explicitly: 'You elegant fowl'
becomes *Noble sieur*, and 'How charmingly sweet you sing' is
raised to *Votre voix est d'une telle élégance*. Even where the translation is simply lexical (*une alliance* for 'a ring'), French ennobles.
Piggy-wig becomes the more adult *cochon de lait*, and does not
merely 'stand in a wood' but emerges from a forest: *Un cochon de
lait surgit d'une forêt*. Steegmuller adroitly echoes the inner rhyme
in 'Dear Pig, are you willing to sell for one shilling . . .' / 'Cochon,
veux-tu bien nous vendre pour un rien . . .', but *un rien* has a
nuance of feline lordliness—the 'letter of credit' theme—entirely
above the source-text. The marriage 'on the hill' is modulated into
the more spacious, rhetorical *sur le mont les unit*. Only in the finale,
and inexplicably, does Steegmuller betray his tactic of elevation.
Once more, the phonetic mimesis is brilliant:

> Et là sur la plage, le nouveau ménage
> Dansa au clair de la lune,
>> La lune,
>> La lune,
> Dansa au clair de la lune.

But *ménage* is irremediably domestic; *la plage* banishes the magic of
Shakespearean reminiscence in 'on the edge of the sand'; and *au*

clair de la lune, no doubt because of the children's song, is oddly flatter than 'the light of the moon'. Transfiguration slips suddenly into diminution.

We have seen that it is the ideal of translation to be neither. This ideal can never be realized totally. No contingent form can be defined as perfect. It is a platitude to say so. But the issue is not entirely trivial. A 'perfect' act of translation would be one of total synonymity. It would presume an interpretation so precisely exhaustive as to leave no single unit in the source-text—phonetic, grammatical, semantic, contextual—out of complete account, and yet so calibrated as to have added nothing in the way of para-phrase, explication or variant. But we know that in practice this perfect fit is possible neither at the stage of interpretation nor at that of linguistic transfer and restatement. The limiting conditions on hermeneutic totality, moreover, are not restricted to translation. We saw at the start of the discussion that there are no perfections and final stabilities of understanding in any act of discourse above the most rudimentary (even there ambiguity may interfere). Understanding is always partial, always subject to emendation. Natural language is not only polysemic and in process of dia-chronic change. It is imprecise, it has to be imprecise, to serve human locution. And although the existence of a 'perfect trans-lation' or 'perfect exchange of the totality of intended meaning' between two speakers is theoretically conceivable, there could be no way of verifying the actual fact. For how would we know? By what means except an alternate formulation and explicative rephrasing could we demonstrate that the case in point was indeed 'perfect'? Yet such demonstration would necessarily reopen the question. In other words: to demonstrate the excellence, the exhaustiveness of an act of interpretation and/or translation is to offer an alternative or an addendum. There are no closed circuits in natural language, no self-consistent axiomatic sets.

But if 'perfect' translation is no more than a formal ideal, and if great translation is rare, there are, none the less, examples which seem to approach the limits of empirical possibility. There are texts in which the initial commitment to the emotional and intel-lectual risks of unmapped, resistant alternity continues vital and scrupulous even to the finished product. There are translations which are supreme acts of critical exegesis, in which analytic understanding, historical imagination, linguistic expertness articulate a critical valuation which is at the same time a piece of totally lucid, responsible exposition. There are translations which not only represent the integral life of the original, but which do so by enriching, by extending the executive means of their own tongue. Lastly, most exceptionally, there are translations which restore, which achieve an equilibrium and poise of radical equity between two works, two languages, two communities of historical experience and contemporary feeling. For a translation to realize all four aspects equally and to the full is, obviously, 'a miracle of rare device'.

No student of the subject will have direct knowledge of more than a small fraction of an immense, somewhat chaotic spectrum. To name a 'short list' of supreme translations would be absurd. There are too many variables in historical circumstance and local purpose. One has competence in far too few languages, literatures, and disciplines. But I would not want to conclude the 'work-shop' section of my argument without citing one or two examples of the 'near-ideal'. The four-stage model I have put forward derives from actual cases such as these.

Though there is, perhaps, a faltering, a nuance of sentimentalization in lines seven and eight, G. K. Chesterton's version of Du Bellay's 'Heureux qui, comme Ulysse . . .' needs no commentary. Far from being a licence, the English sixteen-line form establishes a genuine parity with the French sonnet:

> Heureux qui, comme Ulysse, a fait un beau voyage,
> Ou comme cestuy là qui conquit la toison,
> Et puis est retourné, plein d'usage & raison,
> Vivre entre ses parents le reste de son aage!
> Quand revoiray-je, helas, de mon petit village
> Fumer la cheminee, & en quelle saison
> Revoiray-je le clos de ma pauvre maison,
> Qui m'est une province, & beaucoup d'avantage?
> Plus me plaist le sejour qu'on basty mes ayeux,

Que des palais Romains le front audacieux:
Plus que le marbre dur me plaist l'ardoise fine,
Plus mon Loyre Gaulois que le Tybre Latin,
Plus mon petit Lyré que le mont Palatin,
Et plus que l'air marin la doulceur Angevine.

Happy, who like Ulysses or that lord
    Who raped the fleece, returning full and sage,
With usage and the world's wide reason stored,
    With his own kin can wait the end of age.
When shall I see, when shall I see, God knows!
    My little village smoke; or pass the door,
The old dear door of that unhappy house
    That is to me a kingdom and much more?
Mightier to me the house my fathers made
    Than your audacious heads, O Halls of Rome!
More than immortal marbles undecayed,
    The thin sad slates that cover up my home;
More than your Tiber is my Loire to me,
    Than Palatine my little Lyré there;
And more than all the winds of all the sea
    The quiet kindness of the Angevin air.

My second example or set of examples ought to come under the rubric of impossibility: because of the inherent complexities of the original text, because of the conflicting assumptions regarding permissible syntactic and prosodic experimentations in French and in English. But Pierre Leyris's translations from Gerard Manley Hopkins[1] are among the finest restatements in modern literature and inexhaustibly instructive both in detail and general grasp.

Stanza IV of 'The Wreck of the Deutschland' is characteristic in its sensory exactitude and its involution:

---

[1] Gerard Manley Hopkins, *Reliquiae: Vers, Proses, Dessins réunis et traduits par Pierre Leyris* (Paris, 1957); Gerard Manley Hopkins, *Le Naufrage du Deutschland. Poème traduit par Pierre Leyris* (Paris, 1964).

> I am soft sift
> In an hourglass—at the wall
> Fast, but mined with a motion, a drift,
> And it crowds and it combs to the fall;
> I steady as a water in a well, to a poise, to a pane,
> But roped with, always, all the way down from the tall
> Fells or flanks of the voel, a vein
> Of the gospel proffer, a pressure, a principle, Christ's gift.

Letting 'sift' work on his inner ear, Leyris probably caught the presence of neighbouring 'sieve' and, perhaps, that of Scottish 'siver', the aperture through which liquid drains. Multiple points of reference are now engaged: the motion of sand or water through a pinched channel; the refinement of matter through a strainer at once literal and spiritual; the hourglass used to mark the time for orisons; the entrapment of the *Deutschland* in sanded narrows. Each of these is latent in *Je passe au sas / D'un sablier*—a translation which, seemingly without effort, mimes Hopkins's assonance. *Sas* is a strainer, often of fine linen (a motif taken up later in the poem). It is also the confined section of water between lock-gates in which a vessel is held while the sluices operate. It is conceivable also that Leyris remembered Charles d'Orléans's haunting line: 'Passant mes ennuiz au gros sas', which he would have found in Littré's article on the word. Next he translates 'Fast' by *ferme*:

> D'un sablier—contre la paroi, ferme,
> Mais miné par un mouvement, une coulée,
> Et qui s'ameute et qui se carde vers la chute. . . .

'Fast', of course, compacts two contrary energies: that of speed and that of solidity. But the latter is, at this point, more obvious to Hopkins's design and Leyris rightly opts for it. *Ameuter* is bold and complex. It gathers several vital strands: the theme of 'mutiny' against God's enigmatic, seemingly wasteful purpose; the tumultuous crowding of the passengers evoked in Stanza XVII ('a heartbroke rabble'); and the literal hunting down of the innocent Franciscan nuns through the Falk Laws (the latter connotation is uppermost in *meute*). *Carde*, as it were, goes 'behind' the original: Hopkins's 'combs' being, very likely, a displacement from the less striking possibility 'to card'. The French word also looks back to the element of linen in *sas* and forward to *encordé*. No less than in English, *la chute* carries the needed theological as well as material implications.

As Leyris, who is following the edition and notes of W. H. Gardner points out, the next lines are of exceeding density. They mesh at least two principal veins of imagery, that of the well with its roped bucket descending into and ascending from the deeps, and that of the twined, threading rush of waters down the flank of a tall fell:

> Moi calme comme l'eau d'un puits jusqu'au suspens, jusqu'au miroir,
> Mais encordé—toujours et tout du long des hauts
> A-pics ou flancs de la montagne, d'une veine
> De l'Évangile proposé, pression, principe, don du Christ.

Such is Leyris's interpretative intensity that one almost overlooks those reproductions of internal rhyme which would be the pride of a lesser translation (*eau/hauts*, *suspens/flancs*, *long/don*). *Suspens* beautifully renders 'poise' and prepares, in a way even subtler than Hopkins's for the element of 'poise', of firm stance in *proposé*. But the touch of miracle is *encordé*. The word takes up the entire range of imagery implicit in the spiralling sift of sand, and the part of 'threading' in 'combs' and *carde*. *Cordée* is a miner's term ('but mined with a motion') signifying the time needed to wind the winch which draws up buckets of earth and rubble. *Encordé* (Leyris's own invention?) embeds a vital pun: to proceed *en cordée* is to be 'roped up' while climbing. Pivoting on the word, the stanza modulates from the theme of the hourglass and the carded threads to that of the steep mountains. *Principe* is stronger than it looks. As so often in Descartes and in Pascal, the word is used here with every overtone of inception and radicalism. The Gospel is the start and root of man's meaning.

Equally revealing is Leyris's treatment of the close of Stanza XI:

> Flesh falls within sight of us, we, though our flower the same,
> Wave with the meadow, forget that there must
> The sour scythe cringe, and the blear share come.

Again the translator lets Hopkins's alliterations and assonances guide him into a recreation of the pulse of argument:

> La chair choit sous nos yeux et nous, bien que notre fleur ne soit autre,
> Qu'avec le pré nous ondulions, nous oublions
> Que là doit sévir l'aigre faux, survenir le soc anuiteur.

Chair/fleur is probably triggered by 'flower'/'blear'/'share'. The strident grate of fricatives in the original is perfectly matched by Leyris's last verse. But the stress of understanding is such that it carries the translation almost beyond Hopkins or, more precisely, that it transcends the immediate text to invoke the sum of Hopkins's poetry. *Survenir le soc* recalls 'c'est l'ahan qui fait le soc dans le sillon / Luire', Leyris's earlier rendition of 'sheer plod makes plough down sillion / Shine' from 'The Windhover'. Being so sharply specific, it brushes aside the crucial indistinctness of 'blear share' in which there is, obviously, the 'ploughshare' but also the more diffuse sense of 'lot', 'destined portion'. *Anuiteur*, moreover, a rare, handsome word found in Froissart and Du Bellay, tells us something of the translator's inspired syncretism. The ploughshare brings deadly night, as do the 'shears' of the Fates. In one respect, Leyris is only externalizing the emblematic, personified suggestion of Death's 'sour scythe'; in another, however, he is going past Hopkins to write a line whose intricacy and force of suggestion defeat the original.

One would wish to go on with detailed quotation and the attempt at analysis. As much as any translation I know, Leyris's Hopkins puts the reader on the tantalizing verge of gaining insight into the processes—acoustic, tactile, hermeneutic—whereby the mind can pass from one language into another and then return. The 'permeability' required is, in the present instance, wholly exceptional, but the dynamics are of a general order. Let me conclude by merely quoting Leyris's restatement of 'Pied

Beauty'—an 'impossibility' if ever there was:

> Glory be to God for dappled things—
> *Gloire à Dieu pour les choses bariolées,*
>   For skies of couple-colour as a brinded cow;
> *Pour les cieux de tons jumelés comme les vaches tavelées,*
>   For rose-moles all in stipple upon trout that swim;
> *Pour les roses grains de beauté mouchetant la truite qui nage;*
>   Fresh-firecoal chestnut-falls; finches' wings;
> *Les ailes des pinsons; les frais charbons ardents des marrons chus;*
>   *les paysages*
>   Landscapes plotted and pieced—fold, fallow, and plough;
> *Morcelés, marquetés—friches, labours, pacages;*
>   And áll trádes, their gear and tackle and trim.
> *Et les métiers: leur attirail, leur appareil, leur fourniment.*
> All things counter, original, spare, strange;
> *Toute chose insolite, hybride, rare, étrange,*
>   Whatever is fickle, freckled (who knows how?)
> *Ou moirée, madrurée (mais qui dira comment?)*
>   With swift, slow; sweet, sour; adazzle, dim;
> *De lent-rapide, d'ombreux-clair, de doux-amer,*
> He fathers-forth whose beauty is past change:
>   Praise him.
> *Tout jaillit de Celui dont la beauté ne change:*
>   *Louange au Père!*

Though it is possible to analyse many points of phonetic, grammatical, and semantic detail, and though one can often reconstruct with some confidence the proceedings of trial-and-error, of rejection and amendment which the translator must have followed (the motion from 'rose-moles' to *mouchetant* via an obsolete cosmetic sense being a simple case in point), the underlying facts of language-transfer, of the 'neurophysiology' of bilingualism and 'interlingual thought' entirely escape us (*mais qui dira comment?*)[1] Translation of this distinction does not only penetrate the barrier

---

[1] Cf. the discussion of bilingual interchange in Susan M. Ervin-Tripp, *Language Acquisition and Communicative Choice* (Stanford University Press, 1973), pp. 1–92.

between languages. It seems to break through the barriers of uncertainty which marks any complex speech-act. It arrives at the core, as Matthew Arnold defines it in lines from 'St. Paul and Protestantism':

> Below the surface-stream, shallow and light,
> Of what we *say* we feel—below the stream
> As light, of what we *think* we feel—there flows
> With noiseless current strong, obscure and deep,
> The central stream of what we feel indeed. . . .

Chapter Six

TOPOLOGIES OF CULTURE

THIS study began by trying to show that translation proper, the interpretation of verbal signs in one language by means of verbal signs in another, is a special, heightened case of the process of communication and reception in any act of human speech. The fundamental epistemological and linguistic problems implicit in interlingual translation are fundamental just because they are already implicit in all intralingual discourse. What Jakobson calls 'rewording'—an interpretation of verbal signs by means of other signs in the same language—in fact raises issues of the same order as translation proper. This book has argued, therefore, that a 'theory of translation' (in the 'inexact', non-formalized sense in which I have sought to define this concept) is necessarily a theory or, rather, a historical–psychological model, part deductive, part intuitive, of the operations of language itself. An 'understanding of understanding', a hermeneutic, will include both. It is, consequently, no accident that the methodical investigation of the nature of semantic processes begins with Kant's call for a rational hermeneutic and with Schleiermacher's study of the linguistic structures and translatability of the Hebrew, Aramaic, and Greek scriptures. To study the status of meaning is to study the substance and limits of translation.

These, however, and the philosophic issues which they entail, are not limited to the spoken or written word. The current discipline, if it is that, of semiology addresses itself to every conceivable medium and system of signs. Language, it asserts, is only one among a multitude of graphic, acoustic, olfactory, tactile, symbolic mechanisms of communication. Indeed, urge the semiologist and the student of animal communication ('zoosemiotics'), it is in many respects a restrictive specialization, an evolutionary twist which has assured man's domination over the natural world but which has also insulated him from a much wider spectrum of somatic–semiotic awareness. In this perspective translation is, as we have seen, a constant of organic survival. The life of the individual and of the species depends on the rapid and/or accurate reading and interpretation of a web of vital information. There is a vocabulary, a grammar, possibly a semantic of colours, sounds, odours, textures, and gestures as multiple as that of language, and there may be dilemmas of decipherment and translation as resistant as any we have met. Though it is polysemic, speech cannot identify, let alone paraphrase, even a fraction of the sensory data which man, blunted in certain of his senses and language-bound as he has become, can still register. This is the problem of what Jakobson labels 'transmutation', the interpretation of verbal signs by means of signs in non-verbal sign systems (the curved arrow on the road sign, the 'mantle blue' at the close of 'Lycidas' whose colour encodes 'purity' and 'hope renewed').

But we need not go immediately or entirely outside language. There is between 'translation proper' and 'transmutation' a vast terrain of 'partial transformation'. The verbal signs in the original message or statement are modified by one of a multitude of means or by a combination of means. These include paraphrase, graphic illustration, pastiche, imitation, thematic variation, parody, citation in a supporting or undermining context, false attribution (accidental or deliberate), plagiarism, collage, and many others. This zone of partial transformation, of derivation, of alternate restatement determines much of our sensibility and literacy. It is, quite simply, the matrix of culture. In this closing chapter, I want to apply the notion of 'alternity' and the model of translation put forward in the preceding discussion to the larger question of inherited meaning and culture. To what extent is culture the translation and rewording of previous meaning? Being intermediate and ubiquitous, the great area of 'transformations' and metamorphic repetitions is one in which verbal signs are not necessarily 'transmuted' into non-verbal sign systems. They may, on the contrary, enter into various combinations with such systems. The exemplary case is that of language and music or language in music.

The composer who sets a text to music is engaged in the same sequence of intuitive and technical motions which obtain in translation proper. His initial trust in the significance of the verbal sign system is followed by interpretative appropriation, a 'transfer into' the musical matrix and, finally, the establishment of a new whole which neither devalues nor eclipses its linguistic source. The test of critical intelligence, of psychological responsiveness to which the composer submits himself when choosing and setting his lyric, is at all points concordant with that of the translator. In both cases we ask: 'has he understood the argument, the emotional tone, the formal particularities, the historical conventions, the potential ambiguities in the original? Has he found a medium in which to represent fully and to elucidate these elements?' The means at the composer's disposal—key, register, tempo, rhythm, instrumentation, mode—correspond to the stylistic options open to the translator. The basic tensions are closely analogous. The debate as to whether literalism or recreation should be the dominant aim of translation is exactly paralleled by the controversy, prominent throughout the nineteenth century, as to whether the word or the musical design should be uppermost in the Lied or in opera.

We have seen that the same original text is often translated by several contemporary and subsequent translators, and that such a sequence of alternative versions is an act of reciprocal, cumulative criticism and correction. The musical case is precisely comparable. When Zelter, Schubert, Schumann, and Wolf set the identical Goethe poem to music, when Debussy, Fauré, and Reynaldo Hahn compose music to the same lyrics by Verlaine, when both Berlioz and Duparc write music to Gautier's 'Au cimetière', the contrastive aspects, the problems of mutual awareness and critique are exactly those posed by multiple translation. Has the composer read his poem accurately? Which individual syllables or words, which phrases or prosodic units, does he select for instrumental or vocal emphasis? Does this selection or its converse, the understatement of certain units, fairly enact the poet's intention (is

Schubert right, in setting Schmidt von Lübeck's 'Der Wanderer', when he concentrates the whole meaning of the song on the word *nicht* in the last line, making the word come on a poignant appoggiatura over a strange chord of the sixth)? In what ways are Schumann's, Liszt's and Rubinstein's settings of Heine's 'Du bist wie eine Blume' successive but also divergent commentaries on a deceptively naïve text? To what extent are Wolf's Mörike Lieder an explicit and original act of literary revaluation well before literary critics themselves had recognized the poet's special genius? What brand of Platonism is expressed in Satie's musical setting of passages from the *Symposium* and the *Phaedo* (the analogy with certain of Jowett's edulcorations is striking)? Answers to such questions relate closely to those which turn up in the analysis of literary translation.

Thus there are numerous cases in which the composer simply misreads his text. In all his six settings of Heine, Schubert misconstrues the poet's covert but mordant irony. Often the musician will tamper with the words, altering, omitting or 'improving' on the poem to suit his personal gloss or formal programme (the translator too adds or elides to his own advantage). Mozart tacks on an extra verse to Goethe's 'Veilchen'; wishing to obtain a rise of a full octave on the word, Schubert elides the *e* in *Vögelein* in Goethe's 'Ueber allen Gipfeln'; in Schumann's opus 90, the composer alters Lenau's text, changing words, leaving out several, inserting some of his own (being the most verbally-perceptive of songwriters, Hugo Wolf almost never modifies the lyric).[1] In

---

[1] I owe these three examples to Jack M. Stein, *Poem and Music in the German Lied from Gluck to Hugo Wolf* (Harvard University Press, 1971). Prof. Stein's book is one of the very few extended treatments of the interaction of poetry and musical setting. John Hollander's *The Untuning of the Sky: Ideas of Music in English Poetry 1500–1700* (Princeton University Press, 1961) remains invaluable, but deals only marginally with the actual musical treatment of literary texts. The kind of close investigation made by Vincent Duckles in his article on 'John Jenkins's Settings of Lyrics by George Herbert' (*The Musical Quarterly*, XLVIII, 1962) is somewhat exceptional. The best studies have been elicited by those modern composers who have their own strong views on the relations of word to music. Cf. Wilfrid Mellers, 'Stravinsky's Oedipus as 20th-Century Hero' (*The Musical Quarterly*, XLVIII, 1962); Claudio Spies, 'Some Notes on Stravinsky's Requiem Settings' (*Perspectives of New Music*, V, 1967); and Wolfgang Martin Stroh's important article on

musical translation, no less than in linguistic, there are problems of surpassing. In both *Die Schöne Müllerin* and *Die Winterreise*, Schubert utterly transfigures the feeble poems of Wilhelm Müller, making of them a searching statement on the griefs and doubts of human existence. Little in Chamisso's verse prepares one for the emotional complexity of Schumann's settings in opus 42. Dare one say, in analogy to aspects of 'transfiguration' in the Authorized Version, that there are settings in the *Saint Matthew Passion* (such as the Evangelist's narrative of Calvary and Jesus' last words on the cross) which excel even *their* book, or that Berlioz transfigures and, therefore, in some measure betrays, the Queen Mab speech from *Romeo and Juliet*?

Endowed with a genius for the musical elements and powers of musical suggestion in the oral and written word, involved, at a sovereign, philosophic level, in the whole subject of the transformation of organic and artistic forms, Goethe looked with ambivalence on his musical translators. Yet they swarmed to his writings.[1] Margarethe sings the ballad 'Es war ein König in Tule' (*Faust* I. 2759–82) under circumstances of deep ambiguity. Mephistopheles has put the casket of jewels in her wardrobe; he and Faust are prowling in the garden. Margarethe finds the air strangely heavy. The poem is charged with ironies and menace appropriate to Margarethe's situation but also beyond her con-

---

'Schoenberg's Use of Text: The Text as a Musical Control in the 14th "Georgelied", opus 15' (*Perspectives of New Music*, VI, 1968). A. H. Fox Strangways's 'Song-Translation' (*Music and Letters*, II, 1921) remains the most sensible advocacy yet of the translation of foreign lyrics into English. Herbert F. Peyser's 'Some Observations on Translation' (*The Musical Quarterly*, VIII, 1922) can be seen as a counter-argument. In Peyser's closely-reasoned view, 'the peculiar clang-tint' of each individual language, particularly when it is set to music, makes all but exceptional virtuosities of translation futile. See also the two general articles on word and music by Northrop Frye, 'Introduction: Lexis and Melos' (*English Institute Essays*, New York, 1957); 'Music in Poetry' (*University of Toronto Quarterly*, XI, 1941–2).

[1] The monographic literature is abundant. The three volumes of the Goethe–Zelter correspondence remain our primary source. See also the two editions by Max Friedländer of *Gedichte von Goethe in Kompositionen seiner Zeitgenossen* (*Schriften der Goethe-Gesellschaft*, XI, 1896), and *Gedichte von Goethe in Kompositionen* (*Schriften der Goethe-Gesellschaft*, XXXI, 1916). For a general survey, see the special issue on Goethe and music of *La Revue musicale*, CXXV, 1932.

scious grasp. Goethe's quatrains have a contradictory spell: the short, 'strangled' lines fall numbly, yet the atmosphere is one of indistinct, haunted spaciousness:

> Es war ein König in Tule
> Gar treu bis an das Grab,
> Dem sterbend seine Bule
> Einen goldnen Becher gab.
>
>   Es ging ihm nichts darüber,
> Er leert' ihn jeden Schmaus;
> Die Augen gingen ihm über,
> So oft er trank daraus.
>
>   Und als er kam zu sterben,
> Zählt er seine Städt im Reich,
> Gönnt' alles seinem Erben,
> Den Becher nicht zugleich. . . .

Of innumerable attempts at translation only Nerval's comes within range, and even here the break in the rhyme-scheme saps the original:

> Autrefois un roi de Thulé,
> Qui jusqu'au tombeau fut fidèle,
> Reçut, à la mort de sa belle,
> Une coupe d'or ciselé.
>
> Comme elle ne le quittait guère,
> Dans les festins les plus joyeux,
> Toujours une larme légère
> A sa vue humectait ses yeux.
>
> Ce prince, à la fin de sa vie,
> Légua tout, ses villes, son or,
> Excepté la coupe chérie,
> Qu'à la main il conserve encor. . . .[1]

The original text was set by Zelter, Schumann, and Liszt. Gounod

---

[1] Nerval had several translations of this piece. I quote that of his *Faust* of 1828. In Richault's edition of 1877 (*La Damnation de Faust*), l.2 of the third stanza reads 'ses villes et son or'.

and Berlioz set the French version. Each of these compositions is an act of interpretative restatement in which the verbal sign system is critically illuminated or, as the case may be, misconstrued by a nonverbal sign system with its own highly formal syntax. In other words the musical setting of a poem generates a construct in which the original and its 'translation' (possibly a twofold translation) coexist in active simultaneity.[1]

Zelter's setting is strictly strophic. It is in the key of A minor with a very simple chordal accompaniment. The music is attendant to the authority of the poem in precisely the way Goethe regarded as appropriate. Schumann is far more ambitious. He composed the song in 1849 and published it in volume I of the *Romanzen und Balladen* (opus 67). Margarethe's troubled monody becomes a part-song for solo, either male or female, and five-part mixed chorus. The treatment of the text is straightforward. All stanzas are set without any repeats and in a simple rhythm. There is some modulation, but it is hardly adventurous. The harmonics are essentially 'vertical' (chordal) rather than 'horizontal' (polyphonic). Schumann puts the work squarely in reach of an amateur choral society and seems to underline the folk-element in the poem. But he shows little grasp of what is inherently uncanny in the legend and in the singer's situation. Liszt's reading is far more acute (he first set the words to music in 1843 but revised the score in 1856). It is a reading based on the ambiguity of the narrative, on the tensions between sensuality and death, between fidelity and waste which organize Goethe's treatment and which dramatize Margarethe's unconscious state. The song is set for mezzo and through-composed. The structure is complex and is calculated to enact the plot of the ballad. Lines are repeated; there is a piano introduction and short piano interludes after the fourth and fifth strophe; the last strophe is divided into two sections with a dramatic reprise of the final verse. Liszt's conception is intensely pictorial and romantic. Taking up a motif from a later scene, the composer imagines the heroine spinning as she sings. Thus the

---

[1] In the following analysis I am indebted to a personal communication from Patrick J. Smith, whose *The Tenth Muse: A Historical Study of the Opera Libretto* (New York, 1970) is a pioneering work in the field.

piano begins by imitating the 'churr' of the wheel. The musical phrase closely mimes the appropriate uneven motion (fast-slow-fast). This phrase becomes the motto of the whole ballad, a characteristic Lisztian device. The tempo is 3/4 except for a few bars in 4/4. The basic key is F minor though there is a modulation to the relative major ($A^\flat$) in the third and fourth lines of the third strophe. Liszt treats Goethe's prosody with unworried freedom; the metre is altered in accord with the musical form. Individual words are dramatized and set off pictorially: thus *sinken* is illustrated by descending chromatic octaves. The mood and events of the narrative are explored and rhetorically intensified by the vocal and instrumental translation.

The two settings of 'Es war ein König in Tule' by Berlioz and the setting by Gounod are in the romantic–dramatic vein of Liszt rather than in that of Zelter or Schumann. There is now a dual motion of translation: from German into French, from French into music. All three compositions, moreover, derive from a larger context. They are a part of a general musical treatment of *Faust* I. This means that Gretchen appears before us as a *dramatis persona*; the composition bears both on the ballad and on the key-relations and motifs in the larger design.

Berlioz set the poem in the *Huit Scènes de Faust* of 1829 and in the *Damnation de Faust* of 1846. The earlier version is strophic, the later through-composed. The first setting is in a modal G, the second, whose orchestration is more elaborate, in a modal F. In the *Damnation*, the setting is not strictly modal; modality is used to achieve a feeling of ancientness and distance (Berlioz regards the piece as a 'chanson gothique'). The spinning wheel is, rightly, omitted. In the first version Marguerite sings while undressing— as in Goethe's play. In the second she sings while plaiting her hair. Berlioz's understanding of the problems posed by French translation grows subtler as he proceeds from the first to the second setting. Both are in 6/8 time. The 1829 version begins, rather naïvely, on the strong beat: *Au-tre-fois un roi/ de Thu-lé/*. In the *Damnation* the voice enters on the weak second beat, after an initial rest: *Au-tre-fois/ un roi de Thu-/ lé, Qui jus-qu'au/ tombeau. . . .* The treatment of *Autrefois* as an upbeat wonderfully deepens the effect of archaic remoteness. In both compositions there is an obbligato for viola solo. This, of course, is one of Berlioz's signatures, but it derives ultimately from eighteenth-century opera and adds a hint of classical *bel canto* to a decidedly romantic coloration. The second version ends far more dramatically than the first. Marguerite sings the ballad a second time, pausing between phrases; quite different themes are haunting her mind. She concludes on a profound sigh and the monosyllable *ah!*, followed by a bar of silence and a pizzicato F in the cellos and basses. This coda already appears in the 1829 setting; what is new is the repetition with its forceful suggestion of absentmindedness. The point is one of fundamental interpretation. In Goethe's play the choice of 'Es war ein König in Tule' is, in one sense, fortuitous (an old ballad out of some popular collection or child's book of verse); it is, in another sense, ironically meshed with the dramatic situation and Margarethe's threatened consciousness. The impact stems from the play of seeming irrelevance against dramatic irony. Berlioz psychologizes and simplifies drastically:

Dans l'exécution de cette Ballade, la chanteuse ne doit pas chercher à varier l'expression de son chant suivant les différentes nuances de la poésie; elle doit tâcher, au contraire, de le rendre le plus uniforme possible: il est évident que rien au monde n'occupe moins Marguerite dans ce moment que les malheurs du Roi de Thulé; c'est une vieille histoire qu'elle a apprise dans son enfance, et qu'elle fredonne avec distraction.

The motif of 'distraction' set out in this stage-direction for the 1829 version is further emphasized in the *Damnation*. Though dramatically defensible and musically effective, it flattens the complex shape of the original.

In Gounod's *Faust* (1859), the actual motif of the King of Thule is made wholly accidental. A dotted rhythm in the orchestral introduction simulates the spinning wheel. The ballad itself is set in A minor and the composition is meant to convey an impression of extreme *naïveté*. It is made to sound almost like a nursery-rhyme. Marguerite, moreover, is paying no heed to the words. Her recitative dwells on the handsome cavalier seen at the fair.

Thoughts of him keep interrupting her song. After each inter-jection she repeats the first verse in order to recall the lyric and in order to recapture the appropriate mood. But she fails to do so. The use of the minor key, the hum of the spinning wheel in the orchestral background dramatize her distraction. The idea derives from Berlioz, but the simplicity of Gounod's tune and the device of interruption and reprise give it a peculiar pathos. Goethe's poem, however, is diminished almost to a jingle.

So far as I am able to judge, none of these six transformations is really satisfactory. Liszt's come nearest to the model of equity. It takes liberties, it overdramatizes, but it is attentive to the discipline and secrecy of Goethe's purpose. Zelter's is little more than a musical caption. Berlioz and Gounod exploit the original for their inspired but also peremptory and selective ends. Schumann's vocalization seems oddly irrelevant, like a vague sketch in the margin of the text. 'Goethe est un piège pour les musiciens; et la musique un piège pour Goethe,' wrote André Suarès.[1] But the issue is a general one. The contrastive tonalities, the differing idiomatic habits, the distinct associative contexts which generate resistance and affinity between two different languages are intensified and complicated in the interpenetration of language with music. Both the verbal sign system and the system of musical notation are codes. Both have a grammar, a syntax, a wide di-versity of national and personal styles. Both have their history. Musical analysis is a 'metalanguage' as is formal logic. Yet though the parallels are crucial and in certain respects homologous, they shade quickly into metaphor. Music *is* a language, but in saying so we use 'language' in a peculiarly unstable sense. We may be using it either at the most technical semiotic level (both are 'sequential rule-governed sign systems obeying certain constraints') or in a sense almost too large for proper definition (both can 'com-municate human emotions and articulate states of mind'). Most likely our reference to 'the language of music' points to the special and the general sense simultaneously and in varying proportion. It

[1] André Suarès, 'Goethe et la musique' (*La Revue musicale*, CXXV, 1932), p. 262.

is, therefore, not astonishing that we lack an adequate critical vocabulary in which to analyse or even paraphrase rigorously the phenomenology of interaction between the language of the word and that of music.

But the exponential effect is there. In Gesualdo's *Moro, lasso, al mio duolo* ..., in Schumann's setting of Eichendorff's 'Waldgespräch', in Wolf's rendition of Mörike's 'Der Feuerreiter', word and music perform an action of reciprocal clarification and enrichment in a structure whose centre is neither that of the verbal sign system nor that of the musical notation. As in great translation, so in a great musical setting, something is added to the original text. But that which is added 'was already there'. Put verbally, the assertion is precious and paradoxical. This is not so in execution. One listens to Duparc's setting of the 'Invitation au voyage' and feels precisely in what ways the composer is letting Baudelaire's words be more than themselves and thus entirely themselves. There is then a metamorphosis into an integral but intermediary genre for which we lack a defining term. The intermediacy is at once a crucial and a restrictive condition. The dynamics of preserved identity and temporary fusion—*lexis* and *melos*, to use Northrop Frye's designation, remain themselves while conjoining in a new form—are so complex as to be very fragile. Thus coexistence on a level of genuine parity and interaction tends to be brief. The madrigal, the aria, the Lied, the art-song seem to mark the limits of sustained synthesis. And even here, we have seen, completion is rare. All too often there is cause for Nerval's dictum that only the poet himself can set his own song or for Victor Hugo's decree: 'Défense de déposer de la musique au long de cette poésie.'[1] But the identical motives for rejection apply to much of translation proper. And where the transmutation is accomplished, the two principal grammars of human feeling fuse.

[1] Quoted in René Berthelot, 'Défense de la poésie chantée' (*La Revue musicale*, CLXXXVI, 1938), p. 90.

2

When a text is set to music the words keep their identity though inside a new formal aggregate. When the composer uses a translation, the change effected on the original verbal signs is that of translation proper. But as we move outward from examples of direct transposition and translation, we find innumerable formal possibilities and shadings of change. These extend, as we have noted, from the closest echo to the most remote, often unconscious reference, embedded resonance, or allusion. They range from an interlinear translation of Homer to the Homeric contours in Joyce. But indistinctly and crucially they extend to concentric spaces of recognition far beyond the manifest dependence of *Ulysses* on the *Odyssey*. These spaces will, for instance, comprise the literal and symbolic status of voyages, uncertain homecomings, marital fidelity, survival through cunning, disguise, the reversal of fortune. Transformations can proceed from linguistic to metalinguistic and non-linguistic codes.[1] The Homeric text can be set to music in its original wording or in translation. It can serve as caption to a painting or sculpture which illustrates one or another episode. But the painter, sculptor, or choreographer need not cite his source-text. He can image, reflect, or enact it with greater or lesser fidelity. He can treat it in a limitless variety of perspectives ranging from 'photographic' mimesis to parody, satiric distortion or the faintest, most arcane of allusions. It is up to us to recognize and reconstruct the particular force of relation. (How soon does the generally alert but unaided reader catch the detailed echoes of *David Copperfield* in Dostoevsky's *The Possessed*, or the kinship between the fable of Lear and that of Cinderella, particularly when the latter is cast in the form, say, of a ballet or a pantomime?)

---

[1] Jean Cassou's sequence of thirty-three sonnets written in prison were illustrated by an equal number of lithographs by Jean Piaubert. Six of the sonnets were, in turn, set to music by Darius Milhaud, thus creating a twofold transposition and a threefold reciprocity between a verbal, a pictorial and a musical sign system. I owe this example to Walter Mönch's important article 'Von Sonettstrukturen und deren Uebertragungen' (Karl-Richard Bausch and Hans-Martin Gauger (eds.), *Interlinguistica: Sprachvergleich und Uebersetzung*).

These manifold transformations and reorderings of relation between an initial verbal event and subsequent reappearances of this event in other verbal or non-verbal forms might best be seen as *topological*. By that I mean something quite simple. Topology is the branch of mathematics which deals with those relations between points and those fundamental properties of a figure which remain invariant when that figure is bent out of shape (when the rubber sheet on which we have traced the triangle is bent into conic or spherical form). The study of these invariants and of the geometric and algebraic relations which survive transformation has proved decisive in modern mathematics. It has shown underlying unities and assemblages in a vast plurality of apparently diverse functions and spatial configurations. Similarly, there are invariants and constants underlying the manifold shapes of expression in our culture. It is these which make it possible and, I think, useful to consider the fabric of culture as 'topological'. The constants can be specifically verbal; they can be thematic; they can be formal. Their recurrence and transformations have been studied by such literary scholars as Auerbach, Curtius, Leo Spitzer, Mario Praz, R. R. Bolgar. The history of the *topos*, of the archetype, of the motif, of the genre, is a commonplace in modern comparative literature and stylistics. Iconology, both with reference to verbal content and with reference to the reprise of particular subjects, motifs, landscapes, allegoric devices by successive artists and schools, is one of the main concerns in current art history. The work of Panofsky, of F. Saxl, of Edgar Wind, of E. H. Gombrich and many others has taught us how much of what the painter sees before him is previous painting. We know now how deep is the grip of convention and of traditional codes of identification over reflexes we might have thought spontaneous. I am, therefore, saying nothing new and the examples I will cite are in several cases familiar ones.

What I am suggesting, however, is that they be recognized as part of a topological process. The relations of 'invariance within transformation' are, to a more or less immediate degree, those of *translation*. Viewed in this way, the concepts of 'underlying structure', 'recursiveness', 'constraint', 'rewrite rules', and 'freedom' put forward by transformational generative grammar will take on a larger meaning. And it will be a meaning less in conflict with the realities of natural language and cultural development. Defined 'topologically', a culture is a sequence of translations and transformations of constants ('translation' always tends towards 'transformation'). When we have seen this to be the case, we will arrive at a clearer understanding of the linguistic–semantic motor of culture and of that which keeps different languages and their 'topological fields' distinct from each other.

The distinction between verbal, formal, thematic, or modal constants is bound to be artificial. In any substantive example each will be in play. But it can serve to point up different strategies and ideals of 'rewriting'. Dryden's *Twenty-ninth Ode of the Third Book of Horace; paraphrased in Pindarick Verse, and inscribed to the Right Hon. Laurence, Earl of Rochester* illustrates a crucial mechanism of formal–cultural rewording.[1] The changes of tonal value, of prosodic technique and prosodic identification implicit in the recasting of Alcaic strophes into Pindaric stanzas already create a complex field of innovation in constancy. So does the implied equation of the Earl of Rochester with Maecenas, an equation which allows a wide latitude of literal, neutral, or ironic resolution. But the fundamental 'rewrite rule' is one of aggrandizement, a fact subtly underscored and, it may be, faintly mocked by the notorious elevation of Pindaric verse. Dryden's eighth and ninth stanzas run to twenty-three lines; they are founded on Horace 41–56:

> Happy the Man, and happy he alone,
>> He, who can call to-day his own:
>> He who, secure within, can say,
> To-morrow do thy worst, for I have lived to-day.
>> Be fair, or foul, or rain, or shine,
> The joys I have possest, in spight of fate, are mine.
> Not Heav'n it self upon the past has pow'r;
> But what has been, has been, and I have had my hour.

[1] Cf. the valuable monograph by Bernfried Nugel: *A New English Horace: Die Uebersetzung der horazischen ars poetica in der Restaurationszeit* (Frankfurt, 1971).

> Fortune, that with malicious joy
>   Does Man her slave oppress,
> Proud of her Office to destroy,
>   Is seldome pleas'd to bless:
> Still various, and unconstant still,
> But with an inclination to be ill,
>   Promotes, degrades, delights in strife,
>   And makes a Lottery of life.
> I can enjoy her while she's kind;
>   But when she dances in the wind,
>   And shakes the wings, and will not stay,
>   I puff the Prostitute away:
>   The little or the much she gave, is quietly resign'd:
>   Content with poverty, my Soul I arm;
>   And Vertue, tho' in rags, will keep me warm.

Our analytic means are meagre. We sense plainly enough but cannot tabulate the agencies of visual contrast and correspondence, of what can fairly be called 'iconographic relations' as between Dryden's text and the Latin. The original strophes are visibly trim:

> Fortuna saevo laeta negotio et
> ludum insolentem ludere pertinax
>   transmutat incertos honores,
>   nunc mihi, nunc alii benigna. . . .

The extra syllable (anacrusis) at the beginning of the first three lines of each strophe seems to accentuate the loaded tautness of the measure. Dryden's stanzas have an arboreal undulation and luxuriance. Their punctuation is musical in its regard to pause and lift, to *presto* and ornament. The capitalizations did not, of course, figure in Horace. Set against this fact the Masque-like procession of *Man*, *Heav'n*, *Fortune*, *Office*, *Lottery*, *Prostitute*, *Soul* and *Vertue*. The typographic difference here is one which redirects eye and mind: it ranges from the purely diacritical—the major word set off against the minor—to elaborate conventions of personification and allegory. The sequence *Office*, *Lottery*, *Prostitute* stands out theatrically against the final pairing of *Soul* with *Vertue*. It is, presumably, in order to preserve the vivacity of opposition that Dryden or the printer did not capitalize 'Poverty'.

We feel the intention and complexity of these transformations, but can give them only intuitive markers. Moreover—and this is the issue—it is not the liberty of Dryden's paraphrase which seems to matter so much as it is the sense of confidence, of agreed definition within variation. Dryden's augments are no doubt excessive: *Pauperiem sine dote* is what they lack. Yet the Horatian fabric, the logic and segmentation of feeling in the original are central and obvious in Dryden's inflation. The organizing contiguities and orientations have held. Mastery of self, felicity, consist in a robust presentness; Fortune is mutable but the genius of memory and a man's acceptance of his condition are proof against disaster. The note is that of domestic stoicism, of equability insured by remembrance and disenchantment. It is one of the principal options in Western behaviour and self-representation. Horace's statement of it in the *Odes* is canonic and we have no difficulty in recognizing its authority and economy even where Dryden is freest. To 'puff the Prostitute away' is, in one perspective, sheer Dryden and unmistakably 'period'. The movement is that of a Restoration lyric sliding into satire. But the elements of this motion are all there in *ludum insolentem ludere*. 'The joys I have possest' seems to embroider loosely on *quod fugiens semel hora vexit*. But it is emphatically to the Horatian idiom that we owe the correlation, again canonic in Western sensibility, between the 'fugitive hour' and the assumption of joy. Thus the 'rewrite rules' and procedures of transformation exhibit both constraint and innovation. To use J. B. Leishman's illuminating phrase, Dryden finds in Horace an 'indefinitely expandable formula'.[1] It is on this 'indefinite expansion' of a fairly limited set of 'formulas' that our culture, our capacities for verifiable recollection and response appear to depend almost completely.

---

[1] Leishman's long prefatory essay to *Translating Horace* (Oxford, 1956) is a masterly introduction to the whole problem of the authority and transmission of classic forms in Western literature and feeling. I will be drawing on it at many points in the following discussion.

Translation, in the wider senses which we are now considering, is the primary instrument of formulaic expansion. It transforms the 'deep structures' of inheritance—verbal, thematic, iconographic—into the 'surface structures' of social reference and currency.

Neo-classicism is based on a postulate of timelessness. It posits the constancy of general human traits and, consequently, of expressive forms whether in speech or the plastic arts. All translation from the canon, all imitation, restatement, citation is, therefore, synchronic. Racine summarizes this aesthetic and psychology of invariance in a remark in his preface to *Iphigénie*. He has noted with satisfaction, from the effect produced in the actual theatre by everything which he has transposed from Homer and Euripides, that 'good sense and reason are the same in all centuries. The taste of Paris has shown itself concordant with that of Athens.' Given this normative cohesion of rational and emotive values over two thousand years, the writer, the architect, the painter of public scenes can imitate originally. His translations from past models are at once faithful and new. They are in the full sense—a sense whose contradiction, whose paradoxicality escapes us unless we pause to look hard at the word—*re-creations*. The neo-classical maker assumes that both the original which he is transposing and a straightforward, possibly literal translation or reproduction of this original are readily present to his audience. Their implicit availability defines the extent of thematic variation in his own product. Formal variation generated by, playing against an implicit constant is a central mode of Western art and letters. It causes the vital ambiguity between the 'classical' and the 'neo-classical', between the antique original properly speaking and its reprise which itself can become a 'classic' if it is of great stature and if its Greek or Latin source is no longer immediate to recognition.

Euripides' *Hippolytus* was first performed in 428 B.C. In lines 1173–1255 the Messenger narrates Hippolytus' fatal encounter with 'the monstrous savage bull' sprung from 'the swelling, boiling, crashing surf'. Sent by Poseidon and itself emblematic of Theseus' exploits in Crete which are the ultimate root of the tragedy, the monster maddens Hippolytus' beloved horses (his name embeds his passion). In David Grene's translation:

> Then all was in confusion. Axles of wheels,
> and lynch-pins flew up into the air,
> and he the unlucky driver, tangled in the reins,
> was dragged along in an inextricable
> knot, and his dear head pounded on the rocks,
> his body bruised. He cried aloud and terrible
> his voice rang in our ears: 'Stand, horses, stand!
> You were fed in my stables. Do not kill me!
> My father's curse! His curse! Will none of you
> save me? I am innocent. Save me!'
> Many of us had will enough, but all
> were left behind in the race. Getting free of the reins,
> somehow he fell. There was still life in him.
> But the horses vanished and that ill-omened monster,
> somewhere, I know not where, in the rough cliffs.

We do not know the date of Sophocles' *Electra*. In this play there is also the recital of the cruel death of a young hero trapped between axles, torn reins, and murderous hoofs during a chariot race. The ornate length of the Paedagogue's speech (679–764) is uncharacteristic of the drastic economy of the rest of the play. It may be interpreted as a psychological finesse. The tale is pure invention, for Orestes is alive and near. But Electra's (apparently unaided) hatred and vengeance are to dominate the remaining episodes in the drama. By the graphic detail of his fictive lament the Paedagogue has, in a sense, eliminated Orestes. We never quite visualize or believe in him again. We cannot tell whether Euripides here, as elsewhere, drew on Sophocles or preceded him. But his own treatment of the scene became a source of imitation and illustration to the present day (a recent film version simply transposes the chariot into a sports car). Euripides' composition of mood and motion from the calm of the beginning through the terror of the event to the calm, now desperate, of the coda; his sequence of set pieces—the young men on the seashore,

Hippolytus' prayer, the bestial eruption, the mortal chase, the disappearance of the bull and the horses leaving helpless humanity; the effects of onomatopoeia as in the supernatural thunder rising—

> ἔνθεν τις ἠχὼ χθόνιος ὡς βροντὴ Διὸς
> βαρὺν βρόμον μεθῆκε, φρικώδη κλύειν

(There came a rumbling, deep in the earth, a
muffled growl like Zeus' thunder, terrible to hear.)

(1201–2)

all these served to establish the speech as a canonic text. From it the later dramatist, moralist, allegoric painter, rhetorician could derive an exemplary repertoire of savage, supernatural occurrence tempered by pathos and irony of motive.

Seneca's tragedies, which scholars assign to the 50s of our era, are modulations on Euripides. The dependence is already highly self-conscious and literary. Seneca fixes on Euripides' genius as a rhetorician, as an architect of oration, to produce his own entirely declamatory closet-dramas. Drawing on aspects of technique latent in Euripides, Seneca wholly internalizes the action. His plays become a string of recitations. Being perpetually violent or grotesque, the events are distanced by the ubiquity of static elocution. Seneca makes a change in the relations (topology) of the agents: Phaedra repents and slays herself, falling on Hippolytus' body. But this is only a minor variant on a set theme. The Messenger's recitation runs from line 1000 to line 1113. It is thus thirty-one lines longer than the Greek and we recognize in this increase a characteristic feature of rewording. It is also interrupted once; Theseus asks what shape the monster had, thus directing attention to the ironic appropriateness of the bull ('caerulea taurus colla sublimis gerens').

At all cardinal points, however, the Latin is a partial transformation of a Greek precedent whose stability as a base of form and imaginative logic is assumed. Only word-by-word analysis could show the number and technical status of Seneca's means of dependent innovation. Where Euripides speaks of an 'unearthly' tidal surge and designates those features of the landscape now shrouded in spray, Seneca universalizes hyperbolically: *Non tantus . . . nec tamen*. 'Never' has there been such tumult, 'never' has the Ionian sea witnessed such breakers. Euripides transports the monster to shore on the crest of an uncanny wave. Seneca invents a machine of the kind which was to adorn seventeenth-century opera and masque:

> inhorruit concussus undarum globus,
> solvitque sese, et litore invexit malum
> maius timore.

> (This liquid globe quivers with a terrible clamour,
> breaks open, and spews onto the shore a monster more
> terrifying than any we could have feared.)

$$(1031-3)$$

Euripides does not describe the sea-bull. The dramatic pace and the indirection of confident art allow him to allude to a spectacle 'more hideous than eyes can bear'. Seneca lingers on horror:

> longum rubenti spargitur fuco latus.
> tum pone tergus ultima in monstrum coit
> facies, et urgens bellua immensam trahit
> squamosa partem. . . .

> (His immense flanks are spotted with reddish slime.
> The extremity of his body is made up of a scaly tail
> which the monster drags behind him in writhing
> coils. . . .) $$(1045-8)$$

He does so not merely from native inclination. Where plot, the distribution of essential mass and the ordering of feeling are given, there is only detail to invent. This is a crucial point. Euripides fixes Hippolytus' fall from the chariot in a single word: πίπτει. The metre and placing at the start of the line give it sufficient drama. Seneca expands and complicates by adducing in counter-point another myth of fatal charioteering:

> talis per auras, non suum agnoscens onus,
> solique falso creditum indignans diem,
> Phaethonta currus devio excussit polo.

> (It was thus that the horses of the sun, realizing the
> absence of their accustomed driver, incensed that a
> false hand should be guiding the chariot of day,
> hurled Phaethon down from the heights of heaven.)
>
> (1090–2)

Talis: it was thus that.... This is the key move towards lateral transposition by citation, allusion, simile. The fable of Phaethon falling from his blazing chariot is an innovative reference within a text which is itself referential throughout. It may or may not have the subtle aim of reminding us of Phaedra's descent from the sun-god. In either case, it belongs squarely to that stock of echoes, of formulaic building-blocks with which the recreator goes to work.

Though criticized from the outset for inordinate length, the 'récit de Théramène' in fact comprises only seventy-three lines. After Theseus' first response, already charged with the terror of premonition, the narrative goes on to recount Aricie's arrival on the dread scene. Thus Racine is economical beyond Euripides, and manifestly beyond Seneca. He makes significant changes. Hippolyte partakes of the concatenation of guilt which enmeshes the characters, although the guilt which he incurs in loving Aricie is carefully shown to be minor and, of itself, ennobling. Racine places the *récit* in the mouth not of a messenger but of Théramène, an intimate of both father and son. This gives to the speech an added poignancy and psychological interest. Moreover, Poseidon himself appears to be active in the ferocious assault on the horses, though again, and with a strategy of rationalist in-determination which qualifies the uses of the supernatural throughout the play, Racine leaves the question open:

> On dit qu'on a vu même, en ce désordre affreux,
> Un dieu qui d'aiguillons pressait leur flanc poudreux.

The stylistic energy, the shapely violence, the psychological tension of the narration have often been detailed.[1] But a just sense

[1] Among the best known studies is Leo Spitzer's essay on 'The "Récit de Théramène"' (*Linguistics and Literary History*, Princeton University Press, 1948). Though it has important insights into Racine's technique, the essay is in fact

of Racine's mastery ought not to exclude an equal sense of the relation of the *récit* to its sources. This relation is, very simply, one of causality: Euripides' and Seneca's treatments of the death of Hippolytus are the *raison d'être* of Racine's. Racine can be supremely economical, he can exploit in depth certain discoveries of feeling, precisely because he comes after. He depends on Euripides and Seneca not only for the general scheme of action but for almost every particular touch.

Racine combines. He takes the pastoral and processional flavour of Hippolytus' departure from Euripides. His depiction of the monstrous surge of waters and of the beast itself are a rewording of Seneca. It is, in fact, Seneca's more extravagant turns which tempt Racine to direct transposition. *Undarum globus* becomes

> Cependant, sur le dos de la plaine liquide
> S'élève à gros bouillons une montagne humide,—

a conceit which contemporary critics thought, not unjustly, somewhat inflated. The monster's livid hues and tortuous extremity are transferred almost intact from the Latin:

> Tout son corps est couvert d'écailles jaunissantes;
> Indomptable taureau, dragon impétueux,
> Sa croupe se recourbe en replis tortueux. . . .

The grating sounds, the strange combination of oiliness and rugosity in *squamosa* were very obviously in Racine's ear. Euripides avoids any description of Hippolytus' mangled flesh. Racine's

> De son généreux sang la trace nous conduit:
> Les rochers en sont teints; les ronces dégouttantes
> Portent de ses cheveux les dépouilles sanglantes

with its audacious play on the literal and emotive values of *dégout-*

---

disappointing. There are imprecisions (the play was not, originally, entitled *Phèdre*). The main thrust of the argument, moreover, is dubious. Spitzer sees the key to the *récit* in 'the magic word "baroque" '. He does so mainly because he fails to consider the Senecan text and its role in Racine's reformulation. The traits which he labels as 'baroque' are almost all to be found in the Latin.

tantes, renders Seneca's 1093–6. Proportionally, Seneca's influence over the *récit* is larger than Euripides'.

But the notion of 'influence' is, in such a case, vacuous. We are dealing with a conscious aesthetic and practice of transformation. Racine's ideal and technique can readily accommodate bits of almost literal translation ('Des coursiers attentifs le crin s'est hérissé' only slightly rewords the Greek). Its other parameter is that of thematic variation. In Euripides the crazed horses vanish; in Seneca they would seem to race out of sight; in Racine they come to a halt

> non loin de ces tombeaux antiques
> Où des rois ses aïeux sont les froides reliques.

The touch is inspired. It has psychological and scenic attributes specific to Racine's purpose. Théramène covertly reminds Theseus of the tragic, gratuitous extinction of the royal house. *Tombeaux antiques* and *froides reliques* are marmoreal words, counters of total repose in calculated contrast to the heat and tumult of the preceding action. The effect is exactly that of a Poussin landscape under a retreating storm. But this motif fairly marks the limits of innovation.

Racine is not, of course, seeking to innovate. He takes as given the timeless validity of the Euripidean text, its power to certify the narrative logic and rational stature of his subject. He draws unworriedly on Seneca who is a fellow-artisan, if in respect of taste a dubious one, in a common enterprise and craft of perpetuation. The psychological interests displayed by Racine, the provisional, metaphoric stance of a Christian–Jansenist sensibility towards pagan myths, the changed criteria of theatrical effect— these distance *Phèdre* from both its Greek and Latin cognates. Racine's genius is his own. But it is a genius exercised within deliberate confines of inheritance and ideal contemporaneity. In his preface Racine cites antique authority for even the slightest of his innovations: 'Je rapporte ces autorités, parce que je me suis très scrupuleusement attaché à suivre la fable.' This scruple is no pedantic or conventional gesture. It voices a fundamental conviction about the rule-bound, 'translational' character of civilized art and literature. To Racine creation is, in central respects, re-creation; freedom derives meaning from constraint.

Let us suppose that we can achieve a complete lexical, grammatical semantic, contextual analysis of the three passages. Let us imagine that we can set the narratives of Euripides, Seneca, and Racine next to each other, relating all formal and semantic elements to each other by virtue of derivation, analogy, general similitude, variance, or contrast. I have tried to show that no such complete analysis is possible, that the idea of exhaustive diagnostic formalization in respect of language is a fiction. But let us suppose the thing to be feasible. We would then, I think, have before us an instrument with which to test and elucidate fundamental issues of language, culture, understanding, and imagination. Taking these three orations alone, we would be in a position to say something concrete about the affinities and the differences between Greek and Latin, and about the ways in which these affinities and differences and their mutual relations to his own language were experienced by a master of seventeenth-century French (already the reticulations between variables are more intricate than any we can handle securely, let alone formalize). We could advance substantive hypotheses about the extent to which the recreative merits and defects of a later version reflect back on the source. How are our readings of Euripides now lit or obscured by our knowledge of Seneca and, particularly, of Racine?

We could, in some measure, at least, come closer to a verifiable gradation of the sequence of techniques and aims which leads from literal translation through paraphrase, mimesis, and pastiche to thematic variation. I have suggested that this sequence is the main axis of a literate culture, that a culture advances, spiralwise, via translations of its own canonic past. A single curve of meaning relates *Phèdre* (1677) to *Hippolytus* (428 B.C.). Racine's confidence, the rarefaction of his executive means, derive from the fact that he felt this time-distance to be both real and unreal. Its reality underwrote the majesty, the essential truth of his material. Its unreality allowed him to work with the Greek original active beside him (he speaks of Sophocles and Euripides as his audience and judges). I come back once more to a root sense of 'translation': to move laterally, to proceed point to point on a level plane.

Transference need not be absolute. We can keep an equation in balance by substitution. Like 'transformation' or 'transcription', 'substitution' is one of a number of main concepts and techniques in the general class of ordered metamorphosis. If we care to do so, we may describe every translation as an act of substitution. Equivalence is sought by means of the substitution of 'equal' verbal signs for those in the original. But what I have in mind now is a more special device, though it is one which underlies much of our literate tradition.

Horace's Ode in praise of Lollius (IV. 9) is one of the templates for Western poetry and our image of the poet. Horace affirms that public achievement and heroism survive only through the poet's commemoration. Eros and even the trivial joys sung by Anacreon achieve permanence in verse. This claim has been a talisman for the writer. No reprise has matched Horace's compressed grandeur—

> vixere fortes ante Agamemnona
> multi; sed omnes inlacrimabiles . . .
>
> (Many heroes lived before Agamemnon,
> but all unwept . . .)

but imitations, paraphrase, variants have been legion. Pope's recasting of strophes one, two, three, and seven exactly illustrates what I mean by 'substitution':

> Lest you should think that verse shall die,
>   Which sounds the Silver Thames along,
> Taught, on the wings of Truth to fly
>   Above the reach of vulgar song;
>
> Tho' daring Milton sits sublime,
>   In Spenser native Muses play;
> Nor yet shall Waller yield to time,
>   Nor pensive Cowley's moral lay.
>
> Sages and Chiefs long since had birth
>   Ere Caesar was, or Newton nam'd;

These rais'd new Empires o'er the Earth,
    And Those, new Heav'n's and Systems fram'd.

Vain was the Chief's, the Sage's pride!
    They had no Poet, and they died.
In vain they schem'd, in vain they bled!
    They had no Poet, and are dead.

The 'Silver Thames' stands for 'resounding Aufidus'; Milton is made equivalent to Homer, Spenser is matched against Pindar, 'grave Stesichorus' is identified with Waller, Cowley, it appears, with Alcaeus. Pope expands on Horace's reference to Agamemnon. Characteristically he cites the supreme glory both of statescraft and of the intellect. Characteristically also, he suggests that even the natural sciences depend on the poet for their lasting renown. But this 'doubling' is, at the same time, a stab at equivalence: such is the lapidary elevation of Horace's *ante Agamemnona multi* that Caesar and Newton are required to restore the balance. We saw the same device of substitution at work in Basil Bunting's imitation of Villon. The poet simultaneously denies and telescopes time. Though it was already more cerebral, more calculated than Racine's Atticism, Pope's Augustanism, his identification of eighteenth-century London with imperial Rome was, none the less, strongly felt. His lines derive from Horace but exist in the same temporal dimension; there is a synchronic parallelism between *Maeonius Homerus* and 'daring Milton'. On the other hand, substitution juxtaposes, for purposes of praise or irony, of shock or coherence. It makes for a collage of past and present, revaluing both in complex, unsettling ways.

In Pope's 'versification' of Donne's second Satire the processes of substitution are far more interesting but also more awkward to analyse. The bare term 'versification' is charged with corrective intent. To a greater or lesser degree, the young Pope seems to have shared Warburton's opinion that Donne's lines 'have nothing more of numbers than their being composed of a certain quantity of syllables'. Pope's professed aim is one of drastic improvement. But neither 'elevation' nor 'refinement' covers the case. The moves made between both texts are more intricate. They generate relations which are at once obvious and elusive. Pope's treatment of Donne is cavalier yet dependent. Consider the close which is often cited as an example of Pope's early virtuosity and of a social awareness not always striking in his later works:

> The lands are bought; but where are to be found
> Those ancient woods, that shaded all the ground?
> We see no new-built palaces aspire,
> No kitchens emulate the vestal fire.
> Where are those troops of Poor, that throng'd of yore
> The good old landlord's hospitable door?
> Well, I could wish, that still in lordly domes
> Some beasts were kill'd, tho' not whole hecatombs;
> That both extremes were banish'd from their walls,
> Carthusian fasts, and fulsome Bacchanals;
> And all mankind might that just Mean observe,
> In which none e'er could surfeit, none could starve.
> These as good works, 'tis true, we all allow;
> But oh! these works are not in fashion now:
> Like rich old wardrobes, things extremely rare,
> Extremely fine, but what no man will wear.

These lines are based on an equivalent number in Donne, but in this instance the term 'based on' is decidedly unhelpful:

> But when he sells or changes land, h'impaires
> His writings, and (unwatch'd) leaves out, *ses heires*,
> As slily as any Commenter goes by
> Hard words, or sense; or in Divinity
> As controverters, in vouch'd Texts, leave out
> Shrewd words, which might against them cleare the doubt.
> Where are those spred woods which cloth'd heretofore
> Those bought lands? not built, nor burnt within dore.
> Where's th' old landlords troops, and almes? In great hals
> Carthusian fasts, and fulsome Bachanalls
> Equally I hate; meanes blesse; in rich mens homes
> I bid kill some beasts, but no Hecatombs,

None starve, none surfet so; But (Oh) we allow,
Good workes as good, but out of fashion now,
Like old rich wardrops; but my words none drawes
Within the vast reach of th'huge statute lawes.

What, then, is the relation of change between these two
passages? Pope makes verbal, metrical and semantic substitutions.
At several points he merely expands. Donne's contracted 'not
built, nor burnt within dore' is augmented to a couplet which
elaborates both alternatives. 'Meanes blesse', or 'Meane's blest' as
editions from 1635 to 1669 have it, represents Donne at his most
terse. The general sense is clear enough, but its clarity derives
from context and argumentative progress rather than from the
phrase itself. Pope's 'That both extremes were banish'd from
their walls . . . And all mankind might that just Mean observe' is
explanatory duplication. Yet it is also more than that. Pope gives to
the Aristotelian–Horatian motif of 'the just mean', which is
undoubtedly significant in Donne, a spacious centrality. But why
reverse the sequence of Donne's propositions and rhymes? Pope
is, I think, substituting a characteristic fourfold symmetry—the
reciprocally symmetrical alternance of material reference and
abstract generality in two successive couplets—for Donne's
irregular lunge. 'Some beasts' play against 'whole hecatombs' as
'Carthusian fasts' play against 'fulsome Bacchanals'. In both
couplets the contrast leads to the normative precept: the banishment of extremes, the observance of the mean. This is to draw on
the anatomy of the heroic couplet, with its inherent bias to concordance or contrastive logic precisely as Donne's rhymes, with
their seemingly contingent modification of dramatic blank verse,
do not.

The loss of sinew is unavoidable. Donne's woods literally
clothed lands now bought and stripped. The concreteness of
reference prepares for the 'old rich wardrops' and, ironically I
conjecture, for the 'vast reach' of the final lines ('the vast reach' of
corrupt law as against the live shelter of the 'spred woods'). Pope's
handling of the material is uncertain. How well has he understood

the intricate but malevolently exact focus of Donne's satire, of the special interplay between Popish and judicial cant and rapacity?[1] 'Those ancient woods, that shaded all the ground' is a handsome line. But its note of lost pastoral is beside Donne's incisive point. This unsteadiness of response on Pope's part betrays itself at the close. It is not obvious whether Pope has understood or perhaps rejected as unacceptable the packed syntax of Donne's 'we allow Good workes as good'. His substitution 'These as good works, 'tis true, we all allow' muddles the point. To what does 'These' refer? The couplet following limps irretrievably. It is only padding. Seeking to explicate Donne's deliberately elliptical warning, Pope adds four lines replete with court sycophants, informers, and treason.

One is left with contradictory evidence. Pope's rewording of Donne is manifestly high-handed. It argues a conviction of greatly improved metrical resources. It imposes what are, at this point in the eighteenth century, taken to be self-evident and self-evidently progressive criteria of rhetorical clarity, balance, trimness. Yet at the same time one senses a recurrent discomfort under pressure, as if Pope was aware of facts of reference and facts of sensibility in Donne marginally outside his own grip. Seeing how often he retains Donne's rhymes and how uneasy are his substitutions where Donne is at his most concentrated, one wonders whether Pope fully shared Warburton's dismissive view of Donne's technique. But there may be a supplementary complication. Pope 'versifies' Donne in the light of his own intimate knowledge and imitations of Horace, notably of Epistle II of Book II. Though the directions of argument are opposed—the Epistle mocks those who think that landed possessions insure against levelling death—they are also at several points symmetrical and lead to parallel phrasing. We cannot tell to what extent Pope regarded Donne as being himself an imitator of Horace, but in his own reading of Donne Horace is largely present. The result is, as very often in more complex types of substitution, a 'three-body problem'. This kind

[1] Cf. I. Jack, 'Pope and the Weighty Bullion of Dr. Donne's Satires' (*PMLA*, LXVI, 1951).

of problem will allow of no more rigorous solution in poetics than it does in classical mechanics.

Other cases and modes of substitution would be worth considering to show how ubiquitous the procedure is in our literature. Juvenal's tenth Satire is canonic as an expression of moral censure on the vacant worldliness of political–urban man by the time Dryden writes his adaptation of it. Dr Johnson's 'Vanity of Human Wishes' embodies an Augustan–Christian reading of the Latin, but a reading whose English substitutions have before them the example of Dryden. Robert Lowell's version, which carries the same title as Johnson's, is simultaneously a twentieth-century 'imitation', a treatment after Pound, and a reutilization of both its predecessors. The development of English prosody in Dryden and Johnson and the history of the language so far as it is marked by the two writers figure in Lowell's technique and are brought into relation to the original. In Lowell's Rome, therefore, the inferences of equivalence, the substitutions of mirroring terms, are at least fourfold. At one level the scene is that of Juvenal's imperial city as reconstructed (translated) by modern historical analysis. At a second and third level, it is the imagined Roman world of Dryden and Johnson, that is to say Restoration and Augustan London felt as material and emblematic analogues to (substitutions for) the Rome of Juvenal. At a fourth level, Lowell's metropolis and the predatory empire on which it feeds are New York and an America which Lowell finds blind and destructive of sane values. The intricacy of substitutions is made viable by the underlying solidity and continuity of the model. Each successive version is a rewrite of Juvenal.

We do not know whether there is an earlier source for Asclepiades' epigram in the Greek Anthology bidding young women not to be too coy because

> The joys of the Love-Goddess are to be found only among
> the living,
> girl, and we shall lie as no more than bone and dust in the
> place of Death.

If Asclepiades did indeed 'invent' this line of persuasion, he is one of the principal begetters of Western poetry. The argument is a commonplace when Tasso rephrases it in a renowned chorus in the *Aminta*. Cowley transposes both the antique form of the proposal and its several variants in Jacobean drama—'worms shall feed on that proud flesh, lady'—into his poem 'My Diet'. Cowley's version of the plot leads immediately to Marvell's. There are, to be sure, intrinsic grounds of pungency and of dense economy which determine the genius of 'To his Coy Mistress', which make this particular variation excel above literally hundreds of analogues. But these grounds cannot be rightly gauged if the opportunities of constraint provided by a long tradition, if the inherent features of substitution in Marvell's text are overlooked. The *données* were entirely given and public.

Moving outward concentrically from literal rewording through paraphrase and substitution one would, I think, come next to one or another type of 'permutation'. A thematic constant is kept dominant and visible across a history of changing forms. Once again, the distinction is somewhat arbitrary. 'Substitution' can operate in just this way, maintaining the matter and logic of a theme while altering the expressive convention. But 'substitution' may usefully be regarded as more literal, as nearer to straightforward translation, than is 'permutation'. The one perennially shades into the other, but an example will point to the difference in degree.

We have seen Horace asserting that the poet's work is the sole guarantor of immortality for other men. There is, therefore, a special poignancy in the fact that the poet himself is mortal, that the singer who ensures survival for others should fall prey to death. In William Dunbar's 'Lament for the Makers', which scholars assign to the period between 1510 and 1520, the terror of the theme is unconcealed. Neither clerk nor theologian can escape, and the poets also are doomed:

> I se the makaris amang the laif
> Playis heir ther pageant, sine gois to graif;
> Sparit is nocht ther faculte;
> *Timor mortis conturbat me.*

> He has done petuously devour
> The noble Chaucer, of makaris flowr,
> The Monk of Bery, and Gower, all thre;
> > *Timor mortis conturbat me.*

There follow ten hammering stanzas enumerating other poets gone. Then the vice closes on Dunbar himself:

> Sen he has all my brether tane,
> He will nocht lat me lif alane,
> On forse I man his nixt pray be;
> > *Timor mortis conturbat me.*

The Renaissance takes up the topic but introduces a dialectic of negation: the poet must die, yet either in his own spiritual person or through the poetic lineage of which he is a part, he will know rebirth. In this treatment of the theme there are obvious complications of adjustment to a Christian view. How is Orpheus' return from the underworld, which is used emblematically throughout the whole tradition of elegy and celebration, to be reconciled to the Christian interpretation of death?[1] The conventions of pastoral serve as an ingenious compromise. By transposing into the landscape and idiom of Theocritus and Virgil, the Christian elegist achieves two effects: he gives to the conceit of the poet's immortality an allegoric distance, and he hints subtly at symbolic concordances between the Apollonian–Orphic tradition and that of the Good Shepherd. Pastoral and paschal interact. A number of subsidiary motifs appear in each variant. With the death of the particular poet, the art of the Muses is itself on the point of extinction. The mourning poet, moreover, feels threatened. How much time is there left for *him*? His lament, therefore, has both a public and a private echo. But this lament must cease. The master is not truly gone. The genius of his verse, the reflection of this

---

[1] In his remarkable study of *Orpheus in the Middle Ages* (Harvard University Press, 1970), John Block Friedman has shown how late-antique thought, Neoplatonism, and Christian iconography lead to the gradual evolution of an 'Orpheus–Christus figure'. From the twelfth century on this syncretic conception influences art and literature.

genius, pallid as it may be, in the elegy now being composed, initiate a counter-current of hope. The mourning landscape modulates into spring. These motifs and the general motion of the argument become formulaic. They allow us to read five major English poems as members of a set related by explicit permutations (each poet in turn takes into account the ways in which his predecessors have organized the invariants).

The tension in Thomas Carew's 'Elegy on the Death of Dr Donne' (1640) stems from a need to accord pagan with Christian counters. The need was the more acute because of Donne's ecclesiastical status and the notorious distance between Donne's profane and sacred poetry. The death of the Dean of St Paul's has left poetry 'widowed'. Carew doubts that there is sufficient inspiration left to produce even an adequate lament:

> Have we no voice, no tune? Did'st thou dispense
> Through all our language, both the words and sense?

Donne had found poetry in a barren state:

> So the fire
> That fills with spirit and heat the Delphique quire,
> Which kindled first by thy Promethean breath,
> Glow'd here a while, lies quench't now in thy death. . . .

Through Donne's verse the Muses' garden has been purged of 'Pedantique weedes'. Donne had opened up for English poets a 'Mine of rich and pregnant phansie'. This image of subterranean venture leads naturally to Orpheus. But Carew gives Orpheus' formulaic presence a critical twist. Such was the wealth and masculine energy of Donne's exploitation that even the Thracian singer would have found in Dr Donne an 'Exchequer', a treasure-trove of invention. Donne's merit was the greater as he accomplished these feats in 'our stubborn language' and at a time when the primacy of the classics and the long labours of their imitators had left only 'rifled fields' (the Proserpine theme with its many affinities with that of Orpheus and with the symbolic drama of seasonal change is not far off). Yet although Donne's demise and Donne's own treatments of the topic of universal decay bear witness to 'the death of all the Arts', some impulse to creation remains. Carew's simile is a fine one: a swiftly turning wheel stays in motion for a time even when the hand which spun it is withdrawn. In a final stringency Carew binds together the formulaic strands of classical mythology and Christian vocation. 'Delphique quire' exactly prepares for the necessary conjunction. Donne was

> Apollo's first, at last, the true God's Priest.

This twofold consecration and the ambiguities it entails are, of course, the substance of 'Lycidas' (1645). More readers than Dr Johnson have been left uncomfortable by the poem's uncompromising stylization of grief, by the ways in which mythological–pastoral conventions are made to carry the moral weight and logical progress of Milton's meaning. But this is the point. No major poem in English literature depends more rigorously on implicit citation, on the postulate of a repertoire of allusion, echo, and counterpoint. The flora of the opening lines directs us to Horace's Ode I. 1 and to Spenser's *Shepheard's Calendar* for September and January. 'Hard constraint' (Milton's 'Bitter constraint') had moved Spenser to write his 'Pastoral Eclogue' on Sidney. Lycidas is the name of the shepherd in Theocritus' seventh Idyll and also that of one of the pastoral speakers in the ninth Eclogue of Virgil. Spenser's *Astrophel* and a long-established device of augmentative pathos lie behind Milton's threefold reiteration of Lycidas' name. 'Who would not sing for Lycidas' is a rewording of 'Carmina sunt dicenda; neget quis carmina Gallo' from Virgil's tenth Eclogue (Pope will use the formula in 'Windsor Forest': 'What Muse for Granville can refuse to sing?'). There is hardly a line in 'Lycidas' which does not solicit, and so far as immediacy of effect goes, presume the reader's awareness of relevant classical and Elizabethan constants.

It is Milton's achievement to use the formulaic and the conventional with such control and confident self-projection, that he appears to go behind the conventions, behind the Horatian, Virgilian, Ovidian variants to an original pressure of experience. He intimates, as it were, and brings to bear on personal feeling those facts of death, of desolate and reborn landscape, of the

poet's sense of mystery and doubt as to the nature of his calling,
which underlie, which at some time out of historical reach generated the structure of pastoral. Milton can do so just because
the 'sincerity' of his lament for Edward King is a qualified and
opportunistic one. The anguish of the poem in regard to unfulfilled promise and to the menace of the contemporary political–
religious situation points, obviously, to Milton himself. But this
egotism is, as we noted, a part of the convention; it is a set element
in a poet's mourning for a fellow-poet. The stylized, entirely
expected character of Milton's material everywhere multiplies the
resonance of his statement. Orpheus enters inevitably but to
supreme effect:

> What could the Muse her self that *Orpheus* bore,
> The Muse her self, for her inchanting son
> Whom Universal nature did lament,
> When by the rout that made the hideous roar,
> His goary visage down the stream was sent,
> Down the swift *Hebrus* to the *Lesbian* shore.

The motif of resurrection is present in Carew, but Milton gives it
a new splendour. Welding Orphic and Christian annunciations of
rebirth, 'Lycidas' completes its parabolic motion in joy:

> Weep no more, woful Shepherds weep no more,
> For *Lycidas* your sorrow is not dead. . . .

The paradox is theological but also strictly formulaic. It is first
stated by Pindar, then rephrased by Horace and by Ovid in the
Metamorphoses. The act of poetic lament is itself a proof that poetry
shall endure.

By 1821 the machinery of pastoral was a stale sham. Yet
'Adonais' invests it with a vitality which goes well beyond the
rhetorical flourish, beyond the sheer prosodic drive of the
poem. This is because Shelley's literalism in the handling of
the mythological-antique conventions (at the service, to be sure
of his own highly idiosyncratic allegoric *personae*) is as intense, as
personal as is that of Milton, though in a totally opposed direction

of thought. 'Adonais', writes Harold Bloom, 'is in a clear sense a materialist's poem, written out of a materialist's despair at his own deepest convictions, and finally a poem soaring above those convictions into a mystery that leaves a pragmatic materialism quite undisturbed.'[1] Shelley's despair at Keats's death, at the organic finality of that death, is deliberately in excess of the facts so far as the acquaintance of the two poets goes. But this excess is integral to Shelley's realization—a realization which we know to be formulaic in this pattern of elegies—of his own threatened condition and of the profoundly ambiguous nature of the poet's existence on earth. In a closing movement beyond philosophic or pragmatic evidence, 'Adonais' breaks free of earth and envisions a Platonic–apocalyptic radiance wholly extrinsic to man. The echoes of 'Lycidas', the parallelisms of rhetorical structure are everywhere apparent. But the type of permutation applied to the traditional constants and to Milton's particular format is that of a radical critique. Shelley's text is a rebuttal of Milton's the more focused because it operates by means of intentional echo.

Exactly as in Milton, the name of the dead poet sounds over and over at the start of the lament. And with reference not to Keats but certainly to Lycidas, Shelley hints at a death by drowning:

> Oh, weep for Adonais—he is dead! . . .
> For he is gone, where all things wise and fair
> Descend;—oh, dream not that the amorous Deep
> Will yet restore him to the vital air;
> Death feeds on his mute voice, and laughs at our despair.

From line 19 to line 190 the bleak reality of organic and individual death is reiterated: '*He* will wake no more, oh, never more.' The surge towards transcendence, with its precise echo to Milton, begins with the opening verse of Stanza XXXIX:

> Peace, peace! he is not dead, he doth not sleep—
> He hath awakened from the dream of life. . . .

---

[1] Harold Bloom, 'The Unpastured Sea: An Introduction to Shelley', in H. Bloom (ed.), *Romanticism and Consciousness* (New York, 1970), p. 397.

Orpheus is present though unnamed:

> He is made one with Nature: there is heard
> His voice in all her music, from the moan
> Of thunder, to the song of night's sweet bird. . . .

But the mourner leaves behind earthly reality even though it is now animate with Adonais' genius. The sphere of man is too corrupt a vessel to contain the ultimate energies of poetic–metaphysical vision. The last stanza concentrates a sum of mastered inheritance and self-recognition so great that it erupts—no other word will do—into numbing clairvoyance. Proceeding from a final allusion to 'Lycidas' and the drowning of Edward King via the Platonic and Petrarchan simile, always precious to him, of the soul's bark, Shelley foretells his own death:

> The breath whose might I have invoked in song
> Descends on me; my spirit's bark is driven,
> Far from the shore, far from the trembling throng
> Whose sails were never to the tempest given. . . .

Rejecting both the pastoral and the Christian contract with immortality, yet drawing largely on the formulaic tradition in which both are instrumental, Shelley's lament, like Dunbar's, rounds on its maker.

In 'Thyrsis' (1866) the permutation of canonic features is consciously parasitic. When Matthew Arnold calls on Thyrsis, Corydon, Bion, and their Sicilian 'mates', he does so at second and third hand. The invocation is, patently, one to Milton and Shelley. But the resulting academicism and touch of self-mockery are apt. They communicate the scholastic ambience, the elevated bookish tenor of Arnold's relations with Clough. Fragile as they are, moreover, the pastoral formulas draw a paradoxical integrity from the fact—it is the key fact—that Arnold's sorrow has a private truth present neither in 'Lycidas' nor 'Adonais'. The elegy keeps in delicate poise a self-conscious pathos and gentle irony neither of which cancels out grief or agnosticism. The placing of Orpheus illustrates Arnold's method. A Sicilian shepherd would have followed Thyrsis into the underworld

And make leap up with joy the beauteous head
Of Proserpine, among whose crowned hair
Are flowers first open'd on Sicilian air,
And flute his friend, like Orpheus, from the dead.

Today no such dispensation is allowed. In Clough's early death Arnold, who is at this point entirely formulaic, sees his own prefigured:

Yes, thou art gone! and round me too the night
In ever-nearing circle weaves her shade. . . .

Then, in deliberate rephrasing of Milton and Shelley, the singer turns from desolation:

yet will I not despair.
Despair I will not, while I yet descry
'Neath the mild canopy of English air
That lonely tree against the western sky.

And Thyrsis' voice, here the *genius loci* of the Virgilian eclogue and landscape, confirms:

Why faintest thou? I wander'd till I died.
Roam on! The light we sought is shining still.

That Thyrsis' words contain an allusion to a well-known passage in Clough's own poetry again illustrates the balance between formal convention and intimacy in Arnold's 'monody' (Milton uses the same technical term to designate 'Lycidas').

This elegiac 'set' to which one could add, but only I think with some qualifications, Swinburne's 'Ave atque Vale' and Tennyson's *In Memoriam*, is simultaneously implicit and examined in Auden's 'In Memory of W. B. Yeats' who had died in January 1939. Auden exploits the pathetic fallacy knowing it to be suspect yet fundamental to the interplay of landscape and mourning throughout the pastoral genre:

He disappeared in the dead of winter:
The brooks were frozen, the airports almost deserted,
And snow disfigured the public statues;

The mercury sank in the mouth of the dying day.
O all the instruments agree
The day of his death was a dark cold day.

Orpheus enters. It is not, this time, or in the first instance, the Orpheus of resurrection but as in Milton the singer dismembered: 'Now he is scattered among a hundred cities'. 'The rout which made the hideous roar' in 'Lycidas', the philistine mob which hounded Adonais to his doom, the vulgar positivists who threaten the Parnassus of Thyrsis and the Scholar-Gypsy, are neatly transmuted into brokers 'roaring like beasts on the floor of the Bourse'. But poetry endures:

> it flows south
> From ranches of isolation and the busy griefs,
> Raw towns that we believe and die in; it survives,
> A way of happening, a mouth.

Auden's passage is a permutation, highly personal yet also firmly traditional, of corresponding motifs in Ovid and Milton. It is not poetry as abstraction but Orpheus' head which journeys south 'to the *Lesbian* shore'. It is the slain Orpheus who, as Ovid reminds us, does not cease from song:

> membra iacent diuersa locis, caput, Hebre, lyramque
> excipis: et (mirum!) medio dum labitur amne,
> flebile nescio quid queritur lyra, flebile lingua
> murmurat exanimis, respondent flebile ripae.

> (The poet's limbs lay scattered far and wide. But, Oh
> Hebrus, you received his head and his lyre, and (oh
> miracle!) while they floated in mid-stream, the lyre
> sounded desolate notes, the lifeless tongue murmured
> mournfully, and the river-banks replied sorrowingly.)
>
> *(Metamorphoses*, XI. 50–3)

Finally Auden reflects on the whole exercise of poets mourning poets. He observes its moral ambiguity. He worries over the central paradox of linguistic immortality. There is something strangely disturbing, even distasteful in the fact that

> Time that is intolerant
> Of the brave and innocent,
> And indifferent in a week
> To a beautiful physique,
>
> Worships language and forgives
> Everyone by whom it lives;
> Pardons cowardice, conceit,
> Lays its honours at their feet.

In the scandal of that forgiveness, however, lie the larger obligation and promise. No less than Carew, Milton, Shelley, and Arnold before him, Auden closes bracingly. Orpheus' unconstraining voice must follow man 'To the bottom of the night'. It must persuade us to rejoice even in the black and winter of history. The coda is pure pastoral:

> In the deserts of the heart
> Let the healing fountains start,
> In the prison of his days
> Teach the free man how to praise.

'Permutation' organizes many other 'sets' in Western poetry and poetic drama, as it does also in music and iconography. It enters into play wherever formulaic elements are at once broad enough to shape a literary form and specific enough to produce identifying, lasting verbal expressions peculiar to that form. This is the case in the family of poets' elegies on poets which runs unbroken in English from Sidney and Spenser to Auden. The formulaic elements of pastoral setting, of self-recognition, of transition from despair to hope, were based on the classical idyll and eclogue. They generated stylizations so supple and efficacious as to serve poets of profoundly different temper and outlook over four centuries. Each mourner in turn drew on the formal structure and on the verbal detail of his predecessors' work. It is the constancy not only of verbal turns but of a genre as a unit which makes 'permutation' more comprehensive and wide-ranging than 'substitution' though both are, as we saw, closely related. The line of descent from Cowley's treatment of the 'coy mistress' theme to that of Donne and of Herrick is immediately verbal; it organizes a topic rather than a genre. 'In Memory of W. B. Yeats' marks the further development, with all the stress on organic cohesion which 'development' can carry, or possibly the concluding statement, of a major form.

Let me propose one further heading under the general class of partial transformations; this class extends, as we have seen, from most literal translation to parody and oblique, even unconscious echo or allusion. In 'The Extasie' Donne advanced the thesis that there occurs in the spiritual and carnal union of authentic love a commingling, an osmotic confluence of two souls:

> When love with one another so
> Interinanimates two soules
> The abler soule which thence doth flow
> Defects of loneliness controules.

There is manuscript authority also for a simpler form of the key term: we can read 'interanimates'. And it is this variant which I would use. 'Interanimation' signifies a process of totally attentive interpenetration. It tells of a dialectic of fusion in which identity survives altered but also strengthened and redefined by virtue of reciprocity. There is annihilation of self in the other consciousness and recognition of self in a mirroring motion. Principally, there results a multiplication of resource, of affirmed being. 'Inter-animated', two presences, two formal structures, two bodies of utterance assume a dimension, an energy of meaning far beyond that which either could generate in isolation or in mere sequence. The operation is, literally, one of raising to a higher power.

If we consider these attributes, it will be immediately apparent that they reproduce the terms proposed throughout this study to define and characterize translation itself. Intensely focused penetration, the establishment of mutual identity through con-junction, the heightening of a work's existence when it is con-fronted and re-enacted by alternate versions of itself—these are the structural features of translation proper. Even where it relates works remote from one another in language, formal convention, and cultural context, 'interanimation' will show itself to be one further derivative from, one further metamorphic analogue of translation. If this has not always been obvious, the reason may be that the area of relations covered by this rubric is so immediate to and so ubiquitous in our culture.

One other preliminary is worth noting. Donne's phrase 'defects of loneliness' is acutely suggestive of the condition of feeling and intellect which accompanies the stress of personal invention. The poet in front of the blank page, the painter before the vacant canvas, the sculptor facing the native stone, the thinker in the felt but undeclared proximity of the unthought, are very nearly a cliché for solitude. Even to the agnostic the act of creation of meaning and shape carries archaic intimations of *hubris*. The maker feels himself to be at once the imitator and the rival of a larger making. He is alone with his need and this need, as writers and artists testify, is no comforter (Conrad's *The Secret Sharer* is a perfect allegory of the artist's exposure to a crowding solitude). 'Interanimation', says Donne, controls the privations of singularity. The 'abler soul' enters the work in hand. The new beginning draws on precedent, on canonic models so as to reduce the menacing emptiness which surrounds novelty. This 'transfer of souls' (interanimation) is one which has determined, which has given a logic of form and of locale to a substantial portion of Western literature, plastic art, and philosophic discourse.

The history of Western drama, as we know it, often reads like a prolonged echo of the doomed informalities (literally the failure to define separate forms) between gods and men in a small number of Greek households. The imbroglios suffered by the clan of Atreus were a set theme in epic and lyric poetry by the time Aeschylus, Sophocles, and Euripides gave them theatrical form. After that, echo never ceases. Seneca's *Thyestes* and *Agamemnon* are at the origin of Renaissance verse-tragedy in Italy, France, and England. The line of interanimation is a direct one to Alfieri. Modern drama is steeped in the story: Hofmannsthal, Claudel, O'Neill, Giraudoux, T. S. Eliot, Hauptmann, and Sartre produce some of the more successful variants. If we include musical and choreographic treatments, witness Martha Graham's inspired Clytemnestra, the modern catalogue would double or treble.

Branches from the main stem are equally rich. The Iphigenia chapter is dramatized in a long sequence of plays from Euripides to Racine and Goethe. We know that Aeschylus had staged the catastrophe of the house of Laius before Sophocles' *Oedipus*, and that Euripides' *Phoenician Women* is only one among several Euripidean versions of the Theban cycle (which, of course, extends to the *Bacchae*). Seneca is followed by Corneille and Alfieri. Yeats rephrases *Oedipus at Colonus*. Cocteau's Jocasta daubing cold-cream on her face next to the cradle of her infant son is a continuation, serious yet parodistic, of an unbroken series. In Sophocles, Euripides, Racine, Alfieri, Hölderlin, Cocteau, Anouilh, and Brecht we find dramatizations of the Antigone story and of the fratricidal struggle between Eteocles and Polynices. As we noted earlier, the interanimations of the problem of Antigone in the thought and writings of Hölderlin, Hegel, and Kierkegaard produce one of the most vivid exchanges of feeling and philosophic debate in modern intellectual history. When Giraudoux entitled his play *Amphitryon 38*, he was underestimating the number of his predecessors. Drawing on variants of the tale in Homer, Hesiod, and Pindar, Aeschylus, Sophocles, and Euripides wrote plays, now lost, on the ambiguous good fortune of the Theban general and his divine double. Plautus took up the subject and seems to have initiated the term 'tragicomedy' in order to characterize his interpretation of it. Imitations of Plautus include a Spanish *Amphitryon* by Perez de Oliva, a Portuguese version by Camões, an Italian one by Ludovico Dolce. Molière, Dryden, and Kleist take up and modify the theme. Giraudoux and Georg Kaiser give it contemporary expression seizing on its symbolic equivocation and the bizarre solidity which it gives to the matter of dreams.[1] Euripides' *Medea* lends its 'abler soul' to the Medea-plays of Seneca, Corneille, Anouilh, Robinson Jeffers, and a score of other dramatists, composers, and choreographers. Sophocles' and Euripides' vision of Hercules inspire Seneca, as always the bridge to modern literacy, Wieland, Wedekind, Ezra Pound,

---

[1] Peter Szondi's essay 'Fünfmal Amphitryon' in *Lektüren und Lektionen* (Frankfurt, 1973) offers a characteristically delicate reading of the interanimation of successive versions.

Dürrenmatt. We have seen the interanimation of Euripides' *Hippolytus* with Seneca and Racine. Schiller translates *Phèdre* and the twentieth century will produce numerous transpositions of the myth including novels and films. Prometheus as fire-bringer, revolutionary intellect, martyr is a recurrent *persona* in Western tragedy, art, and music from Aeschylus to Milton, Goethe, Beethoven, Shelley, Gide, and Robert Lowell. There is probably no complete listing of the number of versions of Faust from the medieval puppet-play and Marlowe down to Goethe, Thomas Mann, and Valéry's *Mon Faust*. Estimates run into the hundreds. The cognate theme of Don Juan is dramatized by Tirso de Molina, Molière, Da Ponte, Grabbe, Pushkin, Horváth, Shaw, Frisch, and Anouilh, to name only the most famous examples. Its dissemination in lyric verse, the mock-epic, or the novel multiplies this list a hundredfold.[1] In Shakespeare's *Lear* we find the 'resisted presence' of an earlier *Leir* play and of variants of the plot in Sidney's *Arcadia*, Holinshed's chronicles, and Spenser's *Faerie Queene* ('resisted' because Shakespeare departs violently from the canonic outline at key points). There will, in turn, be interanimations with *Lear* in Pinter's *Homecoming*. But the mechanism of interanimation is by no means restricted to mythical or archetypal subjects. There are in the region of eighty novelistic, lyric, theatrical presentations of the biography of Joan of Arc. The versions by Shakespeare, Schiller, Shaw, Brecht, Claudel, Maxwell Anderson, and Anouilh are simply among the most celebrated. But an inventory of this kind could go on to absurdity.

The family-tree structure and 'translational' continuity in Western epic poetry and drama are a commonplace of literary study. If, as Whitehead pronounced, Western philosophy is a footnote to Plato, our epic tradition, verse theatre, odes, elegies, and pastoral are mainly a footnote to Homer, Pindar, and the Greek tragedians. But 'interanimation' by virtue of a common source and the magnetism of a canonic ideal pertain, fascinatingly, also to the novel. We tend to overlook this point because the fabric

[1] Cf. the exhaustive treatment of the tradition in Gendarme de Bévotte, *La Légende de Don Juan* (Paris, 1911). A third volume would be required to bring his survey up to date.

of prose fiction makes for what Henry James called 'loose, baggy monsters'. Unlike verse or drama, the novel displays principles of coherence so diffuse and many-sided that we often find it difficult to classify them or to keep them in ordered view. Far more than other genres, the novel suggests extreme contingency, an *ad hoc* response to each particular narrative occasion, to the hazards of psychological, social, spatial circumstance in which the narrative is set. It is a form boundlessly available. The claim of the novelist to be 'dealing with real life' in a way more inclusive, more empirical, freer of stylization than either the poet or the playwright has generally been allowed. There are exceptions which declare themselves undeniably. James's shaping of the desolation of Isabel Archer's marriage in *The Portrait of a Lady* refers us, with an inference both of profound indebtedness and critical revision, to the disasters of marriage in George Eliot's *Middlemarch*. And although the very different supremacies of the two books make it difficult to grasp their detailed affinity, there can be no doubt that *Anna Karenina* embodies Tolstoy's close experience and partial denial of the presentation and moral judgement of adultery in *Madame Bovary*. Such cases are less rare than might appear. There are, throughout the development of modern fiction, clusters of mutual cognizance, interactive groupings around the common centre of an 'abler' or exemplary presence.

The power of *La Nouvelle Héloïse* (1761) is intentionally discursive. Rousseau uses the epistolary form, which he derives from Richardson, to develop dramatic and philosophic occasions on a massive, unrealistic scale. The tensions are extreme but buried in a digressive technique whose roots are, as always in Rousseau, a liberal, recollective review of consciousness. Today, the book is very largely unread. This makes it difficult to convey, except by assertion, the depth and dimension of its influence. These were of a degree to alter the style of educated feeling throughout Europe and in literate circles as distant as the Caucasus. The self-consciousness of men and women, so far as it is externalized in scenes of ideal or of drastic occurrence, was imprinted by Rousseau's narrative. Saint-Preux and Julie became public archetypes of possibilities of emotion and posture which every reader felt to be intimately his (the illustrations to the novel, prepared under Rousseau's express guidance, speeded and intensified the reflex of identification). The geography of the book, its scenario of lake, orchard, and alp, constituted a new, yet seemingly definitive, landscape of private sentiment. The diverse aspects of this landscape, its colorations, seasonal attributes, meteorologies acted as graphic objectifications of and incitements to social, philosophic, and erotic modes. If the phrase 'climate of feeling' carries legitimate suggestions of a material setting and counterpart, if modern sensibility records as a commonplace reciprocities or ironic clashes between personal mood and natural terrain, the merit is Rousseau's. Space is picturesque for us, it echoes, as it did not before he compelled on it his pathos and prodigal solitude.

Transpositions of La Nouvelle Héloïse into episodes of private life and into 'non-literary' writings such as letters, journals, memoirs of travel, intimate diaries, effusions en famille, were ubiquitous. In the nature of the case, our evidence is abundant but imprecise. What the literary historian can point to are novels, confessional tales, fictional reminiscences, plays, pastoral entertainments, written in immediate or more or less extensive imitation of Rousseau's book. These variants run into the hundreds. Werther (1774) has its independent genius but belongs to the family. So far as French, English, and Italian romanticism go, Goethe's tragic idyll simply reinforces the emotional and technical authority of Rousseau. It affords an added concision and fatality to the more leisured, philosophically warier themes of La Nouvelle Héloïse. But it is the latter which were seminal.

Structurally, we may see Rousseau's novel as that of the education of a young man through thwarted love of a married woman. The beloved is 'older' either in moral and physical experience or in actual age. Though love is returned in a dialectic of deepening need, adultery is denied. This negation stems from and also brings with it complex relations, partly filial, to the husband of the beloved. At a typologically predictable point in the action, the lover makes an attempt, part vengeful, part therapeutic, to find erotic reward in more accessible quarters. The result is self-loathing. This emotion leads to the realization of ecstasy, of fulfilment in renunciation. The gesture of renunciation is provoked by a highly ambiguous moment of shared peril (a storm on the lake, a dangerous illness, a political threat from the world at large). The lovers part, but there is between them a contract of desolation. They are dead to their own future. Subsidiary to these main motifs is that of the children of the beloved, or of her younger brothers or sisters. The lover's relation to these—didactic, fraternal, conspiratorial—is one of pathos and duplicity. Landscape and solitude in landscape correlate precisely with narrative action and with states of being as yet subconscious. Rousseau, in *La Nouvelle Héloïse*, proved himself both the theoretician and expressive master of this concordance. It embodies as important a step in literary means as did the adaptation of epic plots to direct theatrical utterance in Greek drama.

La Nouvelle Héloïse is pervasive throughout the development of the French novel during the late eighteenth and the nineteenth centuries. But its force of interanimation is, perhaps, most sensible in a particular cluster.

Sainte-Beuve was not a natural novelist. This made his dependence on canonic precedent the more unforced. Yet *Volupté* (1834) is a work of exceptional nervous intelligence. It springs from 'defects of loneliness' in regard both to the author's personal life—his adoration of Adèle Hugo—and in regard to his sense of having failed as a poet and creator in the full romantic guise. Thus Sainte-Beuve invests the theme of renunciation with a peculiar bitterness. The landscape of obsession and abandonment is one of marsh and flat horizons, contrasting deliberately with that of *La Nouvelle Héloïse*. The vein of religiosity so important in Rousseau, where it is however left lyrical and undogmatic, is exploited by Sainte-Beuve. Having lost Mme de Couaen for ever, Amaury enters the Church. The surrounding motifs of husband and children, of sexual temptation, of transfiguration in denial, are placed exactly as Rousseau had instanced. On 15 November 1834, Sainte-Beuve published a somewhat dismissive review of *La Recherche de l'absolu*. This notice irritated Balzac and greatly complicated his attitude towards *Volupté*. The latter troubled him because of its unexpected strength and because it anticipated his own wish to treat the same theme. Now he resolved to drive Sainte-Beuve from the field. *Le Lys dans la vallée* appeared in 1836. Balzac's narrative of the doomed passion of Félix de Vandeness and Mme de Mortsauf (her name, like that of Saint-Preux, contains the novel) is one of the most dramatic, psychologically inventive in modern fiction. The uses of the Angevin setting perfectly illustrate Henry James's remark in his essay on Balzac that there is nothing else which the author of the *Comédie humaine* feels 'with the communicable shocks and vibrations, the sustained fury of perception . . . that *la province* excites in him'. But the book is intimately related to the rival performance of Sainte-Beuve.[1] The relation, moreover, is tripartite. Balzac, as it were, 'rethinks' *La Nouvelle Héloïse*, a novel which he knows in its last detail, via Sainte-Beuve's reading of Rousseau. Fréderic Moreau and Mme Arnoux make up the fourth couple in the set (is there a subtle echo in the choice of names?). *L'Éducation sentimentale*, in its definitive version, appears in 1869. The title itself conveys Flaubert's express realization of the central motif in Rousseau. Numerous touches direct us back to *La Nouvelle Héloïse*. The challenge to Balzac is overt. Flaubert seems to have felt, as did other nineteenth-century readers, that, for all its splendour, *Le Lys dans la vallée* had vulgarized the psychological fineness of the material, that Balzac had, characteristically, injected a dose of melodrama (Lady Dudley and her fierce steeds) into an ambiguous tragedy of private feeling. Hence Flaubert's special alertness to *Volupté*. The melancholy tints of his own novel, its adroit counterpoise of political with familial pressures, show his indebtedness. Sainte-Beuve died on 13 October 1869. The following day Flaubert wrote to his niece: 'In part I had written *L'Éducation sentimentale* for Sainte-Beuve. He will have died without knowing a line of it!'[2]

[1] Cf. Maurice Allem, *Sainte-Beuve et 'Volupté'* (Paris, 1935), pp. 265–74, for a general discussion of the relation to Balzac. M. Le Yaounc's edition of *Le Lys dans la vallée* (Paris, 1966) singles out numerous verbal and thematic imitations of Sainte-Beuve in Balzac's text.

[2] The background material is to be found in R. Dumesnil's edition of *L'Éducation sentimentale* (Paris, 1942).

Only an intensive comparative recension of the four texts, set out in parallel together with the relevant drafts, letters, and critical statements, could demonstrate the extent, the vitality of 'inter-animate' relation. (Proust's reprise of the two themes of a young man's education of sensibility through love of an older woman and of a lover's complex relationship to the child of a former beloved, are clearly in the tradition, but no longer a direct variant. The link with Rousseau and Flaubert is 'collateral'.) *La Nouvelle Héloïse* generates, serves as focus for, a 'topological space' of mutual readings and challenges. It is within this space that we can best locate, in relation to a common centre and to one another, Sainte-Beuve's *Volupté*, Balzac's immediate riposte, and Flaubert's master-piece. R. P. Blackmur would have spoken of 'reticulation', of a network whose threads take on different hues, different meshings and tensions, as each new work enters the pattern. Donne's term, on the other hand, reminds us of the solitude which nags even the major artist at the start of invention. The 'abler soul' of the great precedent, the proximity of the rival version, the existence, at once burdensome and liberating, of a public tradition, releases the writer from the trap of solipsism. A truly original thinker or artist is simply one who repays his debts, in excess.

'Substitution', 'permutation', 'interanimation' are no more than awkwardly abstract, elusive terms in a sequence of metamorphic relations and possibilities of relation. The guide in the crypt at Chartres informs us that the edifice towering above encases, is literally a product of, six preceding cathedrals, each *imbriquée* in the next. We look at the raw idiosyncrasy of Soutine's painting of 'The Skate' only to realize that the details of spatial arrangement, of colour-contrast, are a deliberate restatement of Chardin's still-life with the same title. We recall the conceit in Nerval's *Filles du feu* whereby all books are hidden repetitions of each other in a chain of metempsychosis which stretches back, as in Plato's *Ion*, to an initial mystery of divine vocation. The 'rewrite rules' vary widely from period to period, from genre to genre. Tennyson does not imitate or translate as did Pope. Picasso's variations on Velasquez have a somewhat different aesthetic from Manet's uses of Goya. But the central point is that all these metamorphic relations have as their underlying deep structure a process of translation. It is this process, and the continuum of reciprocal transformation and decipherment which it ensures, that determine the code of inheritance in our civilization.

One may celebrate this fact as does Leishman when he speaks 'of the continuity of Western European culture and civilization, of the endless possibilities of individual difference within that great identity, and of the perfect freedom that is possible within that service'.[1] Or one can find this 'translational' condition maddeningly oppressive, as did the poets of Dada, as did D. H. Lawrence in his essay on 'The Good Man': 'This is our true bondage. This is the agony of our human existence, that we can only feel things in conventional feeling-patterns. Because when these feeling-patterns become inadequate, when they will no longer body forth the workings of the yeasty soul, then we are in torture.' But whether we experience it as a source of strength or of suffocation, the fact itself remains. No statement starts completely anew, no meaning comes from a void:

Even the greatest artist—and he more than others—needs an idiom to work in. Only tradition, such as he finds it, can provide him with the raw material of imagery which he needs to represent an event or a 'fragment of nature'. He can re-fashion this imagery, adapt it to its task, assimilate it to his needs and change it beyond recognition, but he can no more represent what is in front of his eyes without a pre-existing stock of acquired images than he can paint it without the pre-existing colours which he must have on his palette.[2]

Western art is, more often than not, about preceding art; literature about literature. The word 'about' points to the crucial ontological dependence, to the fact that a previous work or body of work is, in some degree, the *raison d'être* of the work in hand. We have seen that this degree can vary from immediate reduplication to tangential allusion and change almost beyond recognition. But the dependence is there, and its structure is that of translation.

[1] J. B. Leishman, *Translating Horace*, p. 105.
[2] E. H. Gombrich, *Meditations on a Hobby Horse and other Essays on the Theory of Art* (London, 1963), p. 126.

3

We are so much the product of set feeling-patterns, Western culture has so thoroughly stylized our perceptions, that we experience our 'traditionality' as natural. In particular, we tend to leave unquestioned the historical causes, the roots of determinism which underlie the 'recursive' structure of our sensibility and expressive codes. The problem of origins is one of extreme difficulty if only because the accumulated pressures from the past, embedded in our semantics, in our conventions of logic, bend our questions into circular shapes. The themes of which so much of our philosophy, art, literature are a sequence of variations, the gestures through which we articulate fundamental meanings and values are, if we consider them closely, quite restricted. The initial 'set' has generated an incommensurable series of local variants and figures (our 'topologies'), but in itself it seems to have contained only a limited number of units. How is one to think of these? The concept of 'archetypes' is seductive. Robert Graves's assurance 'To Juan at the Winter Solstice' that 'There is one story and one story only / That will prove worth your telling' sets echo going. Great art, poetry that pierces, are *déjà-vu*, lighting for recognition places immemorial, innately familiar to our racial, historical recollection. We have been there before; there is a genetic code of transmitted consciousness. Until now, however, no biological mechanism is known which could make the persistence and reduplication of archetypes, especially at the level of specific images, episodes, scenes, at all plausible. There is a more naïve objection as well. Given our common neurophysiological build, archetypal images, sign systems ought to be demonstrably universal. Those stylizations and continuities of coding which we can verify are, however, cultural specific. Our Western feeling-patterns, as they have come down to us through thematic development, are 'ours', taking this possessive to delimit the Graeco-Latin and Hebraic circumference.

This suggests an alternative source of constancy. It may be that the Mediterranean achievement proved inescapable. Sixty years after *Lear*, Milton, in his prefatory note to *Samson Agonistes*, spoke of Greek tragic drama as the timeless model 'unequalled yet by any'. To the Renaissance, to Winckelmann, the whole issue seemed straightforward. Granted the fact that fundamental intellectual insights and psychological attitudes are of a limited order, the Greeks had found for both means of plastic and verbal expression which were supreme and which had exhausted the likely possibilities. What came after was variation, adjustment to local context, and critique (the critique of the canonic being *the* modern and ontologically inferior mode). Yielding to intuitive conviction, and in patent rebuke to his own construct of history, Marx proclaimed that Greek art and literature would never be surpassed. They had sprung from a concordance, by definition unrepeatable, between 'the childhood of the race' and the highest levels of technical craft. For Nietzsche the record of the species after the ruin of the antique *polis* was one of progressive diminution. All renascences were only partial, strained spurts of nostalgia for a lost mastery over intellectual and aesthetic expression. Even as the history of religion in the West has been one of variations on and accretions to the Judaic–Hellenistic canon, so our metaphysics, visual arts, humanities, scientific criteria, have reproduced, more or less designedly, the Platonic, Aristotelian, Homeric, or Sophoclean paradigm. The novelty of content and of empirical consequence in the natural sciences and technology have obscured the determinist constancy of tradition. But in philosophic discourse and the arts, where novelty of content is at best a problematic notion, the impulse to repetition, to organization via backward reference, is sovereign. Testimony from an unexpected quarter makes the point exhaustively. Civilization, as we know and pursue it, writes Thoreau in *Walden* (III. 6) is transcription:

Those who have not learned to read the ancient classics in the language in which they were written must have a very imperfect knowledge of the history of the human race; for it is remarkable that no transcript of them has ever been made into any modern tongue, unless our civilization itself may be regarded as such a transcript. Homer has never yet been printed in English, nor Aeschylus, nor Virgil even, works as refined, as solidly done, and as beautiful almost as the morning itself; for later writers, say what we will of their genius, have rarely, if ever, equalled the elaborate beauty and finish and the lifelong and heroic literary labours of the ancients.

This view may or may not be adequate to the facts. It may apply only to certain great currents of high culture and conservatism. It may underestimate the element of genuine discovery or rediscovery in what seems inherited. But the sense of a persistent authority of the classical and Hebraic precedent has been one of the principal forces—perhaps the principal force—during some two millennia of Western sensibility. It has largely determined the Western image of reason and of form. The new design, the new utterance, are tested within and against the exemplary legacy. We move forward from quotation, explicit or not, of the classic formula. The actual metaphor which D. H. Lawrence uses to voice his iconoclasm, 'the workings of the yeasty soul', is an echo of an Orphic and Platonic simile.

This does not signify immobility. We have seen that the diachronic reality of language is one of incessant change. Great mutations of feeling, of cognitive and perceptual frameworks, do occur. The meshing of individual temperament with landscape dramatized by Rousseau is a case in point. Yet language is, nevertheless, inherently conservative. Vocabulary and grammar embed the past. The contrast with other media of expression is instructive. The Renaissance discovery of perspective altered the visual arts and the relations of our optic and tactile sensibility to the material context. The evolution of chordal harmony transformed the texture and conventions of music. Language, particularly written language, is, by comparison, stable (the constance of the principal literary modes since high antiquity being, as we have noted, a direct consequence). At this point, again, the transformational generative model needs amendment. Chomsky's emphasis on the innovative character of human speech, on the ability of native speakers to formulate and interpret correctly a limitless number of previously unspoken, unheard sentences, served as a dramatic rebuttal to naïve behaviourism. It demonstrated the inadequacy of the stimulus–response paradigm in its

Pavlovian vein. Chomsky's observation, moreover, has had notable consequences for education and speech-therapy. But looked at from a semantic point of view, the axiom of unbounded innovation is shallow. An analogy with chess may clarify the issue. It is estimated that the number of possible board-positions is of the order of 10^{43} and that there are, within the constraint of accepted rules, some 10^{125} different ways of reaching these. Until now, it is thought, men have played fewer than 10^{15} games. There is, therefore, no practical limit to the previously untried moves still to be made, or to the number which the opponent can understand and reply to. But despite this boundless potential for novelty, the occurrence of genuinely significant innovation, of inventions which in fact modify or enlarge our sense of the game, will always be quite rare. It will always be in a minuscule proportion to the totality of moves played or playable. The man who has something really new to say, whose linguistic innovation is not merely one of *saying* but of *meaning*—to poach on H. P. Grice's distinction—is exceptional. Culture and syntax, the cultural matrix which syntax maps, hold us in place. This, of course, is the substantive ground for the impossibility of an effective private language. Any code with a purely individual system of reference is existentially threadbare. The words we speak bring with them far more knowledge, a far denser charge of feeling than we consciously possess; they multiply echo. Meaning is a function of social-historical antecedent and shared response. Or in Sir Thomas Browne's magnificent phrase, the speech of a community is for its members 'a hieroglyphical and shadowed lesson of the whole world'.

Will this 'dynamic traditionality' so distinctive of Western literacy persist? There are indications that we have become acutely conscious of the question. We know now that the modernist movement which dominated art, music, letters during the first half of the century was, at critical points, a strategy of conservation, of custodianship. Stravinsky's genius developed through phases of recapitulation. He took from Machaut, Gesualdo, Monteverdi. He mimed Tchaikovsky and Gounod, the Beethoven piano sonatas, the symphonies of Haydn, the operas of Pergolesi and Glinka. He incorporated Debussy and Webern into his own idiom. In each instance the listener was meant to recognize the source, to grasp the intent of a transformation which left salient aspects of the original intact. The history of Picasso is marked by retrospection. The explicit variations on classical pastoral themes, the citations from and *pastiches* of Rembrandt, Goya, Velasquez, Manet, are external products of a constant revision, a 'seeing again' in the light of technical and cultural shifts. Had we only Picasso's sculptures, graphics, and paintings, we could reconstruct a fair portion of the development of the arts from the Minoan to Cézanne. In twentieth-century literature, the elements of reprise have been obsessive, and they have organized precisely those texts which at first seemed most revolutionary. 'The Waste Land', *Ulysses*, Pound's *Cantos* are deliberate assemblages, in-gatherings of a cultural past felt to be in danger of dissolution. The long sequence of imitations, translations, masked quotations, and explicit historical painting in Robert Lowell's *History* has carried the same technique into the 1970s. The apparent iconoclasts have turned out to be more or less anguished custodians racing through the museum of civilization, seeking order and sanctuary for its treasures, before closing time. In modernism *collage* has been the representative device. The new, even at its most scandalous, has been set against an informing background and framework of tradition. Stravinsky, Picasso, Braque, Eliot, Joyce, Pound—the 'makers of the new'—have been neo-classics, often as observant of canonic precedent as their seventeenth-century forbears.

A second symptom points to our heightened awareness of traditionality, of the symbolic and expressive constraints encoded in our culture. The modern attention to myth and ritual has transformed anthropology. We are being taught to look on the 'stasis', on the myth-bound structure of primitive societies with an entirely new understanding and intuition of analogy. Had he not been conscious of the constraints, of the conservatism inherent in our own language habits and behavioural format, Lévi-Strauss could never have explored the determinism, the normative reciprocities of speech and myth, of myth and social practice in Amerindian civilizations. Long persuaded of the privileged dynamism of Western ways, of the presumably unique factor of iconoclasm and futurism operative in Western science and technology, we are now experiencing a subtle counter-current, a new understanding of our confinement within ancient bounds of mental habit. We too are creatures of fable and recursive dreams.

Does this reflexive use of the cultural past, this recognition of how much is 'translational' in our field of reference, point to a real crisis? Do those whose antennae are most alert, who, in the words of the Russian poetess Tsvetaeva, have 'perfect pitch for the future' really anticipate the end of the linguistic-cultural continuum? And if so, what evidence is there to support their terror, their flight to the *musée imaginaire*? I have sought to discuss the issue elsewhere.[1] The flowering of a sub- and semi-literacy in mass education, in the mass media, very obviously challenges the concept of cultural canons. The discipline of referential recognition, of citation, of a shared symbolic and syntactic code which marked traditional literacy are, increasingly, the prerogative or burden of an élite. This was always more or less the case; but the élite is no longer in an economic or political position to enforce its ideals on the community at large (even if it had the psychological impulse to do so). There is no doubt that patterns of articulate speech, reading habits, fundamental legacies of grammaticality, are under pressure. We read little that is ancient or demanding; we know less by heart. But although the inroads of populism and technocracy on cultural coherence have been drastic, the scale, the depth of penetration of the phenomenon are very difficult to assess. The outward gains of barbarism which threaten to trivialize our schools, which demean the level of discourse in our politics, which cheapen the human word, are so strident as to make deeper currents almost impalpable. It may be that cultural traditions are more firmly anchored in our syntax than we realize, and that we shall continue to translate from the past of our individual and social being whether we would or not.

The threat of dispersal, of a crisis in the organic coherence

[1] Cf. Chapter IV of *In Bluebeard's Castle: Some Notes Towards the Re-definition of Culture* (London, 1971), and 'Do Books Matter?' in *Do Books Matter?*, edited by B. Baumfield (London, 1973).

between language and its cultural content, could stem from another and paradoxical direction. Here the argument bears crucially on English.

'At countless points on the earth's surface, English will be the most available language—English of some sort.'[1] I. A. Richards's prediction, made in 1943, has proved accurate. Like no other tongue before it, English has expanded into a world-language. It has far outstripped its potential competitors. A large part of the impulse behind the spread of English across the globe is obviously political and economic. In the aftermath of the Second World War, and building on earlier colonial–imperial foundations, English acted as the vulgate of American power and of Anglo-American technology and finance. But the causes of universality are also linguistic. There is ample evidence that English is regarded by native speakers of other languages whether in Asia, Africa or Latin America, as easier to acquire than any other second language. It is widely felt that some degree of competence can be achieved through mastery of fewer and simpler phonetic, lexical, and grammatical units than would be the case in North Chinese, Russian, Spanish, German, or French (the natural rivals to world status). Today, English is being taught as a necessary skill for modern existence not only throughout continental Europe, but in the Soviet Union and China. It is the second language of Japan, and of much of Africa and India. It is estimated that 88 per cent of scientific and technical literature is either published in English initially or translated into English shortly after its appearance in such languages as Russian, German, and French. The novelist, the playwright, whether his native tongue be Swedish, Dutch, Hebrew, Hungarian, or Italian, looks to English translation for his window on the world. Though figures are very uncertain, the community of English-speakers has been reckoned at 300 million, and is growing rapidly. But statistics, however dramatic, do not make the main point. In ways too intricate, too diverse for socio-linguistics to formulate precisely, English and American-English seem to embody for men and women throughout the world—and

[1] I. A. Richards, *Basic English and its Uses* (London, 1943), p. 120.

particularly for the young—the 'feel' of hope, of material advance, of scientific and empirical procedures. The entire world-image of mass consumption, of international exchange, of the popular arts, of generational conflict, of technocracy, is permeated by American-English and English citations and speech habits.

Doubtless there are opposing trends. Threatened at their most vulnerable point of self-definition, other language communities are resisting the Anglo-Saxon tide. Witness the politically organized struggle of French to maintain itself in the Middle East and French Africa, and to halt the inroads of *franglais* at home. There is evidence also that the very pressures for social, technological uniformity generated by the Anglo-American model are producing reactions. The bitter struggles between Walloons and Flemings, the language riots which plague India, the resurgence of linguistic autonomy in Wales and Brittany point to deep instincts of pre-servation. Norway now has two standard languages where it had only one at the turn of the century. Dialect and variant forms of speech are tending towards autonomy. Nevertheless, English dominates as a world-language whose reach far exceeds that of Latin in the historical past, and whose efficacy has all but nullified such schemes as Esperanto.

The consequences lie outside the scope of this study. They are, at many points, contradictory. American English, West Indian English, the idiom of Australia, of New Zealand, of Canada, the varieties of English spoken and written in West Africa have immensely enriched the total spectrum of the mother-tongue. It can fairly be argued that the energies of innovation, of linguistic experiment, have passed from the centre. Has there been an 'English English' author of absolutely the first rank after D. H. Lawrence and J. C. Powys? The representative masters of litera-ture in the English language, since James, Shaw, Eliot, Joyce, and Pound have been mainly Irish or American. Currently, West Indian English, the English of the best American poets and novelists, the speech of West African drama demonstrate what can be called an Elizabethan capacity for ingestion, for the enlistment of both popular and technical forms. In Thomas Pynchon, in Patrick White, the language is fiercely alive. The metropolitan response has been, in several respects, one of fastidious re-trenchment. Much of contemporary verse, drama, fiction written in England is spare, minimalist, and thoroughly distrustful of verbal exuberance. The techniques of Philip Larkin, Geoffrey Hill, Harold Pinter, and David Storey enact a hoarding of old treasures by means of incisive austerity. It is too early to tell. But the question of the future influence of English at large on English 'at home' is one of the most interesting to face the linguist and historian of culture.

If there is enrichment, moreover, there is also loss. 'English of some sort' said Richards, meaning a basic, orthographically rationalized version. But the simplifications may be of an even more damaging order. The externals of English are being acquired by speakers wholly alien to the historical fabric, to the inventory of felt moral, cultural existence embedded in the language. The landscapes of experience, the fields of idiomatic, symbolic, communal reference which give to the language its specific gravity, are distorted in transfer or lost altogether. As it spreads across the earth, 'international English' is like a thin wash, marvellously fluid, but without adequate base. One need only converse with Japanese colleagues and students, whose technical proficiency in English humbles one, to realize how profound are the effects of dis-location. So much that is being said is correct, so little is right. Only time and native ground can provide a language with the interdependence of formal and semantic components which 'translates' culture into active life. It is the absence from them of any natural semantics of remembrance which disqualifies artificial languages from any but trivial or *ad hoc* usage.

The internationalization of English has begun to provoke a twofold enervation. In many societies imported English, with its necessarily synthetic, 'pre-packaged' semantic field, is eroding the autonomy of the native language-culture. Intentionally or not, American-English and English, by virtue of their global diffusion, are a principal agent in the destruction of natural linguistic diversity. This destruction is, perhaps, the least reparable of the ecological ravages which distinguish our age. More subtly, the modulation of English into an 'Esperanto' of world-commerce, technology, and tourism, is having debilitating effects on English proper. To use current jargon, ubiquity is causing a negative feedback. Again, it is too soon to judge of the dialectical balance, of the reciprocities between profit and loss which accrue to English as it becomes the lingua franca and shorthand of the earth. If dissemination weakened the native genius of the language, the price would be a tragic one. English literature, the penetrating yet delicate imprint of a uniquely coherent, articulate historical experience on the vocabulary and syntax of English speech, the supple vitality of English in regard to its unbroken past—these are one of the excellences of our condition. It would be ironic if the answer to Babel were pidgin and not Pentecost.

AFTERWORD

THIS book has applied poetics, literary criticism, and the history of cultural forms to aspects of natural language. Its focus throughout has been on the act of translation. Translation is fully implicit in the most rudimentary communication. It is explicit in the coexistence and mutual contact of the thousands of languages spoken on the earth. Between the utterance and interpretation of meaning through verbal sign systems on the one hand, and the extreme multiplicity and variety of human tongues on the other, lies the domain of language as a whole. I have argued that these two ends of the spectrum—elementary acts of speech and the paradox of Babel—are closely related, and that any coherent linguistics must take both into account.

Only the professional linguist and logician are competent to assess fully the results achieved by formal and meta-mathematical analyses of language. Of these transformational generative grammars are currently the most prestigious, but by no means the only embodiment. This study has testified to the intellectual fascination of contemporary technical linguistics, and to the fact that the formal approach has helped to bring the investigation of language into a central position in philosophy, psychology, and logic. At the same time, I have expressed the conviction that models such as that put forward by Chomsky drastically schematize their material, and that they neglect, often to the point of distortion, the social, cultural, historical determinants of human speech.[1] By divorcing itself from that intimate collaboration with poetics which animates the work of Roman Jakobson, of the Moscow and Prague

[1] In recent papers, Chomsky himself has been modifying his standard theory. He now allows that rules of semantic interpretation must operate on surface structures as well as deep structures. He is also prepared to shift key morphological phenomena from the grammatical model, whose power may have been exaggerated, to the lexicon. Developed further, both these modifications would bring transformational generative grammars nearer to sociolinguistic and contrastive approaches.

language-circles, and of I. A. Richards, formal linguistics has taken an abstract, often trivialized view of the relations between language and mind, between language and social process, between word and culture.

This reductionism has been most dramatic in regard to the issue of linguistic diversity and of the nature of universals. When I began this book the question of Babel, and the history of that question in religious, philosophic, and anthropological thought were hardly respectable among 'scientific' linguists. Now, only four years later, one of the foremost comparative linguists concludes that

the discovery of putative universals in linguistic structure does not erase the differences. Indeed, the more one emphasizes universals, in association with a self-developing, powerful faculty of language within persons themselves, the more mysterious actual languages become. Why are there more than one, or two, or three? If the internal faculty of language is so constraining, must not social, historical, adaptive forces have been even more constraining, to produce the specific plenitude of language actually found? For Chinookan is not Sahaptin is not Klamath is not Takelma is not Coos is not Siuslaw is not Tsimshian is not Wintu is not Maidu is not Yokuts is not Costanoan. . . . The many differences do not disappear, and the likenesses, indeed are far from all Chomskyan universals. . . . Most of language begins where abstract universals leave off.[1]

This last point is decisive, and I have underlined it throughout my argument. Whether attempts at a comprehensive anatomy of language by formal and logical means are more than an intellectual exercise, often illuminating on the level of the ideal, remains a moot question.[2] This study has sought to show that other approaches may have much to contribute.

[1] Dell Hymes, 'Speech and Language: On the Origins and Foundations of Inequality Among Speakers' (*Daedalus*, issued as the *Proceedings of the American Academy of Arts and Sciences*, CII, 1973), p. 63.

[2] For the most recent attempt to apply formal logic to vagueness, context dependence, metaphor, and polysemy in natural language, cf. M. J. Cresswell, *Logics and Languages* (London, 1973). Nothing in this acute treatment seems to overcome Wittgenstein's admonition against the derivation of systematic logic from ordinary language or Tarski's theorem that 'there can be no general criterion of truth for sufficiently rich languages'—*all* natural languages being 'sufficiently rich'.

In particular, I have put forward the hypothesis that the proliferation of mutually incomprehensible tongues stems from an absolutely fundamental impulse in language itself. I believe that the communication of information, of ostensive and verifiable 'facts', constitutes only one part, and perhaps a secondary part, of human discourse. The potentials of fiction, of counterfactuality, of undecidable futurity profoundly characterize both the origins and nature of speech. They differentiate it ontologically from the many signal systems available to the animal world. They determine the unique, often ambiguous tenor of human consciousness and make the relations of that consciousness to 'reality' creative. Through language, so much of which is focused inward to our private selves, we reject the empirical inevitability of the world. Through language, we construct what I have called 'alternities of being'. To the extent that every individual speaker uses an idiolect, the problem of Babel is quite simply, that of human individuation. But different tongues give to the mechanism of 'alternity' a dynamic, transferable enactment. They realize needs of privacy and territoriality vital to our identity. To a greater or lesser degree, every language offers its own reading of life. To move between languages, to translate, even within restrictions of totality, is to experience the almost bewildering bias of the human spirit towards freedom. If we were lodged inside a single 'language-skin' or amid very few languages, the inevitability of our organic subjection to death might well prove more suffocating than it is.

There is no greater virtuoso of strangulation than Beckett, no master of language less confident of the liberating power of the word. Hamm says in *Endgame*:

I once knew a madman who thought that the end of the world had come. He was a painter—and engraver. I had a great fondness for him. I used to go and see him, in the asylum. I'd take him by the hand and drag him to the window. Look! All that rising corn! And there! Look! The sails of the herring fleet! All that loveliness! He'd snatch away his hand and go back into his corner. Appalled. All he had seen was ashes. He alone had been spared. Forgotten. It appears the case is ... was not so ... so unusual.

Beckett translates himself, or perhaps interleaves as he composes:

J'ai connu un fou qui croyait que la fin du monde était arrivée. Il faisait de la peinture. Je l'aimais bien. J'allais le voir, à l'asile. Je le prenais par la main et la traînais devant la fenêtre. Mais regarde! Là! Tout ce blé qui lève! Et là! Regarde! Les voiles des sardiniers! Toute cette beauté! Il m'arrachait sa main et retournait dans son coin. Épouvanté. Il n'avait vu que des cendres. Lui seul avait été épargné. Oublié. Il paraît que le cas n'est . . . n'était pas si . . . si rare.

The transfer is flawless (except for that enigmatic addition or omission, depending on which text came first, of the engraver). Yet the differences in cadence, in tone, in association are considerable. The English slopes to a dying fall via long *o* sounds; the French spirals to a final nervous pitch. Set the two passages side by side, and a curious effect follows. Their claustral bleakness remains, but the measure of distance between them is sufficient to create a sense of liberation, of almost irresponsible alternative. 'That rising corn' and 'ce blé qui lève' speak of worlds different enough to allow the mind both space and wonder.

The Kabbalah, in which the problem of Babel and of the nature of language is so insistently examined, knows of a day of redemption on which translation will no longer be necessary. All human tongues will have re-entered the translucent immediacy of that primal, lost speech shared by God and Adam. We have seen the continuation of this vision in theories of linguistic monogenesis and universal grammar. But the Kabbalah also knows of a more esoteric possibility. It records the conjecture, no doubt heretical, that there shall come a day when translation is not only unnecessary but inconceivable. Words will rebel against man. They will shake off the servitude of meaning. They will 'become only themselves, and as dead stones in our mouths'. In either case, men and women will have been freed forever from the burden and the splendour of the ruin at Babel. But which, one wonders, will be the greater silence?

SELECT BIBLIOGRAPHY

THE following is a check-list of material which the student of translation will find of particular use. It is set out chronologically, and begins with Schleiermacher's essay of 1813. As is pointed out in Chapter Four, this text initiates the modern approach to translation as part of a larger theory of language and understanding. Works marked with an * themselves contain important bibliographies.

1813

Friedrich Schleiermacher, 'Ueber die verschiedenen Methoden des Uebersetzens' reprinted in Hans Joachim Störig (ed.), *Das Problem des Übersetzens* (Darmstadt, 1969)

1816

Wilhelm von Humboldt, Preface to *Aeschylos' Agamemnon metrisch übersetzt* (Leipzig, 1816)

Mme de Staël, 'De L'Esprit des traductions', first published in an Italian newspaper, then included in the volume entitled *Mélanges* (Brussels, 1821)

1819

J. W. v. Goethe, 'Uebersetzungen' in 'Noten und Abhandlungen zu besserm Verständnis des west-östlichen Divans', *West-Östlicher Divan* (Stuttgart, 1819)

1861–2

Matthew Arnold, 'On Translating Homer' (Arnold's articles are gathered into a book of this title edited by W. H. D. Rouse, London, 1905)

Francis W. Newman, *Homeric Translation in Theory and Practice* (London, 1861)

1863

É. Littré, *Histoire de la langue française* (Paris, 1863), I, pp. 394–434

1881

Herbert A. Giles, 'The New Testament in Chinese', *The China Review*, X (1881)

1886

Tycho Mommsen, *Die Kunst des Uebersetzens fremdsprachlicher Dichtungen ins Deutsche* (Frankfurt am Main, 1886)

1892

J. Keller, *Die Grenzen der Uebersetzungskunst* (Karlsruhe, 1892)

1904

Ludwig Fulda, 'Die Kunst des Uebersetzens', in *Aus der Werkstatt* (Stuttgart, 1904)

1908

Rudolf Borchardt, 'Dante und deutscher Dante', in *Süddeutsche Monatsheften*, V (1908)

1914

W. Fränzel, *Geschichte des Uebersetzens im 18. Jahrhundert* (Leipzig, 1914)

1917–18

Ezra Pound, 'Notes on Elizabethan Classicists', reprinted in *Literary Essays of Ezra Pound* (London, 1954)

R. L. G. Ritchie and J. M. Moore, *Translation from French* (Cambridge University Press, 1918)

1919

F. Batjuškov, K. Čukovskij, and N. Gumilev, *Principy xudožestvennogo perevoda* (*Principles of Artistic Translation*) (Petrograd, 1919)

1920

F. R. Amos, *Early Theories of Translation* (New York, 1920)

G. Gentile, 'Il torto e il diritto della traduzione' in *Rivista di cultura*, I (1920)

Ezra Pound, 'Translators of Greek: Early Translators of Homer', reprinted in *Literary Essays of Ezra Pound*

1922

Ferdinand Brunot, *La Pensée et la langue. Méthode, principe et plan d'une théorie du langage appliquée au français* (Paris, 1922)

J. B. Postgate, *Translation and Translations, Theory and Practice* (London, 1922)

1923

Walter Benjamin, 'Die Aufgabe des Übersetzers', introduction to a translation of Charles Baudelaire, *Tableaux parisiens* (Heidelberg, 1923)

1925

Ulrich von Wilamowitz-Moellendorff, 'Was ist "Übersetzen"?', in *Reden und Vorträge* (Berlin, 1925)

1926

B. Croce, *Estetica* (Bari, 1926)

Franz Rosenzweig, *Die Schrift und Luther* (Berlin, 1926)

Karl Wolfskehl, 'Richtlinien zur Übersetzung von de Costers "Ulenspiegel"', in *Ein Almanach für Kunst und Dichtung* (Munich, 1926)

1927

C. H. Conley, *The First Translators of the Classics* (Yale University Press, 1927)

Eva Fresel, *Die Sprachphilosophie der deutschen Romantik* (Tübingen, 1927)

Franz Rosenzweig, Postscript to *Jehuda Halevi, Zweiundneunzig Hymnen und Gedichte* (Berlin, 1927)

Wolfgang Schadewaldt, 'Das Problem des Uebersetzens' in *Die Antike*, III (1927)

1928

Albert Dubeux, *Les Traductions françaises de Shakespeare* (Paris, 1928)

1929

Hilaire Belloc, 'On Translation' in *A Conversation with an Angel and other Essays* (London, 1929)

A. F. Clements, *Tudor Translations* (Oxford University Press, 1929)

Marcel Granet, *Fêtes et chansons anciennes de la Chine* (Paris, 1929)

Ezra Pound, 'Guido's Relations', *The Dial*, LXXXVI (1929)

1930

Marc Chassaigne, *Étienne Dolet* (Paris, 1930)

Roman Jakobson, 'O překládání veršâ' ('The Translation of Verse'), *Plan*, II (Prague, 1930)

Karl Wolfskehl, 'Vom Sinn und Rand des Übersetzens', in *Bild und Gesetz* (Berne, Zürich, 1930)

1931

A. Barthélémy, *Saint-Évremond* (Lyons, 1931)

Hilaire Belloc, *On Translation* (Oxford University Press, 1931)

F. O. Matthiessen, *Translation: An Elizabethan Art* (Harvard University Press, 1931)

1932

I. A. Richards, *Mencius on the Mind: Experiments in Multiple Definition* (London, 1932)

E. Horst von Tscharner, 'Chinesische Gedichte in deutscher Sprache: Probleme der Übersetzungskunst', *Ostasiatische Zeitschrift*, XVIII (1932)

C. B. West, 'La Théorie de la traduction au XVIIIᵉ siècle', *Revue de littérature comparée*, XII (1932)

1933

H. B. Lathrop, *Translations from the Classics into English from Caxton to Chapman 1477–1620** (University of Wisconsin Press, 1933)

1934

André Thérive, *Anthologie non-classique des anciens poètes grecs* (Paris, 1934)

M. Toyouda, 'On Translating Japanese Poetry into English', *Studies in English Literature*, XIV (Tokyo, 1934)

Arthur Waley, Introduction to *The Way and Its Power, A Study of the Tao Tê Ching* (London, 1934)

Frances A. Yates, *John Florio* (Cambridge University Press, 1934)

1935

Georges Bonneau, *Anthologie de la poésie japonnaise* (Paris, 1935)

C. W. Luh, *On Chinese Poetry* (Peiping, 1935)

1936

G. Bianquis, 'Kann man Dichtung übersetzen?', *Dichtung und Volkstum*, XXXVII (1936)

E. R. Dodds (ed.), *Journal and Letters of Stephen MacKenna* (London, 1936)

1937

José Ortega y Gasset, 'Miseria y Esplendor de la Traducción', reprinted in book form in 1940, and included in *Obras completas* (Madrid, 1947)

1939

Jules Legras, *Réflexions sur l'art de traduire* (Paris, 1939)

V. Weidlé, 'L'Art de traduire', *Nouvelles littéraires*, XXX (1939)

1941

David Daiches, *The King James Version of the English Bible* (University of Chicago Press, 1941)

A. Fedorov, *O xudožestvennom perevode (Artistic Translation)* (Leningrad, 1941)

Vladimir Nabokov, 'The Art of Translation', *The New Republic*, CV (1941)

1943

E. S. Bates, *Intertraffic. Studies in Translation* (London, 1943)

J. McG. Boothkol, 'Dryden's Latin Scholarship', *Modern Philology*, XL (1943)

Alexandre Koyré, 'Traduttore—traditore: à propos de Copernic et de Galilée', *Isis*, XXXIV (1943)

1944

Paul Valéry, Preface to a translation of the *Cantiques spirituels de Saint Jean de la Croix*, included in *Variétés*, V (Paris, 1944)

1945

H. Bernard, 'Les Adaptations chinoises d'ouvrages européens', *Monumenta Sinica*, X (1945)

André Gide, 'Lettre-Préface' to the bilingual edition of *Hamlet* in Gide's translation (New York, 1945)

J. Urzidil, 'Language in Exile' in *Life and Letters To-day*, XLVIII (1945)

O. Weissel, *Dolmetsch und Übersetzer* (Geneva, 1945)

1946

Valery Larbaud, *Sous l'invocation de Saint Jérome* (Paris, 1946)
G. Panneton, *La Transposition* (Montreal, 1946)

1947

Eugene A. Nida, *Bible Translating, An Analysis of Principles and Procedures** (New York, 1947)
J. G. Weightman, 'The Technique of Translation', in *On Language and Writing* (London, 1947)

1948

Herbert Grierson, *Verse Translation* (Oxford University Press, 1948)

1949

Ronald Knox, *Trials of a Translator* (New York, 1949)
R. E. Teele, *Through a Glass Darkly: A Study of English Translations of Chinese Poetry* (University of Michigan Press, 1949)

1951

Douglas Knight, *Pope and the Heroic Tradition* (Yale University Press, 1951)
D. D. Paige (ed.), *The Letters of Ezra Pound 1907–1941*, notably the letters to W. H. D. Rouse on Homeric translation (London, 1951)

1952

J. Herbert, *Manuel de l'interprète* (Geneva, 1952)
J. P. Vinay, 'Traductions', in *Mélanges offerts en mémoire de Georges Panneton* (Montreal, 1952)

1953

Yehoshua Bar-Hillel, 'The Present State of Research on Mechanical Translation', *American Documentation*, II (1953)
A. Fedorov, *Vvedenie v teoriju perevoda (Introduction to the Theory of Translation)** (Moscow, 1953; second, revised edition, 1958)
J. W. MacFarlane, 'Modes of Translation', *Durham University Journal*, XLV (1953)
Paul Valéry, 'Variations sur les Bucoliques', in *Traduction en vers des Bucoliques de Virgile* (Paris, 1953)

1954

Yehoshua Bar-Hillel, 'Can Translation be Mechanized?', *The American Scientist*, XLII (1954)
Olaf Blixen, *La traducción literaria y sus problemas* (Montevideo, 1954)
Martin Buber, *Zu einer neuen Verdeutschung der Schrift*, issued as a Supplement to Martin Buber and Franz Rosenzweig, *Die Fünf Bücher der Weisung* (Cologne, 1954)
Jackson Mathews, 'Campbell's Baudelaire', *Sewanee Review*, LXII (1954)

1955

E. Betti, *Teoria generale della interpretazione* (Milan, 1955)

P. Brang, 'Das Problem der Übersetzung in sowjetischer Sicht', *Sprachforum*, I (1955). This paper is reprinted with addenda in H. J. Störig (ed.), *Das Problem des Übersetzens*

Hermann Broch, 'Einige Bemerkungen zur Philosophie und Technik des Übersetzens', in *Essays*, I (Zürich, 1955)

E. Fromaigeat, *Die Technik der praktischen Übersetzung* (Zürich, 1955)

W. Frost, *Dryden and the Art of Translation* (Yale University Press, 1955)

W. N. Locke and A. D. Booth, *Machine Translation of Languages* (New York, 1955)

Georges Mounin, *Les Belles infidèles* (Paris, 1955)

Vladimir Nabokov, 'Problems of Translation: *Onegin* in English', *Partisan Review*, XXII (1955)

W. Schwarz, *Principles and Problems of Biblical Translation: Some Reformation Controversies and their Background* (Cambridge University Press, 1955)

1956

R. G. Austin, *Some English Translations of Virgil* (Liverpool University Press, 1956)

E. Cary, *La Traduction dans le monde moderne* (Geneva, 1956)

B. Croce, *Critica e Poesia* (Bari, 1956)

J. R. Firth, 'Linguistic Analysis and Translation', in *For Roman Jakobson* (The Hague, 1956)

J. B. Leishman, *Translating Horace* (Oxford, 1956)

G. F. Merkel (ed.), *On Romanticism and the Art of Translation. Studies in Honor of E. H. Zeydel* (Princeton University Press, 1956)

P. Myami, 'General Concepts or Laws in Translation', *Modern Language Journal*, XL (1956)

Allardyce Nicoll, 'Commentaries' to *Chapman's Homer*, edited by Allardyce Nicoll (New York, 1956)

N. Rescher, 'Translation and Philosophic Analysis' in the *Journal of Philosophy*, LIII (1956)

K. Thieme, A. Hermann, and E. Glässer, *Beiträge zur Geschichte des Dolmetschens** (Munich, 1956)

1957

E. Cary, 'Théories soviétiques de la traduction', *Babel*, III (1957)

R. Fertonani, 'A proposito del tradurre', *Il Ponte*, XIII (1957)

Martin Heidegger, *Der Satz vom Grund* (Pfullingen, 1957)

K. Horalek, *Kapitoly z teorie překládání* (*Chapters from a Theory of Translation*) (Prague, 1957)

Ronald Knox, *On English Translation* (Oxford University Press, 1957)

Jiří Levý (ed.), *České teorie překladu* (*Czech Theories of Translation*) (Prague, 1957)

Pierre Leyris, introduction to Gerard Manley Hopkins, *Reliquiae. Vers, proses, dessins* (Paris, 1957)

R. Poncelet, *Cicéron traducteur de Platon* (Paris, 1957)

T. H. Savory, *The Art of Translation* (London, 1957)

Wolfgang Schadewaldt, 'Hölderlins Übersetzung des Sophokles' in Sophokles, *Tragödien, deutsch von Friedrich Hölderlin* (Frankfurt am Main, 1957)

Arno Schmidt, review of George Goyert's translation of James Joyce's *Ulysses* in the *Frankfurter Allgemeine Zeitung*, 26 October 1957

B. Terracini, *Conflitti di lingue e di culture* (Venice, 1957)

1958

É. Benveniste, 'Catégories de pensée et catégories de langue', *Les Études philosophiques*, IV (1958)

Eric Jacobsen, *Translation: A Traditional Craft* (Copenhagen, 1958)

Boris Pasternak, 'Translating Shakespeare', *20th Century*, CLXIV (1958)

Wolfgang Schadewaldt, 'Die Wiedergewinnung antiker Literatur auf dem Wege der nachdichtenden Übersetzung', *Deutsche Universitätszeitung*, XIII (1958)

A. H. Smith (ed.), *Aspects of Translation* (London, 1958)

J.-P. Vinay and J. Darbelnet, *Stylistique comparée du français et de l'anglais* (Paris, 1958)

J. Wirl, *Grundsätzliches zur Problematik des Dolmetschens und des Übersetzens* (Vienna, 1958)

1959

O. Braun and H. Raab, *Beiträge zur Theorie der Übersetzung* (Berlin, 1959)

Reuben A. Brower (ed.), *On Translation** (Harvard University Press, 1959)

F. Flora, 'L'Unità delle lingue e le traduzioni', *Letterature moderne*, IX (1959)

O. Koundzitch, V. Stanevitch, E. Etkind, *et al.*, *Masterstvo perevoda* (*The Art of Translation*) (Moscow, 1959)

W. Widmer, *Fug und Unfug des Übersetzens* (Cologne and Berlin, 1959)

1960

A. G. Oettinger, *Automatic Language Translation** (Harvard University Press, 1960)

Willard V. O. Quine, *Word and Object* (M.I.T. Press, 1960)

1961

William Arrowsmith and Roger Shattuck (eds.), *The Craft and Context of Translation: A Critical Symposium* (University of Texas Press, 1961)

G. Barth, *Recherches sur la fréquence et la valeur des parties du discours en français, en anglais et en espagnol* (Paris, 1961)

Karl Dedecius, 'Slawische Lyrik—übersetzt—übertragen—nachgedichtet', *Osteuropa*, XI (1961)

Dell Hymes, 'On the Typology of Cognitive Styles', *Anthropological Linguistics*, III (1961)

G. Steiner, 'Two Translations', *Kenyon Review*, XXII (1961)

K. D. Uitti, 'Some Linguistic Aspects of Translation', *Romance Philology*, XIV (1961)

1962

W. H. Auden, 'On Goethe: For a New Translation', *Encounter*, XIX (1962)

J. Brooke, 'Translating Proust', *London Magazine*, I (1962)

Die Kunst des Übersetzens, Proceedings of the Bavarian Academy of Fine Arts (no ed.), (Munich, 1962)

Harry Levin, *Refractions* (Oxford University Press, 1962)

1963

L. Bonnerot, *Chemins de la traduction* (Paris, 1963)

C. Chadwick, 'Meaning and Tone', *Essays in Criticism*, XIII (1963)

B. Etkind, *Poezija i perevod* (*Poetry and Translation*) (Moscow and Leningrad, 1963)

Fritz Güttinger, *Zielsprache, Theorie und Technik des Übersetzens* (Zürich, 1963)

Alfred Malblanc, *Stylistique comparée du français et de l'allemand* (Paris, 1963)

Georges Mounin, *Les Problèmes théoriques de la traduction** (Paris, 1963)

1964

Émile Delaveney (ed.), *Traduction automatique et linguistique appliquée* (Paris, 1964)

Georges Mounin, *La Machine à traduire** (The Hague, 1964)

Eugene A. Nida, *Toward a Science of Translation: With Special Reference to Principles and Procedures in Bible Translating** (Leiden, 1964)

Aleksandr Pushkin, *Eugene Onegin. Translated from the Russian, with a Commentary by Vladimir Nabokov* (New York, 1964)

Josef Vachek (ed.), *A Prague School Reader in Linguistics* (University of Indiana Press, 1964)

1965

Anthony Burgess, 'Pushkin and Kinbote', *Encounter*, XXIV (1965)

J. C. Catford, *A Linguistic Theory of Translation* (Oxford University Press, 1965)

H. Friedrich, *Zur Frage der Übersetzungskunst* (Heidelberg, 1965)

Robert Graves, 'Moral Principles in Translation', *Encounter*, XXIV (1965)

Joseph Needham, 'Notes on the Chinese Language', in *Science and Civilisation in China*, I (Cambridge University Press, 1965)

J. P. Sullivan, *Ezra Pound and Sextus Propertius: A Study in Creative Translation* (London, 1965)

W. Tosh, *Syntactic Translation* (The Hague, 1965)

1966

Alexander Gerschenkron, 'A Manufactured Monument', *Modern Philology*, LXIII (1966)

Erica and Alexander Gerschenkron, 'The Illogical Hamlet: A Note on Translatability', *Texas Studies in Literature and Language*, VIII (1966)

Helmut Gipper, *Sprachliche und geistige Metamorphosen bei Gedichtübersetzungen: eine sprachvergleichende Untersuchung zur Erhellung deutsch-französischer Geistesverschiedenheit* (Düsseldorf, 1966)

Paul Selver, *The Art of Translating Poetry* (London, 1966)

G. Steiner, Introduction to *The Penguin Book of Modern Verse Translation* (London, 1966)

1967

Donald Davie, 'The Translatability of Poetry', *The Listener*, LXXVIII (1967)

Rolf Kloepfer, *Die Theorie der literarischen Übersetzung. Romanisch-deutscher Sprachbereich** (Munich, 1967)

Maynard Mack, Introduction to *The Iliad of Homer* in *The Poems of Alexander Pope*, VII (London and Yale University Press, 1967)

W. Sdun, *Probleme und Theorien der Übersetzung in Deutschland vom 18. bis 20. Jahrhundert* (Munich, 1967)

1968

Charles J. Fillmore, 'Lexical Entries for Words', *Foundations of Language*, IV (1968)

H. P. Grice, 'Utterer's Meaning, Sentence-Meaning, and Word-Meaning', *Foundations of Language*, IV (1968)

Jiří Levý, 'Translation as a Decision Process', in *To Honor Roman Jakobson* (The Hague, 1968)

1969

Ward Allen, *Translating for King James. Notes Made by a Translator of King James's Bible* (Vanderbilt University Press, 1969)

K. Čukovski, *Vysokoye iskusstvo* (*The High Art*) (second, revised edition, Moscow, 1969)

Jiří Levý, *Die literarische Übersetzung. Theorie einer Kunstgattung** (Frankfurt am Main, 1969)

A. Ljudskanov, *Traduction humaine et traduction automatique* (Paris, 1969)

Eugene A. Nida and Charles R. Taber, *The Theory and Practice of Translation** (Leiden, 1969)

H. Orlinsky, *Notes on the New Translation of the Torah* (Philadelphia, 1969)

H. J. Störig (ed.), *Das Problem des Übersetzens* (Darmstadt, 1969)

Mario Wandruszka, *Sprachen vergleichbar und unvergleichbar* (Munich, 1969)

Ralph Rainer Wuthenow, *Das fremde Kunstwerk. Aspekte der literarischen Übersetzung* (Göttingen, 1969)

Wai-Lim Yip, *Pound's Cathay* (Princeton University Press, 1969)

J.-M. Zemb, *Les structures logiques de la proposition allemande* (Paris, 1969)

1970

Émile Benveniste, *Le Vocabulaire des institutions indo-européennes* (Paris, 1970)

C. Day-Lewis, *On Translating Poetry* (Abingdon-on-Thames, 1970)

J. S. Holmes (ed.), *The Nature of Translation* (The Hague, 1970)

1971

Karl-Richard Bauch and Hans-Martin Gauger (eds.), *Interlinguistica. Sprachvergleich und Übersetzung* (Tübingen, 1971)

Ernst Leisi, *Der Wortinhalt. Seine Struktur im Deutschen und Englischen* (fourth edition, revised), (Heidelberg, 1971)

Mario Praz, 'Shakespeare Translations in Italy' and 'Sul tradurre Shakespeare', in *Caleidoscopio shakespeariano* (Bari, 1971)

Annelise Senger, *Deutsche Übersetzungstheorie im 18. Jahrhundert 1734–1746** (Bonn, 1971)

1972

Velimir Chlebnikov, *Werke*, edited by P. Urban (Harmburg, 1972). Cf. particularly II, pp. 597–606

A. S. Dil (ed.), *The Ecology of Language: Essays by Einar Haugen* (Stanford University Press, 1972)

H. A. Mason, *To Homer Through Pope* (London, 1972)

Morris Swadesh, *The Origin and Diversification of Languages* (London, 1972)

1973

Robert M. Adams, *Proteus: His Lies, His Truth: Discussions of Literary Translation* (New York, 1973)

Henri Meschonnic, 'Poétique de la traduction', in *Pour la poétique II* (Paris, 1973)

A. C. Partridge, *English Biblical Translation* (London, 1973)

Jacqueline Risset, 'Joyce traduit par Joyce', *Tel Quel*, LV (1973)

F. D. Spark, *On Translations of the Bible* (London, 1973)

1974

C. Allen, *The Greek Chronicles. The Relation of the Septuagint of I and II Chronicles to the Massoretic Text*, part 1. The Translator's Craft (Supplementum to *Vetus Testamentum*) (Leiden, 1974)

M. E. Coindreau, *Mémoires d'un traducteur. Entretiens avec Christian Guidicelli* (Paris, 1974)

A. Fedorov, 'The Problem of Verse Translations', in *Linguistics*, CXXXVII (1974)

La Traduction en jeu (Paris, 1974)

On Language, Culture and Religion: In Honour of E. A. Nida (The Hague and Paris, 1974)

Translators and Translating. Selected Essays of the American translators association. Summer Workshops, 1974 (Binghamton, NY, 1974)

1975

L. S. Baruxadov, *Jazyk i perevod. Voprosy obščej i častnoj teorii perevoda* (*Language and Translation. Problems of general and specific theory of translation*) (Moscow, 1975)

H. Broch, 'Einige Bemerkungen zur Philosophie und Technik des Übersetzens', in *Schriften zur Literaturtheorie*, II (Baden-Baden, 1975)

R. B. Harrison, *Hölderlin and Greek Literature* (Oxford, 1975)

G. Jäger, *Translation und Translationslinguistik* (Halle (Saale), 1975)

A. Lefevere, *Translating Poetry: Seven Strategies and a Blueprint* (Assen and Amsterdam, 1975)

A. Neubert and R. Růžička, *Verständlichkeit, Verstehbarkeit, Übersetzbarkeit. Sprachwissenschaft und Wissenschaftssprache* (East Berlin, 1975)

Modern Language Notes, VI (1975)

E. A. Nida, *Language Structure and Translation. Essays* (with bibliography of E. A. Nida's works) (Stanford University Press, 1975)

A. Popovič, *Teória umeleckého prekladu. Aspekty textu a literárnej metakomunikácie (The Theory of Artistic Translation. Aspects of the Text and of Literary Metacommunication)* (Bratislava, 1975)

A. Popovič, *Dictionary for the Analysis of Literary Translation* (University of Alberta Press, 1975)

T. R. Steiner, *English Translation Theory, 1650–1800* (Assen and Amsterdam, 1975)

1976

R. W. Brislin, *Translation* (New York, 1976)

L. A. Černjaxovskaja, *Perevod i smyslovaja struktura (Translation and Meaning-Structure)* (Moscow, 1976)

A. R. Chakraborty, *Translational Linguistics of Ancient India* (Calcutta, 1976)

H. W. Drescher and S. Scheffzek (eds.), *Theorie und Praxis des Übersetzens und Dolmetschens. Referate und Diskussionsbeiträge des internationalen Kolloquiums am Fachbereich Angwandte Sprachwissenschaft der Johannes Gutenberg-Universität Mainz in Germersheim (2.–4. Mai 1975)* (Frankfurt am Main, 1976)

G. Mounin, *Les Problèmes théoriques de la traduction* (Paris, 1976)

Problemas de la traducción (Sur, rev. semestrial, enero-dic.) (Buenos Aires, 1976)

N. Rudd, *Lines of Enquiry: Studies in Latin Poetry* (Cambridge University Press, 1976)

G. Steiner, 'Aspects du langage et de la traduction. Entretien avec Jacques De Decker', in *Cahiers internationaux de Symbolisme*, XXXI (1976)

P. Valesio, 'The Virtues of traducement. Sketch of a Theory of Translation', in *Semiotica*, I (1976)

R. W. Brislin (ed.), *Translation: Applications and Research* (New York, 1976)

J. Vincent, 'On Translation: A First Approximation', in *Annali anglistica* (1976)

T. Webb, *The Violet in the Crucible. Shelley and Translation* (Oxford, 1976)

1977

A. K. France, *Boris Pasternak's Translations of Shakespeare* (University of California Press, 1977)

J. Grayson, *Nabokov Translated: A Comparison of Nabokov's Russian and English Prose* (Oxford University Press, 1977)

A. Lefevere, *Translating Literature. The German Tradition from Luther to Rosenzweig* (Assen and Amsterdam, 1977)

M. G. Rouse (ed.), *Translation in the Humanities* (Binghamton, NY, 1977)

G. Vázquez-Ayora, *Introducción a la traductología: curso básico de traducción* (Georgetown University Press, 1977)

W. Wilss, *Übersetzungswissenschaft. Probleme und Methoden* (Stuttgart, 1977)

1978

R. de Beaugrande, *Factors in a Theory of Translating* (Assen and Amsterdam, 1978)

J. Belitt, *Adam's Dream: A Preface to Translation* (New York, 1978)

Colloque sur la traduction poétique. Sorbonne. Nouvelle Paris III les 8–10 déc. 1972 (Paris, 1978)

D. Constantine, 'Hölderlin's Pindar: The Language of Translation', in *Modern Language Review* (1978)

H. Diller and J. Kornelius, *Linguistische Probleme der Übersetzung* (Tübingen, 1978)

F. Guenther and M. Guenther-Reutter (eds.), *Meaning and Translation: Philosophical and Linguistic Approaches* (London, 1978)

H. Meschonnic, 'Traduction restreinte, traduction généralisée', in *Pour la poétique* (Paris, 1978)

K. Reiss, *Möglichkeiten und Grenzen der Übersetzungkritik: Kategorien und Kriterien für eine sachgerechte Beurteilung von Übersetzungen* (Munich, 1978)

A. Schroeter, *Geschichte der deutschen Homer-Übersetzung im 18. Jahrhundert* (Hildesheim, 1978)

Theory and Practice of Translation. Nobel Symposium 39. Stockholm, Sept. 6–10, 1976 (Berne, 1978)

A. F. Tytler (Lord Woodhouselee), *Essay on the Principles of Translation* (new edition, Amsterdam, 1978)

M. Yassin and F. Aziz, 'Translation between Two Models: The Whorfian Hypothesis and the Chomskyan Paradigm', in *Incorporated Linguist*, XVII (1978)

1979

M. Levin, 'Forcing and the Indeterminacy of Translation', in *Erkenntnis*, XIV (1979)

L. G. Kelly, *The True Interpreter. A History of Translation Theory and Practice in the West* (Oxford, 1979)

J.-R. Ladmiral, *Traduire: Théorèmes pour la traduction* (Paris, 1979)

J.-C. Margot, *Traduire sans trahir. La théorie de la traduction et son application aux textes bibliques* (Lausanne, 1979)

B. M. Snell, *Translating and the Computer. Proceedings of a Seminar, London, 14th November, 1978* (Amsterdam, New York, 1979)

D. West and T. Woodman (eds.), *Creative Imitation and Latin Literature* (Cambridge University Press, 1979)

1980

S. Bassnett-McGuire, *Translation Studies* (London, 1980)

A. Bonino, *Il traduttore. Fondamenti per una scienza della traduzione.* Vol. I (Turin, 1980)

N. Hofmann, *Redundanz und Äquivalenz in der Literarischen Übersetzung, dargestellt an fünf deutschen Übersetzungen des Hamlet* (Tübingen, 1980)

V. N. Komisarov, *Lingvistika perevoda (Linguistics of Translation)* (Moscow, 1980)

Literary Communication and Reception, Communication littéraire et reception, Literarische

Kommunikation und Rezeption. Proceedings of the IXth Congress of the International Comparative Literature Association, Innsbruck 1979 (Innsbruck, 1980)

A. Newman, *Mapping Translation Equivalence* (Leuven, 1980)

S. O. Poulsen and W. Wilss, *Angewandte Übersetzungswissenschaft. Internationales Übersetzungswissenschaftliches Kolloquium an der Wirtschaftsuniversität Århus/ Dänemark, 19.–21. Juni 1980*

D. Stein, *Theoretische Grundlagen der Übersetzungswissenschaft* (Tübingen, 1980)

'Theory Serving Practice/La Théorie au service de la pratique, Colloque, Collège Glendon, York University', in *Meta*, XXV. 4 (1980)

G. Toury, *In Search of A Theory of Translation* (Tel Aviv, 1980)

W. Wilss, *Semiotik und Übersetzen* (Tübingen, 1980)

O. Zuber, *The Languages of Theatre—Problems in the Translation and Transposition of Drama* (Oxford, 1980)

1981

J. House, *A Model for Translation Quality Assessment* (Tübingen, 1981)

G. M. Hyde, *D. H. Lawrence and the Art of Translation* (London, 1981)

Masterstvo perevoda 1979 (The Art of Translation 1979) (Moscow, 1981)

G. T. Kühlwein and W. Wilss (eds.), *Kontrastive Linguistik und Übersetzungs-wissenschaft. Akten des Internationalen Kolloquiums Trier/Saarbrücken, 25.–30. 9. 1979* (Munich, 1981)

P. Newmark, *Approaches to Translation* (Oxford and New York, 1981)

K. Reiss, 'Type, Kind and Individuality of Text. Decision-Making in Translation', in *Poetics Today: Theory of Analysis of Literature and Communication*, II. 4 (1981)

M. G. Rose (ed.), *Translation Spectrum. Essays in Theory and Practice* (State University of New York Press, 1981)

P. Rónai, *A tradução vívida* (Rio de Janeiro, 1981)

'Translation in the Renaissance/La Traduction à la Renaissance', in *Canadian Review of Comparative Literature*, VIII. 2 (1981)

1982

F. Apel, *Sprachbewegung: eine historisch-poetologische Untersuchung zum Problem des Übersetzens* (Heidelberg, 1982)

E. Etkind, *Un art en crise. essai de poétique de la traduction poétique* (Lausanne, 1982)

W. Frawley (ed.), *Translation: Literary, Linguistic and Philosophical Approaches* (University of Delaware Press, 1982)

V. García Yerba, *Teoría y práctica de la traducción* (Madrid, 1982)

D. Lehmann, *Arbeitsbibliographie des Übersetzens: interdisziplinäre Aspekte der Sprach-und Übersetzungswissenschaft sowie der Übersetzungspraxis* (Trier, 1982)

A. L. Willson, *Übersetzen als Hochstaplerei* (Wiesbaden, 1982)

1983

G. I. Anzilotti, *Four English/Italian Stories: Experiments in Translation* (Lake Bluff, Ill., 1983)

R. Opioli (ed.), *Tradurre poesia* (Brescia, 1983)

N. Briamonte, 'Tradizione, traduzione, traducibilità', in *Paragone*, XXXIV (1983)

V. García Yerba, *En torno a la traducción: teoría, crítica, historia* (Madrid, 1983)

W. Koller, *Einführung in die Übersetzungswissenschaft* (Heidelberg, 1983)

C. Picken (ed.), *The Translator's Handbook* (London, 1983)

B. Terracini and B. Mortara Garavelli (eds.), *Il problema della traduzione* (Milan, 1983)

H. J. Vermeer, *Aufsätze zur Translationstheorie* (Heidelberg, 1983)

1984

R. Apter, *Digging for the Treasure. Translation After Pound* (Berne, 1984)

A. Berman, *L'Épreuve de l'etranger. Culture et traduction dans l'Allemagne romantique* (Paris, 1983)

N. Briamonte, *Saggio di Bibliografia sui problemi storici, teorici e pratici della traduzione* (Naples, 1984)

D. Davidson, *Inquiries into Truth and Interpretation* (Oxford University Press, 1984)

J. Holz-Mänttäri, *Translatorisches Handeln. Theorie und Methode* (Helsinki, 1984)

H. G. Hönig and P. Küssmaul, *Strategie der Übersetzung* (Tübingen, 1984)

Premières assises de la traduction littéraire (Arles 1984)

G. P. Norton, *The Ideology and Language of Translation in Renaissance France and Their Humanist Antecedents* (Geneva, 1984)

K. Reiss and H. J. Vermeer, *Grundlegung einer allgemeinen Translationstheorie* (Tübingen, 1984)

J. Stackelberg, *Übersetzungen aus zweiter Hand: Rezeptionsvorgänge in der europäischen Literatur vom 14. bis zum 18. Jahrhundert* (Berlin and New York, 1984)

G. Steiner *Antigones* (Oxford, 1984)

1985

H. Aris, 'De Bagdad à Tolède: le rôle des traducteurs dans la transmission des patrimoines culturels grec et arabe à l'Occident' (Master's thesis, University of Ottawa)

J. F. Graham (ed.), *Difference in Translation* (Cornell University Press, 1985)

H. Grassegger, *Sprachspiel und Übersetzung: eine Studie anhand der Comics-Serie 'Asterix'* (Tübingen, 1985)

Th. Hermans (ed.), *The Manipulation of Literature. Studies in Literary Translation* (London and Sydney, 1985)

E. Honig, *The Poet's Other Voice. Conversations on Literary Translation* (University of Massachusetts Press, 1985)

I. D. Levin, *Russkie perevodčiki XIX veka i razvitie xudožestvennogo perevoda* (*Russian Translators of the 19th Century and the Evolution of Artistic Translation*) (Leningrad, 1985)

A. Neubert and G. Jäger, *Text and Translation* (Leipzig, 1985)

1986

E. Cary, *Comment faut-il traduire? Introduction, bibliographie et index de Michel Ballard* (Presses Universitaires de Lille, 1986)

J. Felstiner, 'Mother Tongue: Holy Tongue: On Translating and Not Translating Paul Celan', *Comparative Literature*, XXXVIII. 2 (1986)

Y. Gambier, *Trans* (Turun Yliopisto, Kaantajankoulutus Laitos, 1986)

R. Kirk, *Translation Determined* (Oxford University Press, 1986)

F. Paepke, *Im Übersetzen Leben: Übersetzen und Textvergleich* (Tübingen, 1986)

D. Seleskovitch and M. Lederer, *Interpréter pour traduire* (Publications de la Sorbonne, 1986)

1987

J. Albrecht, H. W. Drescher, H. Göhring, and H. Salnikow (eds.), *Translation und interkulturelle Kommunikation* (Berne, 1987)

Bibliographie du traducteur/Translator's Bibliography (University of Ottawa Press, 1987)

I. D. K. Kelly and D. J. Wigg, *Computer Translation of Natural languages* (Wilmslow, 1987)

S. Nirenburg, *Machine Translation. Theoretical and methodological Issues* (Cambridge University Press, 1987)

W. Radice and B. Reynolds (eds.), *The Translator's Art. Essays in honour of Betty Radice* (Harmondsworth, 1987)

R. V. Schoder, *The Art and Challenge of Translation* (Oak Park, Ill., 1987)

B. Schultze (ed.), *Die literarische Übersetzung. Fallstudien zu ihrer Kulturgeschichte* (Göttingen, 1987)

1988

J. A. Holmes, *Translated! Papers on Literary Translation and Translation Studies* (Amsterdam, 1988)

B. Hollander (ed.), *Translation Tradition: Paul Celan in France* (*Acts*, VIII/IX) (1988)

H. Kittel (ed.), *Die literarische Übersetzung. Stand und Perspektiven ihrer Erforschung* (Göttingen, 1988)

P. Newmark, *A Textbook of Translation* (New York and London, 1988)

C. Nord, *Textanalyse und Übersetzen. Theoretische Grundlagen, Methode und didaktische Anwendung einer übersetzungsrelevanten Textanalyse* (Heidelberg, 1988)

B. Raffel, *The Art of Translating Poetry* (Pennsylvania State University Press, 1988)

R. M. Rosini, *Questioni traduttive* (Udine, 1988)

M. Shell-Hornby, *Translation Studies: An Integrated Interpretation* (Amsterdam, 1988)

H. Schot, *Linguistics, Literary Analysis, and Literary Translation* (University of Toronto Press, 1988)

J. Slocum (ed.), *Machine Translation Systems* (Cambridge University Press, 1988)

W. Wilss, *Kognition und Übersetzen: zu Theorie und Praxis der menschlichen und der maschinellen Übersetzung* (Tübingen, 1988)

1989

A. Benjamin, *Translation and the Nature of Philosophy* (London and New York, 1989)

J. Biguenet and R. Schulte (ed.), *The Craft of Translation* (University of Chicago Press, 1989)

F. Buffoni (ed.), *La traduzione del testo poetico* (Milan, 1989)

R. Ellis (ed.), *The Medieval Translator. The Theory and Practice of Translation in the Middle Ages* (Cambridge University Press, 1989)

R. Larose, *Théories contemporaines de la traduction* (Presses de l'Université de Québec, 1989)

R. Warren (ed.), *The Art of Translation: Voices from the Field* (Northeastern University Press, 1989)

1990

E. Barilier, *Les Belles fidèles: petit essai sur la traduction* (Lausanne, 1990)

S. Bassnett and A. Lefevere (eds.), *Translation, History, and Culture* (London, 1990)

Masterstvo perevoda 1985 (*The Art of Translation 1985*) (Moscow, 1990)

S. Olofsson, *God is my Rock: A Study of Translation Technique and Theological Exegesis in the Septuagint* (Stockholm, 1990)

1991

E. Cheyfitz, *The Poetics of Imperialism: Translation and Colonization from The Tempest to Tarzan* (New York, 1991)

R. Copeland, *Rhethoric, Hermeneutics, and Translation in the Middle Ages* (Cambridge University Press, 1991)

G. Folena, *Volgarizzare e tradurre* (Milan, 1991)

E. A. Gutt, *Translation and Relevance. Cognition and Context* (London, 1991)

J. Sailhamer, *The Translational Technique of the Greek Septuagint for the Hebrew Verbs and Participles in Psalms 3–41* (New York, 1991)

The student of translation will also want to familiarize himself with the publications of the proceedings of the International Federation of Translators (FIT) founded in Paris in 1953. Notable among these are E. Cary and R. W. Jumpelt (eds.), *Quality in Translation* (Oxford, London, New York, Paris, 1963), and I. J. Citroën (ed.), *Ten Years of Translation* (Oxford, London, New York, Paris, 1967). First issued in Paris in 1932, and taken over by UNESCO in 1947, the annual *Index Translationum* is an indispensable guide to trends and areas of concentration in world translation. The *Yearbook of Comparative and General Literature* (1952–) contains an annual review of work in and about translation. Particular emphasis is placed on works in the theory of translation which are not listed in the Bibliography of *General and Comparative Literature*.

The number of journals in the field is increasing. Several are concerned almost exclusively with professional and technical aspects of the art. These include *Traducteur* (Montreal, 1939–), *Babel** (1955–), the *Journal des traducteurs*, later *Meta* (Montreal, 1956–), and *Der Übersetzer* (Neckarrems, 1964–). Important statistical information can be found in the *Translation Monthly* first issued in 1955 by the University of Chicago, and then taken over by the Department of Commerce, Washington, D.C. Since 1954, *Mechanical Translation* (Cambridge, Mass.) has been the senior journal in a rapidly expanding discipline. See also *La Traduction automatique* (The Hague, 1960–). Numerous important papers on the theory and practice of translation, though with an obvious special focus, have appeared in *The Bible Translator* (London, 1949–).

Journals in the general field of linguistics and comparative philology often include articles on translation. This is true notably of the *Revue des langues vivantes* (Brussels, 1932–), *Die Sprache* (Vienna, 1949–), *Sprachforum* (Münster, Cologne, 1955–), *Langues et styles* (Paris, 1959–), *Language Research* (Washington, D.C., 1965–), *Language Sciences* (Bloomington, Indiana, 1968–), *Sprachkunst* (Vienna, 1970–). In 1967, the journal *Sprache im technischen Zeitalter* (Berlin) produced two special numbers on translation (21, 24).

Nine (Venice, 1949–), *Stand* (London, 1952–), *Agenda* (London, 1959–) and *L'Éphémère* (Paris, 1967–72) have been among the literary and 'little' magazines most active in the field of poetic translation. Appearing since 1965, *Modern Poetry in Translation* (London) has been devoted entirely to the publication of foreign verse in English translations. Appearing from 1968 to 1971, the six issues of *Delos* (University of Texas at Austin) constitute the most distinguished and influential effort so far to create a journal concerned exclusively with the theory, history, and art of translation. A very interesting new journal entirely dedicated to the theory and practice of translation is *Testo a fronte*, published by Guerini in Milan. Although Soviet scholars had always been very prolific in the study of the different aspects of the art of translation, it seems that at the moment there is no special Soviet publication specifically dealing with that discipline. Students who are interested in Soviet works on translation may consult the journals *Inostrannaja literatura* (Foreign Literature) and *Družba narodov* (Fraternity of Nations), where short reviews and articles on translation sometimes appear, or, in a more academic key, the Tartu journal *Semiotike*.

INDEX

Aarsleff, Hans, 177n., 209n.

Achilles, 22, 230, 370, 417–19, 421, 422

Act without Words (S. Beckett), 194

Ada (V. Nabokov), 127

Adam, 60, 61, 62, 64, 129, 185, 213, 324, 499

Adamov, Arthur, 22

Addison, Joseph, 11

'Adonais', 470–2 *passim*

Adorno, T. W., 168, 240

Advancement of Learning, The (F. Bacon), 208

Aeneid, The (Virgil), 269

Aeschylus, 22, 157, 329–32 *passim*, 384, 477, 478, 479

Aesop, 369

Africa (Petrarch), 138

Agamemnon, 157, 331, 461

Agamemnon (Aeschylus trans. R. Browning), 329

Agamemnon (Seneca), 477

Âge d'homme, L' (M. Leiris), 207

Agrippa von Nettesheim, Henry Cornelius, 64

Ajax, 348

Akin, J. *et al.*, 302n.

Alajouanine, T., 298n.

Alazraki, Jaime, 70n.

Alcaeus, 461

Alcestis, 44

Alfieri, Vittorio, 477, 478

Alice in Wonderland (Lewis Carroll), 37

Allem, Maurice, 483n.

Allemann, Beda, 340n.

Allen, Ward, 288n., 366n.

Allgemeine Naturgeschichte und Theorie des Himmels (I. Kant), 160–1

Aminta (Tasso), 466

Ammerman, R. R., 217n.

Amores (Ovid), 423

Amos, 154

Amphitryon, 478 & n.

Amphitryon (F. Perez de Oliva), 478

Amphitryon 38 (J. Giraudoux), 478

Amyot, Jacques, 248, 260, 265, 288 & n., 354, 375

Anacreon, 460

Anderson, Maxwell, 479

Anderson, Neils, 126n.

Anderson, Robin, vii

Andromaque (J. Racine), 46

'Angelus Silesius' (Johann Scheffler), 64, 65

Anglo-Saxon Dictionary (J. Bosworth), 25

Anna Karenina (Leo Tolstoy), 480

Anouilh, Jean, 478, 479

Anscombe, G. E. M., 140n.

Anthropological Linguistics, 250

Anthropologie structurale, L' (C. Lévi-Strauss), 319

Antigone, 23, 42, 344n., 346, 348, 349, 478

Antigone (Sophocles trans. J. C. F. Hölderlin), 67, 344 & n., 345, 346, 347, 349, 350

Antony (Mark Antony, the Triumvir), 387, 389

Antony and Cleopatra, 383, 387, 388, 389

Apel, Karl-Otto *et al.*, 256n.

Apollinaire, Guillaume, 318

Apostles, The: descent of gift of tongues on, 61, 65

Aquinas, Thomas, 148, 150 & n.

Arabian Nights, The, 379

Arbogast, H., 200n.

Arcadia (Sir P. Sidney), 479

Archilochus, 269

Ardener, Edwin, 126n.

d'Argental, Comte, 384

Argot ancien, L' (P. Champion), 25

Aricie, 456

Ariosto, Lodovico, 7, 272
Aristophanes, 263
Aristophanes' Apology (R. Browning), 329
Aristotle, xv, 148, 149 & n., 267, 284, 287, 319, 324; influence of, 79, 85, 259, 260, 267, 463, 487
Aristotelian Society, Proceedings of the, 221
Armorial Families (A. C. Fox-Davies), 25
Arndt, Hans Werner, 211n.
Arnold, Matthew, xv, 249, 362, 418, 435, 472, 473, 474
Arp, Hans, 202, 203 & n.
Arrowsmith, William, vii; and Roger Shattuck, 250, 286, 287, 289
Ars Magna (R. Lully), 209
Ars poetica (Horace), 248, 268n., 327, 328
Ars Signorum, vulgo Character Universalis et Lingua Philosophica (G. Dalgarno), 78, 210
Artaud, Antonin, 31
de Arte Combinatoria (G. W. Leibniz), 210
Artmann, Hans Carl, 370
Asclepiades, 465
Ashton, E. B., 255n.
Aspects of the Theory of Syntax (N. Chomsky), 99n., 106n., 111
Astrophel (E. Spenser), 469
As You Like It, 423
Atemwende (P. Celan), 167
Atreus, king of Mycenae, 477
Auberique, P., 109n.
Auden, W. H., 375, 473–5
Auerbach, F., 161n., 448
Aufgabe des Uebersetzers, Die (W. Benjamin), 66 & n.
Augustine, St. (Aurelius Augustinus), 148, 149, 150, 153, 159n., 228, 229, 230, 314
Austen, Jane, 8–11 *passim*, 26
Austin, J. L., 141 & n., 143, 216, 217, 218, 219, 222, 224 & n., 225, 228
Autret, Jean, 335n.
Auvray, Lucien, 355n.
Aveling, Eleanor Marx, 395–7

Awkward Age, The (H. James), 37
Ayer, A. J., 169n., 172, 217 & n., 219n.
Ayers, M. R., 225n.

Bach, Emmon, 106 & n.; and R. T. Harms, 106n.
Bacon, Francis, 208, 284
Bacon, Roger, 77, 98
Ball, Hugo, 201n., 202–4 *passim*
Balzac, Guez de, 20
Balzac, Honoré de, 386, 482, 483 & n., 484
Bar-Hillel, Yehoshua, 117n., 326n.
Barker, S. F. and P. Achinstein, 164n.
Basilides, 72
Bates, E. S., 289n.
Baudelaire, Charles Pierre, 287, 386, 422, 446
Baudissin, Wolf Heinrich, Count, 401
Baumfield, B., 491n.
Bausch, Karl-Richard and Hans-Martin Gauger, 447n.
Beardsley, Aubrey, 15
Beattie, James, 99
Beaufret, Jean, 340n.
Becher, J. J., 210
Beckett, Samuel, 48, 194, 289, 498
Beethoven, Ludwig van, 28, 163, 479, 489
Befristeten, Die (E. Canetti), 146
Beissner, Friedrich, 340n.
Bender, M. Lionel, 20n.
Benjamain, Walter, ix, 22, 66–7, 68, 249, 250, 257, 262, 273, 283, 290n., 313, 324, 332, 339, 340n., 344n.
Benn, M. B., 340n., 343n.
Benveniste, Emile, 109n., 137
Beowulf, 71, 319
Bérénice (J. Racine), 46, 391
Bergson, Henri, 148, 151, 152
Berlioz, Hector, 384, 438, 440, 442, 443, 445
Bernstein, Basil, 34n.
Bertaux, Pierre, 340n.
Berthelot, René, 446n.
Bethge, Hans, 379
Bever, T. G. and W. Weksel, 130n.
Bévotte, Gendarme de, 479n.

Beyond the Pleasure Principle (S. Freud), 167

Bible, The: Anchor Bible, 422; Authorized Version (King James VI), 13, 259, 288, 322, 323, 335, 364, 366 & n., 367, 375, 422, 440; Bishops' Bible, 366; code of history, 165; German Bible, 323; Gospels, The, 143, 257, 366; Luther Bible, 259, 272, 280, 315; New English Bible (N.E.B.), 322; New Testament, 259, 363n.; Old Testament, 23, 153, 155, 257; translation of, 366 & n.; Vulgate, 322, 334; *see also* Coverdale, Matthew, Tyndale, Whitchurch

Bible of Amiens, The (J. Ruskin), 335

Billington, James, 159n.

Binswanger, L., 205n.

Biological Foundations of Language, The (E. H. Lenneberg), 295

Bisterfeld, J. H., 210

Black, M., 140n.

Black, Max, 217n., 220n., 223 & n.

Blackmur, R. P., 484

Blake, William, 39, 72, 79

Bleak House (C. Dickens), 26

Bloch, Ernst, 168, 219, 227, 228

Bloch, Marc, 143, 144n.

Bloom, Harold, 471 & n.

Bloomfield, Leonard, 19, 128

Boas, Franz, 89

Boccaccio, Giovanni, 413

Böhme, Jakob, 65 & n., 71

Bois, John, 366n.

Boisjermain, Luneau de, 334

Bolgar, R. R., 448

Boltzmann, Ludwig, 162 & n.

Boman, Thorlief, 165n.

Bonner, Anthony, 74 & n., 76

Boole, George, 70, 216

Borchardt, Rudolf, 356–9 *passim*, 370

Borges, Jorge Luis, xx, 70–3 *passim*, 74, 352, 356

Borst, Arno, 59n., 62n.

Bosch, Hieronymus, 28

Bossuet, Jacques Benigne, 385

Bowra, C. M., 373n.

Bradley, F. H., 220

Brahmin linguistic mythology, 62

Braque, Georges, 490

Brecht, Bertolt, 359, 378, 478, 479

Brillouin, Léon, 163n.

Brink, C. O., 268n.

Broad, C. D., 134, 148

Broca, Paul, 131, 297

Broca's area of the brain, 297

Broch, Hermann, 336–8

Brod, Max, 68

Bronowski, J. and Ursula Bellugi, 130n.

Brooke, Tucker, 403

Brothers Karamazov, The (F. Dostoevsky), 37

Brotheryon, B., 232n.

Brower, Rueben A., 250, 274n., 326n., 329n.

Brown, R. L., 89n.

Brown, R. W., 108

Browne, Sir Thomas, 319, 489

Browning, Robert, 15, 190, 329–32 *passim*

Brunel, P., 386n.

Bruner, Jerome S., 134, 223n.

Brunhes, B., 161n.

Bruni, Leonardo, 248, 276, 311

Bruno, Giordano, 261

Buber, Martin, 154 & n., 155n.

Büchner, Georg, 13, 120

Buffon, George Louis Leclerc, Comte de, 160

Bull, William E., 139n.

Bültmann, Rudolf, 143

Bunting, Basil, 367, 461

Burgess, Anthony, 289

Burke, Edmund, 286

Burke, Kenneth, x, 35, 80

Burling, Robbins, 34n.

Butor, Michel, 289

Bynner, Witter, 376

Byron, George Gordon, 6th Baron, 328

Byronism, 285

Byzantine rhetoric and theology, 230

Caesar, Gaius Julius, 259, 461

Cain, mark of, 64

Calderon de la Barca, Pedro, 270, 272

Cammaerts, Emile, 196n.

Camões, Luis de, 478
Campbell, Roy, 287, 327
Canetti, Elias, 146, 147, 163
Cantos (E. Pound), 386, 490
Carew, Thomas, 468–9, 470, 475
Carlyle, Thomas, 262
Carnap, R., 106, 141, 212, 216–20
  passim
Carnot, N. L. Sadi, 160–3 *passim*
Carnot's Theory, An Account of
  (W. Thomson), 161–2
Carroll, John B., 93n.
'Carroll, Lewis' (Charles Lutwidge
  Dodgson), 37, 196, 197 & n., 222
Cartesianism, 79, 81, 91, 96, 119, 215,
  228, 229, 255, 298, 372, 385, 390
Casaubon, Isaac, 278
Cassandra, 44
Cassirer, Ernst, 89, 97
Cassou, Jean, 447n.
Cathay (E. Pound), 375, 376, 377n.,
  378, 412
Catullus, 24, 370, 371n.
Cavalcanti, Guido, 353, 354 & n., 370
Caxton, William, 3
Celan, Paul, 167, 191–2, 197, 318,
  370, 400, 409–11, 412, 425–6,
  426n.
'Céline, Louis-Ferdinand'
  (L.-F. Destouches), 385
Cellini, Benvenuto, 270
Cervantes (Saavedra), Miguel de, 72,
  74–5, 272, 285
Cézanne, Paul, 490
Chamisso, Adelbert von, 440
Champion, P., 25
Chanson du vieux marin, La (V.
  Larbaud), 365
Chapman, George, 248, 260, 284,
  418–19
Chappell, V. C., 171n.
Char, René, 385
Characteristica universalis (G. W.
  Leibniz), 73
Character, pro notitia linguarum universali
  (J. J. Becher), 210
Chardin, Jean Simeon, 484
Charterhouse of Parma, The (Stendhal),
  47

Chassaigne, Marc, 277n.
Chateaubriand, François René,
  Vicomte de, 139, 333–5 *passim*
Chaucer, Geoffrey, 4, 5, 29, 269
Cheke, Sir John, 278
Chesterton, G. K., 288, 429
chinesische Flöte, Die (H. Bethge), 379
Choerilus of Athens, 329
Chomsky, Noam, vii, xiv, 62, 77, 99n.,
  103–14 *passim*, 116, 119, 177 & n.,
  178n., 219, 245, 302–4, 488, 489,
  496 & n.
Christian: church in West, 258;
  culture, 45, 467–8; doctrines, 165,
  467 & n., 472; early churches, 149,
  157; Jews, 158; view, 467–8
Christianity, 258; early, 230
Cicero, Marcus Tullius, 157, 248, 251,
  272, 278, 292, 324
Cimetière marin, Le (P. Valéry), 73
Cinderella fable, 447
Clapeyron, B. P. E., 161
Claudel, Paul, 324, 385, 477, 479
Claudius, Matthias, 283
Clausius, Rudolf Julius Emmanuel,
  161, 162
Clough, Arthur Hugh, 18, 472, 473
Cocteau, Jean, 478
Cohen, Jonathan, 208n., 211 & n.
Cohn, Albert, 400n.
Coleridge, Samuel Taylor, xv, 26, 64,
  72, 83, 85, 364
Collectanea etymologica, 211
Collins French Phrase Book, 319
Comédie humaine (H. de Balzac), 483
Comenius, Johann Amos, 209 & n.
Comte, Auguste, 266
Confessions (St. Augustine), 149, 153
Congreve, William, 16
Conrad, Joseph, 335, 477
Convivio (Dante), 253
Cook, Raymond, 204n.
Cooper, James Fenimore, 285
Copernicus, Nicolaus, 60, 218, 285
Corinthians, Epistles to the, 251
Coriolanus (W. Shakespeare), 5, 44, 187
Corneille, Pierre, 384, 478
Cornford, F. M., 156n., 363 & n.
Corsen, Meta, 340n.

Courier, Paul-Louis, 353
Couteras, Helen and Sol Saporta, 302n.
Couturat, L., 211n.; and L. Leau, 208n.
Coverdale, Miles, 366
Coward, Noël, 16–17
Cowley, Abraham, 248, 267–8, 343, 461, 466, 475
Cowper, William, 420
Craft and Context of Translation: A Critical Symposium, The (W. Arrowsmith and R. Shattuck), 250, 253n.
Cratylus (Plato), 255
Creation, The (F. J. Haydn), 322
Cresswell, M. J., 497n.
Cribb, T., 254
Critique of Linguistic Philosophy, A (C. W. K. Mundle), 173
Croce, Benedetto, 140, 249, 256 & n., 264
Crow (T. Hughes), 370
cummings, e. e., 236
Curtius, Ernst Robert, 358, 448
Cusanus, Nicholas, 65, 84
Cymbeline, 1–8 *passim*, 11, 27–8
Cyrano de Bergerac, Savinien, 129

'Dada' movement, 24, 201–5 *passim*, 485
Daiches, David, 366 & n.
Dalgarno, George, 78, 210
Dali, Salvador, 28
Damnation of Faust, The (H. Berlioz), 443, 444
Daniel, Samuel, 124, 261
Dante Alighieri, 13, 49, 168, 185, 186n., 253, 339, 353–9 *passim*
Dante and His Circle (D. G. Rossetti), 353
Dante Deutsch (R. Borchardt), 357 & n., 358, 359
Da Ponte, Lorenzo, 479
Darwin, Charles, xiii, 56–7, 372
David Copperfield (C. Dickens), 447
Davie, Donald, 375
Death of Virgil, The (H. Broch), 336–8

Deborah, Song of, 24
Debussy, Claude, 438, 490
Decadent movement, 15, 202
De Cecco, John P., 103n.
Decembrio, Pierro Candido, 259
Declaration of the Word As Such (A. Kručenyx), 194
De compositione verborum (Dionysius of Halicarnassus), 347
De Divinatione (Cicero), 157
De Fato (Cicero), 157
Defaucompret, Auguste-Jean-Baptiste, 285
Défence et illustration de la langue française (J. Du Bellay), 253
Defoe, Daniel, 18
De Interpretatione (Aristotle), 148
De interpretatione (P.-D. Huet), 276, 279
De interpretatione recta (L. Bruni), 248
Delafosse, M., 33n.
De L'Allemagne (Mme de Staël), 82
Delille, Jacques, 272
Delos, A Journal on and of Translation, 289
Delphi, oracle at, 156
De mendacio (St. Augustine), 228
Denison, N., 126n.
De optimo genere interpretandi (P.-D. Huet), 248
De Quincey, Thomas, 72
Deregowski, Jan B., 223n.
Derrida, Jacques, 110n., 381 & n.
Descartes, René, 73, 79, 103, 209, 255, 284, 432
Description of the Empire of China, A (J. B. Du Halde), 376
Deuteronomy, 153
Dewitz, Hans-Georg, 357n.
Dialects: Alemannic, 357; Alpine, 357; Atakama, 56; Gueno, 56; Khalka, 32; Kung, 100; Puelče, 56
Dialogues (Plato), 362
Dichtung (M. Heidegger), 205
Dickens, Charles, 26, 285
Dickinson, Emily, 368
Dictionary (S. Johnson), 254
Dictionary, A Middle English (Kurath and Kuhn), 25

Dictionary of Early English, A (Shipley), 25

Dictionary of Gardening, A (Royal Horticultural Society), 25

Dictionary of Naval Equivalents (Admiralty), 26

Diderot, Denis, 36, 253, 270, 401

Diebold, A. R., 126n.

Dilthey, Wilhelm, 262

Diodorus Chronos, 149

Dionysius of Halicarnassus, 347

Divina Commedia (Dante), 339, 354, 355, 356

'*Divan*' (*Westöstlicher Divan*) (J. W. von Goethe), 66

Dobschütz, Ernst von, 159n.

Doctor Faustus (C. Marlowe), 4

Dodd, C. H., 159 & n.

Dodds, E. R., 156 & n., 256n., 282 & n.

Dohl, Reinhard, 201n., 202n.

Dolan, John M., 310n.

Dolce, Ludovico, 478

Dolet, Étienne, 276, 277 & n., 280

Doll's House, The (H. J. Ibsen), 395

Dombey and Son (C. Dickens), 26

Don Carlos (J. C. Schiller), 83

Don Giovanni (W. A. Mozart), 46, 47

Don Juan (J. Molière), 388

'Don Juan' legend, 479 & n.

Donne, John, 15, 191, 424, 461–4, 468, 469, 476–7, 484

Don Quixote (M. de Cervantes), 73–5, 285

Dostoevsky, F., 37, 147, 285, 447

Dowson, Ernest, 15

Drapers' Dictionary, The (S. William Beck), 25

Drummond of Hawthornden, William 327

Dryden, John, 15, 18, 21, 29, 248, 267–70 *passim*, 274, 283, 324, 327, 328, 351, 360, 423, 449–51, 465, 478

Du Bellay, Joachim, 248, 253–5 *passim*, 264, 288, 429, 433

Dubeux, Albert, 386n.

Duchamp, Marcel, 202

Duckles, Vincent, 439n.

Du Halde, J. B., 376

Duine, F., 354n.

Dumesnil, R., 483n.

Dummett, Michael, 223–4 & n., 310n.

Dunbar, William, 466–7, 472

Duparc, Henri, 438, 446

Dupront, A., 257n.

Durandin, Guy, 232n.

Dürrenmatt, Friedrich, 479

Early Italian Poets, The (D. G. Rossetti), 353

Eckhart, Johannes ('Meister Eckhart'), 64, 65

Eclogues (Virgil), 365

Eddington, Sir Arthur Stanley, 148

Éducation sentimentale, L' (G. Flaubert), 483 & n.

Eichendorff, Joseph, Freiherr von, 446

Einführung in die Metaphysik (M. Heidegger), 241

Einiges über die neuen Uebersetzer-fabriken (J. J. Hottinger), 280

Einstein, Albert, 164n.

Eins und Alles (J. W. Goethe), 273

Einundzwanzig Sonette (W. Shakespeare trans. P. Celan), 409

Eisenhower, D. D., 421

Electra, 453

Electra (Euripides), 453

Electra (Sophocles), 453

Elefantenkarawane (H. Ball), 204

Elements (Euclid), 295

Eliade, Mircea, 159n.

'Eliot, George' (Mary Ann Evans), 480

Eliot, T. S., xv, 15, 237, 377, 477, 490, 493

Éluard, Paul, 31

Empson, William, x, 213, 222

Encyclopédistes, The, 211

Ende aller Dinge, Das (I. Kant), 151

Endgame (S. Beckett), 498

Enfer mis en vieux langage François, L' (E. Littré), 355, 356

Enneades (Plotinus), 281

Enquiry (D. Hume), 150

Enzensberger, Magnus, 370

Erasmus, Desiderius, 258, 259, 366

Ernst, Max, 202

Ervin-Tripp, Susan M., 434n.

Eschenbach, Wolfram von, 200

Essay on Criticism (A. Pope), 328

Essay on the Principles of Translation (A. F. Tytler), 249

Essays on Truth and Reality (F. H. Bradley), 220

Essay towards a real character and a philosophical language (J. Wilkins), 73, 78, 210

Esther, Book of, 281

Eteocles, 157–8, 478

Etymological Dictionary, An (W. W. Skeat), 25

Euclid, 118, 144, 279, 295

Eugene Onegin (trans. V. Nabokov), 289, 315, 331

Euripides, 44, 281, 329, 340, 452–9 *passim*, 477, 478

Evans, H. M., 162n.

Évolution créatrice (H. Bergson), 152

Exhortations to the Diligent Study of Scripture (Erasmus), 258

Exodus, Book of, 64

Exstasie (J. Donne), 424, 476

Extraterritorial: Papers on Literature and the Language Revolution (G. Steiner), vii, 186n.

Ezekiel, Book of, 155, 262

Fables (La Fontaine), 368

Faerie Queene, The (E. Spenser), 5, 353, 360, 479

Fang, Achilles, 376n., 377n.

Fann, K. T., 169n.

Faulkner, William, 416

Fauré, Gabriel, 438

Faust (J. W. von Goethe), 65, 83, 440–5 *passim*

Faust (C. F. Gounod), 440–5 *passim*

'Faust' theme, 479

Fawn (J. Marston), 7

Fedorov, Andrei, 249

Fenollosa, E. F., 375, 377 & n.

Ferguson, Charles A., 100n., 126n.

Ficino, Marsilio, 30, 259, 260

Ficker, Ludwig, 192

Filles du feu (G. de Nerval), 484

Finnegans Wake (J. Joyce), 190, 199, 289

Firth, J. R., 212n., 214

Fishman, J. A., 126n.

FitzGerald, Edward, 375

Fitzgerald, Robert, vii, 289

Flak, Otto, 201n.

Flaubert, Gustave, 184, 359, 391–7 *passim*, 412, 483, 484

Flew, A. N., 217n.

Florio, John, 124, 248, 261, 280, 281, 284

Flucht aus der Zeit, Die (H. Ball), 203

Focillon, Henri, 159

Fontenelle, Bernard le Bovier de, 160

Ford, Ford Madox, 377

Formalist movement, 249

Forrest, David V., 205n., 206n.

Forster, Leonard, 126n., 198

Fox Strangways, A. H., 440n.

Fragmente (J. von Herder), 82

Fraisse, Simone, 318n.

François de Sales, St., 73

Franzos, Karl Emil, 120

Fraser, J. T., 148n.

Freeman, K., 263n.

Frege, Gottlob, 140, 148, 175 & n., 212, 216, 217, 310n.

Freud, Sigmund, 30, 167

Frey, Hans, 340n.

Friedman, John Block, 467n.

Frisch, Max, 479

Frisk, Hjalmar, 232n.

Froissart, Jean, 433

Frost, Robert, 21

Frost, W., 267n.

Frye, Northrop, 440n., 446

Futurism, 202

Gadamer, Hans-Georg, 109n., 143n., 149n., 191n., 211n., 250, 256n., 285

Gadda, Carlo Emilio, 21

Galatians, Epistle to The, 364

Gale, Richard M., 148n.

Galen, 287

Galileo Galilei, 152, 260, 285

Gallic War (Caesar), 259

Gardner, Allen R. and Beatrice T., 240n.

Gardner, W. H., 432

Garnett, Constance, 285

Garver, N., 172n.

Garvin, Paul L., 326n.

Gate of Tongues Unlocked and Opened, The (trans. of Comenius), 209

Gaudier-Brzeska, Henri, 276n.

Gautier, Judith, 378

Gautier, Théophile, 423, 438

Gazzaniga, M. S., 298n.

Geertz, Clifford, 34n.

Geissler, H., 209n.

Geist der Utopie (E. Bloch), 228

Gellius, Aulus, 311

Genée, Rudolf, 400n.

Genesis, 62, 69, 322, 363n.

Genet, Edmond Charles, 25, 35

Gentile, G., 264 & n.

George, Stefan, 190, 200–1, 287, 397, 404–8 *passim*, 411

Germ, The, 12

Gerschenkron, Alexander, 332n.

Gershman, Herbert S., 201n.

Gesammelte Werke (H. Broch), 337

Geschichte der alten und neuen Literatur (F. Schlegel), 401

Geschwind, Norman, 295n.; and Walter Levitsky, 295n.

Gesualdo, Don Carlos, 446, 489

Ghil, René, 239, 244

Gibson, James J., 223n.

Gide, André, 336, 383–5 *passim*, 387–90 *passim*, 479

Gilbert, W. S. and A. Sullivan, 25

Gilson, Étienne, 150n.

Ginsberg, Allen, 21

Gipper, Helmut, 109n.

Giraudoux, Hippolyte Jean, 477, 478

Glinka, Michael Ivanovich, 489

Glossary of Tudor and Stuart Words (Skeat & Mayhew), 25

Gluck, Christoph Willibald, 439n.

Gödel, Kurt, 220

Godolphin, Sidney Godolphin, Earl of, 269

Goethe, Johann Wolfgang von, 65, 66, 79, 83, 84, 186, 190, 249, 261, 270–4, 285, 288, 340, 346, 375, 401, 438–45, 478, 479, 481

Golding, Arthur, 259

Goldsmith, Oliver, 11

Gombrich, E. H., 448, 485n.

Goodman, Nelson, 164 & n., 178 n.

Gorgias of Leontini, 263

Gosse, Edmund, 186n.

Gottscheid, Johann Christoph, 356

Gounod, Charles François, 441, 443, 444, 489

Goya y Lucientes, Francisco, 484, 490

Grabbe, Christian Dietrich, 400, 479

Graham, Martha, 477

Granville-Barker, Harley, 6

Grassi, B., 206n.

Graves, Robert, 42, 486

Gray, Nicolette, 354n.

Greek: culture, 30, 143n., 155, 156 & n., 195, 229–30, 268, 331, 346, 416, 477–8, 479, 486, 487; mythology, 24, 230

Greeks, the, 156, 487

Green, William Chase, 156n.

Greenberg, J. H., 102 & n., 104, 107n.

Gregor, Joseph, 400n.

Grene, David, 453

Grice, H. P., 489

Grundzüge der Phonologie (N. S. Trubetskoy), 99

Guarino da Verona, 259

Guillaume, Gustave, 139n.

Guitton, Jean, 149n.

Gulliver, 230

Gulliver's Travels (J. Swift), 37

Gumperz, John J., 126n.; and Charles A. Ferguson, 34n.; and D. Hymes, 126n.

Gundolf, Friedrich, 200n., 401 & n., 402

Haack, R. J. and Susan, 219n.

Haas, Mary R., 41n., 147n.

Hafiz (Persian singer), 273, 375

Hagen, Einar, 126n.

Hahn, Reynaldo, 438

Halborow, L. C., 170n.

Hall, Robert A., Jr., 109n., 111, 112 & n.

Hamann, J. G., 60, 76, 79–81 *passim*

Hamlet, 25, 384, 386n., 387

Hamm, 498

Hampshire, Stuart, 79n., 225n.

Handke, Peter, 182

Hardin, C. L., 173 & n.

Hardy, Thomas, 364
Harrington, Sir John, 7
Harris, E. E., 128
Hartmann, Nicolai, 262
Hartshorne, Charles, 225n.
Hasidism, 63, 67, 68
Hatsopoulos, G. N. and J. H. Keenan,
    162n.
Hauptmann, Gerhart, 401, 477
Hauvette, Henri, 355n.
Haydn, Franz Joseph, 322, 489
Hebb, D. O., W. E. Lambert and E. R.
    Tucker, 301n.
*hebraischen Synonyma der Zeit und
    Ewigkeit genetisch und
    sprachvergleichend dargestellt, Die*
    (C. von Orelli), 164–5
Hegel, Georg Wilhelm Friedrich, 88,
    137n., 216, 227, 256, 266, 285, 313,
    317, 344n., 410, 416, 478
Heidegger, Martin, xx, 143n., 192,
    205, 219, 241, 250, 313, 315, 317,
    339, 340n., 341, 344n., 362, 395,
    408
Heine, Heinrich, 252, 439
Hélie, Pierre, 98
Hellenism, 155, 184, 487
Hellingrath, Norbert von, 340n., 347
Hemingway, Ernest, 17
Hempel, C. G., 220
Hemphill, R. E., 306n.
Heracles (Euripides), 329
Heraclitus, 18, 24, 157
Herbert, George, 439n.
Hercules, 478
Herder, Johann Gottfried von, 20, 81,
    82, 85, 195, 279, 280, 341, 353
Hermeneutik (F. Schleiermacher), 263
Hermes Trismegistus, 62
Herodotus, 165, 353
Herrick, Robert, 476
Hesiod, 478
Hesse, Hermann, 358
Hewes, Gordon W., 241n.
Hexter, J. H., 141n.
Higman, B., 212n.
Hill, Geoffrey, 494
Hints from Horace (Lord Byron), 328
Hippias minor (Socrates), 230

Hippolytus (Euripides), 136, 281,
    452–8 *passim*, 459, 479
History (R. Lowell), 490
Hobbes, Thomas, 400, 418–19
Hobson-Jobson (Sir H. Yule and A. C.
    Burnell), 25
Hockett, C. F., 101
Hofmannsthal, Hugo von, 192–3, 194,
    358, 477
Hofmiller, Josef, 357
Hoijer, H., 45n.
Hölderlin, Johann Christian Friedrich,
    26, 30, 66, 67, 83, 185, 192, 222,
    248, 280, 283, 339–42, 400, 478
Holinshed, Raphael, 479
Holland, Elizabeth, 390
Holland, Philemon, 259
Hollander, John, 439n.
Homage to Sextus Propertius (E. Pound),
    308
Homecoming (H. Pinter), 479
Homer, 22, 23, 185, 187, 195, 248,
    260, 266, 269, 280, 284, 288 & n.,
    289, 318, 329, 340, 360 & n., 361,
    380, 417–22, 447, 461, 478, 479,
    487
Hood, Thomas, 422
Hopkins, Gerard, 396–7
Hopkins, Gerard Manley, 15, 289,
    430–4
Horace, 24, 248, 267–9 *passim*, 272,
    276, 327–9, 340, 342 & n., 450–1,
    460–1, 464, 466, 469, 470, 485n.
Horn-Monval, M., 386n.
Horváth, János, 479
Hosea, 155
Hottinger, J. J., 280
Hough, E., 354n.
Housman, A. E., 26, 278
Houyhnhnms, the, 229, 233, 236, 237
Howard, Richard, 284n.
Hudson, W., 223n.
Huelsenbeck, Richard, 202
Huet, Pierre-Daniel, 248, 257 & n.,
    276, 278–80 *passim*, 283
Hughes, Ted, 370
Hugo, Adèle, 482
Hugo, Victor, 186, 365, 384, 446
Huit Scènes de Faust (H. Berlioz), 443

Humboldt (Friedrich Henrich), Alexander, 84
Humboldt (Karl), Wilhelm von, 59, 82–91 *passim*, 98, 101, 104, 106, 181, 249, 250, 279
Hume, David, 144, 148, 150–1, 215, 225, 227, 229
Humphrey (Humfrey), Lawrence, 277–80 *passim*
Husserl, Edmund, 219, 292
Hydén, Holger, 302
Hymes, Dell, 51, 126n., 497n.
Hynd, James and E. M. Valk, 66n.
Hyppolite, Jean, 416
Hyslop, A., 310n.

Ibsen, Henrik Johan, 167, 238, 395, 416
Igitur (S. Mallarmé), 202
Iliad (Homer), 22, 195, 230, 248, 259, 260, 265, 272, 354, 360, 370, 417, 418, 420, 421
Illuminations (J. A. Rimbaud), 186
Il pleut doucement sur la ville (J. A. Rimbaud), 321
Index translationum (UNESCO), 284
Inferno (Dante), 167, 356, 358
Ingres, Jean Auguste Dominique, 12–14
In Memoriam (Lord Tennyson), 473
Institutiones oratoriae (Quintilian), 265
Interlinguistica, Festschrift for Professor Mario Wandruszka (ed. by Karl-Richard Bausch and Hans-Martin Gauger), 250
Interpretatio linguarum etc. (L. Humphrey), 277–8
Introduction to Semantics (R. Carnap), 220
Introduction to the Theory of Translation (A. Fedorov), 249
Ion (Plato), 484
Ionesco, Eugène, 194
Iphigenia legend, 478
Iphigénie (J. Racine), 46, 452, 478
Irby, James E., 74 & n., 76
Irenaeus, Bishop of Lyons, 159n.
Isaiah, 23, 154, 384
Ishaq, Hunain ibn, 287

Isou, Isidore, 204–5, 206
Italiander, R., 125n.
Izvestia, 35

Jack, I., 464n.
Jacob, 153, 155
Jacob, André, 139n.
Jager, Ronald, 219n.
Jakobson, Roman, x , 83, 92, 99–100, 130 & n., 245, 274–5, 293, 302n., 422, 436, 437, 496
James, Henry, 9, 37, 397, 480, 483, 493
James, William, 49, 75, 134, 221
Janik, Allan and Stephen Toulmin, 374n.
Jankélévitch, Vladimir, 232n., 236
Janua linguarum reserata (J. A. Comenius), 209
Japanese haiku, 333, 377n., 379
Jarry, Alfred, 31
Jay, P. C., 240n.
Jeffers, Robinson, 478
Jenkins, John, 439n.
Jeremiah, Book of, 154–5
Jerome, St. (Eusebius Sophronius Hieronymus), 248, 253, 257, 262, 264, 275, 281, 283, 292, 314
Jesus Christ, 38, 257, 384, 401, 426, 440
Joan of Arc, 479
Job, 23
Job, Book of, 422
Jocasta, 157, 478
John, Gospel according to St., 158
John (of the Cross), St., 327
Johnson, Lionel, 15
Johnson, Samuel, xv, 26, 254, 264, 267, 423, 465, 469
Jonah, Book of, 154, 155
Jones, O. K., 170n.
Jones, Sir William, 82
Jonson, Ben, 30, 43, 248, 267, 268 & n., 327–9
Journal des débats, 354
Jowett, Benjamin, 362–4, 439
Joyce, James, 199, 266, 297, 447, 490, 493
Judaism, 45, 62, 79–80, 155, 158, 252, 487

Juvenal, 43, 269, 465

Kabbalism, 60–5 passim, 70, 72, 76, 79, 93, 129, 257, 313, 499
Kafka, Franz, 36, 68–70 passim, 179, 194, 378
Kahn, David, 176n.
Kaiser, Georg, 478
Kaldor, Susan and Ruth Snell, 126n.
Kandinsky, Wassily, 202
Kant, Immanuel, xvi, 80, 85, 94, 96, 132n., 148, 151, 160, 215, 218, 255, 284, 285, 338, 410, 436
Kaplan, B. and S. Wagner, 130n.
Katz, Jerrold J., 217
Kazantzakis, Nikos, 416
Keats, John, 14, 42, 284, 353, 412, 471
Keesing, Felix M. and Marie M., 34n.
Kelletat, Alfred, 191n.
Kempter, Lothar, 340n.
Kenner, Hugh, 377n., 378n.
Kepler, Johann, 60, 65, 160
Khlebnikov, Velimir, 190, 204, 242, 246, 370
Kierkegaard, Sören Aaby, 76, 145, 165, 344n., 416, 478
King, Edward, 470, 472
King, Hugh R., 149n.
Kipling, Rudyard, 25
Kircher, Athanasius, 209, 210
Kirk, R., 310n.
Klänge (W. Kandinsky), 202
Klanggedichte (H. Ball), 206
Kleist, Bernd Heinrich Wilhelm von, 20, 83, 478
Kloepfer, Rolf, 276n.
Klopstock, Friedrich Gottlieb, 263, 341, 342 & n.
Kloss, C. B., 373n.
Knox, Ronald A., 251 & n.
Koenig, F. O., 161n.
Koestler, Arthur and J. R. Smythies, 119n.
Kolakowski, L., 200n.
Koyré, Alexandre, 65n., 160n., 260, 285
Kraus, Karl, 22, 203, 287, 407–8, 411
Kristeva, J., 304n.
Kroesch, Samuel, 232n.
Kručenyx, Alexei, 194

Labé, Louis, 314, 423–5
Labor, William, Paul Cohen, and Clarence Robbins, 34n.
Lacan, Jacques, 140n.
La Fontaine, Jean de, 67, 368–9
Laius, King of Thebes, 478
Lakoff, George, 113
Lambert, W. E., 127n.; see also 302n.
Lamennais, Hughes Felicité Robert de, 354
Lancaster, J. B., 240n.
Language (L. Bloomfield), 19
languages: Aba, 56; African group, 61, 96, 493; Ainu, 379; Algonkian group, 96; American Indian group, 36, 41, 45, 53, 55–6, 91, 96, 163; Anglo-Norman, 198; Anglo-Saxon, 25, 72, 375; Apache, 97; Arabic, 74, 265, 287, 379; Aramaic, 24, 65, 436; Araucanian, 102; Arči, 54; Armenian, 279; Armoric, 199; Aztec, 93; Bengali, 352; Bergamasque, 32; Bikol group, 58; Breton-Celtic, 334; Cahita group, 55; Cantonese, 32; Carib, 42, 61; Castilian, 74, 352; Catalan, 198, 265; Celtic, 62; Chabokano, 58; Cherokee, 42; Chichewa, 96; Chinese, 24, 30, 32, 78, 93, 107, 208, 211, 243, 279, 333, 375–80 passim, 382, 492; Cœur d'Alène, 96; Coptic, 279; Cuna, 24; Daghestan group, 103; Dido, 54; Dutch, 352, 492; Elizabethan English, 28; Ermitano, 58; Eskimo group, 42, 57; Esperanto, 62, 211, 212, 214, 493; European group, 96; Finno-Ugaritic, 56, 107; Fortran (artificial), 226; French, 139, 145, 179, 198, 319–25, 353, 355; Galician-Portuguese, 198; German, 272–3, 280, 281, 285, 321–3, 337–8, 340, 342, 349–50, 356–9 passim, 379, 381, 400 & n., 401, 404, 409, 412, 492; Greek, 24, 42, 67, 79, 87, 93, 148, 157, 165 & n., 199, 252, 255, 259, 260, 268, 270, 278, 279, 281, 282, 318, 329, 332, 341, 343–5 passim, 349–50, 364, 367, 375, 418, 421, 436, 454, 458, 459; Hebrew, 7, 24, 62–5, 67, 68, 79, 93, 107, 121,

languages (*cont.*):
153, 155, 164, 165 & n., 195, 261, 278, 323, 367, 422, 436, 486, 492; Hitchiti, 42; Hopi group, 55, 93–5; Huite (or Yecarome), 55; Hungarian, 252, 492; Ido (artificial), 211; Indo-European group, 94, 96, 98, 107, 137, 163, 165, 319; Italian, 198, 200, 270, 276, 289, 322–3, 334, 353, 379, 405, 413, 414, 492; Japanese, 7, 42, 93, 333, 352, 375, 379; Kakoma, 55; Kamtchadal, 54; Khoisan group, 100; Koasati (Muskogean group), 41; Kot (or Kotu), 54; Kota, 93; Kučarete, 55; Kupeño, 55; Lapp, 107; Latin, 87–8, 93, 95, 112, 198–9 *passim*, 208–10 *passim*, 255, 259, 260, 270, 275, 278, 279, 282, 308, 311, 320, 322, 323, 328, 340n., 367, 370, 413–15 *passim*, 450, 454, 457n., 458, 459, 486, 493; Latine sine flexione (artificial), 211; Lithuanian, 319; Manchu, 32; Mandarin, 32; Matagalpa, 55; Middle English, 366, 367; Middle High German, 270, 341; Milanese, 32; Mongol, 32; Mongolian, 375; Muskogean, 41; Nawa, 55; Neapolitan, 32; Nootka, 102; Novial (artificial), 211, 214; Occidental (artificial), 211; Old English, 404; Old French, 199, 319, 334, 369; Old High German, 341; Old Norse, 320, 404; Old Scottish, 199; Oubykh, 54; Palaeosiberian group, 54; Persian, 270, 273, 333; Polish, 375; Portuguese, 376, 381; Provençal, 198, 319, 370; Qapuči, 54; Quileute, 101; Russian, 93, 322, 332n., 352, 375, 409, 426n., 492; Salishan, 101; Samoan, 107; Samoyed, 287; Sanskrit, 42, 95, 145, 320; Semitic groups, 165, 166; Shawnee, 93; Slavic group, 270; Spanish, 55, 270, 492; Standard Average European (SAE), 96; Swabian, 341; Swahili, 58, 107, 382; Swedish, 492; Syriac, 279; Tagalog, 58; Tarascan, 24; Thai, 42; Tibetan, 32; Tomateka, 55; Tubatulabal, 55; Urdu, 287, 379; Uto-Aztec group, 55; Venetian, 32; Volapük (artificial), 211; Walloon, 334; Welsh, 319; West African English, 493; West Indian English, 493; Wishram, 100; Wraywaray, 58; Xwarši, 54; Yiddish, 68; Zulu, 96; Zuni Indian, 32; Zyriene, 56

Language, Thought and Reality (B. L. Whorf), 91
La Place, Pierre-Antoine de, 384
Larbaud, Valery, 249, 284, 289, 365
Larkin, Philip, 494
Latinism, 276, 334, 415
Lattimore, Richmond, 330–1, 420–1
Lawrence, D. H., 485, 488, 493
Lawrence, Gertrude, 17
Lawrence, T. E. ('T. E. Shaw'), 361
Lear, fable of, 447
Lear, King, 36, 233, 404, 479, 486
Lear, Edward, 196–7, 359, 426–7
Lees, R. B., 20n.
Le Gallienne, Richard, 15
Lehrer, K., 225n.
Leibniz, Gottfried Wilhelm, 60, 73–82 *passim*, 84, 90, 91, 98, 148, 177 & n., 210–11 & n., 212, 213, 220, 378
Leiris, Michel, 33n., 207
Leishman, J. B., 451 & n., 485 & n.
'Lenau, Nikolaus' (Nikolaus Franz Niembsch von Strehlenau), 439
Lenneberg, E. H., 45, 97, 108, 295
Lenin, Nicolai, 287
Leonardo da Vinci, 12, 21
Leopardi, Giacomo, Count, 255n., 353
Leopold, W., 301n.
Le Page, R. B., 126n.
Le Roy, Louis, 260
Les Bonnes (E. C. Genet), 34
Le Senne, René, 232n.
Lessing, Gotthold Ephraim, 401
Lettres philosophiques (P.-A. de La Place), 384
Lettre sur les sourds et muets (D. Diderot), 253
Lévinas, Emmanuel, 292, 293n.
Lévi-Strauss, Claude, vii, 30, 47, 52, 80, 86, 95, 107, 137n., 163, 319, 490
Leviticus, 28, 153

Lévy-Brühl, Lucien, 91
Lewis, M. M., 301n.
Le Yaounc, M., 483n.
Leyris, Pierre, 289, 386n., 390, 430–4
Libellus de optimo genere oratorum (Cicero), 248
Lichtenberg, Georg Christoph, 222
Liddell, H. G. and R. Scott, 282
Liberman, Philip H., 240n.; and Edmund S. Crelin, 234n.; and Edmund S. Crelin and Dennis H. Klatt, 130n.
Liede, Alfred, 196n.
Lied von der Erde, Das (G. Mahler), 379
Lifton, Robert, 168
Linacre, Thomas, 278
Linnaeus (Carl von Linne), 288
Linsky, L., 212n.
Lipmann, Otto and Paul Blaut, 232n.
Leconte de Lisle, Charles Marie René, 318
Liszt, Franz, 439, 441–3 *passim*
Littlewood, J. E., 179
Littré (Maximilian Paul), Emile, 24, 334, 354–6, 358–9
Liu, James J. Y., 376n.
Livius Andronicus, 286
Livre de Jade, Le (Judith Gautier), 378
Livy (Titus Livius), 259
Locke, D., 172n., 175n.
Locke, John, 10, 284, 286
Logical Positivists, 216, 220
Logische Syntax der Sprache (R. Carnap), 220
Logopandecteision (Sir T. Urquhart), 209
Logue, Christopher, 265, 370, 380, 421, 422 & n., 423
Lohmann, Johannes, 109n., 256n.
Loki, 235
Lollius, Marcus, 460
Longfellow, Henry Wadsworth, 421
Longinus, Cassius, 284
Longus, 353
Lorenzaccio (A. de Musset), 388
Lorenz, Konrad, 119
Lounsbury, F. G., 45
Lowell, Amy, 376
Lowell, Robert, 268, 287, 376, 391, 423, 465, 479, 490

Lubac, Henri de, 362n.
Lubeck, Schmidt von, 439
Lucan (Marcus Annaeus Lucanus), 340
Lucretius (Titus Lucretius Carus), 81, 334
Lully, Raymond, 209
Luria, A. R., 298n.
Luther, Martin, 67, 89, 248, 257, 259, 264, 272, 280, 283, 315, 341, 356
Luther, Wilhelm, 109n., 232n., 265
'Lycidas' (J. Milton), 469–73 *passim*
Lydgate, John, 5
Lys dans la vallée, Le (H. de Balzac), 483 & n.

Macbeth, 44
Machaut, Guillaume de, 489
Machiavelli, Niccolò, 35, 285
MacKenna, Stephen, 256n., 281–3, 315, 400
Mackey, W. F., 126n.
McKinnon, Donald, 151n.
McTaggart, J. E., 148, 151, 152 & n.
Madame Bovary (G. Flaubert), 391–7 *passim*, 480
Mahler, Gustav, 379
De Maistre, Joseph, 22
Malcolm, N., 169, 172 & n., 207
Malebranche, Nicholas, 148
Malherbe, François de, 20
Mallarmé, Stéphane, 28, 66, 72, 185–7 *passim*, 190–1, 192, 200, 202, 239, 244, 289, 385, 402
Malraux, André, 227
Malthus, T. R., 11
Mandelbaum, D., 91n.
Mandelstam, Osip Emilyevich, 168
Manet, Edouard, 484, 490
Manichaeism, 71
Manière de bien traduire d'une langue en aultre (E. Dolet), 276
Manilius, 279
Mann, Thomas, 479
Manual of Phonology (C. F. Hockett), 101
Manzoni, Alessandro, F. T. A., 288
Mardrus, J. C., 379
Marinetti, Filipo Tommaso, 202

Marivaux, Pierre Carlet de Chamblain de, 16, 352
Marlowe, Christopher, 4, 423, 479
Marriage of Figaro, The (W. A. Mozart), 46–7
Marsh, John, 165n.
Marston, John, 7
Martinet, A., 126n.
Marvell, Andrew, 21, 25, 466
Marx, Karl Heinrich, 195, 284, 395, 487
Marx, O., 404n.
Marxism, xvii, 137n., 221, 227, 304n.
Mason, H. A., 360n.
Massignon, L., 415
Mathers, Edward Powys, 379
Matthew Passion, St. (J. S. Bach), 440
Matthew, Thomas, 366
Mauthner, Fritz, 175n., 181 & n., 187
Maxwell, James Clark, 164n.
Mayo, Bernard, 225n.
Mazon, P., 331, 346
Medea legend, 478
Medea (Euripides), 478
Megillath Taanith, 252
Meillet, A., 54; and M. Cohen, 54n.
Meinecke, Dietlind, 189n.
'Meister Eckhart', *see* Eckhart, Johannes
Mellers, Wilfred, 439n.
Mémoire (B. P. E. Clapeyron), 161
Menard, Pierre, 73–6
Mencius, 256n., 376n.
Mendel, Johann Gregor, 288
Menninger, Karl, 163n.
Merkabah mysticism, 63
Merleau-Ponty, M., 116–17, 134 & n., 292
Merrill, Stuart, 376
Merry Wives of Windsor, The, 3, 25, 423
Mersenne, Marin, 209
Merz, J. T., 161n.
Meschonnic, Henri, xx, 323n:, 416n.
Mesnard, A., 354
Messias (F. G. Klopstock), 263
Metamorphoses (Ovid), 414–15, 470
Metaphysics (Aristotle), 148
Michelet, Jules, 353
Michels, Gerd, 200n.

Mickiewicz, Adam, 375
Middlemarch (G. Eliot), 480
Middleton, Christopher, vii, 289
Middleton, Thomas, 3
Midsummer Night's Dream, A, 407
Milhaud, Darius, 447n.
Miller, Robert L., 89n.
Milton, John, 13, 21, 44, 190, 261, 333–5, 419, 420, 461, 469–75 *passim*, 479, 486
Miner, Earl, 377n.
Minima Moralia (T. W. Adorno), 240
Minnis, N., 195n.
Minsky, M., 212n.
Miseria y esplendor de la traducción (J. Ortega y Gasset), 264
Mithraic cults, 158
Moby Dick: or The White Whale (H. Melville), 289
Mohr, J. C. B., 340n.
Molière (Jean-Baptiste Poquelin), 34, 384, 388, 478, 479
Molina, Tirso de, 479
Mönch, Walter, 447n.
Mon Faust (P. Valéry), 479
Monod, Jacques, 133 & n., 166 & n.
Montaigne, Michel de, 28, 125, 248, 260, 261, 281, 284, 383, 385
Montesquieu, Charles Louis de Secondat, Baron de la Brede et de, 85
Monteverdi, Claudio, 489
Moore, G. E., 216, 219, 222
Moore, George, 395
Moore, Marianne, 67, 368–9
More, Sir Thomas, 278
Moreau, Fréderic, 483
Morick, H., 170n., 172n.
Mormon Church, 200
Moro, lasso, al mio duolo . . . (D. C. Gesualdo), 446
Morris, William, 15, 361
Morse, R. J., 354n.
Morstein, P. von., 170n.
Morwitz, Ernst, 200n.
Mosaic Law, 70
Moses, Books of, 62
Moses und Aron (A. Schoenberg), 184
Motherwell, R., 201n.

Mots, Les (J.-P. Sartre), 184
Motteux, Pierre Antoine, 285
Mozart, Wolfgang Amadeus, 46–7, 439
Much Ado About Nothing, 3, 4, 6
Mueller, Friedrich Max, 164
Muir, Edwin and Willa, 336
Müller, Wilhelm, 440
Mundle, C. W. K., 173 & n.
Musset, Alfred de, 338
Muttersprache und Geistesbildung (L. Weisgerber), 90

Nabokov, Vladimir, 77, 127, 252, 254, 264, 289, 315, 328, 331, 332n., 400
Nachdichtung (K. Kraus), 407
Narski, I. S., 221n.
Nazism, vocabulary of, 35
Neanderthal man, 153n., 234 & n.
Needham, Joseph, 279, 373 & n.
Neo-classicism, 184
Neoplatonism, 406, 467n.
Nerval, Gérard de, 185, 441 & n., 446, 484
Nestor, 22
Neveu de Rameau (D. Diderot), 270
Newton, Sir Isaac, 94, 160 & n., 163, 461
Nicholas of Cusa, *see* Cusanus
Nicolas V, Pope, 259
Nidditch, P. H., 164n.
Nietzsche, Friedrich Wilhelm, 168, 216, 232–3, 237–9, 243, 260, 283, 285, 487
Nimrod, 60, 68
Nims, John Frederick, vii, 289
Nizan, Paul, 289
Noctes (Aulus Gellius), 311
Nolan, Rita, 219n.
North, Sir Thomas, 259, 260, 284, 315, 375
Noss, Richard B., 117n.
Nosworthy, J. M., 2
Notitias super lingua internationale (G. Peano), 211
Nouvelle Héloïse, La (J.-J. Rousseau), 480–4
'Novalis' (Friedrich Leopold, Freiherr von Hardenberg), 250, 283, 356

Noyes, G. R., 375
Nugel, Bernfried, 449n.

Oberon (C. M. Wieland), 272
Ockham, William of, 148, 150 & n.
Odes (Horace), 342 & n., 451, 469
Odes et ballades (V. Hugo), 365
Ödipus der Tyrann (J. C. F. Holderlin), 344
Odysseus, 22, 229, 230, 235, 362
Odyssey (Homer), 22, 195, 200, 201, 230, 272, 286, 289, 360, 361, 422, 447
Oedipus, 30, 157, 340n., 348, 439n.
Oedipus Coloneus (Sophocles), 44, 478
Oedipus Rex (Sophocles), 345, 349, 478
Oettinger, A. G., 326n.
Ogden, C. K., 86, 141, 213
Old Believers in Russia, 159
d'Olivet, Fabre, 93
O'Neill, Eugene Gladstone, 477
On Translation (ed. R. A. Brower), 250
Opie, Iona and Peter, 36
Opitz von Boberfeld, Martin, 356
Orbis sensualium pictus (or *Comenius's Visible Word* etc.) (Comenius), 209
Ordinary Language (G. Ryle), 134
Orelli, C. von, 164
Oresteia (Aeschylus), 23
Origen, 278
Origin and Diversification of Language, The (M. Swadesh), 51
Orlando Furioso (L. Ariosto), 7, 14
Orléans, Charles d', 431
Orpheus, 140, 384, 467–75 *passim*, 488
Ortega y Gasset, José, 48, 236, 249, 264, 283, 314
Orwell, George, 22, 35
Osgood, Charles, 109–10
Othello, 7, 48, 387, 403
Otway, Thomas, 391
Ovid (Publius Ovidus Naso), 260, 267, 269, 340, 413–15, 423, 469, 470, 474
d'Ovidio, Francesco, 355n.
Ovid's Epistles, Translated by Several Hands (J. Dryden), 267

Oxford English Dictionary (O.E.D.), 3, 11, 13, 25, 320

Palmer, Richard, E., 143n.
Panofsky, Erwin, 448
Pan Tadeusz (A. Mickiewicz), 375
Paolucci, A., 354n.
Pap, A., 220n.
Paracelsus, Theophrastus Bombast von Hohenheim, 59, 64, 84, 200
Paradise Lost (J. Milton), 7, 333, 334, 420
Paradiso (Dante), 265, 339
Paradis perdu (F. R. Chateaubriand), 335
Parattore, E., 413n.
Parerga und Paralipomena (A. Schopenhauer), 281
Parmenides (Plato), 255, 362
Parry, Milman, 185, 421
Partridge, A. C., 366, 367n.
Partridge, Eric, 25, 197 & n.
Pascal, Blaise, 160, 265, 285, 432
Pascal, Roy, 400n.
Pasternak, Boris, 375, 423
Patmos (J. C. F. Holderlin), 350
Patristic philosophy, 229, 257
Paul, St., 251, 364, 435
Paulhan, Jean, 134
Pavlov, Ivan Petrovitch, 296, 489
Payne, Robert, 376n.
Paz, Octavio, vii, 190n., 198, 247 & n., 289
Peano, Giuseppi, 211, 216
Pears, D., 225n.
Péguy, Charles, 318 & n.
Pei, Mario Andrew, 58
Peirce, C. S., 148, 221, 274
Pena, Jean, 279
Penguin Book of Modern Verse Translation, The (ed. G. Steiner) vii
Pensée sauvage, La (C. Lévi-Strauss), 95
Pentateuch, The, 362
Pepys, Samuel, 352
Percy, Thomas, 376
Perez de Oliva, F., 478
Pergolesi, Giovanni Batista, 489
Pericles, 142
Perkins, Moreland, 172n.

Perotti, Niccolò, 259
Persian philosophy, 155
personae (E. Pound), 276
Petrarch (Francesco di Petracco), 138, 404, 406, 411, 423, 472
Peyser, Herbert, F., 440n.
Phädra (J. C. F. Schiller), 287, 479
Phaedo (Plato), 439
Phaedra, 454, 456
Phaedrus (Plato), 156
Phaethon, 456
Phèdre (J. Racine), 46, 391, 458, 459, 479
Phenomenologie (G. Hegel), 416
Philologische Einfälle und Zweifel (J. G. Hamann), 81
Philosophical Investigations (L. Wittgenstein), 62, 91, 169–70
Philosophical Review, 169
Philosophy and Logical Syntax (R. Carnap), 141
Philosophy of Symbolic Forms (E. Cassirer), 89
Phoenician Women (Euripides), 478
Physics (Aristotle), 148
Piaget, Jean, 38, 119, 134; and Barbel Inhelder, 132n.
Piaubert, Jean, 447n.
Picabia, Francis, 202
Picasso, Pablo, 484, 490
Pichot, Amédée, 285
Pietism, 61, 76, 210, 341
Pimsleur, Paul and Terence Quinn, 126n.
Pindar, 23, 67, 185, 195, 248, 268, 340 & n., 342–4, 390, 449, 461, 470, 478, 479
Pindarique Odes (A. Cowley), 267
Pinter, Harold, 46, 182, 194, 479, 494
Piranesi, Giovanni Battista, 72
Piro, S., 206n.
Pitcher, George, 219n., 224n.
Planck, Max, 162n.
Plato, 23, 27, 30, 41, 57, 83, 107, 130, 132, 141, 156, 215, 255, 259, 260, 268 & n., 280, 282, 285, 331, 362–4, 401, 404–6 *passim*, 424, 439, 471, 472, 479, 484, 487, 488
Plautus, Titus Maccius, 478

Plotinus, 255, 281, 283, 315, 323, 400
Plutarch of Cheronea, 248, 259, 260, 265, 284, 288, 315, 375
Po, Li, 377
Pocock, J. G. A., 286 & n.
Poe, Edgar Allan, 186 & n.
Poem into Poem (ed. G. Steiner), vii
Poems (D. G. Rossetti), 13
Poems in Translation (J. F. Nims), 289
Poetics (Aristotle), 284
Poet's Tongues, The (L. Forster), 198
Polybius, 259, 278
Polygraphia Nova et Universalis (A. Kircher), 210
Pons, C., 386n.
Ponzio, Augusto, 206n., 304n.
Pope, Alexander, 248, 288 & n., 328, 360 & n., 419–20 *passim*, 460–3 *passim*, 469, 484
Pope, M. H., 422
Popper, Karl, 234n.
Portrait of a Lady, The (H. James), 480
Possessed, The (F. Dostoevsky), 447
Pouillon, Jean, 139n.
Pound, Ezra, 15, 27, 249, 266, 268, 276 & n., 283, 289, 308, 354 & n., 368, 370, 375, 377 & n., 378, 380, 386, 412, 423, 465, 478, 490, 493
Poussin, Nicolas, 30, 458
Powys, J. C., 493
Praz, Mario, 448
Prelatical Episcopacy (J. Milton), 13
Pre-Raphaelites, 15, 30
Preston, M. S. and W. E. Lambert, 302n.
Pride and Prejudice (J. Austen), 9
Principia Mathematica (Bertrand Russell and A. N. Whitehead), 220
Principii d'una scienza nuova (G. B. Vico), 78
Principles of English Etymology (W. W. Skeat), 25
Prinzip Hoffnung (E. Bloch), 228
Prior, A. N., 152n., 212, 225n.
Private Lives (N. Coward), 16–17
Proceedings of the Aristotelian Society, 221
Proceedings of the Colloquium on Translation, of the Bavarian Academy of Fine Arts 1962, 280n.

Prometheus myth, 241, 479
Prometheus (Aeschylus), 331
Prometheus Unbound (P. B. Shelley), 241
Propertius, Sextus, 266, 370
Proust, Marcel, 33, 139, 285, 335 & n., 386, 484
Provinciales (B. Pascal), 265
Psalms, The, 67, 195, 366, 422
Psycho-Biology of Language, The, 250
Ptolemy (Claudius Ptolemaeus), 259
Puech, Henri-Charles, 159n.
Puhvel, J., 127n.
Pupil, The (H. James), 37
Purgatorio (Dante), 13, 358
Pushkin, Aleksandre, 254, 262, 332n., 333, 400, 479
Putnam, Hilary, 178n.
Pynchon, Thomas, 493
Pyritz, Hans *et al.*, 270n.
Pythagoras, 60, 65

Quasimodo, Salvatore, 413–15
Quevedo y Villegas, Francisco Gomez de, 73
Quine, Willard van Orman, ix, 125, 212, 217, 221, 249, 283, 291 & n., 292, 294, 310n., 371, 372n., 374
Quintilian (Marcus Fabius Quintilianus), 248, 251, 265, 267

Rabelais, François, 129, 185, 259, 260, 288, 354, 359, 385, 418
Rabinowitz, Isaac, 346n.
Racine, Jean, 28, 46, 287, 384, 385, 390–1, 452, 456–9, 461, 478, 479
Racine et Shakespeare (Stendhal), 384
Ramanujan, Srinivasa, 179
Ramsey, F. P., 221
Ray, Alain, 354n.
Ray, Gordon, vii
Rayfield, J. R., 126n.
Recherche de l'absolu, La (H. de Balzac), 482
Réflexions sur la puissance motrice du feu et les moyens propres à la développer (S. Carnot), 160
Reichenbach, Hans, 106, 161n.
Reiff, Arno, 268n.

Reinhardt, Karl, 340n., 344n., 350n.
Reinsch, Hugo, 268n.
Rembrandt (Rembrandt Harmens Van Rijn), 490
Renaissance, The, 286, 334, 467, 477, 487, 488
Renaissance in Italy, The (J. A. Symonds), 259
Renfrew, Colin, 147n.
Republic (Plato), 260, 331
Restoration comedy, 10, 25
Retz, Jean François Paul de Gondi, 321
Reuchlin, Johann, 366
Revelation of St. John the Divine, The, 158, 366n.
Revolutionary Immortality (R. Lifton), 168
Revzin, I.I., 117n.
Rexroth, Kenneth, 376
Rhees, R., 169n., 172n.
Ricardo, David, 11
Richards, I. A., vii, 50 & n., 92, 113n., 141, 213, 245n., 249, 256n., 376n., 492 & n., 494, 497
Richardson, Samuel, 10, 272, 480
Richardson, Tony, vii
Richter, Hans, 201n.
Ricœur, Paul, 140n., 313n.
Rilke, Rainer Maria, 253, 314, 424–5
Rimbaud, Jean Arthur, 185–6 *passim*, 200, 321, 386
Rime of the Ancient Mariner, The (S. T. Coleridge), 364
Roberts, W. Rhys, 347
Robespierre, Maximilien François Marie Isidore de, 142
Robinson Crusoe, 172
Rochefort, Henri, Marquis de Rochefort-Luçay, 248
Roger Délivrant Angélique (J. A. D. Ingres), 14
Rollins, C. D., 170n.
Roman literature, 268
Romans, Epistle to the, 364, 366n.
Romanticism, 184
Romanzen und Balladen (R. Schumann), 442
Romeo and Juliet, 401, 440
Ronconi, Alessandro, 139n.

Ronsard, Pierre de, 383
Rorty, Richard, 217 & n., 219
Rose, Steven, 302
Rosenzweig, Franz, 257, 283, 290n.
Ross, Donald Carne, vii
Rossetti, Dante Gabriel, 12–15, 25, 28, 353–4
Rossi, Paolo, 211n.
Rossi-Landi, F., 304n.
Roubaud, Jacques, 198, 247n.
Rouse, W. H. D., 289
Rousseau, Jean-Jacques, 284, 480–4, 488
Rubáiyát of Omar Khayyám (E. FitzGerald), 375
Rubinstein, Anton Grigorovich, 439
Ruskin, John, 335 & n.
Russell, Bertrand, 212, 216, 217, 219
Russell, Bertrand, and A. N. Whitehead, 212, 216, 220
Russell, Claire and W. M. S. Russell, 195n.
Ryle, G., 134, 187n., 218

Sade, Donatien Alphonse François, Marquis de, 41
Sainte-Beuve, Charles Augustin, 482–4
Saint-Maur, Dupré de, 334
Saint-Simon, Claude Henri de Rouvroy, Duc de, 321
Salinger, J. D., 38
Salomé (G. Apollinaire), 318
Samson Agonistes (J. Milton), 420, 486
Sanguineti, Edoardo, 198, 247n.
Santayana, George, 423
Sapir, Edward, 81, 89, 91 & n., 98, 106
Sartre, Jean-Paul, 137n., 184, 219, 289, 477
Satie, Erik Leslie, 439
Šaumjan, S. K., 117n.
Saussure, Ferdinand de, 32, 83, 206
Saxl, F., 448
Scève, Maurice, 191
Schadewaldt, Wolfgang, 281, 340n., 344n., 345
Schelling, Felix E., 268n.
Schelling, Friedrich Wilhelm Joseph von, 216

Schiller, F. C. S., 221
Schiller, Johann Christoph Freidrich von, 83, 85, 89, 114, 195, 287, 340, 346, 479
Schilpp, P. A., 234n.
Schlegel, August Wilhelm von, 87, 249, 272, 280, 356, 365, 401, 423
Schlegel, Freidrich von, 82, 87, 401
Schleiermacher, Friedrich Daniel Ernst, 143, 249, 263, 265, 279, 280, 283, 318, 436
Schlesinger, G., 152n.
Schleyer, J. D., 232n.
Schleyer, J.-M., 211
Schlick, Moritz, 173, 220
Schoenberg, Arnold, 184, 440n.
Scholem, Gershom, 63n., 130n.
Schöne Müllerin, Die (F. Schubert), 440
Schopenhauer, Arthur, 249, 255, 281, 344n.
Schubert, Franz Peter, 438–40 passim
Schumann, Robert Alexander, 438–42 passim
Schwitters, Kurt, 202
Scientific Thought (C. D. Broad), 134
Scott, Sir Walter, 25, 30, 285
Scott, Wilson L., 161n.
Scott Moncrieff, C. K., 285
Seaman, P. David, 126n.
Searle, J. R., 175n., 177n.
Sebeok, Thomas, vii, 2, 54
Secret Sharer, The (J. Conrad), 477
Selden, John, 366
Sellin, Ernst, 154n.
Sendbrief vom Dolmetschen (M. Luther) 248
Seneca, Lucius Annaeus, 72, 260, 334, 370, 454–9 passim, 477, 478
Sense and Sensibility (J. Austen), 8–12
Sentimental Journey through France and Italy, The (L. Sterne), 11
Seven Against Thebes (Aeschylus), 157
Sewell, Elizabeth, 196 & n., 197n.
Seyssel, Claude de, 260
Shakespeare, William, 2–8 passim, 25–6, 28, 44–5, 48, 163, 185, 187–9, 260, 267, 272, 280, 284, 287, 320, 365, 383–90 passim, 400–12 passim, 420, 423, 427, 479

Shakespeare und der Deutsche Geist (F. Gundolf), 401 & n.
Shakespeare und kein Ende (J. W. von Goethe), 401
Shattuck, Roger, see Arrowsmith, William
Shaw, George Bernard, 479, 493
'T. E. Shaw', see Lawrence, T. E.
Shelley, Percy Bysshe, xv, 30, 241, 470–3 passim, 479
Shepheard's Calendar, The (E. Spenser), 469
Sidgwick, Henry, 18
Sidney, Sir Philip, 469, 475, 479
Sièyes, Emmanuel Joseph, 139
Silent Woman, The (B. Jonson), 43
Simenon, Georges, 426n.
Simonini, R. C., 354n.
Skeat, W. W., 25
Skelton, John, 25
Skinner, B. F., 103n., 304n.
Skinner, C. A., 154n.
Skinner, Quentin, 142n., 143n.
Slakta, Denis, 304n.
Sleepwalkers, The (E. & W. Muir), 336
Sly, Christopher, 408
Smerud, Warren B., 170n.
Smith, Miles, 366
Smith, Patrick, J., 442n.
Smollett, Tobias, George, 285
Smyth, Herbert Weir, 330–1
Socrates, 230
Solomon, The Song of, 44
Sonnet en 'ix' (S. Mallarmé, trans. O. Paz), 289
Sophocles, 66, 248, 280, 318, 340 & n., 344 & n., 345, 347, 350, 400, 453, 459, 477, 478, 487
Sörbom, Göran, 268n.
Sordello (R. Browning), 190
Sorescu, Marin, 254
Sous l'invocation de Saint Jérome (V. Larbaud), 249
Soutine, Chaim, 484
Speak Memory (V. Nabokov), 127
Spencer, Herbert, 282
Spenser, Edmund, 352, 353, 461, 469, 475, 479
Spiegel, Shalom, 154n.

Spies, Claudio, 439n.
Spindler, Robert, 329n.
Spinoza, Benedictus de, 64, 155
Spitzer, Leo, 448, 456n.
Staël, Mme de (Anne Louise Germaine Necker, Baronne de Staël-Holstein), 82
Stalinism, 35, 146, 231
Statius, Publius Papinius, 269
Steegmuller, Francis, 426–7
Steel, T. B., 212n.
Stegmüller, W., 220n.
Stein, Gertrude, 202, 338
Stein, Jack M., 439n.
Steiner, George, 120–3
Steiner, Mrs. (née Franzos) (mother of author), 120
Steinke, G. E., 201n.
Steinthal, H., 84n., 89
'Stendhal' (Marie Henri Beyle), 47, 384
Sterne, Laurence, 11
Stevens, Wallace, 413
Stocker, Michael A. G., 170n., 173n.
Stocks, J. L., 149n.
Stoicism, 148, 149, 157, 230
Storey, David, 494
Strabo, 259
Strassburg, Gottfried von, 199
Strauss, Bruno, 276n.
Stravinsky, Igor Fedorovich, 439n., 489, 490
Strawson, P. F., 216–21 passim, 224
Strich, Fritz, 270n.
Strindberg, Johan August, 46, 416
Stroh, Wolfgang Martin, 439n.
Sturel, René, 288n.
Suarès, André, 445 & n.
Supervielle, Jules, 425–6
Sur la pluralité des mondes (B. de Fontenelle), 160
Surrealist literature, 201n.
Suter, R., 149n.
Swadesh, Morris, 20n., 51, 58
Swift, Jonathan, 18, 229–30, 233, 236
Swinburne, Algernon Charles, 15, 363, 473
Sylvae (J. Dryden), 269
Symbolists, 202

Symposium (Plato), 439
Symonds, John Addington, 259
Symons, Arthur, 15
Szilard, Leo, 163n.
Szondi, Peter, 256n., 409n., 478n.

Tagliacozzo, G., 79n.
Tagliaferro, R. Catesby, 363n.
Tagliavini, C., 107n.
Talleyrand-Perigord, Charles Maurice de, 236
Talmud, The, 63, 68
Taming of the Shrew, The, 25, 408
Tanburn, N. P., 172n.
Tantalus, 59
Tarn, Nathaniel, vii
Tarski, A., 212, 216, 220 & n., 223, 497n.
Tartakower, S. G., 21
Tasso, Torquato, 272, 413, 466
Taylor, Thomas, 363 & n.
Tchaikovsky, Peter Ilich, 489
Teele, Roy Earl, 376n.
Tempest, The, 25, 390
Tennyson, Alfred Tennyson, 1st Baron, 15, 361, 473, 484
Thackeray, William Makepeace, 33
Theocritus, 467, 469
Thessalonians, The Epistles to the, 364
Third Anniversary Discourse on the Hindus (Sir W. Jones), 82
Thom, A., 153n.
Thom, René, 309n.
Thomson, J. F., 170n., 174 & n.
Thomson, W. (Lord Kelvin), 162
Thoreau, Henry David, 487
Thucydides, 23, 143, 157, 166, 259, 260, 400, 418
Thyestes (Seneca), 477
Tieck, Dorothea, 401
Tieck, Johann Ludwig, 272, 280, 285, 365, 423
Timaeus (Plato), 362–4
Timber, or Discoveries made upon Men and Matters (B. Jonson), 269, 327
Timon of Athens, 188
Todd, W., 173n.
Tolmer, Léon, 279n.

Tolstoy, Leo N., Count, 480
Tomlinson, Charles, 198, 247n.
Tonkin, Elizabeth, 126n.
Torah, The, 61, 63
Toulmin, Stephen and June Good-
  field, 160n.; *see also* 374n.
Tractatus (L. Wittgenstein), 192, 213,
  217, 219, 227, 236, 338
Traité du Verbe (R. Ghil), 239
Translators, International Federation
  of, 284
Treatise (D. Hume), 144, 225
Trier, Jost, 90, 91
Tripolitus, Theodosius, 279
Troilus and Criseyde (G. Chaucer), 5
Troy Book (J. Lydgate), 5
Trubetskoy, Prince Nikolai S., 99, 100,
  302n.
Trudgill, Peter, 34n.
Tseveteva, M. I., 491
Turandot (G. Puccini), 236
Turn of The Screw, The (H. James), 37
Twain, Mark, 38
*Twenty-ninth Ode of the Third Book of
  Horace* etc. (J. Dryden), 449
Two Gentlemen of Verona, 4, 423
Tyler, S., 126n.
Tyndale, William, 258, 366
Typhoon (J. Conrad), 335
Tytler, Alexander Fraser (Lord
  Woodhouselee), 249
Tzara, Tristan, 202

Uccello, Paolo, 14
Ueber die Sprache und Weisheit der Indier
  (F. von Schlegel), 82
*Ueber die verschiedenen Methoden des
  Uebersetzens* (F. Schleiermacher), 249
*Ueber die Verschiedenheit des menschlichen
  Sprachbues* etc. (W. von Humboldt),
  84, 86
Ugolini, A. F., 413n.
Ullendorff, Prof. Edward, x
Ullmann, Stephen, 107n.
Ulysses, 355, 358, 359
Ulysses (J. Joyce), 190, 289, 297, 447,
  490
UNESCO, 284
Untermeyer, Jean Starr, 336-7

Urquhart, Sir Thomas, 209-10, 259,
  288

Valéry, Paul, 73, 249, 283, 365, 479
Valla, Lorenzo, 259
Van Gennep, A., 33n.
Velasquez, Diego Rodriquez de Silva,
  484, 490
Verdenius, W. J., 268n.
Verkauf, Willy, 201n.
Verlaine, Paul, 321, 425, 438
Vermischte Anmerkungen (J. G.
  Hamann), 80
Versuch über eine akademische Frage
  (J. G. Hamann), 79
Vico, Giovanni Battista, 78-83 *passim*,
  92, 107, 199
Vie de Rancé, La (F. Chateaubriand),
  139
Vildomec, V. 126n.
Villon, François, 25, 359, 367, 461
Vinay, Jean-Paul, 126n.
Virgil (Publius Vergilius Maro), 158,
  269, 270, 334, 337, 340, 355, 356,
  365, 419, 469, 473, 487
Vollrath, Ernst, 149n.
Voltaire, François Marie Arouet de,
  384, 385
Volupté (C. A. Sainte-Beuve), 482, 483
  & n., 484
Vom Weltbild der Deutschen Sprache
  (L. Weisgerber), 90
Voss, Johann Heinrich, 83, 272, 280,
  318
Vossler, K., 358
Voynich manuscript, 176 & n.
Voznesensky, Andrey, 375
Vroon, Ronald, 204n.
Vygotsky, L. S., 125

Waddell, Helen, 376
Walden (H. D. Thoreau), 487
Waley, Arthur, 285, 333, 375-8 *passim*
Waller, Edmund, 269, 461
Wallis, R., 137n.
Wandruszka, Mario, 250, 323n.
Wange-Wei, 379
Warburton, William, 461, 464
Webern, Anton von, 490

Wedekind, Frank, 401, 478
Weiler, Gershon, 175n.
Weinrich, Harald, 139n., 189n., 232n.
Weinreich, Uriel, 126n.
Weisgerber, Leo, 90
Weiss, Paul A., 132n., 140n.
Wernicke, Carl, 282, 297
Wernicke area of brain, 297, 298
Werther (Die Leiden des jungen Werthers) (J. W. von Goethe), 83, 285, 352, 481
Wesley, John, 11
West-Östlicher Divan (J. W. von Goethe), 270, 272
What Maisie Knew (H. James), 37
Whitchurch, Edward, 366
White, Alan R., 219n.
White, Patrick, 493
Whitehead, Alfred North, 479
Whiteley, W. H., 126n.
Whorf, Benjamin Lee, 81, 89, 91–8 *passim*, 106, 109n., 111, 137, 243, 292
Widmer, Walter, 284n., 333n.
Wieland, Christoph Martin, 83, 272, 280, 365, 401, 478
Wiener, Norbert, 163n.
Wilamowitz-Moellendorff, Ulrich von, 281
Wilbur, Richard, 71, 287
Wilde, Oscar, 232
Wilkins, John, 73, 78, 210, 211, 212
Willis, Thomas, 80
Will to Power, The (F. W. Nietzsche), 237, 238
Wilson, J. Dover, 403
Winckelmann, Johann Joachim, xv, 195, 487
Wind, Edgar, 448
Winterreise, Die (F. Schubert), 440

Winter's Tale, The, 3, 7
Wittenberg, A. I., 132n.
Wittgenstein, Ludwig, 8, 41, 62, 64, 91, 92, 97, 105, 116, 149, 169–77, 178, 192, 203, 207, 212–13, 216–18 *passim*, 222, 223, 227, 228, 236, 290 & n., 310 & n., 338, 374n., 497n.
Wodehouse, P. G., 33, 34
Wolf, Hugo, 438, 439 & n., 446
Wolkenstein, Oswald von, 198
Word and Object (W. van O. Quine), 125, 249
Wordsworth, William, 373
Worsley, P. H., 360
Wörterbuch (J. & W. Grimm), 25
Wörterbuch (E. Littré), 25
Wozzeck (G. Büchner), 13, 120
Wright, Arthur, F., 50n.
Wright, G. H. von, 164n.
Wycliffe, John, 366
Wyle, Nicholas von, 275, 276n.

Xenophon, 259, 324

Yeats, William Butler, 14, 157, 473, 476, 478
Yip, Wai-lim, 377n., 378n.
Yuille, J. C., A. Paivio and W. E. Lambert, 302n.

Zamenhof, L. L., 211
Zangwill, O. L., 298n.
Zelter, K. F., 438, 440n., 442, 443, 445
Zibaldone (G. Leopardi), 255n.
Ziff, Paul, 219n.
Zohar, 63
Zuberbühler, Rolf, 340n., 341n.
Zukofsky, Celia and Louis, 370, 371n.
Zuntz, Günther, 340n.